AF182763

Eutrophication of Shallow Lakes with Special Reference to Lake Taihu, China

Developments in Hydrobiology 194

Series editor

K. Martens

Eutrophication of Shallow Lakes with Special Reference to Lake Taihu, China

Edited by

B. Qin[1], Z. Liu[1,2] & K. Havens[3]

[1] *Nanjing Institute of Geography and Limnology, Chinese Academy of Sciences, Nanjing, China*
[2] *Institute of Hydrobiology, Jinan University, Guangzhou, China*
[3] *Department of Fisheries and Aquatic Sciences, University of Florida, Gainesville, USA*

Reprinted from Hydrobiologia, Volume 581 (2007)

Springer

Library of Congress Cataloging-in-Publication Data

A C.I.P. Catalogue record for this book is available from the Library of Congress.

ISBN-13 978-90-481-7552-9 e-ISBN-13 978-1-4020-6158-5

Published by Springer,
P.O. Box 17, 3300 AA Dordrecht, The Netherlands

Cite this publication as Hydrobiologia vol. 581 (2007).

Cover illustration: Lake Taihu, a large shallow lake in the Changjiang River delta (China). Satellite image: China Center for Resource Satellite Data and Applications.

Printed on acid-free paper

TABLE OF CONTENTS

Hydrobiologia (2007) 581:1–2
DOI 10.1007/s10750-006-0522-4

EUTROPHICATION IN LAKES

Preface

Boqiang Qin · Zhengwen Liu · Karl Havens

Eutrophication and algal blooms are worldwide environmental issues in lakes. The eutrophication process and forming mechanisms of algal blooms are particularly complicated in shallow lakes due to the strong lake–land, air–water and water–sediment interactions.

In China there are more than 2,700 lakes, with a total area of 91,000 km^2, and one third of them are shallow lakes. In recent years, many shallow lakes in China have displayed rapid eutrophication and suffered from algal blooms. These problems have resulted in a shortage of drinking water supply and degradation of lake ecosystems. The control of eutrophication in shallow lakes is one of the main issues with which the local people and Chinese governments are concerned today.

Lake Taihu is the third largest freshwater lake in China, with an area of about 2,338 km^2 and a mean depth of 1.9 m. Now its main function is supplying drinking water for the surrounding cites such as Shanghai, Suzhou and Wuxi. It is also very important in tourism, culture fisheries and navigation. However, with economic development and increased population in the lake basin, Lake Taihu is suffering from serious eutrophication, so research work on ecosystem restoration in the lake is receiving increasing attention.

The 'International Symposium on the Eutrophication Process and Control in Large Shallow Lakes—With Special Reference to Lake Taihu, a Shallow Subtropical Chinese Lake' was held in Nanjing during April 22–26, 2005, and provided a forum for discussion and exchange of ideas and information by scientists and environment managers on the issues of eutrophication process, control, and management of shallow lakes in general and Lake Taihu in particular. This special issue collects papers presented at the symposium. The editors and the authors of the papers in this special issue hope that it will make a significant contribution to the understanding and management of eutrophication in shallow lakes.

Guest editors: B. Qin, Z. Liu & K. Havens
Eutrophication of shallow lakes with special reference to Lake Taihu, China

B. Qin (✉)
Nanjing Institute of Geography & Limnology, Chinese Academy of Sciences, 73 East Beijing Road, Nanjing 210008, P.R. China
e-mail: qinbq@niglas.ac.cn

Z. Liu
Institute of Hydrobiology, Jinan University, Guangzhou 510632, P.R. China

K. Havens
Department of Fisheries and Aquatic Sciences, University of Florida, Gainesville, Florida 32653, USA

Acknowledgements

The papers in this volume have all gone through the usual review process with at least two referees for each. As editors, we would therefore like to express our sincere thanks to the following persons for their cooperation and valuable comments on one or several manuscripts:

Reviewers

Alois Herzig, Akio Imai, Alan J. Lewitus, Amy M. Kamarainen, Andrzej Gorniak, Asa Danielsson, Barbara Robson, Binhe Gu, Bob Foy, Boping Han, Boqiang Qin, Brain Moss, Christian E. W. Steinberg, Christopher W. K. Chow, Claire L. Schelske, Colin S. Reynolds, Curtis J. Richardson, David Hamilton, Delphine Latour, Dietrich Soyez, Dmitry Lajus, Duran Robert, Edward Santoro, Elly Spijkerman, Erik Jeppesen, Fernando Unrein, Figen ERKOC, Francois G Schmitt, Gea-Jae Joo, Genxing Pan, Giovana O. Fistarol, H.J. (Marieke) de Lange, Harvey Shear, Heinz Löffler, Herman J. Gons, Hijran Yavuzcan Yildiz, Huayu Lu, Hui Li, Isabel Ramirez, James W. McClelland, Jaroslav Vrba, Jianchao Lee, Jidong Gu, Jingsong Yan, John Smol, Judith Wolf, Kaiqin Xu, Karl E. Havens, Kelly A. Rusch, Kohji Michioku, Koren R. Nydick, Lianchun Fan, Liping Zhou, Liuyan Yang, Marc Schallenberg, Mark Honti, Mark J. McCarthy, Marley Waiser, Martin T. Dokulil, Morton Siyuan Chen, Moshe Gophen, Nathan Hawley, Nicola Fohrer, Nimal Chandrasena, Olav Slaymaker, Peeter Nõges, Ping-Mei LIEW, Piotr Dawidowicz, Rajendra Kumar Gupta, Raoul Henry, Robert G. Wetzel, Robert Portielje, Rosaluz Tavera, Ryo Matsumoto, S.S.S. Sarma, Sam Lau, Scotta A. Lowe, Shahamat U. Khan, Shigeo Tsujimura, Suwanchai Nitisoravut, Sven Erik Jorgensen, Takashi Asaeda, Tara Reed, Thomas L. Crisman, Tiina Nõges, Ulrike-G. Berninger, Vera Istvánovics, Vitor Vasconcelos, Walter Dodds, Wayne S. Gardner, William R. Herb, Xinde (Rocky) Cao, Zhengwen Liu, Zoltan Vekerdy

We are also indebted to Henri Dumont for his valuable suggestions and help throughout the process of producing this special issue, Liancong Luo and Liuyan Yang for their excellent editorial assistance, Mark McCarthy for the language checking and Hongwei Yang for providing the cover illustration. We acknowledge the Chinese Academy of Sciences, National Natural Science Foundation of China for their financial support (No. KXCZ1-SW-12 and 40514034) for this symposium and the editorial affairs of this volume.

Hydrobiologia (2007) 581:3–14
DOI 10.1007/s10750-006-0521-5

EUTROPHICATION IN LAKES

Environmental issues of Lake Taihu, China

Boqiang Qin · Pengzhu Xu · Qinglong Wu · Liancong Luo · Yunlin Zhang

Abstract Lake Taihu is characterized by its shallowness (mean depth = 1.9 m) and large surface area (2,338 km^2). Runoff sources are mostly from the mountainous west and southwest, and outflows are located throughout East Taihu. This causes shorter retention times in the south. In contrast, urban pollutants discharge into northern Taihu and result in poor water quality. Non-point pollution from rural areas and sewage wastewater is the primary pollution source. Water current velocity ranges from 10–30 cm s^{-1}, and surface currents normally follow wind direction. Bottom currents appear to be a compensation flow. Most wave heights are less than 40 cm, and underwater irradiance correlates to seston in the water column. Lacustrine sediment is distributed in littoral zones, mostly along the western shoreline, with almost no accumulation in the lake center. Intensive aquaculture in East Taihu caused eutrophication and hampered water supply in surrounding areas. In addition, development of marshiness in the eastern littoral zones and East Taihu has occurred. The function of flood discharging of East Taihu has been limited by flourishing macrophytes. The problems facing in Lake Taihu will be alleviated by improving the management of nutrient sources into the lake.

Keywords Lake Taihu · Hydrography · Water quality · Eutrophication · Aquaculture

Introduction

Lake Taihu is situated in the Changjiang (Yangtze) delta, the most industrialized area in China with high population density, urbanization, and economic development. Although Taihu Basin occupies only 0.4% of China's territory and 3% of China's population, it contributes about 10% of Gross National Product (GNP) and 15% of province revenues (NSB, 2000; SBSC, 2000; SBZP, 2000; SBJP, 2000). Lake Taihu is important for water supply, flood control, tourism and recreation, shipping and aquaculture. It is the drinking water source for several cities, such as Shanghai, Suzhou, Wuxi, and Huzhou. Since 1980's, however, economic development has resulted in pollutants being produced and discharged into rivers and the lake. With the deterioration of water quality, eutrophication and algal blooms (*Microcystis* spp.) have occurred. Recently, the algal bloom has extended its

Guest editors: B. Qin, Z. Liu & K. Havens
Eutrophication of shallow lakes with special reference to Lake Taihu, China

B. Qin (✉) · P. Xu · Q. Wu · L. Luo · Y. Zhang
Nanjing Institute of Geography & Limnology, Chinese Academy of Sciences, 73 East Beijing Road, 210008 Nanjing, P.R. China
e-mail: qinbq@niglas.ac.cn

coverage and persists throughout the summer, which affects the function of the lake as a drinking water supply. In 1996, the Chinese government targeted pollution and eutrophication control before 2000 and 2010. Factories within the watershed now must treat wastewater before it is discharged to rivers. Also, rivers flowing into Lake Taihu should meet grade III standards (National Surface Water Quality Standard GB3838-2002) before 2000, and lake water should be clarified before 2010. However, lake water quality has not improved to date, and, in 2001, pollution control work was reevaluated and redirected. In fact, problems facing Lake Taihu are similar to most freshwater lakes in eastern China. Based on field investigations and literature review, this paper attempts to review and address main eco-environmental issues in Lake Taihu.

General description of the lake

Lake Taihu is the third largest freshwater lake in China and is located between 30°55′40″–31°32′58″ N and 119°52′32″–120°36′10″ E. Lake length (from north to south) is 68.5 km and width (from east to west) is 56 km. Mean depth is 1.9 m, and maximum depth is 2.6 m corresponding to an elevation of 3.0 m above sea level. The lake bottom features flat terrain with an average topographic gradient of 0°0′19.66″ and elevation of 1.1 m above sea level. Shallow-water area with mean depth <1.5 m is about 452 km^2, mostly in East Taihu, accounting for 19.3% of the total surface area. Deepest areas (>2.5 m) are in the north and west occupying 197 km^2 or 8.4% of the total lake area.

Origin of the lake

Many researchers have discussed the origin and history of Lake Taihu (Wang & Ding, 1936; Wissman, 1941; Chen et al., 1959; NIGLAS, 1965; Yang et al., 1985; Yan & Xu, 1987; Sun et al., 1987; Sun & Huang, 1993; Huang, 2000; Sun, 2004). In general, Lake Taihu probably was an ancient lagoon, as suggested by its location in the Changjiang delta where solid particles have been transported and deposited in the transitional estuarine zone. Evidence of marine sediment materials found in drill cores supported this assumption (Chen et al., 1959). Continental shelf extension resulted in enclosing of the lagoon and it eventually became a freshwater lake (Chen et al., 1959; Yang et al., 1985; Yan & Xu, 1987). Recently, new evidence from stratigraphy and archaeological sites suggest that Taihu originated from extreme flooding. Flood water discharged by man-made channels was retained in the depression (Sun & Huang, 1993; Sun, 2004).

Hydrology of the lake

Lake Taihu has a complicated river and channel network. There are 172 rivers or channels connecting to the lake (Xu & Qin, 2005). The total length of rivers in Taihu basin is ca. 12,000 km, i.e., about 3.24 km km^{-2}. During flooding season (spring and summer), flooding runoff from the west or southwest goes through lake to the east and empties into the East China Sea (Fig. 1). In the dry season (autumn and winter), however, water flow can reverse, especially in rivers in the east. Generally, rivers in the west of the lake are defined as upstream and vice versa. Runoff is defined as positive or negative. Table 1 shows the runoff of main rivers flowing into or out of the lake from May 2001 to April 2002. The greatest inflow rivers are Chendonggang, Xitiaoxi, and Yincungang. Three are located in the west or southwest watershed. The main outflow rivers are Taipu, Xinyunhe, and Xijiang located in the southeast (Table 1). Table 2 shows the monthly distribution of runoff from May 2001 to April 2002. Some rivers flow into the lake year-round, some flow out all year, and some reverse flow direction in different seasons (Table 2). There is no definitive estimate of water balance yet because of the complicated hydraulic connections between rivers and the lake. Table 3 is a rough estimate of water balance based on surveys of 115 rivers from May 2001 to April 2002. Flowing water from these rivers account for 90% of total runoff input and output, but the measuring frequency was insufficient because measurements were taken once in normal months and twice in

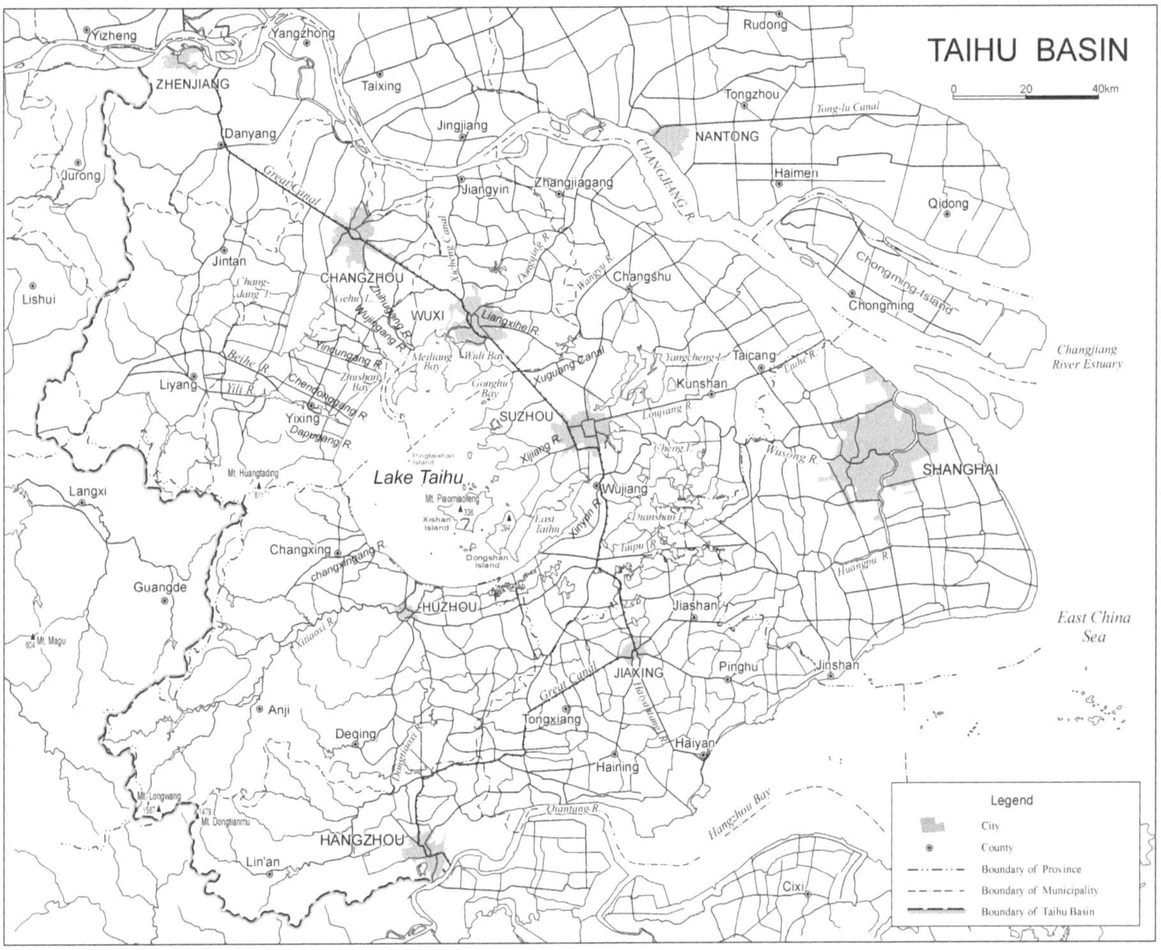

Fig. 1 Sketch map of Taihu Basin

flooding months. Annual runoff input, output, and water balance from May 2001 to April 2002 is estimated based on the measurement (Table 3). This water balance represents more flooding years versus the multi-year averaged water balance (Mao et al., 2004). Water retention time is about 5 months, but it is shorter in the south since most runoff water is discharged via the Taipu River, in the southeast. Water quality, therefore, is better in the south than in the north.

Hydrodynamics

There still is not a complete understanding of water currents in Lake Taihu. Earliest measurements were made in 1960, when two boats cruised around the lake at the same speed (NIGLAS, 1965). Other measurements were made in 1986–1987 (Sun & Huang, 1993) and summer in 1997 (Qin et al., 2000). These measurements were incomplete and incomparable since records were collected on different dates or at different sites. In summer, prevailing winds from the southeast or southwest generate a counter-clockwise water current in western Taihu centered near Pingtaishan Island (Fig. 2). Current velocity ranged from 10 cm s^{-1} to 30 cm s^{-1} (Qin et al., 2000). There was a stable southward current along the western coast and a weak and unstable current in Meiliang Bay (Fig. 2). This current pattern was reproduced by a numerical simulation under the same conditions (Luo & Qin, 2004; Hu et al., 2004). Regardless of season, there is a weak counter-clockwise current in northern Meiliang Bay (Qin et al., 2000; Luo et al., 2004a, b), which may explain

Table 1 Inflow and outflow of main rivers around Lake Taihu during May 2001 to April 2002 (if the sum of number of days of inflow and outflow less than 365, the remainder is days of no flow)

River	Inflow				Outflow				Annual runoff (10^8 m^3)
	Days	Mean daily inflow rate (m^3 s^{-1})	Mean inflow rate (m^3 s^{-1})	Total inflow (10^8 m^3)	Days	Mean daily outflow rate (m^3 s^{-1})	Mean outflow (m^3 s^{-1})	Total outflow (10^8 m^3)	
Xitiaoxi R.	234	273.0	34.9	7.05	125	–57.8	–12.79	–1.38	5.67
Dongtiaoxi R.	118	400.0	71.5	7.29	247	–122.0	–45.37	–9.68	–2.39
Changxingang R.	301	61.0	6.8	1.76	64	–10.4	–2.21	–0.12	1.64
Chendonggang R.	338	147.0	34.2	9.98	26	–38.7	–10.07	–0.23	9.76
Yincungang R.	347	72.9	18.5	5.56	18	–14.8	–5.47	–0.09	5.47
Wujingang R.	339	49.8	6.6	1.95	26	–7.1	–2.89	–0.06	1.88
Zhihugang R.	244	78.1	8.3	1.75	27	–14.1	–5.56	–0.13	1.62
Liangxihe R.	174	60.7	11.8	1.78	60	–23.3	–9.58	–0.50	1.28
Wangyu R.	116	240.0	82.8	8.30	249	–177.0	–44.10	–9.49	–1.19
Xinyun River	32	33.1	15.4	0.43	333	–183.0	–31.47	–9.05	–8.63
Dapugang R.	14	4.2	2.7	0.03	348	–14.2	–5.78	–1.74	–1.71
Xijiang R.	7	16.5	8.7	0.05	358	–44.3	–18.72	–5.79	–5.74
Xuguang Canal	38	7.7	3.3	0.11	290	–27.3	–8.03	–2.01	–1.91
Taipu River	8	73.2	27.5	0.19	357	–296.0	–119.47	–36.85	–36.66

Table 2 Monthly runoff rate of rivers around Lake Taihu during May 2001 to April 2002 (m^3 s^{-1}) (inflow is defined as positive and outflow is defined negative)

River	2001								2002				Yearly Average
	May	Jun	Jul	Aug	Sep	Oct	Nov	Dec	Jan	Feb	Mar	Apr	
Xitiaoxi R.	8.8	49.3	22.0	57.0	–9.4	–1.7	–6.3	6.8	7.9	–4.1	28.5	55.6	17.9
Dongtiaoxi R.	–2.2	80.6	–5.9	8.2	–76.7	–45.6	–54.5	–20.3	–15.1	–31.0	9.6	61.1	–7.7
Changxingang R.	3.7	10.7	6.6	10.3	1.1	0.1	1.0	3.6	3.1	1.2	9.8	11.0	5.2
Chendonggang R.	15.7	44.8	33.1	67.1	31.1	26.4	24.7	32.0	22.2	12.7	37.4	22.3	30.8
Yincungang R.	14.8	28.8	11.9	37.2	18.9	15.4	14.0	19.3	10.0	7.1	18.4	11.6	17.3
Wujingang R.	4.5	11.2	4.9	11.7	5.5	4.9	5.8	4.2	3.1	4.2	7.0	4.6	6.0
Zhihugang R.	5.7	13.4	0.1	12.9	5.4	3.9	2.9	2.1	0.9	2.9	9.0	2.4	5.1
Liangxihe R.	5.9	14.1	–5.8	2.6	3.6	10.3	5.8	10.4	1.6	0	0	0	4.0
Wangyu R.	12.0	–12.4	–87.4	–48.5	–41.6	15.0	–12.8	–18.0	–37.2	115.0	116.0	–36.5	–3.0
Xinyun R.	–23.4	–47.4	–34.3	–24.8	–22.0	–31.3	–26.6	–28.7	–30.9	–21.8	–15.8	–21.1	–27.3
Dapugang R.	–2.7	–2.0	–5.9	–5.2	–5.5	–5.3	–6.5	–8.4	–9.5	–5.5	–5.8	–2.4	–5.4
Xijiang R.	–10.7	–9.3	–27.4	–27.9	–25.2	–17.6	–17.0	–15.3	–17.4	–14.0	–18.5	–17.6	–18.2
Xuguang Canal	–1.6	0.2	–10.5	–12.4	–10.2	–4.7	–6.1	–3.9	–7.3	–2.2	–7.7	–5.6	–6.0
Taipu R.	–49.6	–84.9	–132.0	–135.0	–125.0	–85.4	–107.0	–127.0	–118.0	–116.0	–160.0	–156.0	–116.3

Table 3 The water balance of Lake Taihu during May 2001 to April 2002

Input runoff (10^8 m^3)	Output runoff (10^8 m^3)	Precipitation (10^8 m^3)	Evaporation (10^8 m^3)	Change in capacity (10^8 m^3)	Inflow from uncontrolled areas and river net (10^8 m^3)
80.11	–96.67	27.19	–17.71	–5.38	12.46

why this area is favorable for concentrating algal blooms (Qin et al., 2000).

In July 2003, an Acoustic Doppler Current Profiler was deployed at the center of Meiliang Bay, and 10-min interval current velocity and direction data from 11 layers were collected. Surface currents followed the direction of wind forcing (Fig. 3a), but current velocity attenuated

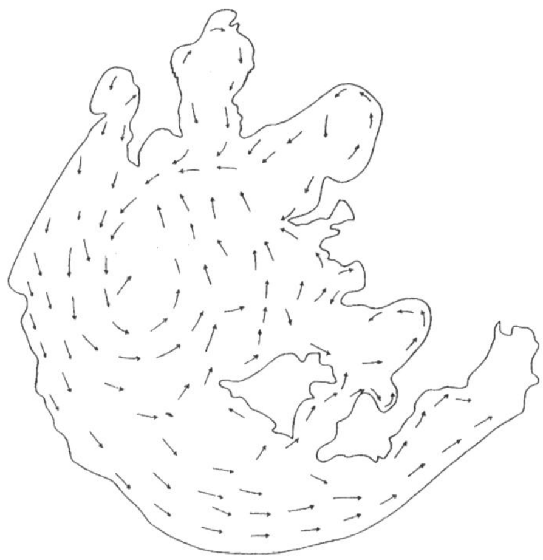

Fig. 2 The prevailing current pattern during the summer, the velocity of current ranged from 10 cm s^{-1} to 30 cm s^{-1}

with depth and reversed direction at the bottom (Fig. 3b). This indicates that there is a compensation current at the bottom opposite wind forcing on the surface (Fig. 3b).

Wave action is a key factor in interactions between the sediment and water (Qin et al., 2004).

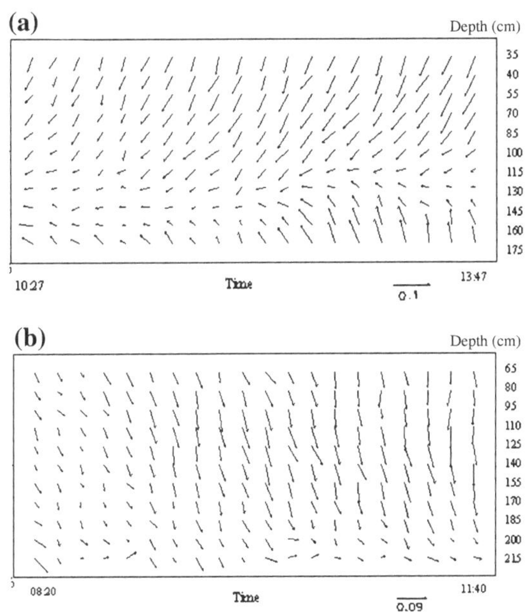

Fig. 3 (**a**) Current vectors at 35, 40, 55, 70, 85, 100, 115, 130, 145, 160, 175 cm below surface (water depth is 2.4 m) at 10-min intervals in Meiliang Bay during 10:27–13:47 on 14 July 2003. (**b**) Current vectors at 65, 80, 95, 110, 125, 140, 155, 170, 185, 200, 215 cm below surface (water depth is 2.4 m) at 10-min intervals in Meilaing Bay during 8:20–11:40 on 17 July 2003

In August 1986, a continuous observation of waves was made in the northwest (Wu et al., 1987). Statistical analyses with cumulative frequency of one-third highest wave height ($H_{1/3}$) and maximum wave height (H_{max}) showed that most waves were low in height with 80% of $H_{1/3} < 8$ cm, 98% of $H_{1/3} < 20$ cm, and 5% > 40 cm (Wu et al., 1987). From field observations, the empirical relationship among average wave height ($\overline{H}$), wind speed (W), wind fetch (F), and period (T) was derived as follows (Hu et al., 2004):

$$\frac{g\overline{H}}{W^2} = 0.00168\left(\frac{gF}{W^2}\right)^{0.46}$$

If water depth was assumed not to change, then wave height was proportional to wind speed (Hu et al., 2004):

$$\overline{H} = 5.497W^{0.5} - 4.516$$

Similarly, the relationship between maximum wave height (H_{max}) and wind speed can be expressed as (Wu et al., 1987):

$$\frac{gH_{max}}{W^2} = 0.1094\left(\frac{gD}{W^2}\right)^{0.7508}$$

$$H_{max} = 5.237 \times 10^{-2}\ W$$

Underwater light climate

Lake Taihu is a shallow and eutrophic lake, and Secchi depths are normally low, i.e., about 30–40 cm in the north and 40–50 cm in the south (Qin et al., 2004; Zhang et al., 2004a). Based on water quality monitoring conducted by Taihu Laboratory for Lake Ecosystem Research (TLLER), transparency is related to seston abundance (Fig. 4):

$$S^{1/4} = 8.103 - 5.847 \ln ST$$

$$(R = 0.87,\ SD = 0.32,\ N = 135,\ P < 0.0001)$$

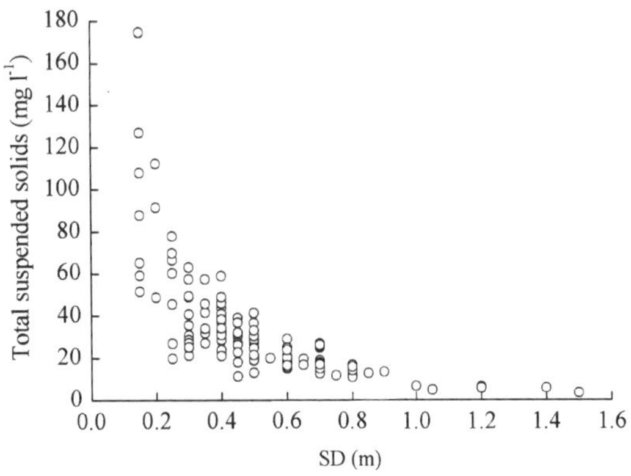

Fig. 4 The relationship between the solid suspension and secchi depth (SD) in Lake Taihu

where S is seston abundance (mg l^{-1}), ST is secchi depth (m). Correlation and regression analysis based on measurements of light attenuation and seston, including inorganic and organic seston, and chlorophyll-*a* showed that inorganic solids were important in autumn, whereas organic solids were important in summer, especially in the north (Zhang et al., 2004b). Correlation analysis also indicated that chlorophyll was not a large factor on light attenuation (Zhang et al., 2004b).

Underwater irradiance measurements show that daily light distribution at specific depth followed solar radiation, i.e., increase in the morning and decrease in the afternoon (Fig. 5) (Zhang et al., 2004b). Irradiance at 1 m was about 10% of that at the surface (Fig. 5). Ultraviolet (UV) radiation, which is harmful for phytoplankton growth, was attenuated rapidly, i.e., UV-B at 20 cm was about 1% of surface levels (Zhang et al., 2004b). Short-wave irradiance, therefore, is attenuated faster than long-wave irradiance in this lake.

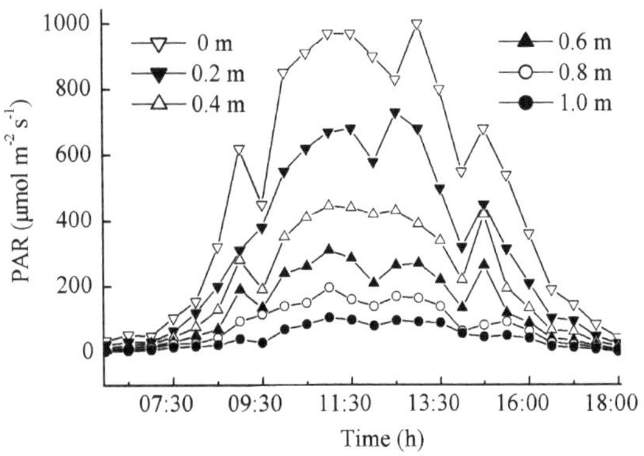

Fig. 5 Daily variations of underwater irradiance at different depth (0.0, 0.2, 0.4, 0.6, 0.8, 1.0 m under the surface) at the site nearby Taihu Laboratory for Lake Ecosystem Research (TLLER) on May 17, 1998

Sediment in Lake Taihu

Spatial and vertical distribution of lacustrine sediment was investigated at 723 sites in Lake Taihu. Successive correction schemes and geostatistical methods were used to determine sediment thickness in the 69 × 69 whole lake grids. Spatial statistical analysis revealed that sediment covers about 1,100 km^2 of lake bottom, which accounts for about 47.5% of lake area (Fig. 6) (Luo et al., 2004a, b). Total volume of lacustrine sediment is estimated to be 1.86×10^9 m^3 (Luo et al., 2004a, b), and, in most regions, lacustrine sediment depth was 0.5–2.0 m (Luo et al., 2004a, b). Most lacustrine sediments are distributed along the western shoreline from Meiliang Bay to the south. There is little lacustrine sediment in the center of the lake (Fig. 6).

Dating sediment at different depths from drill cores showed that sedimentation rates range from 0.6 mm $year^{-1}$ to 3.6 mm $year^{-1}$ with an average of 2 mm $year^{-1}$ (Fan et al., 2004). Bulk density of surface sediment is 1.2 g cm^{-3} with lowest bulk density from East Taihu and highest from the western lake (Fan et al., 2004). Organic matter content ranges from 0.13% to 16.6% with maxima from East Taihu and minima from the center and west (Fan et al., 2004). TN and TP in sediments, range from 0.01% to 0.66% and 0.01% to 0.015%, respectively (Fan et al., 2004).

Pollution and eutrophication

In 1960, Lake Taihu was categorized as oligotrophic because total inorganic nitrogen (TIN), total inorganic phosphorus (PO_4^{3-}-P), and chemical oxygen demanded (COD_{Mn}) were low (Table 4) (NIGLAS, 1965). TIN increased dramatically until 1981 to about 18 times greater than that in 1960, COD_{Mn} increased by 49%, and PO_4^{3-}-P remained stable (Table 4) (Sun & Huang,

Fig. 6 Distribution of lacustrine sediment in Lake Taihu

Table 4 Change of water quality indexes since 1960's

Year	TIN (mg l^{-1})	TN (mg l^{-1})	PO_4^{3-}-P (mg l^{-1})	TP (mg l^{-1})	COD_{Mn} (mg l^{-1})	Sources
1960	0.05		0.02		1.90	NIGLAS (1965)
1981	0.894	0.9	0.014		2.83	Sun & Huang (1993)
1988	1.115	1.84	0.012	0.032	3.3	Sun & Huang (1993)
1994	1.135	2.05	0.010	0.086	5.77	SEPA (2000)
1995	1.157	3.14	0.011	0.111	5.53	SEPA (2000)
1998[a]	1.582	2.34	0.007	0.085	5.03	TLLER
1999[a]	1.79	2.57	0.004	0.105	4.99	TLLER

[a] Data provided by TLLER

1993). From 1981 to 1998, TP and COD_{Mn} increased by 2.7 and 1.5 times, respectively (Table 4). Total nitrogen (TN), total phosphorus (TP), and chemical oxygen demand (COD_{Mn}) increased from 1.8, 0.03, and 3.3 mg l^{-1} in 1988 to 2.3, 0.085, and 5.03 mg l^{-1} in 1998, respectively (Table 4).

Table 5 shows quantities of pollutants which rivers transport into the lake. Most pollutant inflow rivers are located in the west or northwest, whereas pollutant outflow rivers are located in the south or southeast. Based on studies in 2002 and 2003, about 30–40% of nitrogen and phosphorus is retained inside the lake (Table 5).

Table 5 Nutrient flux of inflow and outflow of Lake Taihu during 2001–2002

	COD_{Mn} (ton)	NH_4^+-N (ton)	NO_2^--N (ton)	NO_3^--N (ton)	TN (ton)	TP (ton)
Inflow	37,571.06	12,431.70	1,007.86	10,469.12	28,657.97	1,029.37
Outflow	–35,431.36	–4,864.99	–379.77	–5,862.28	–14,599.90	–668.27
Inflow–outflow	2,139.70	7,566.71	628.09	4,606.83	14,058.07	361.10
Retention rate	5.7%	60.9%	62.3%	44.0%	49.1%	35.1%

Wastewater produced within the basin amounted to 2.25×10^5 t year^{-1} of COD_{Cr}, 4.13×10^4 t year^{-1} of TN and 3.8×10^3 t year^{-1} of TP in the 1980's (Sun & Huang, 1993), increasing to 2.82×10^5 t year^{-1} of COD_{Cr} and 8.0×10^4 t year^{-1} of TN and 5.66×10^3 t year^{-1} of TP in the early 1990's (SEPA, 2000), and increasing further to 7.46×10^5 t year^{-1} of COD_{Cr}, 1.09×10^5 t year^{-1} of TN and 1.56×10^4 t year^{-1} of TP in the late 1990's (Huang, 2004). However, pollutants discharged into the lake were less than the pollutants produced within the basin. About 1.06×10^5 t year^{-1} of COD_{Cr}, 2.0×10^4 t year^{-1} of TN and 1.55×10^3 t year^{-1} of TP in the 1980's were discharged into the lake (Huang et al., 2001). During the early 1990's, total pollutants discharged increased to 1.31×10^5 t year^{-1} of COD_{Cr}, 3.1×10^4 t year^{-1} of TN and 1.75×10^3 t year^{-1} of TP (SEPA, 2000). In the late 1990's, the COD_{Cr} empting into the lake increased to 1.71×10^5 t year^{-1}, but TN and TP decreased to 2.5×10^4 t year^{-1} and 1.3×10^3 t year^{-1}, respectively (Huang, 2004). During the early 2000's, inputs of COD_{Mn}, TN, and TP were 3.76×10^4 t year^{-1}, 2.87×10^4 t year^{-1} and 1.03×10^3 t year^{-1}, respectively (Table 5) (Xu & Qin, 2004), which are much lower than earlier estimates. These results varied among different investigators, but pollutants empting into the lake probably increased in the early 1990's and decreased in the late 1990's. This change may relate to increased wastewater treatment in surrounding areas.

Analysis of pollutant composition in wastewater in 1994 showed that 16% of total nitrogen (TN) was from industrial sources, 25% from domestic sewage, and 38% from non-point sources (agriculture production) (SEPA, 2000). For total phosphorus (TP), 10% was from industry, 60% from sewage, and 15% from non-point sources (SEPA, 2000). For COD, 39% was from industry, 42% from sewage, and 10% from non-point sources (SEPA, 2000) (Table 6). In 1998, industrial pollution to the lake, decreased dramatically either in total COD, or TN and TP (Table 6), whereas the contribution from agriculture to COD and TN increased significantly and sewage pollution contributed 70% of TP (Table 6). This change in pollution composition reflects rapid economic growth and land use alteration.

Based on data from 1998, pollutants produced by industry were mainly from the east or southeast of the lake basin, such as Suzhou, Wuxi, and Jiaxing (Huang, 2004). Pollutants from agriculture (cropping, rice growing, etc.) accounted for 37% of COD, 49.5% of TN and 48% of TP produced in the basin (Huang, 2004). These non-point pollutants came mostly from the west. In addition, poultry culture scattered in the rural region accounted for 5% of COD, 6% of TN, and 18% of TP (Huang, 2004). Fish culture made up 5.5% of COD, 5.2% of TN, and 4.1% of TP. With strengthening control of point sources, the contribution from non-point sources, i.e., agriculture and animal culture, likely will increase with time.

Aquaculture in the lake

The main form of aquaculture in Lake Taihu is pen-fish-culture. Aquaculture has been limited to East Taihu, a macrophyte-dominated bay in the southeast part of the lake with an area of

Table 6 Comparison of composition of different pollution sources in COD, TN, and TP in 1994 and 1998

Year	Pollution source	COD_{Cr}		TN		TP	
		Amount (ton)	Percentage (%)	Amount (ton)	Percentage (%)	Amount (ton)	Percentage (%)
1994	Industry	111,061	39	12,544	16	591	10
	Domestic	119,029	42	19,948	25	3,394	60
	Agriculture	28,138	10	29,842	38	852	15
	Aquaculture			13,195	17	533	9
1998	Industry	93,822	26	2,686	4	191	2
	Domestic	96,782	27	16,232	25	6,510	70
	Agriculture	91,582	25	17,815	28	1,362	15
	Aquaculture	49,627	14	17,750	28	733	8

126 km^2. Although the documented area of pen-fish-culture was 2,830 ha (Chen & Wu, 2004), field observations reveal that almost the entire water surface of East Taihu is occupied by net pens. Fish culture in this area has raised some environmental problems. East Taihu is a shallow bay with 95% coverage by macrophytes. In 1993, annual macrophyte production in East Taihu was 1.12×10^6 t, which accounts for 3,920 tons of nitrogen and 496 tons of phosphorus (Wu et al., 1995; Yang & Li, 1996), corresponding to 58% and 95%, respectively, of total external loading to East Taihu. About 60×10^4 t of macrophytes were harvested for fish farming resulting in removal of 1,890 t of nitrogen and 296 t of phosphorus, corresponding to 28% and 57%, respectively, of total external loading to East Taihu (Wu et al., 1995; Yang & Li, 1996). Harvesting of macrophytes now is blocked due to the proliferation of net pens. Consequently, the lake lost an important pathway for nutrient export. This loss has promoted sedimentation and eutrophication in East Taihu (Table 7). Aquaculture activities also contribute to increases in nutrient loading and eutrophication. It is estimated that producing 1 ton of fish requires 141 kg nitrogen and 14 kg phosphorus (Yang & Li, 1996; Wu et al., 1995; Li, 1998), and about 65% of each is not assimilated and is subsequently released into the overlying water (Yang & Li, 1996). In pen-fish-culture areas, increased nutrient loading leads to rapid growth of phytoplankton, zooplankton, and bacteria. After one year of fish farming, the phytoplankton abundance was three times higher than in non-culture areas, and heterotrophic bacteria abundance increased 3- to 4-fold (Wu et al., 1995). Total organic carbon, total nitrogen, and total organic

Table 7 Changes in water quality parameters in East Taihu in 1990's

Year	TN (mg l^{-1})	NH_4^+-N (mg l^{-1})	NO_3^--N (mg l^{-1})	TP (ug l^{-1})	COD_{Mn} (mg l^{-1})
1991	1.01	0.14	0.32	43	5.5
1993	1.04	0.20	0.25	75	5.6
1997	1.39	0.19	0.10	31	5.5
1999	1.65	0.475	0.23	40	6.6
2000	1.08	0.26	0.30	100	6.4

nitrogen in surface sediments increased by 141, 87.5, and 86%, respectively, after 2 years of fish culturing (Li, 2004). Recently, fish culturing has been replaced by the more profitable freshwater crab culturing, which will increase the input of feed, and further increase the deposit of organic materials from the remnants of feed. Because fish culture will make use of macrophytes for fish food, the lake will output the nutrients and retard the development of marsh in this way. In turn, the recent aquaculture aggravates the deterioration of water quality and degradation of ecosystem.

Marsh development

Marsh development mainly takes place in the eastern shoreline areas, i.e., East Taihu, Xikou Bay, and Gonghu Bay. Figure 7 shows the expansion of spatial macrophyte distribution (TBA & NIGLAS, 2000). From Fig. 7, macrophytes have extended from East Taihu in the 1960's (Fig. 7a) to Xikou Bay and Gonghu Bay in the 1980's (Fig. 7b). Marsh development in East Taihu is more serious in East Taihu because it also functions as a flood discharge area and drinking water supply for Suzhou and Shanghai. These functions, however, are hampered by marsh development. Generally, marsh development in lakes results from organic or inorganic material deposition, which elevates the lake bottom, decreases water capacity, and eventually leads to elimination of the lake (Li, 2004).

According to field investigations and TM satellite image data in 1997, 95% of East Taihu is covered by aquatic vegetation. Dominant species include *Potamogeton maackianus, Vallisneria natans*, and *Zizania latifolia*. Average macrophyte biomass in East Taihu is about 3.8 kg m^{-2} with a range of 2.0–5.6 kg m^{-2} (Li, 2004). Total standing biomass of aquatic plants in Lake Taihu is about 500,000 t, and *Potamogeton maackianus* accounted for 42%, *Zizania latifolia* 29.5% and *Vallisneria natans* 0.2% (Li, 2004). Floating-leaf and emergent vegetation, such as *Zizania latifolia*, indicate accelerated marsh development in East Taihu (Li, 2004).

Since East Taihu is the major outlet of Taihu Lake, particles are transported and deposited in

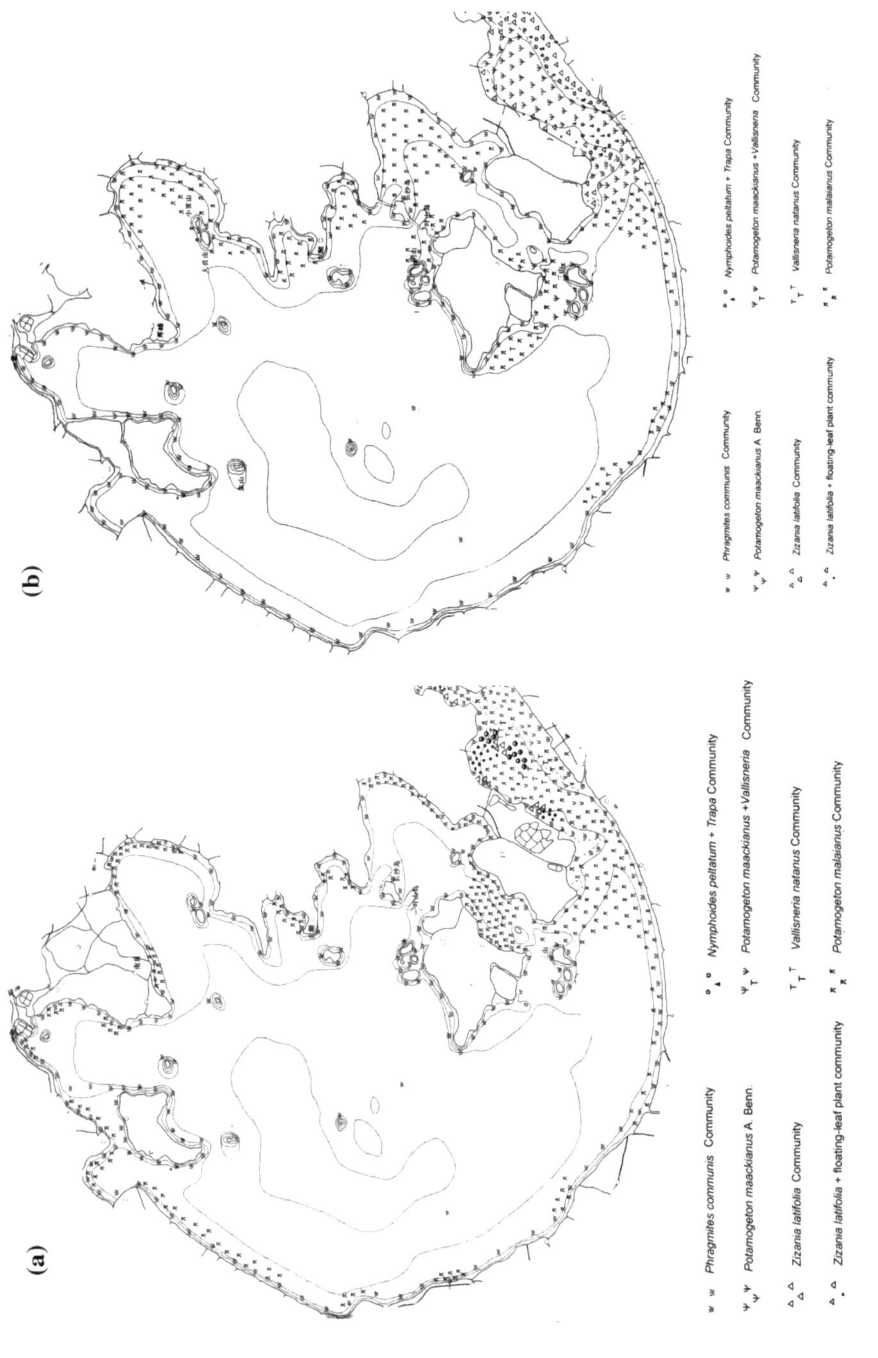

Fig. 7 Spatial distribution of macrophytes in 1960's (**a**) and in 1980's (**b**)

East Taihu. Macrophytes reduce water flow velocity and, hence, increase silting. Investigation of sediment distribution in East Taihu showed that the top 1 m of sediment was soft and composed of sand (98%) and organic materials (1.5%) (Li, 2004). Sediment composition showed that most deposit materials were allocthonous. Thus, increased sedimentation rates and marsh development may be attributed to pen-fish-culture, since it limited macrophyte harvest, promoted succession from submerged to emergent and floating-leaf vegetation, inhibited water movement, and augmented deposition of macrophyte detritus and other particles.

Management

Lake Taihu has many functions, including drinking water supply, flood water retention, fisheries, tourism and shipping. These varied functions have resulted in multi-purpose-oriented management without sufficient coordination. Over-exploration and over-utilization caused degradation of the lake ecosystem and deterioration of water quality, since some functions are contradictory, such as water supply and fish culture. Development of a centralized-management system, therefore, should be considered to decelerate eutrophication and marsh development, restore the ecosystem, and improve water quality. Regulations and administrative policies should be enacted, e.g., reclamation should be forbidden and fish-culture should be limited in spatial distribution. A long-term plan for pollution and eutrophication control should be based on accurate prediction of economic development and contaminant sources. Watershed management should be planned and implemented aiming at pollution source control. Scientific research should be emphasized and focus on internal loading from sediment, water and nutrient budgets, and ecological restoration in the littoral zone.

Conclusion

1. Lake Taihu has a complicated inflow and outflow system. Water retention time is shorter and water quality is better in the south or southeast, whereas retention time is longer and water quality is worse in the north or northwest.
2. Major pollutants are mostly from the western watershed. Most COD input is balanced by river output, and about 30–40% of nutrient input is retained in the lake. Most pollution is from non-point agriculture production and domestic sewage.
3. Light attenuation is associated with suspended solids, which is related to wave forcing. Most sediments are distributed in the littoral zone, especially along the western shoreline.
4. Aquaculture in East Taihu increases nutrient concentrations in water and sediments, which accelerates eutrophication and marsh development. Swamping in East Taihu results from macrophyte success, which increases sedimentation rate, reduces the ability of flooding water passing through the system and quickens the tempo of extinguishment of the lake.

Acknowledgements This work was funded by Chinese Academy of Sciences (No. KZCX1-SW-12), Ministry of Science and Technology of China (No. 2002AA601011), and National Natural Science Foundation of China (No. 40501078). We thank Prof. Karl Havens, Prof. Wayne Gardner, and Prof. Koren Nydick for their careful reviews and constructive suggestions. Thanks are also to Dr. Mark McCarthy for his help with English polishing.

References

Chen, J., Z. Yu & C. Yun, 1959. Morphology of Changjiang Delta. Acta Geographica Sinica 25: 201–220 (in Chinese).

Chen, W. & Q. Wu, 2004. Structure and evolution of ecosystem in Lake Taihu. In Qin, B., W. Hu & W. Chen (eds), Process and Mechanism of Environmental Changes of Lake Taihu. Science Press, Beijing: 289–297 (in Chinese).

Fan, C., G. Zhu, W. Li, S. Sun, C. Hu & B. Qin, 2004. Exchange between the water–sediment interface and its environmental effects. In Qin, B., W. Hu & W. Chen (eds), Process and Mechanism of Environmental Changes of Lake Taihu. Science Press, Beijing: 170–186 (in Chinese).

Hu, W., B. Qin & L. Luo, 2004. The hydrodynamic processes and the environmental effects in Lake Taihu. In Qin, B., W. Hu & W. Chen (eds), Process and Mechanism of Environmental Changes of Lake Taihu. Science Press, Beijing: 107–161 (in Chinese).

Huang, W., 2004. The pollutant budget in Lake Taihu. In Qin, B., W. Hu & W. Chen (eds), Process and Mechanism of Environmental Changes of Lake Taihu. Science Press, Beijing, 21–28 (in Chinese).
Huang, X., 2000. Layout and Integrated Management of Taihu Basin. Press of Water & Electricity, Beijing, 1–20 (in Chinese).
Huang, Y., C. Fan, P. Pu, J. Jiang & Q. Dai, 2001. Aquatic Environment and Pollution Control of Lake Taihu. Science Press, Beijing, 187–206 (in Chinese).
Li, W., 2004. Tendency and reason for evolution toward mash in the East Taihu. In Qin, B., W. Hu & W. Chen (eds), Process and Mechanism of Environmental Changes of Lake Taihu. Science Press, Beijing, 33–51 (in Chinese).
Li, W., 1998. Utilization model of aquatic macrophyte for fishery in Chinese shallow lakes. Ecological Engineering 11: 61–72.
Luo, L. & B. Qin, 2004. Numerical simulation based on a three-dimensional shallow-water hydrodynamic model-current circulations in Lake Taihu with prevailing wind-forcing. Journal of Hydrodynamics (Serial B) 16(3): 341–349.
Luo, L., B. Qin, G. Zhu & Y. Zhang, 2004a. The current pattern of Meiliang Bay in the winter time. Journal of Lake Sciences 16(1): 73–76 (in Chinese).
Luo, L., B. Qin & G. Zhu, 2004b. Sediment distribution pattern maped from the combination of objective analysis and geostatistics in the large shallow Lake Taihu, China. Journal of Environmental Sciences 16(6): 908–911.
Mao, R., P. Xu & J. Gao, 2004. Hydrography and water balance of Lake Taihu. In Qin, B., W. Hu & W. Chen (eds), Process and Mechanism of Environmental Changes of Lake Taihu. Science Press, Beijing, 9–18 (in Chinese).
Nanjing Institute of Geography, Academia Sinica (NIGLAS), 1965. Investigation of Lake Taihu. Science Press, Beijing, 1–84 (in Chinese).
National Statistics Bureau (NSB), 2000. Statistics Almanac of China. China Statistics Press, Beijing (in Chinese).
Qin, B., W. Hu, W. Chen, J. Ji, C. Fan, Y. Chen, X. Gao, L. Yang, G. Gao, W. Huang, J. Jiang, S. Zhang, Y. Liu & Z. Zhou, 2000. Studies on the hydrodynamic processes and related factors in Meiliang Bay, Northern Taihu. Journal of Lake Sciences 12(4): 327–334 (in Chinese).
Qin, B., W. Hu, G. Gao, L. Luo & J. Zhang, 2004. The dynamics of resuspension and conceptual mode of nutrient releases from sediments in large shallow Lake Taihu, China. Chinese Sciences Bulletin 49(1): 54–64.
State Environmental Protection Administration (SEPA), 2000. Water Protection in the Three Rivers and Three Lakes of China. Chinese Environmental Press, Beijing: 151–175 (in Chinese).
Statistics Bureau of Jiangsu Province (SBJP), 2000. Jiangsu Statistics Almanac. China Statistics Press, Beijing (in Chinese).
Statistics Bureau of Shanghai City (SBSC), 2000. New Shanghai Statistics Almanac of National Economic and Social History in Past 50 Years. Statistics Bureau of Shanghai City (in Chinese).
Statistics Bureau of Zhejiang Province (SBZP), 2000. New Zhejiang Statistics Almanac. China Statistics Press, Beijing (in Chinese).
Sun, S., 2004. Topographic features and origin of Lake Taihu and its catchment. In Qin, B., W. Hu & W. Chen (eds), Process and Mechanism of Environmental Changes of Lake Taihu. Science Press, Beijing, 1–9 (in Chinese).
Sun, S. & Y. Huang, 1993. Lake Taihu. China Ocean Press, Beijing, 23–89 (in Chinese).
Sun, S., Y. Wu & B. Dong, 1987. The Bottom Configuration and Recent Deposition of Lake Taihu. Memoirs of Nanjing Institute of Geography and Limnology, Academia Sinica, No. 4. Science Press, Beijing, 1–14 (in Chinese).
Taihu Basin Authority (TBA) & Nanjing Institute of Geography & Limnology (NIGLAS), 2000. Atlas of Eco-environment of Lake Taihu. Science Press, Beijing, 1–118.
Wang, H. & W. Ding, 1936. Formation and shrinkage of Lake Taihu. Monthly of Water Resources 11(6): 8–13 (in Chinese).
Wissman, F., 1941. Geographical problem in the lower delta of Changjiang River (translated by Wang, D.). Geography 1(1): 74–77 (in Chinese).
Wu, J., X. Han & J. Jiang, 1987. A study of the wind waves at the platform and resuspension of the sediments in Taihu Lake. In ISTB'87 – International Symposium on Lake Taihu and Lake Biwa, Nanjing, 8/1–8/6.
Wu, Q., K. Chen & G. Gao, 1995. Effects on aquatic environment of large-scale inning for aquaculture. Journal of Aquaculture 19: 343–349 (in Chinese).
Xu, P. & B. Qin, 2005. Water quantity and pollutant fluxes of the surrounding rivers of Lake Taihu during the hydrological year of 2001–2002. Journal of Lake Sciences 17(3): 213–218 (in Chinese).
Yan, Q. & S. Xu, 1987. Recent Yangtze Delta Deposits. East China Normal University Press, Shanghai, 68–75 (in Chinese).
Yang, H., Z. Xie & D. Yang, 1985. Sea-level change in holocene and evolution of Lake Taihu. In Yang, H. (ed.), Contribution to the Quaternary Glaciology and Quaternary Geology. Geological Publishing House, Beijing, 49–64 (in Chinese).
Yang, Q. & W. Li, 1996. Fishery impacts on aquatic plants and possible strategy. Journal of Applied Ecosystem 7: 83–88 (in Chinese).
Zhang, Y., B. Qin, W. Chen & L. Luo, 2004a. A study on total suspended matter in Lake Taihu. Resources and Environment in the Changjiang River 13(3): 266–271 (in Chinese).
Zhang, Y., Q. Cai & P. Yang, 2004b. Light transmission in the water. In Qin, B., W. Hu & W. Chen (eds), Process and Mechanism of Environmental Changes of Lake Taihu. Science Press, Beijing, 53–96 (in Chinese).

Hydrobiologia (2007) 581:15–24
DOI 10.1007/s10750-006-0524-2

EUTROPHICATION IN LAKES

The art and science of lake restoration

Brian Moss

Abstract Science and art encompass understanding, through human hand and mind; both are applied in the form of creation by human hand and mind. Like Janus, the Roman god of doorways and bridges, they are two faces of the same thing. Great art requires great understanding and great understanding is wasted unless it is applied to create meaningful art, whether this is in the form of written works or paintings. These principles apply not least to the restoration of lakes and whole landscapes from the human-induced damage that is the antithesis of significant art and fundamental science. Because they lie at the lowest points of hydrologic basins, lakes reflect the balance of human activities in their catchments. If these activities are carried out well, they contribute greatly to the landscape and to the welfare of the human population. If carried out badly in the sole interests of exploitation of resources, both human and otherwise, they may become problems reflected, for example, in eutrophication, algal blooms and fish kills. Lake restoration involves first an understanding in a reductionist (scientific) way of the processes that drive lake ecosystems, but then of the ultimate reasons that create these proximate problems. Ultimate reasons lie in a much wider sphere of human nature and the organisation of society. Lake restoration, in its most trivial form, may be simply a form of gardening to allay the symptoms of problems and create the illusion of a solution. Lake restoration in its most profound form involves an understanding of cultural significance and the workings of human societies and forms an epitome for the solution of much greater, global problems. Only in this form does it become truly creative. Approval must come from both of the faces of Janus.

Keywords Eutrophication · Shallow lakes · Culture · China

Introduction

There has arisen a divide in the way that knowledge is perceived. It has been called the two cultures by the English novelist and physicist. Snow (1959) and it concerns the gulf between scientific perception and perception by artists, linguists, historians, social scientists and all those loosely grouped into the subject areas of the humanities. Science is perceived as objective, collects facts from survey and experiments, and

Guest editors: B. Qin, Z. Liu & K. Havens
Eutrophication of shallow lakes with special reference to Lake Taihu, China

B. Moss (✉)
School of Biological Sciences, Biosciences Building, University of Liverpool, Liverpool L69 3BX, UK
e-mail: brmoss@liverpool.ac.uk

interprets them in a way that is alleged to be value-free and objective. Scholars in the humanities also collect facts. The methodology of a historian is no different from that of an ecologist describing past and present ecosystems, but controlled experiments are not conceivable in the arts and thus determination of mechanism has to be left to the perception of the observer and therefore is regarded as subjective rather than objective. This simplified overview disguises the fact that there can be a great deal of subjectivity in the interpretation of information collected by scientists and that there are insights available to other scholars that are too readily rejected by some scientists because they are not amenable to controlled experiment. A better way of dividing approaches to understanding is to distinguish the reductionist from the holistic. Scientists and those in the humanities use both approaches, but current practice is for scientists to be more reductionist and humanities' scholars more holistic.

Where ecological restoration is concerned, there can be a harmony between the two approaches and the lessons of the humanities, indeed of fine art, can be applied to guide the process. Ecological restoration, for example, involves several stages. First there is a problem. A habitat has to be restored because it has been degraded to some extent. The reasons for the damage need first to be understood or otherwise restoration may be misguided. Secondly, the extent of the damage must be assessed and thirdly a target towards which restoration should be directed must be decided. Then, fourthly, the scientific understanding of the system is brought into play to determine how to achieve the target and lastly it is likely that maintenance of this new status will require management of some sort. The lessons of the humanities are applicable at all but the fourth stage of technical restoration.

The parallels with the conception and creation of fine art are considerable. Art exists to interpret society as much as to decorate it. Science likewise diagnoses problems as well as helps create the artefacts of comfortable life styles. The power with which artworks are expressed mirrors the degree of the problem just as scientific approaches assess the degree of damage. Restoration involves design and the parallels are obvious here and management may be both a scientific and aesthetic process if the results of the restoration are to be permanent and significant.

The existence of the problems

Lakes lie at the bottoms of catchment basins. Simply because of the operation of gravity, they will reflect activities in their catchments, manifested in changed water quality, but also in alterations to watercourses for the purposes of water storage, flood control, pollutant disposal and farming. A catchment covered by pristine ecosystems will pose no problems for the freshwater systems that drain it and indeed there is a seamless connectance between the terrestrial and freshwater systems. Intact ecosystems are parsimonious with nutrients and very little available nutrient leaches from the land to the water. Most key nutrients are held and recycled within the terrestrial biomass. Pristine freshwater systems thus have extremely low nitrogen and phosphorus concentrations, even when the catchment soils are made fertile from easily weathered rocks.

Pristine freshwater systems will also have considerable structure. They are not just masses of mixed water. Rivers have structures created by geomorphological processes (sand bars, riffles, floodplains, levees) and by woody debris entering from the immediate catchment. Lakes have littoral zones with a rich diversity of emergent and submerged plants. Maintenance of this structure may depend on the intactness of the diversity of the ecosystem. Top predators, wolves and bears, salmonids and other piscivorous fish and birds may exert considerable top-down effects (Ripple & Beschta, 2004a, b) that maintain both this structure and the efficiency of cycling of nutrients in a system where nutrients are naturally extremely scarce. There will also be a great connectivity between systems. A lake does not exist as an isolated system. It depends on organic matter washed in from the catchment through the inflow rivers for part of its productivity (Hanson et al., 2004). The surrounding wetlands may be crucial for fish and invertebrate life histories and in the interesting case of the functioning of northern rivers, nutrients brought from the ocean in the bodies of salmonid fish are recycled through the faeces of their brown bear predators into the riparian woodland. In turn tree debris retains carcases after spawning to provide the nutrients needed by the periphyton and micro invertebrates that sustain the hatched fish once they have used up their yolk sacs. Movements of pike (*Esox lucius*) between lakes and wetlands perform parallel roles (Klinge et al., 1995).

Parsimony of nutrients, preservation of structure, including appropriate biodiversity, and maintenance of connectivity are the three key elements that constitute the 'gold standard' for any ecosystem restoration. A system with these three characteristics intact is self-maintaining. Scientists comment on deviation from this state by recording the impacts on freshwater systems that damage these characteristics. Nutrient parsimony is eroded by eutrophication, which may also result in a loss of structure. Catchment management for efficient forestry may deny to rivers the structure provided by forest debris (Harmon et al., 1986) and connectance is easily lost through the construction of dams and the channelisation and embankment of lowland rivers so that they lose hydrological connection with their floodplains. Essentially, a long catalogue of impacts that damage freshwater systems ultimately boils down to damage to the three key elements.

The problems of freshwaters are ultimately the problems of how landscapes are managed and therefore of the way that human societies are controlled and managed. In the developed world the current philosophy is of maximising resource use in the interests of profit with rather little consideration of the sustainability of the resource. Early ecology contented itself with description of the workings of intact systems; now it is equally concerned with documenting the damage. Likewise fine art once celebrated the contemplative delights of pristine or lightly used landscapes. The traditions of Chinese scroll painting very much manifest this. But currently much of it is dedicated to protest and comment. Pablo Picasso's painting of *Guernica*, a horrific view of the bombing of a Spanish town in support of extreme right wing policies is a famous example. Peter Scott's painting. 'The Natural Environment of Man' (Fig. 1) uses art as a means of expressing environmental controversy.

Fig. 1 The Natural World of Man. Peter Scott, 1961. The painting is a comment on the dilemma faced by humans in contemplating the advances of technology, to the right of the picture and the consequent losses to ecosystems, represented by a selection of endangered mammals and birds to the left. The figure stands apart from the scene, representing philosophies that regard the natural world as something to be controlled by humans in their own interests but is also transparent and melds into the natural world. Humans are as dependent on the biosphere as all other creatures. The figure also holds his or her arms in a gesture of uncertainty, not knowing quite what to do about the problems being caused by human activities. By permission of Lady Scott, Wildfowl and Wetlands Trust, Slimbridge

The six artistic principles of Xi He and their equivalents in lake restoration

It is in the design of targets for restoration, however that the parallel approaches between art and science become most interesting. In the sixth century (Northern and Southern Dynasty) Xie He (Hsieh Ho), in his *Ku-hua p'inlu* (Acker, 1954; Tregear, 1997) posited, for the judgement of Chinese paintings, six principles, which have been so influential that they influenced Chinese painting into the twentieth century and may still influence it. These were:

Q'i yun sheng tung: The painter should endow his work with life and movement through harmony with the spirit of nature

Ku fa yung pi: Refers to structural power and tension of the brush stroke

Ying wu hsiang hsing: Fidelity to the object in portraying forms

Sui lei fu ts'ai: Conforming to kind in applying colours

Ching ying wei chih: Plan and design, place and composition

Ch'uan I mu hsieh: Transmission of ancient models by copying

Gombrich (1995) talks of a wonderful restraint in Chinese painting (Fig. 2). There is an economy of line, a sense of contemplation and inner feeling for the subject rather than a precise elaboration of details. Technique in brushwork is economical and firm and painted scrolls are usually adorned with philosophic comment, also elegantly executed in text characters. The long tradition of Chinese painting, adopted also by Japanese painters, has its dangers in that rigid tradition inhibits individual inspiration and innovation, but a balance between tradition and innovation often gives the best of both worlds and in such a balance lies the relevance of art to lake restoration.

Fig. 2 A characteristic Chinese painting by Sheng Mou, 1330–1369. Waiting for the Ferry on an autumn river (1351). Hanging scroll on paper. Palace Museum, Beijing

With some contraction of the third and fourth principles, which concern fidelity of form and colour, there are equivalents to Xie Hi's principles in the guidance that might be given for effective lake restoration, especially in the complex case of shallow lakes. There should be skill in carrying out the work and economy in resources and effort; there should be fidelity to the structures and nature of the system; there should be taste in reflecting the nature of the landscape; there should be wisdom in understanding the essence of the relationships by reference to pristine examples; and there should be harmony with the spirit of wise and sustainable use. It is not known whether there was particular significance in the order of Xie Hi's strictures on painting, except that the first was the most important. The logical parallels for lake restoration however might re-order the equivalents in terms of: Wisdom, Skill, Fidelity, Taste and Harmony.

Wisdom in understanding the essence of the relationships

Understanding of the functioning of shallow lake systems has advanced greatly since 1970. The natural state of all lakes is to have very low available nutrient concentrations, as discussed above. If the natural catchment vegetation is destroyed for urbanisation or agriculture of any kind, these mechanisms are destroyed and leaching losses increase greatly. Such eutrophication has been associated with the loss of clear water and submerged plant communities from shallow lakes and their replacement with filamentous algae and phytoplankton.

The idea that this was a process entirely driven by nutrient increase is still held by some workers, but was challenged by Balls et al. (1989) and Irvine et al. (1989) who found that adding large amounts of nutrients to such plant communities in experimental ponds did not displace the plants. Indeed they grew better. A switch to phytoplankton only occurred if the plants were independently damaged by mechanical removal. In subsequent years a range of switch mechanisms has been recognised, each of which, especially but not necessarily, in tandem with nutrient enrichment may displace the plants (Moss et al., 1996). These include herbicide application, grazing by introduced birds and fish like common carp (*Cyprinus carpio* L.), pesticide leachate and increasing salinity (both of which disturb the zooplankton grazer interactions that maintain clear water), rise in water level and violent storms. Thus over a large range of nutrient loadings, dependent on whether such a switch has occurred, shallow lakes may have alternative states of clear water and plant dominance and turbid water with phytoplankton dominance (Fig. 3). Sometimes, turbidity may be created through suspended sediment raised by wind or benthic fish in plant-devoid waters. This understanding has come from a range of approaches, including analysis of very large data sets (Jeppesen et al., 1999), mesocosm (Moss et al., 2004) and pond experiments (Balls et al., 1989; McKee et al., 2003), whole-lake experiments (Meijer et al., 1999) and theoretical treatments (Scheffer et al., 1993). The general model applies to small and moderate sized lakes (perhaps up to the order of 10^2–10^3 km^2 but is relevant also to very large shallow lakes though with very large wind fetches, the more exposed portions of these may behave differently.

Part of the understanding has also come from attempts to restore the plant communities to

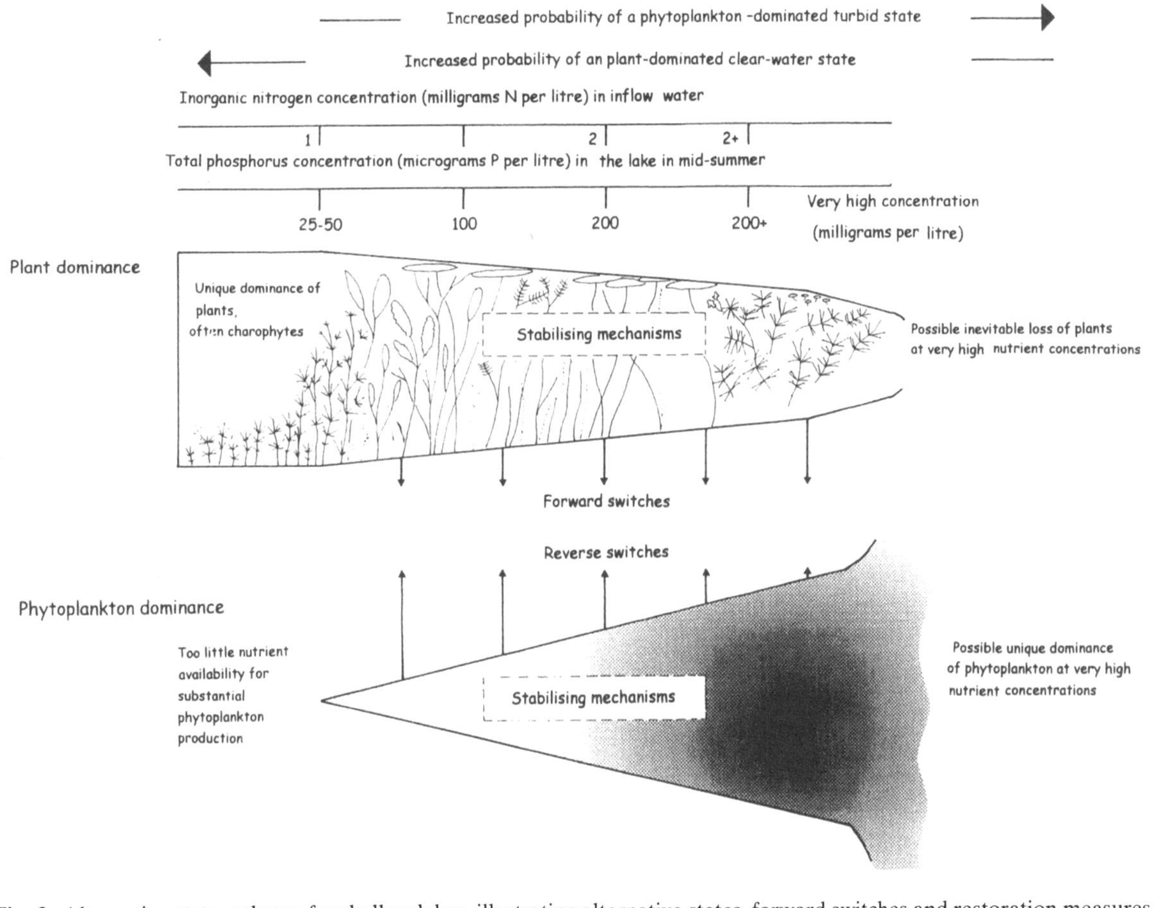

Fig. 3 Alternative states scheme for shallow lakes, illustrating alternative states, forward switches and restoration measures in the context of a range of nutrient loadings or concentrations (from Moss, 2002)

turbid lakes. Some cultures, not least those that have large populations to feed, may prize the turbid state for its ease of fishing and productivity of fish for food. The general presumption, however, is that the more desirable state of a shallow lake is to be plant-dominated, because this state is associated with the greatest biodiversity and aesthetic value and the least problems of smell, fish kills and toxicity of algal blooms. Plant communities use a number of mechanisms to maintain clear waters, including competition for nitrogen with algae, allelopathy, and reduction of eddy disturbance necessary for phytoplankton to remain suspended. Indirect, but equally important, mechanisms include provision of refuges against fish predators for zooplankton grazers on phytoplankton or macroinvertebrate grazers on periphyton (Moss et al., 1996; Scheffer, 1998).

Restoration attempts initially centred on reinstating the role of grazers by temporary removal of their fish predators, a process called biomanipulation. This resulted in clear water in which plants could grow, but protection against damage from grazing birds was sometimes necessary to establish self-sustaining communities. The communities were, however, usually of low diversity and vulnerable to slight disturbance that led to return of turbid conditions. Few restorations have been entirely successful in restoring diverse, stable plant communities (Moss et al., 1996; Meijer et al., 1999).

The reasons have perhaps been because the scale of restoration attempts has been too limited, that nutrients have not been adequately controlled or that only one of two important nutrients has been controlled. Shallow lakes have strong links in their natural states with surrounding wetlands that provide habitat for piscivorous fish that control zooplanktivorous fish and hence result in large zooplankton populations (Klinge et al., 1995). Extensive restoration of whole floodplains rather than just isolated lake basins is needed. Although alternative states exist over a wide range of nutrient loadings, it appears that the threshold for switches to act to remove plants is lower when nutrients are high and that biomanipulation is more effective if nutrients are reduced and that perhaps nutrient control should involve both phosphorus and nitrogen. Phosphorus may control algal production but the diversity of the plant community and hence its stability seems to be determined by nitrogen (James et al., 2005). Nutrient control is difficult especially when diffuse sources dominate and generally only phosphorus is controlled because point sources (for example, waste water treatment works) tend to be prominent and phosphorus can easily be precipitated at them.

Moss et al. (1996) developed a technical strategy for restoration, which forms an idealised basis for restoration attempts. First the target for restoration should be determined. The nature of the plant community is linked to the nutrient loading and generally diverse charophyte or isoetid communities require very low nutrient concentrations; mixed pondweed communities will persist over a middle range; and nymphaeid communities will still grow at very high concentrations and even when conditions are switching to turbid conditions. Once the target is decided, any existing or potential forward switches should be removed. There can be no success if this is not done. Thus, damaging fish or unusually high numbers of grazing birds need to be removed. Usually these are of introduced species and have built up to large numbers in the absence of natural predators. Risks from pesticide and herbicide contamination must be reduced and any damaging management such as mechanical clearance eliminated.

Nutrients should then be reduced, including both phosphorus and nitrogen. A range of methods exists for controlling point and diffuse sources but generally it is difficult to reduce concentrations to those that are likely to favour successful restoration. Restorations are most likely to be successful if total phosphorus concentration is reduced to winter values of less than 0.05 mg l^{-1} and total nitrogen to less than 1 mg l^{-1}. These suggestions are subject to much variability dependent on local circumstances however and emulation of the pristine condition of very low available nutrients should be the guiding principle. The fourth step is that of biomanipulation and a common source of failure is for too few fish to be removed. Successful restoration will be more likely if stocks are reduced to lower than 2 g m^{-2} and if some piscivores are left in the lake. This will allow daphnids and other zooplankters to increase and will clear the water, although if retention times are long and favourable to Cyanobacteria, there may be difficulties (Moss et al., 1991). Cyanobacteria can be controlled by grazing, but generally only if the proportion of such algae is low. Thick, surface blooms generally cannot. In the clear water, plants may spontaneously return from residual fragments but may have to be reintroduced. New plantings are vulnerable to bird damage and may need to be protected by caging. Thereafter, once plants have established vigorous beds, perhaps after two years, a fish community can be reintroduced, but should reflect the native community for the lake, should not include exotics and should always include piscivorous fish. Finally restoration projects should always be carefully monitored for several years afterwards. Many are not and valuable lessons as to why they succeed or fail will be lost. And in all things, patience is a virtue. It is not possible to replace over a year what may have been lost over a century.

Beyond this catalogue of necessary technicalities, however, is a philosophical basis for restoration and it is in this that the parallels with the principles of Xie He for painting become very relevant. Wisdom, skill, fidelity, taste and harmony are further contracted here into a hierarchy of skill, fidelity and taste, harmony and wisdom.

Skill in carrying out the work with economy of resources and effort

Lake restoration is often carried out by contractors, for whom it is simply a means of making profit. The results can then fall far short of the exquisiteness that Xie He would have expected of his painters and that sensitive ecologists would expect in the final result of a restoration. Mass-produced art is rarely prized. Skill comes in appreciating details and in minimising costs without compromise on the final result.

In one restoration attempt in which I was involved, at a lake in Wales (Moss et al., 2002), a detailed specification of what should be done was drawn up, including a strong proviso that certain plant species, like *Ceratophyllum demersum* L. should not be used, for they were likely to create dense beds that would interfere with angling. The use of plant stock bought from garden suppliers, which might include exotics and cultivars, was also precluded. In the event, the work was carried out by contractors who took the cheapest options of putting in any plant they could find, which included the very common *Ceratophyllum*, and at least effort, which meant they bought water lily stock from garden suppliers. The result was a lake choked by *Ceratophyllum* and fringed with an inappropriate, for the site, and probably non wild-type nymphaeid, *Nymphoides peltata* L. There was public protest at the *Ceratophyllum* choking and the result was that the local authority treated the lake with herbicide and effectively reversed the entire restoration after spending of a considerable sum of money ultimately to achieve nothing.

This however, is a small example of this particular precept. Skill involves choosing the right level of approach, and economy has to be considered in the long, not the short term. There are three levels of approach to restoration that correspond to different levels of cause of the problem. These are immediate, proximate and ultimate. For a shallow lake that has switched to phytoplankton dominance with troublesome, perhaps toxic cyanobacterial blooms, the immediate cause of the problem is the bloom. The immediate treatment is to use mixing methods or algicide. Both are cheap, temporary and ultimately ineffective in solving the problem for they treat only symptoms and the money spent on them is ultimately wasted.

The next level of treatment is to carry through the technical strategy outlined above. This deals with the proximate causes such as increased nutrient loading in the catchment, introduction of exotic fish or profligate use of herbicides and pesticides. It will cost much more but will be effective for a longer term. It may not be ultimately effective, however, for continuous control of nutrients will be needed and should funds become unavailable or should inappropriate fish be reintroduced for some new commercial intention, the money will have been wasted. What is really needed in this, as in all environmental problems, is a profound enquiry into the ultimate causes, which will be political, social and cultural. Treatment of the ultimate cause of the problem may have to be a change in agricultural policy so that large areas of the catchment are used as buffer zones to allow intensive agriculture to continue in the remainder. Or they may mean an enquiry into the relative rights of some parts of society to exploit the resources of the lake (for a fishery for introduced, very productive species for example), and the rights of other parts of society not to tolerate this in the interests of an environment that preserves biodiversity. Skill may come in realising these levels of approach and in aiming to tackle the problem as high up the hierarchy of causation as is possible.

Fidelity and taste in reflecting the nature of the landscape

Lakes that have been damaged are frequently incapable of being restored to anything remotely like their pristine equivalents. Shorelines often have been developed and damaged and may support building development that cannot be removed. Restoration may have to take on board such changes. A parallel aspect of restoration is re-creation of new lakes from quarries, gravel excavations and other such engineering activities. Pond creation is a widespread activity in Europe (Andrews & Kinsman, 1990; Williams, 1999; Moore, 2002) where former small ponds have often been filled in as a result of agricultural intensification to ease the use of large agricultural machines. Taste and design are very important in both these activities.

The guidance that Xie He would doubtless have given to lake restorers would have been that if natural structures such as emergent wetland margins cannot be restored, and hard edges have to be maintained, these should be of natural materials, appropriate to the area, especially its local geology, and not concrete or building stone of colours and textures that do not fit into the surrounding landscape. He would have spurned the artificiality of structures like fountains and artificial rafts for bird roosting. Fountains are particularly favoured by local government authorities, partly because the idea is rife that water needs continually oxygenating and partly by a perception that moving waters are tourist attractions. Water does not need artificially oxygenating, unless there has been such severe organic pollution that it is in danger of becoming permanently anaerobic and fountains are not good oxygenating devices anyway. Sub-surface circulation using powerful pumps is needed in such situations. Fountains may be appropriate to man-made pools in urban situations but not to natural or large lakes.

In designing new lakes, there is a tendency also to ignore the tenets of art in favour of the conveniences of engineering. Newly excavated lakes tend to have either geometric shapes, square or oblong, or exaggerated sinuosity in an attempt to avoid geometricity. Both are equally offensive to the eye. In natural lake basins the shape and outline are determined by the local geology and geomorphology. No general prescription can be given except that design should take account of the subtleties of the landscape and that this takes different talents from those needed for the engineering of earth movement and water retention. An appreciation of the nature of unmodified local lake basins and what gives them their particular characteristics is crucial to this. In an aerial photograph the ultimate results of lake creation or restoration should blend with the landscape and not stand out prominently as clearly different. One group of man-made lakes, the Norfolk Broads (Moss, 2002) formed in the 14th century in the United Kingdom, fits so well into the landscape that for many years their anthropogenic origin was not even realized.

Harmony with the spirit of wise and sustainable use

Xie He's first and most important stricture for painters is ultimately the most important one in lake restoration also. Problems in lakes reflect problems in society. An algal bloom is a symptom of a society that has not satisfactorily dealt with its immediate problems of waste disposal and food production in the best possible way. It reflects a dominance of some interests over others and generally these will have been the richest and most powerful interests. The current dominant philosophy of the human population of this planet is one of short-term exploitation of resources with rather little thought for ultimate consequences. Were this not the case we would not be facing severe climate change, heavily overfished seas, severe water shortages and polluted lakes.

There is acknowledgement of the problems in many political quarters but a policy of acknowledging problems but doing little and hoping that solutions will emerge spontaneously is a dangerous one. High quality lakes are set in high quality landscapes. High quality landscapes reflect sustainable values in society. Xie He taught that great art reflects great understanding, fundamental values and worthy objectives. Great lake restoration likewise reflects understanding and objectives worthy of a society that can be sustained, that reflects the needs of all people and that is not just a commercial activity. And if a Chinese art critic can be seen to offer guidance in lake restoration, perhaps ancient Chinese philosophy (Lao Tzu, approx. 500BC) can provide a solid basis for creating the sort of society that will sustain itself and its associated natural world, not least its freshwater resources. It defies all pleas above for connectedness, and a holistic and profound approach to quote a single passage to illustrate an entire philosophy, but Verse 67 will have to suffice:

Some say that my teaching is nonsense.
Others call it lofty but impractical,
But to those who have looked inside themselves,
This nonsense makes perfect sense.
And to those who put it into practice,
This loftiness has roots that go deep.

I have just three things to teach:
Simplicity, patience, compassion.
These three are your greatest treasures.
Simple in actions and in thoughts,
You return to the source of being.
Patient with both friends and enemies,
You accord with the way things are.
Compassionate towards yourself,
You reconcile all beings in the world.
(Translated by S. Mitchell, 1999)

References

Acker, W. B. R., 1954. Some T'ang and pre T'ang texts on Chinese paintings. Leiden University press, Leiden.

Andrews, J. & D. Kinsman, 1990. Gravel pit restoration for wildlife. Royal Society for the Protection of Birds, Sandy.

Balls, H., B. Moss & K. Irvine, 1989. The loss of submerged plants with eutrophication. I. Experimental design, water chemistry, aquatic plant and phytoplankton biomass in experiments carried out in ponds in the Norfolk Broadland. Freshwater Biology 22: 71–87.

Gombrich, E. H., 1995. The Story of art. 16 edn. Phaidon Press, London.

Hanson, P. C., A. I. Pollard, D. L. Bade, K. Predick, S. R. Carpenter & J. A. Foley, 2004. A model of carbon evasion and sedimentation in temperate lakes. Global Change Biology 10: 1285–1298.

Harmon, M. E., J. F. Franklin, F. J. Swanson, P. Sollins, S. V. Gregory, J. D. Lattin, N. H. Anderson, S. P. Cline, N. G. Aumen, J. R. Sedell, G. W. Lienkaemper, J. Cromak & K. W. Cummins, 1986. Ecology of coarse woody debris in temperate ecosystems. Advances in Ecological Research 15: 133–302.

Irvine, K., B. Moss & H. R. Balls, 1989. The loss of submerged plants with eutrophication II Relationships between fish and zooplankton in a set of experimental ponds, and conclusions. Freshwater Biology 22: 89–107.

James, C. S., J. Fisher, V. Russell, S. Collings & B. Moss, 2005. Nitrate availability and hydrophyte species richness in shallow lakes. Freshwater Biology 50:1049–1063.

Jeppesen, E., M. Sondergaard, B. Kronvang, J. P. Jensen, L. M. Svendsen & T. L. Lauridsen, 1999. Lake and catchment management in Denmark. Hydrobiologia 395/396: 419–432.

Klinge, M., M. P. Grimm & S. H. Hosper, 1995: Eutrophication and ecological rehabilitation of Dutch lakes: presentation of a new conceptual framework. Water Science, Technology 31: 207–218.

Lao Tzu C5 BC approx. Tao Te Ching, 1999. An illustrated journey (translated by Mitchell, S. Tao Te Ching). Frances Lincoln, Ltd, London.

McKee, D., D. Atkinson, S. E. Collings, J. W. Eaton, A. B. Gill, I. Harvey, K. Hatton, T. Heyes, D. Wilson & B. Moss, 2003. Response of freshwater microcosm communities to nutrients, fish, and elevated temperature during winter and summer. Limnology & Oceanography 48: 707–722.

Meijer, M.-L., I. de Boois, M. Scheffer, R. Portielje & H. Hosper, 1999. Biomanipulation in shallow lakes in the Netherlands: an evaluation of 18 case studies. Hydrobiologia 408/409: 13–30.

Moore, N. W., 2002. Oaks, dragonflies and people. Harley Books, Colchester.

Moss, B., 2002. The Broads: the people's wetland. Harper Collins, London.

Moss, B., J. Madgwick & G. L. Phillips, 1996. A guide to the restoration of nutrient enriched shallow lakes.

Environment Agency, Broads Authority & European Union Life Programme, Norwich.
Moss, B., D. Stephen, D. M. Balayla, E. Becares, S. E. Collings, C. Fernandez-Alaez, M. Fernandez-Alaez, C. Ferriol, P. Garcia, J. Goma, M. Gyllstrom, L. A. Hansson, J. Hietala, T. Kairesalo, M. R. Miracle, S. Romo, J. Rueda, V. Russell, A. Stahl-Delbanco, M. Svensson, K. Vakkilainen, M. Valentin, W. J. Van de Bund, E. Van Donk, E. Vicente & E. G. Villena, 2004. Continental-scale patterns of nutrient and fish effects on shallow lakes: synthesis of a pan-European mesocosm experiment. Freshwater Biology 49: 1633–1649.
Moss, B., J. Stansfield & K. Irvine, 1991. Development of daphnid communities in diatom- and cyanophyte-dominated lakes and their relevance to lake restoration by biomanipulation. Journal of Applied Ecology 28: 586–602.
Moss, B., L. Carvalho & J. Plewes, 2002. The Lake at Llandrindod Wells – a restoration comedy ? Aquatic Conservation: Marine & Freshwater 12: 229–245.
Ripple, W. J. & R. I. Beschta, 2004a. Wolves, elk, willows, and tropic cascades in the upper Gallantin Range of South-western Montana, USA. Forest Ecology and management 200: 161–181.
Ripple, W. J. & R. L. Beschta, 2004b. Wolves and the ecology of fear: can predation risk structure ecosystems? BioScience 54: 755–766.
Scheffer, M., S. H. Hosper, M. L. Meijer, B. Moss & E. Jeppesen, 1993. Alternative equilibria in shallow lakes. Trends in Ecology & Evolution 8: 275–279.
Scheffer, M., 1998. Ecology of shallow lakes. Chapman & Hall, London.
Snow, C. P., 1959. The two cultures. Cambridge University Press, Cambridge.
Tregear, M., 1997. Chinese art. Thames & Hudson, London.
Williams, P., 1999. The pond book: a guide to the management and creation of ponds. The ponds conservation trust, Oxford.

Hydrobiologia (2007) 581:25–42
DOI 10.1007/s10750-006-0502-8

EUTROPHICATION IN LAKES

Phosphorus dynamics at multiple time scales in the pelagic zone of a large shallow lake in Florida, USA

Karl E. Havens · Kang-Ren Jin · Nenad Iricanin · R. Thomas James

Abstract Phosphorus (P) dynamics in large shallow lakes are greatly influenced by physical processes such as wind-driven sediment resuspension, at times scales from hours to years. Results from long-term (30 year) research on Lake Okeechobee, Florida (area 1,730 km^2, mean depth 2.7 m) illustrate key features of these P dynamics. Variations in wind velocity result in changes in water column transparency, suspended solids, and total P (TP). In summer there are diurnal changes in TP associated with afternoon winds, and in winter, when strong winds occur for multiple days, monthly average TP remains high compared to summer. The magnitude of daily and seasonal TP changes can exceed 100 μg l^{-1}. Hurricanes and tropical storms also cause extreme changes in TP that are superimposed on seasonal dynamics. When a hurricane passed 80 km south of the lake in October 1999, mean pelagic TP increased from 88 to 222 μg l^{-1}. During large resuspension events, light attenuation is substantially increased, and this influences the biomass and spatial extent of submerged plants, as well as water column TP. In Lake Okeechobee, TP concentrations typically are ~20 μg l^{-1} when submerged plants are dense, and soluble reactive P concentrations are reduced below detection, perhaps by the periphyton and plant uptake and by precipitation with calcium at high pH. In contrast, TP exceeds 50 μg l^{-1} when submerged plants and periphyton are absent due to prolonged deep water, and phytoplankton biomass and algal bloom frequency both are increased. In Lake Okeechobee and other large shallow lakes, complex models that explicitly consider wind-wave energy, hydrodynamics, and sediment resuspension, transport, and key biological processes are needed to accurately predict how lake water TP will respond to different management options.

Keywords Phosphorus · Shallow lakes · Sediment resuspension · Wind effects · Physical processes · Diffusion · Eutrophication · Long-term studies

Guest editors: B. Qin, Z. Liu & K. Havens
Eutrophication of shallow lakes with special reference to Lake Taihu, China

K. E. Havens (✉)
Department of Fisheries and Aquatic Sciences, University of Florida, Gainesville, FL 33458, USA
e-mail: khavens@ufl.edu

K.-R. Jin · N. Iricanin · R. T. James
South Florida Water Management District, West Palm Beach, FL 33406, USA

Introduction

The seasonal and year-to-year dynamics of TP in deep dimictic lakes have been studied for decades, and they follow generally predictable patterns (e.g., Wetzel, 2001). During summer stratification, dissolved P is incorporated into the biomass of plankton, which subsequently settles from the epilimnion, or is incorporated into higher trophic levels of the food web. With the exception of buoyant cyanobacteria or other algae that can actively maintain their position in the upper water (Reynolds, 1994), there is a net downward movement of P from the epilimnion to the hypolimnion and sediments. Once particles have settled, the only time they return to surface waters in a deep stratified lake may be at fall overturn. Various authors have examined and summarized these patterns and their underlying processes (e.g., Hutchinson, 1975; Wetzel, 2001) that are largely based on work done decades ago by Mortimer (1941), Rigler (1956, 1964, 1973), and Lean (1973a, b).

In contrast to their stratified counterparts, shallow lakes may display highly irregular seasonal and year-to-year variations in TP, largely driven by stochastic processes that are not as readily predicted. These lakes also are considerably less studied, although in recent years that tendency has changed (Scheffer, 1998). At any given time, the physical transport of particulate and interstitial P into the water column of a shallow lake by wind energy may be orders of magnitude greater than other inputs of P, including diffusion, bioturbation, and external loading. As long as a concentration gradient exists between sediment-bound P and dissolved P in the water ($P_{water} > P_{sediment}$), downward transport of P with sediment particles may greatly exceed other loss processes, such as algal settling (e.g., James et al., 1997). In shallow lakes where wind strongly influences sediments, physical resuspension by wind-waves dominates the upward transport of P, and settling of particles dominates the downward flux (Fig. 1). In lakes not as heavily influenced by wind, other processes in the P cycle, including bioturbation, uptake of allochthonous P by plankton, plants, and periphyton may represent downward fluxes of greater magnitude.

There are several ways to predict whether P dynamics of a lake will be strongly driven by wind-waves. These predictions range from complex hydrodynamic models dealing with conservation of momentum and mass (e.g., Jin et al., 2000, 2002) to calculations of intermediate complexity based on wavelength and water depth (Caper & Bachmann, 1984) to simple index values. An example of an index is the dynamic ratio of Hakanson (1982), calculated as the square root of lake surface area (km^2) divided by mean depth (m). This index was developed to identify the relative portion of a lake bottom likely to be influenced by wind-driven sediment resuspension. Using data from 36 Florida lakes, Bachmann et al. (2000) showed that when the dynamic ratio is above 0.8, 100% of the lake bottom is subject to resuspension, whereas at ratios below 0.8, the risk of resuspension rapidly declines in a linear manner. This method can be used to characterize the influence of wind-generated waves on large lakes of the world (Fig. 2). Lakes are distributed according to their mean depth and area, with the diagonal line indicating the critical depth for

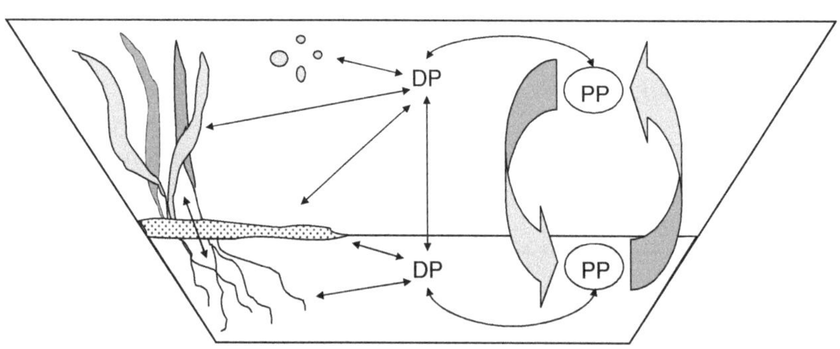

Fig. 1 Conceptual diagram of phosphorus (P) dynamics in a shallow lake that is heavily influenced by wind-induced waves. The arrows indicate fluxes of P between plants and periphyton, benthic algal mats, plankton, sediments, and water/sediment dissolved (DP) and particulate (PP) forms. In this example, resuspension and settling of PP dominates the lake's P dynamics

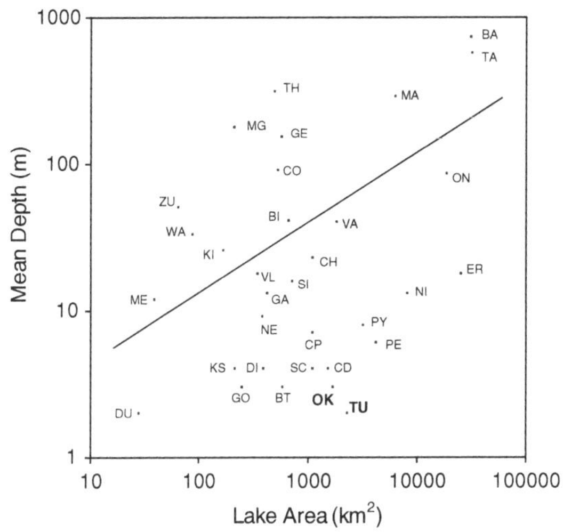

Fig. 2 Large lakes of the world, plotted as a function of their mean depth and surface area. The diagonal line indicates the mean depth for any particular surface area where the dynamic ratio (Hakanson, 1982) equals 0.8. Lakes lying below this line have higher dynamic ratios, and as such, are expected to be increasingly influenced by wind-induced waves. Symbols: Balaton (BT), Constance (CO), Biwa (BI), Chad (CD), Champlain (CH), Chapala (CP), Dianchi (DI), Donghu, (DU), Gatiun (GA), George (GO), Kasumigaura (KA), Kinneret (KI), Malawi (MA), Maggiore (MG), Neagh (NE), Peipsi (PE), Poyang (PY), Saint Clair (SC), Simcoe (SI), Tahoe (TH), Taupo (TZ), Valencia (VL), Vattern (VA), Baikal (BA), Erie (ER), Mendota (ME), Nicaragua (NI), Okeechobee (**OK**), Ontario (ON), Taihu (**TU**), Tanganyika (TA), Washington (WA), and Zurich (ZU). Data were obtained from the World Lake Database of the International Lake Environment Committee (www.ilec.or.jp)

any given area where the dynamic ratio is exactly 0.8. Lakes lying far below the 0.8 line are expected to be more influenced by wind-induced waves and sediment resuspension. Examples of these types of lakes are Okeechobee (OK), Taihu (TU), Dianchi (DI), Balaton (BT), George (GO), Kasumigaura (KS), Saint Clair (SC), Chad (CD), Chapala (CP), and Piepsi (PE). These lakes also fall into the range of large size and low maximal depth that indicates a predominance of random redistribution of sediments by wind-induced waves (Hilton, 1985; Douglas et al., 2002). Results presented in this paper can be most readily generalized to these other large lakes, as well as to smaller lakes with shallower depth.

Qin et al. (2006) describe the processes involved in P flux from sediments to water in Lake Taihu, P.R. China. Lake Taihu is very similar in size and depth to a lake in the USA – Lake Okeechobee, Florida. In this paper, we examine the P dynamics of Lake Okeechobee as another representative case study for large shallow lakes, and focus in particular on how P dynamics vary at different temporal scales in these highly dynamic ecosystems.

Materials and methods

Characteristics of the study site

The physical, chemical, and biological characteristics of Lake Okeechobee have been extensively documented (e.g., Aumen, 1995; James et al., 1995a, b; Havens et al., 1996; Steinman et al., 2002a). Therefore, only a brief overview is provided here except for attributes specifically related to the topic of this paper. The lake has a nutrient residence time of 3.4 years (James et al., 1995a) and a high rate of external P loading (>0.3 g m^{-2} y^{-1}). That loading is substantially higher than a background load of 0.08 g m^{-2} y^{-1} estimated from sediment accumulation rates (Brezonik & Engstrom, 1998). Pelagic TP concentration now averages 110 μg l^{-1}, the ratio of total nitrogen to total P (TN:TP ratio, by mass) is 12:1, the ratio of dissolved inorganic N to soluble reactive P (DIN:SRP ratio, by mass) is 6:1, and cyanobacteria dominate the phytoplankton (Cichra et al., 1995; Havens et al., 2003). The lake is N-limited with surplus SRP in the water (Aldridge et al., 1995; Phlips et al., 1997). Despite N limitation and low N:P ratios, non N-fixing cyanobacteria (*Oscillatoria* and *Lyngbya*) are most common in the plankton because low underwater irradiance restricts the occurrence of N-fixing taxa (*Anabaena*, *Aphanizomenon*) to infrequent calm summer days (Havens et al., 1998, 2003).

The lake bottom is relatively flat and is 60–90 cm above sea level (Fig. 3a). An underwater reef of limestone near the southern end separates the central pelagic zone from shallower south and southwestern regions. On the western side of the lake, there is a relatively steep bottom slope at the edge of a low-gradient littoral marsh that supports

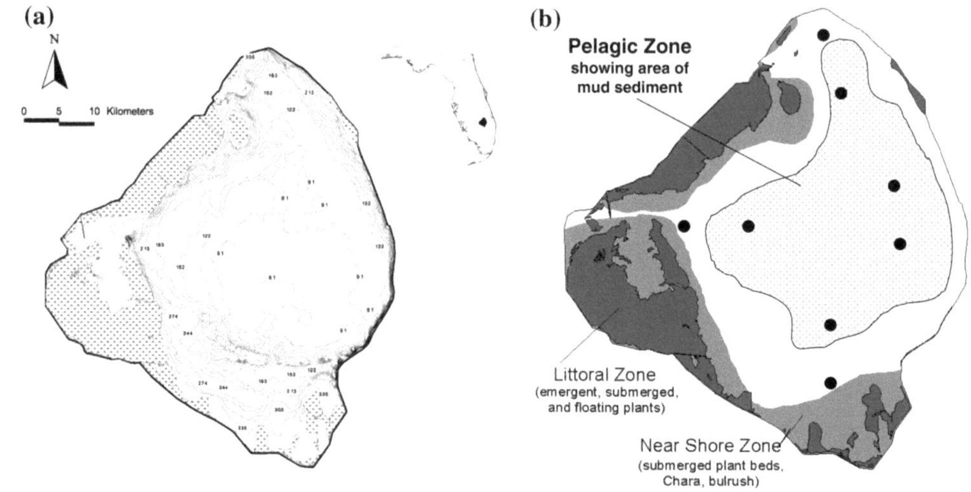

Fig. 3 (**a**) Bathymetric map of Lake Okeechobee, showing bottom elevations in cm above NGVD (National Geodetic Vertical Datum, formerly mean sea level). The lake's littoral marsh, a region of emergent vegetation, is indicated by the stippled area. The inset map indicates the lake's location in Florida, USA. (**b**) Map of Lake Okeechobee, showing three distinct habitats (pelagic, littoral, and near-shore), as well as location of fine mud sediments on the lake bottom and the eight long-term water quality monitoring stations of the South Florida Water Management District (black circles)

emergent vegetation over an area of 400 km^2. The entire lake, including the littoral zone, is enclosed by a dike constructed in the mid-1900s by the United States Army Corps (USACE) of Engineers.

As a result of two factors—water depth and sediment type—there also are three distinct zones (Fig. 4) in the lake that support different communities of primary producers and have markedly different water chemistry. A central pelagic zone is the deepest region of the lake. It is underlain by fine-grained mud sediments (Fisher et al., 2001) that are readily resuspended by wind-induced waves (Jin & Ji, 2001) as in other large shallow lakes (Otsubo & Muraoka, 1987; Luettich et al., 1990; Kristensen et al., 1992; Qin et al., 2006). The water column in this zone has high concentrations of P (Phlips et al., 1995), frequent light limitation of phytoplankton (Phlips et al., 1997), and it does not support vascular plants or benthic algae. Phytoplankton chlorophyll *a* concentrations do not correlate with TP (Havens et al., 1999) because most of the P is associated with resuspended sediment particles (details below). In sharp contrast, TP concentrations in the interior littoral zone are frequently below 10 μg l^{-1} (Hwang et al., 1998) because the main source of water input for this zone is rainfall. Little transport of nutrients occurs between the pelagic and littoral zones except when water level is very high (Havens, 1997). Lying between the pelagic and littoral zones is an area of intermediate depth, turbidity, and nutrient enrichment (Hwang et al., 1998; Havens & Walker, 2002). This region is referred to as the near-shore zone (Havens, 2002). When the lake experiences prolonged high water, as in the late 1990s, the near-shore zone becomes dominated by phytoplankton (Havens et al., 2001a). However, under shallow conditions it develops a dense community of submerged plants and clear water (Havens, 2003). The relationship between chlorophyll *a* and TP in the near-shore zone is similar to what is found in deeper Florida and North America lakes (Mazumder & Havens, 1998).

Data sources

This paper synthesizes prior work dealing with P dynamics in Lake Okeechobee, and aims to bring this information together in a manner that allows for comparisons of processes occurring at different temporal scales. Citations are provided to the original published sources of the information, where details can be found for field sampling,

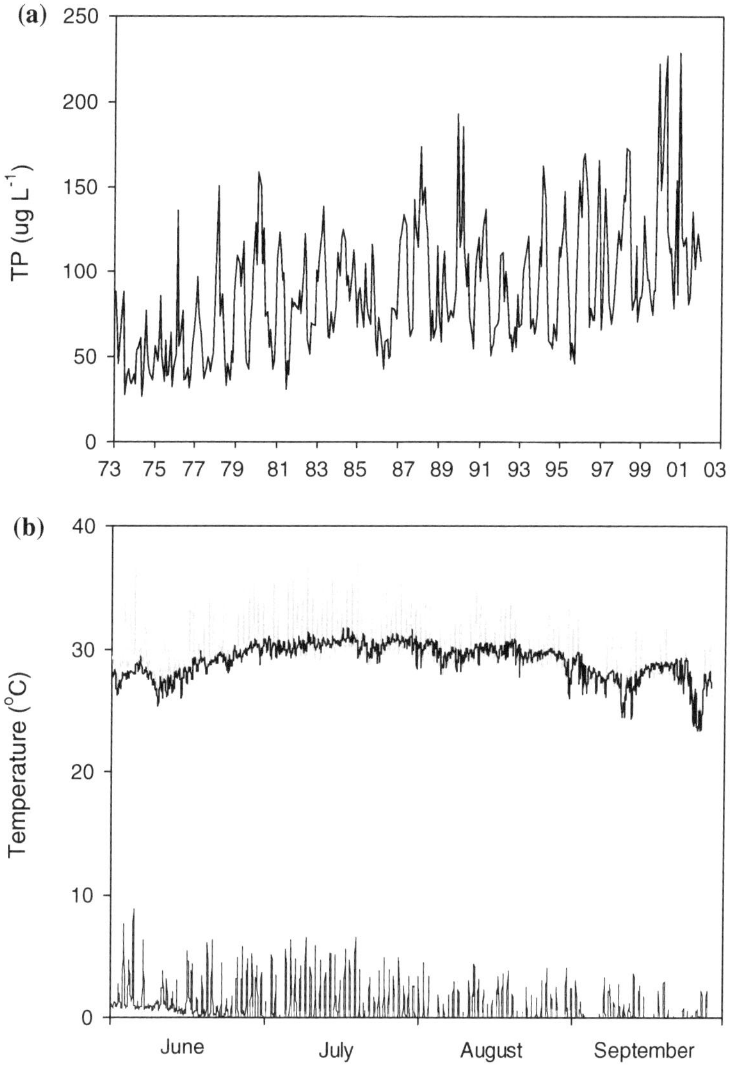

Fig. 4 (**a**) Historical record (1973–2002) of pelagic total phosphorus (TP) concentrations in Lake Okeechobee, based on monthly data collected at eight stations shown in Fig. 3. (**b**) Representative data regarding water temperatures near the surface (upper grey trace) and at the bottom (black trace) of the mid-lake water column in Lake Okeechobee. These are hourly data collected during summer 1999. The lower data set represents the difference between top and bottom temperature

laboratory methods, and experimental protocols. Only abbreviated information is presented here, coincident with presentation of the particular results and conclusions of each study. Most of the data were collected in the pelagic zone where the South Florida Water Management District (SFWMD) conducts monthly or semi-monthly water quality assessment at eight stations, which have existed since 1973 (Fig. 3b). Data from the near-shore zone, which has been sampled since the mid-1980s, also are presented, along with results from studies done by the University of Florida, under contract with the SFWMD. The littoral zone is not addressed here because it typifies a wetland more than a lake ecosystem. All of the water quality sampling and analyses carried out by the South Florida Water Management District (SFWMD), and the agencies and universities working under contract to the SFWMD, follows standard quality control standards that have been accepted by the State of Florida (SFWMD, 1998). Unless otherwise indicated, statistical significance is defined in this paper at $P = 0.05$.

Results

Greater than 30 years of data indicate that temporal dynamics in pelagic TP in Lake Okeechobee occur at seasonal, yearly, multi-year, and multi-decadal time scales (Fig. 4a). Considering first the short-term dynamics, we examined the hypothesis of Moore & Reddy (1994) that large fluxes of P from lake sediments might occur during brief periods of anoxia associated with transitory thermal stratification. During 1997–1999 we measured water temperature and dissolved oxygen (DO) at surface and bottom depths every 30 min with a YSI multi-parameter sonde. Representative data (Fig. 4b) for temperature indicate that in summer daily thermal stratification does occur, forming in morning and breaking up in the afternoon. However, anoxia never was observed (data not shown), with 95% of DO measurements exceeding 5 mg L^{-1} in the study period. These data do not support a major role for anoxia in stimulating sediment P release in this lake.

At the seasonal time scale, there is a regular pattern in water column TP (Fig. 5) that closely tracks seasonality of wind velocity. Concentrations of TP are highest in winter, when prevailing winds are strongest and lowest in summer when winds are relatively weak. The exception to this pattern is when hurricanes pass close to the lake, as occurred in October 1999 when a Category I storm crossed the Everglades (Fig. 6) and produced maximal sustained winds of 90 km h^{-1} over the center of Lake Okeechobee.

Results from the Lake Okeechobee Ecosystem Model (LOEM, Jin et al., 2000, 2002) indicate that the hurricane generated currents as strong as 50 cm s^{-1} over a large portion of the lake, and that large amounts of mud sediment were suspended from the center of the lake and transported to shoreline areas (Fig. 7). Field observations confirmed that fine mud sediment was deposited into bays in the south and west

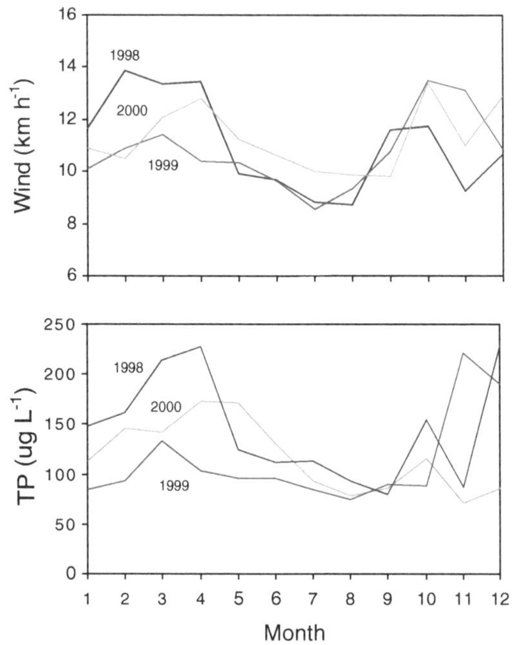

Fig. 5 Representative data regarding wind velocity at mid-lake and pelagic mean total phosphorus (TP) from the eight stations shown in Fig. 3. The three selected years are 1999, 2000, and 2001

regions of the lake where prior to the storm the sediments were comprised of peat and sand, respectively. Water quality monitoring data indicate that the hurricane resulted in lake-wide increases of TP from an average of 88 $\mu g\ l^{-1}$ before the storm to 222 $\mu g\ l^{-1}$ afterwards (Fig. 8).

Year-to-year variation in TP in Lake Okeechobee is quite pronounced and some of the variation can be explained by changes in water level (Fig. 9). A long-term increasing trend in water column TP (Fig. 10) coincides with increasing inflow TP from 1973 to 1985, but then contrasts with decreasing inflow TP afterwards.

Discussion

Pelagic phosphorus dynamics at hourly time scales – effects of wind and diffusion

In lakes where wind-generated waves and circulation are major factors controlling pelagic P dynamics, large changes in TP concentration can occur in short time intervals. Kristensen et al.

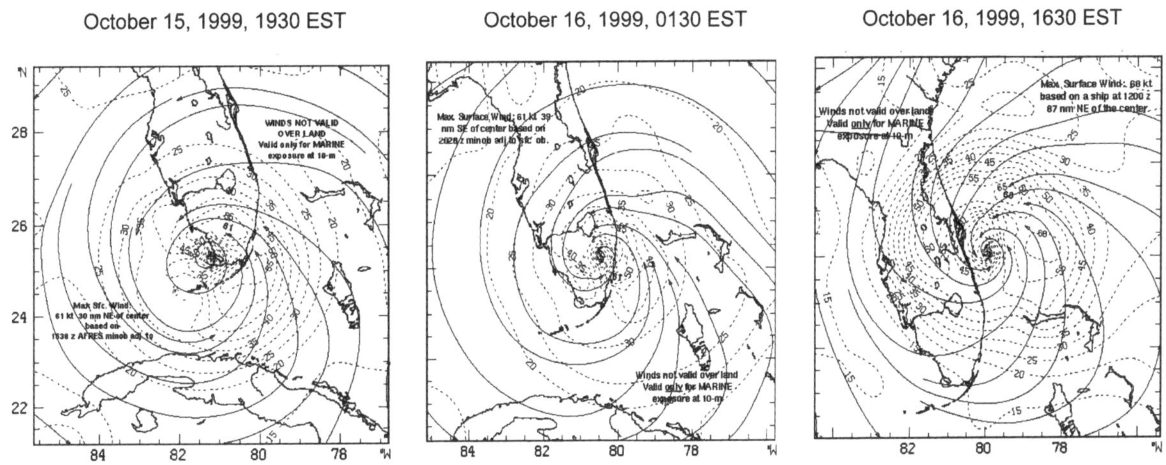

Fig. 6 Graphics from the National Hurricane Center, showing the movement of Hurricane Irene across the Florida peninsula, from 1930 h on October 15 to 1630 h on October 16, 1999. These figures are presented to illustrate the location of the eye of the storm as it moved south of Lake Okeechobee, as well as the wind field patterns. As indicated on the panels, the specific wind velocities are not valid over land

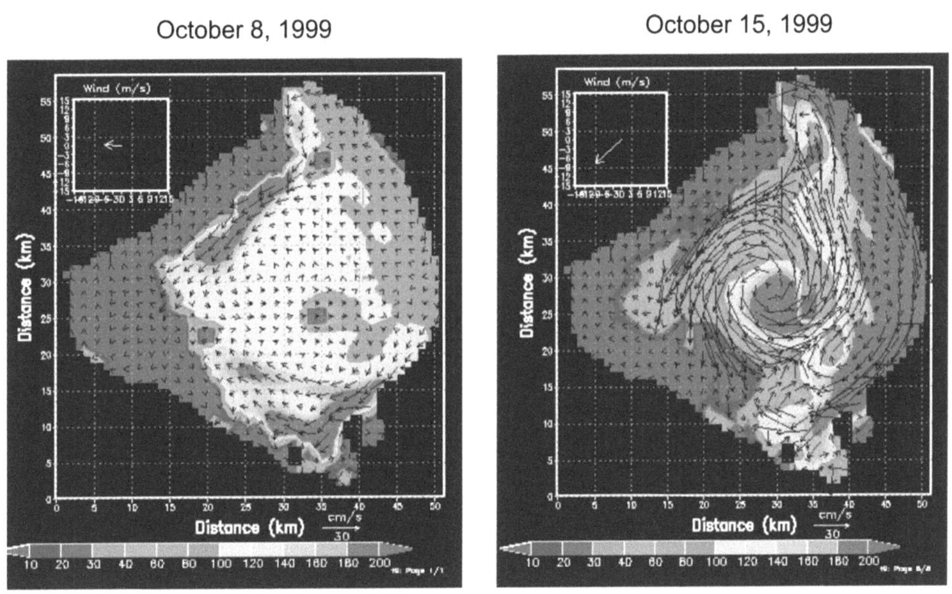

Fig. 7 Results of the Lake Okeechobee hydrodynamic and sediment transport model (Jin et al., 2000, 2001; Jin & Ji, 2003), showing surface current velocity vectors (black arrows) and resuspended sediment concentrations (grey scale) on October 8, before Hurricane Irene, and October 15, 1999, during the onset of the storm. The inset panel indicates wind velocity and direction recorded at the meteorological station at mid-lake

(1992) evaluated changes in TP in Lake Arreso, Denmark (mean depth 2.9 m) and documented increases of 100–400 μg P l^{-1} over time periods less than 12 h. Large increases in TP followed short-lasting periods of increased wind velocity with rapid declines in TP concentration after the winds had subsided. Similarly, Hamilton & Mitchell (1997) examined TP concentration on calm vs. windy days in seven lakes in New Zealand (mean depth < 1.5 m) and estimated that daily variations of TP were as great as 700 μg P l^{-1}. Sub-daily data are not available for Lake Okeechobee.

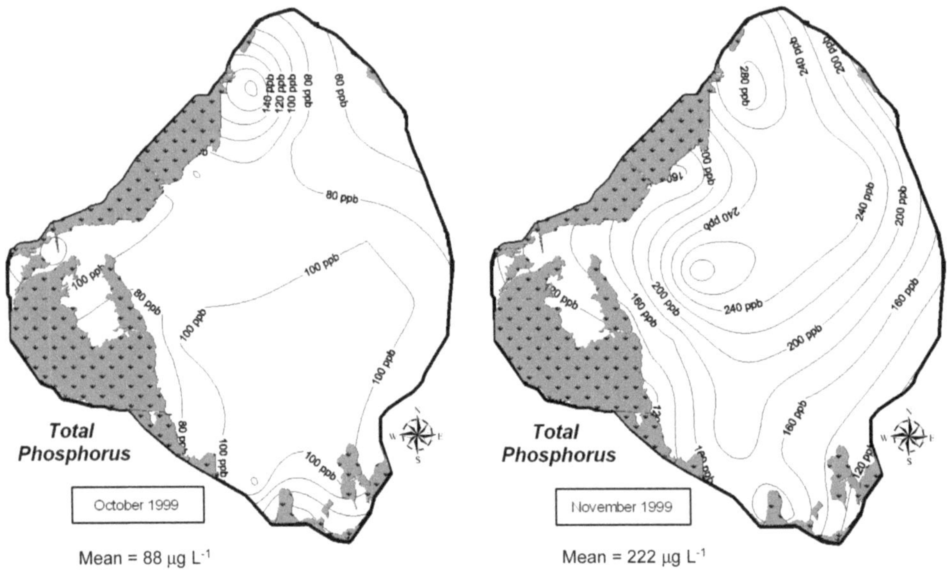

Fig. 8 Total phosphorus (TP) concentrations in the open water regions (pelagic and near-shore zones) of Lake Okeechobee, during October 1999, two weeks before Hurricane Irene, and November 1999, two weeks after the storm. Contours of TP concentration were interpolated from point data using Surfer® Software.

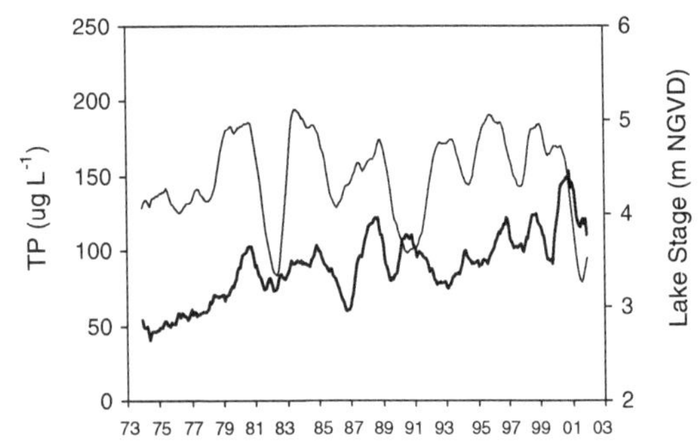

Fig. 9 Pelagic total phosphorus (TP) concentrations and stage in Lake Okeechobee, shown as 12-month rolling averages from 1973 to 2002

However, correlations exist between total suspended solids, vertical attenuation coefficient (K_0) and TP (Reddy et al., 1995; Hanlon et al., 1999; Sheng, 1999), allowing for an estimated occurrence of wind-induced high TP events. During summer, wind velocities are maximal in afternoon (coinciding with thunderstorm activity) and again in the evening due to winds off the Atlantic Ocean. Hanlon et al. (1999) found large increases in K_0 (from below 2 up to as high as 9 m^{-1} and lasting for just 1–2 h) within 2 h of maximal wind occurrence. Output from the LOEM indicate a similar time lag in Lake Okeechobee (Jin et al., 2002). The predicted time lag does not appear to be associated with energy transfer to the lake bottom, which occurs within 30 minutes or less. Rather, it seems to be associated with vertical transport of fine sediment particles to the upper water column. Sheng (1993) estimated that wind events of 30 km h^{-1} could increase water column TP by as much as 180 μg P l^{-1} within approximately the same period of time as estimated by the LOEM.

Wind-induced sediment resuspension is estimated to transport 6–18 times the amount of P to the water column as diffusive flux in Lake Okeechobee (Sheng, 1993; Moore & Reddy, 1994), and up to 6 times the amount of P found in external loads, when the data are compared on a yearly basis. This compares with the estimate by Qin et al. (2006) that TP loading due to resuspension in Lake Taihu is 5–10 times the external load. However, Reddy et al. (1995) noted that the amount of SRP released to the water column in a typical diurnal event is only a small fraction of the TP that is resuspended in a shallow lake. They concluded that the increase in SRP is limited by the high Fe content of this lake's sediments, and their low equilibrium P concentration, which

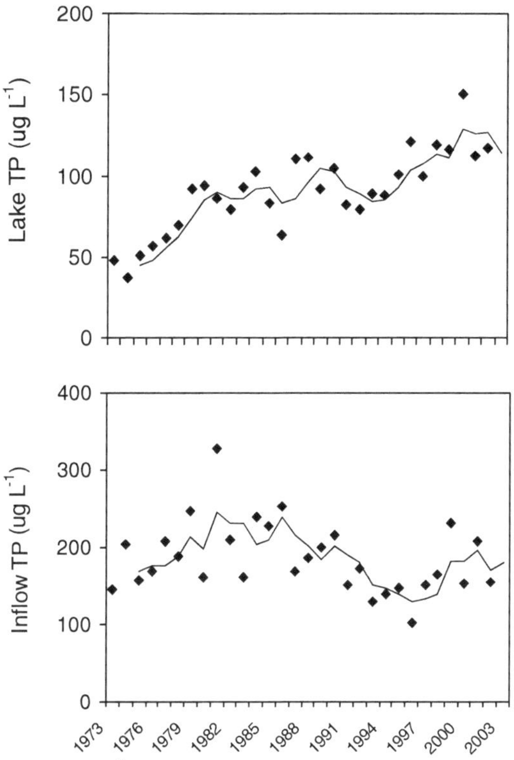

Fig. 10 Monthly mean concentrations of total phosphorus (TP) for Lake Okeechobee pelagic zone and inflows, from 1973 to 2002. The line indicates a long-term trend in the data for pelagic TP. The inflow data do not display a linear trend

favors sorption over desorption on sediment particles in the water column. Not all lakes may behave in this manner. Sediment resuspension may be a source, a sink, or have no effect on water column SRP depending on sediment geochemistry and other factors (e.g., Andersen, 1974; Peters & Cattaneo, 1984; Sondergaard et al., 1992). An opposite extreme from Lakes Okeechobee and Taihu is Lake Arreso, Denmark, where Sondergaard et al. (1992) documented a large net transport of SRP into the water column during resuspension events (20–30 times the SRP release from undisturbed sediments). They concluded that this massive SRP release occurred because (1) a large portion of sediment P in Lake Arreso occurred in a loosely-adsorbed organic form; (2) the sediments had a low ratio of Fe to P; and (3) there was a high equilibrium P concentration (50 to 180 μg l^{-1}). In contrast, Lakes Okeechobee and Taihu have high Fe:P ratios in their sediments, so that desorption of SRP during resuspension is not favored (Moore & Reddy, 1994; Qin et al., 2006).

Sondergaard et al. (1992) described a process whereby the combination of SRP desorption from resuspended sediment particles and rapid uptake of SRP by phytoplankton acts like a "pump" for transferring P into the water column of Lake Arreso. In that lake, light availability in the mixed layer is sufficient to allow a high biomass and productivity of phytoplankton to occur. In Lake Okeechobee, phytoplankton is severely light limited (Phlips et al., 1997) in the pelagic zone, has a low biomass, and is less likely to have this impact on P dynamics. Hence, while generalizations may be made about effects of wind resuspension on TP in large shallow lakes, the effects on SRP cannot be predicted without information regarding such factors as sediment biogeochemistry and mean irradiance in the water column.

Hourly variations of TP concentration in Lake Okeechobee are most pronounced in summer when there are large-scale changes in wind velocity between late morning (low wind) and afternoon-evening (higher wind, due to thunderstorms). During winter, when convective thunderstorms do not form around the lake and daily variations in wind from the ocean are less pronounced, regional low versus high pressure systems generally determine wind velocities over the lake. In that case it is common to have periods lasting for two or more days with strong sustained winds from a particular direction. As a result, the concentration of TP is more likely to remain elevated over multiple days, with less intra-daily variation.

Passive diffusion of P from lake sediments to the water column also can be important in shallow eutrophic lakes, as in deeper systems. However, when Fe chemistry regulates P release, the flux will depend on the occurrence of calm conditions of a sufficient duration to allow thermal stratification and bottom water anoxia. In profiles of intact sediments from Lake Okeechobee, soluble reactive P increased from near 0.1 mg l^{-1} at the sediment-water interface to 1.1 mg l^{-1} at sediment depths below 10 cm (Moore & Reddy, 1994). The authors concluded that large fluxes of P from the sediment could occur if the water column were to experience low dissolved oxygen levels, but as shown here that is unlikely to occur on a regular basis. Thus a relatively low rate of P release under oxidized conditions dominates the diffusive release of P to the water column in this lake. Similar results were obtained for Lake Taihu, where Qin et al. (2006) note that "constant dynamic disturbance frequently transports oxygen to the sediment-water interface" such that variations in dissolved oxygen and Eh are quite small.

Although diffusive P release may be small compared to the P transport associated with sediment resuspension, it still could play an important role in providing readily available P to the water column (Moore & Reddy, 1994). The effects on water quality then will depend on whether there are adequate light conditions for net phytoplankton growth during or subsequent to the resuspension event.

The results regarding P dynamics at hourly to daily time scales in Lake Okeechobee also are consistent with the findings of Kristensen et al. (1992), who concluded that in the shallow Danish lake (Lake Aresso, area 41 km^2, mean depth 3 m), the input of P to the lake water is dominated by resuspension. The same result is found in Lake Kasumigaura, Japan, where despite high rates of P release by diffusion (Ishikawa & Tanaka, 1993), the ratio of TP supply to the water column from resuspension vs. diffusion is 10:1 (Havens et al., 2001b). This is similar to the ratio estimated for Lake Okeechobee.

Seasonal dynamics of pelagic phosphorus

As a result of seasonal variation in wind velocity, and its influence on duration and magnitude of sediment resuspension (Jin & Ji, 2001), a predictable seasonal pattern exists for TP in Lake Okeechobee. Phlips et al. (1995) identified this seasonal pattern using contour maps of winter vs. summer TP content based on four years of integrated water samples from 84 locations in the lake. In winter, TP ranged from 50 to 110 $\mu g\ l^{-1}$, while in summer the range was between 30 and 80 $\mu g\ l^{-1}$. We found that in any given year, over 70% of the monthly variation in pelagic mean TP could be explained by wind velocity.

The response of phytoplankton to the winter increase in pelagic TP is minimal because most of that P is associated with abiotic seston particles. Those particles attenuate light to the extent that the algae are light limited during most of the winter months, when there is complete mixing of the water column (Aldridge et al., 1995; Phlips et al., 1995). One might consider that lower water temperatures and shorter photoperiod in winter also could limit phytoplankton biomass. However, this is unlikely because chlorophyll *a* concentrations in near-shore regions not influenced by sediment resuspension often are maximal during winter. That differing seasonality has been attributed to a reduced competition with plants and periphyton for water column phosphorus during winter months (Phlips et al., 1993).

Hurricane effects on pelagic phosphorus

In regions of the world that are influenced by tropical storms (hurricanes and cyclones), these extreme and unpredictable events may have strong impacts on P dynamics of lakes, yet those impacts are little studied. A major limitation is lack of substantive data prior to and after the event, which reflects the unpredictable timing and location of these severe storms.

There is evidence that a tropical storm contributed to a large decline in submerged plants in Chesapeake Bay in the 1970s and accompanying changes in water quality (Bayley et al., 1978), and it has been suggested that a hurricane caused the large-scale loss of plants and development of dense phytoplankton blooms in Lake Apopka, Florida in the 1940s (Clugston, 1963; Bachmann et al., 1999). We documented major impacts of a Category I hurricane in 1999 on sediment suspension and transport and water column TP in Lake Okeechobee.

When wind velocities decline abruptly, as they did after the 1999 hurricane passed beyond Florida, strong currents persist in the lake as residual circulation from the previous wind forcing event (Jin & Ji, 2001). Hence, sediment particles may be transported long distances from where they are resuspended, in particular if they are very fine-grained like the mud sediments that predominate in the center of Lake Okeechobee.

The modeling scenario illustrated here (Fig. 7) indicated that over 10,000 metric tons of fine-grained mud sediment was uplifted during the storm. Concentrations of TP remained high in the lake for two months after the hurricane, and displayed a second large increase (to above 230 $\mu g\ l^{-1}$) when another wind storm passed over the lake in late winter 2001.

During the last century, the Florida peninsula experienced landfall by a named tropical storm, on average, once per year, and landfall by a hurricane (wind velocity >120 km h^{-1}) every 2–3 years. A major hurricane (wind velocity >180 km h^{-1}) impacted the peninsula every 5–6 years. Wind velocities decline as a tropical storm moves over land, so sustained winds over the lake in any given storm might be considerably lower than at the time of landfall. Nevertheless, Lake Okeechobee is located in a position that has a high probability for tropical storm impacts. This also may be true for some large shallow lakes in lowland areas of Asia. Given that Hurricane Irene had low wind velocities for a tropical storm and did not pass directly over Lake Okeechobee, the impacts summarized here may be in the lower range of what can occur when a tropical storm passes over a large shallow lake. In August and September 2004, for example, three major hurricanes passed over the Florida peninsula, all causing substantial storm surges and resuspension events in Lake Okeechobee. We are in the process of analyzing the impacts, and will report those results in a future publication.

Of additional interest is the expected influence of long-term trends in ocean currents on future hurricane frequencies. During the period from 1995 to 2000, the Atlantic Ocean experienced the highest activity of hurricanes ever measured, and this has been linked with a warm phase of the Atlantic Ocean Thermohaline Current (AOTC), which is projected to last 20–40 years (Landsea, 2001). Thus the P dynamics of a lake impacted by hurricanes may be influenced by environmental changes occurring at a much greater spatial scale than the lake's own watershed. While limnologists have become attuned to the need to manage lakes in a watershed context, an even broader context may sometimes be required.

Inter-annual variations in pelagic phosphorus

Canfield & Hoyer (1988) suggested that the variation in yearly mean TP of the lake's pelagic zone was due to variations in water level ($r = 0.70$, 1973–1984). Havens (1997) confirmed that a statistical relationship existed for the years 1973–1995, albeit with a lower correlation ($r = 0.50$) than in Canfield & Hoyer's data set, and he summarized the various mechanisms that might explain the correspondence between lake stage and TP. Four hypotheses have been proposed: (1) littoral flooding occurs at high stage, and is followed by P release from dead plants (Canfield & Hoyer, 1988); (2) increased horizontal transport of resuspended sediment occurs when water levels are high (Maceina, 1993); (3) greater uptake of P by submerged plants and periphyton occurs when water levels are low (Phlips et al., 1993); and (4) co-variation of wind and water level occurs on a yearly basis, but wind actually is the driving factor (Maceina & Soballe, 1990) and water level only is a covariate, not a causal factor.

The first hypothesis is rejected based on several lines of evidence (Havens, 1997). The lake's littoral zone is nutrient-poor compared to the pelagic, and taken together, the biomass of all submerged and emergent plants and periphyton in the littoral and near-shore zones contain just 0.8% of the lake's P, compared to 97.1% in the sediments (Steinman et al., 1999). Dierberg (1992) and Harris et al. (1995) carried out decomposition experiments to quantify P release from submerged and emergent vegetation, respectively, in this lake. They concluded that the potential for increased TP in the pelagic zone due to P release from dead plants was on the order of 3–11 $\mu g\ l^{-1}$ much lower than the year-to-year variation in TP that has been observed with rising and falling lake water levels. Therefore, we focus here on the other three hypotheses, which seem more likely to be responsible for the TP variability at an intermediate time scale.

Water level, phosphorus, and horizontal mixing

Maceina (1993) suggested that the bottom topography of Lake Okeechobee, in particular the limestone reef near the southern and southwestern area, influences horizontal transport of sediment particles that are resuspended by wind-induced waves at mid-lake. He further suggested that at high lake stage (surface elevation > 4.9 m mean sea level) sediments could readily be transported over the top of the reef, whereas at low water levels (surface elevation < 4.6 m) this transport would be partially blocked. As a result, the lake-wide average TP, which is based on eight stations, two of which occur in the near-shore zone, would be higher at higher lake stages. To support this proposed mechanism, Maceina (1993) showed that the difference between pelagic and near-shore TP concentration declined with increasing lake water level, i.e., the lake became more homogenized. We examined this relationship using recent data from the lake, including data from the eight long-term stations, plus data from 30 stations located in the near-shore zone that have been sampled since 1986. The result was consistent with what Maceina (1993) observed. In the pelagic zone, TP increased only slightly with increasing lake stage, whereas in the near-shore zone, TP increased much more rapidly, from TP near 40 $\mu g\ l^{-1}$ at stage <4 m to TP near 80 $\mu g\ l^{-1}$ at stage ~5 m. As a result the difference in TP concentration between pelagic and near-shore zones dropped from near 60 $\mu g\ l^{-1}$ at low stage to near 20 $\mu g\ l^{-1}$ at high stage. The two pelagic stations located furthest south and west (Fig. 3b) respond to stage in a similar manner to the near-shore stations, and are largely responsible for the slight decline in TP at lower stage when the eight-station pelagic network is averaged.

In addition to the empirical relationships, hydrodynamic modeling (Sheng, 1993) supports the concept of greater horizontal mixing of the lake at high stage. Sheng concluded that "during low lake stage, wind-driven circulation gyres are generally confined to the open water (pelagic) zone, with little mixing into the vegetation (near-shore) zone." The LOEM gave a similar result when it is run under low vs. high stage conditions, with all other factors (e.g., wind velocity) being held constant.

In this particular lake it is important to understand why the near-shore zone displays a strong correspondence between TP and lake stage. That zone includes drinking water intakes for several municipalities, it is the main location of recreational fishing, and it provides critical vegetative habitat for fish, wading birds, and other wildlife (Havens, 2002). Hence, changes in TP concentration of the near-shore zone have a greater potential to impact ecological and societal values of the lake than changes in TP of the pelagic zone. This situation likely exists in other large shallow lakes.

Water level, phosphorus, and submerged plants

Another explanation for the reduced lake-wide TP at lower stage is that submerged plants and algae sequester large amounts of P from the water column, either by direct uptake (Burkholder et al., 1990; Hansson, 1990) or other processes associated with dense plant beds, including increased particle settling (Kufel & Kufel, 2002), sediment stabilization (Vermaat et al., 2000), and co-precipitation of P with calcium at high pH (Murphy et al., 1983). In Lake Okeechobee, it has been established that the biomass and spatial extent of submerged plants and algae is controlled by water level (Steinman et al., 2002b; Havens, 2003). In the late 1980s, following a drought and low lake stage, submerged plants covered between 18,000 and 21,000 hectares of the near-shore zone (Zimba et al., 1995), and nutrient-poor water and low biomass of phytoplankton coincided with seasonal maxima of submerged plants (Phlips et al., 1993). As noted above, this generally happens in winter. From 1995 to 1999, water levels in the lake were high, and the spatial extent of submerged plants declined to below 2,000 ha. Water in the near-shore zone displayed high TP concentrations and was extremely turbid (Havens et al., 2001a). During 2000–2001, another drought resulted in record low water levels in the lake, and the spatial extent of submerged plants again increased, to over 17,000 hectares (Havens et al., 2002). Havens (2003) found that the average biomass of submerged plants in the near-shore zone could be predicted from water depth and non-volatile suspended solids.

These results demonstrate correlation between plants and clear water, but cannot tell us whether plants are responsible for the improved water quality, or were simply responding to clear water caused by some other processes. Even if plants have local impacts on water quality, it is important to determine the spatial scale at which they influence TP in such a large lake, where much of the bottom (due to soft mud sediments) is not suitable for their colonization. At their maximal spatial extent in 2002, submerged plants covered just 13% of the combined surface area of the pelagic and littoral zones. It therefore is unlikely that they would substantially control TP concentration at a lake-wide scale. The vegetation might influence the 8-station average calculated in the long-term monitoring program, because 2 of those stations occur in areas that can support plants, but this effect on 25% of the stations should be limited. Only if there are strong currents passing through the near-shore zone would submerged plants be expected to influence lake-wide TP, and as discussed, water and nutrient transport to the near-shore zone is reduced when lake stage is low and plants are at their greatest density (Sheng, 1993). Submerged plants in Lake Okeechobee likely have a localized effect that somewhat dampens seasonal and year-to-year variations of TP, but physical processes are of considerably greater importance. Considering data from a seasonal survey program for submerged plants in Lake Okeechobee (Havens, 2003), there is a synchronized rise and fall in water depth, non-volatile solids (indicative of inorganic sediments), and phytoplankton chlorophyll *a* between 1999 and 2002 (Havens et al., 2004). This pattern occurs both at sites with and without submerged plants.

Wind and phosphorus at a yearly time scale

Wind has been identified as the main driving force behind P dynamics at hourly, daily, and seasonal time scales, and it also may be the main factor influencing year-to-year variations in pelagic TP. Maceina & Soballe (1990) examined this relationship, using pelagic TP data from 1975 to 1988. They found that yearly-averaged TP and stage were correlated ($r = 0.61$), but noted that the correlation between TP and wind velocity was stronger ($r = 0.78$). Havens (1997) examined a longer dataset (1973–1995) and obtained similar results (TP vs. stage, $r = 0.50$; TP vs. wind, $r = 0.64$).

We cannot identify the causal mechanisms behind year-to-year variations in TP based on field observations and correlations, and it is not possible to perform controlled experiments at a scale that is relevant to what occurs in the >1,000 km^2 pelagic zone of this lake. Furthermore, simple empirical models cannot effectively predict temporal changes in water column TP, chlorophyll *a*, and other attributes. In the case of Lake Okeechobee, our approach to understanding the driving factors for variations in TP and reasonably predicting future trends has been to develop the LOEM. Similar tools are used for management of large shallow coastal systems, such as the Chesapeake Bay (Cerco & Cole, 1994) and Tampa Bay, Florida (Yassuda & Sheng, 1997), and in other large shallow lakes including Lake Taihu.

Phosphorus dynamics over multiple decades

Underlying the seasonal and year-to-year variation in pelagic TP of Lake Okeechobee is a 30-year trend of increasing concentrations. This increase is approximately 20 $\mu g\ l^{-1}$ per decade ($P < 0.01$, seasonal Kendall's Tau test). The minima, maxima, and central tendency of the data all display a systematic increase, and a non-linear pattern can be seen in the smooth curve fit to these data. Paleolimnological data (Brezonik & Engstrom, 1998) indicate a trend of increasing P deposition rate in the lake that extends back as far as 1910.

The trend in pelagic TP documented during the last 30 years may be a result of several factors, including: (1) substantial changes in the TP concentration of water flowing into the lake; (2) a delayed response of the lake to recent reductions in TP input due to P recycling from sediments; and (3) reduced P assimilation by the ecosystem due to saturation effects. The concentration of TP in water flowing into the lake was near 150 $\mu g\ l^{-1}$ in the early 1970s, increased to over 200 $\mu g\ l^{-1}$ in the early 1980s, and then declined back to near 150 $\mu g\ l^{-1}$ in the early 1990s. Two recent years (1999 and 2001) are outliers, perhaps reflecting effects of (1) intense runoff during and after the hurricane in 1999, and (2) flushing of the watershed after a 2-year drought in 2001. Otherwise, there has not been any upward or downward trend in inflow TP during the last decade.

Declining inflow TP concentration after the 1980s has been attributed to an array of P management programs that were implemented in the watershed (Flaig & Havens, 1995; Havens et al., 1996). Coincident increases in lake water TP from the 1970s to 1980s, and a plateau of lake water TP from the 1980s to 1990s, are typical response patterns for shallow eutrophic lakes with high internal recycling of P (Sas, 1989). Concentrations of TP in the water column increase with increasing external inputs, but then remain high because the surface sediments (the source of internal P loading) reflect the past loading conditions until they are buried by new sediment material with a lower concentration of P. The sediment accrual rate in Lake Okeechobee is ~0.1 cm y^{-1} with the upper 5 cm of sediments interacting with the overlying water column through diffusion and resuspension (Reddy et al., 1995). As a result, the lake response time to changes in external P inputs is expected to be on the order of decades. For predicting response time of other shallow lakes to P load reductions, deposition rate and depth of active sediment layer are of particular importance, yet they are difficult to determine. Estimating net sediment deposition rate, for example, requires paleolimnological analysis and undisturbed sediment cores (Brezonik & Engstrom, 1998). In large shallow lakes that are heavily influenced by wind, it is important to confirm that sediments preserve an undisturbed historic record before using profiles to estimate process rates or to infer past conditions. In some lakes, this may not be the case. Depth of the active layer also can be influenced by bioturbation of macroinvertebrates (Van Rees et al., 1996), and in some lakes with reduced P loading, SRP migrates from deeper sediment layers as surface sediment P content is depleted through diffusion (Sondergaard et al., 2001).

There also is evidence that the sediments of Lake Okeechobee have become saturated with P over recent decades, and that this may contribute in part to the lake's poor response to reduced P inputs. Fisher et al. (2001), for example, documented that porewater SRP concentration in the lake's surface sediments increased significantly between 1988 and 1998, suggesting a saturation of binding sites. Further, the P mass balance for Lake Okeechobee indicates that the net P sedimentation rate declined four-fold from the 1970s to the 1990s (Havens & Schelske, 2001). This apparent P saturation led resource managers to consider sediment treatment, with calcium or alum, as a way to help speed recovery of the lake. However, due to the large size of Lake Okeechobee, its extreme shallow depth, and a variety of other engineering, ecological, and economic issues, this active management was determined to be non-feasible. Smaller lakes may be more suitable for this type of management (e.g., Cooke et al., 1993).

Conclusions

A substantial portion of the global storage of surface freshwater occurs in large shallow lakes (Herdendorf, 1984). These lakes typically are located in lowland areas (Fig. 11) that are readily used for agriculture and other human activities. As such, large shallow lakes are likely to receive substantial inputs of non-point source agricultural pollution, as well as solids from soil erosion in upstream watersheds. Unless they are rapidly flushed, their sediments become natural sinks for accumulation of P, nutrients, and other pollutants (Havens & Steinman, 1995). Based on the information presented in this case study of Lake Okeechobee, the following conclusions are arrived at regarding large shallow lakes, with emphasis on lakes that have a dynamic ratio substantially greater than 0.8 and occur in coastal areas influenced by sea breeze winds and/or tropical storms.

(1) Physical processes of sediment resuspension and horizontal circulation dominate ecosystem P dynamics.

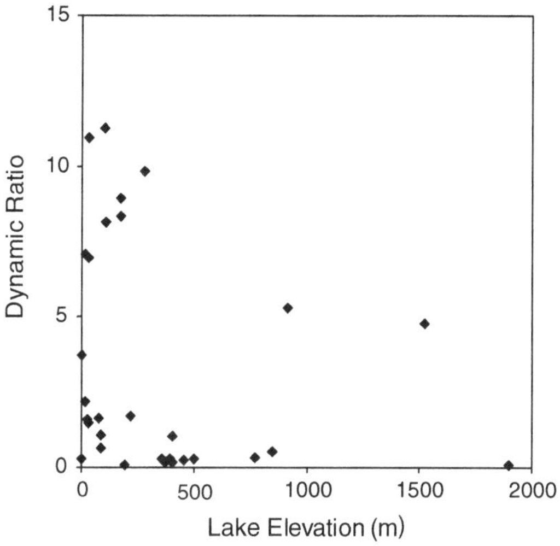

Fig. 11 Dynamic ratio of large lakes (>500 km^2) plotted as a function of their elevation above sea level. The dynamic ratio, calculated as the square root of surface area divided by mean depth, indicates a lake's susceptibility to wind effects. Ratios above 0.8 indicate high susceptibility (Bachmann et al., 2000). This figure illustrates that large lakes with high dynamic ratios occur in lowland regions of the world. The lakes plotted here are from the same dataset used to generate Fig. 2

(2) Rapid increases in P concentration can occur at time scales from days to weeks, and at a magnitude that is considerably greater than at longer time scales.
(3) The potential to modify trophic state by manipulating the food web is minimal because physics overwhelms biology in terms of controlling P dynamics, light availability, and other key attributes in this ecosystem.
(4) One exception is that by controlling water level, there might be potential to influence TP by promoting growth of submerged aquatic vegetation. However, depending on sediment type and bottom profile, this has a variable influence on lake-wide TP.
(5) Responses of lake water TP concentration to external P load reduction are significantly delayed by internal recycling, as indicated by Sas (1989), and this situation may be exacerbated where lakes are influenced periodically by tropical storms.
(6) Simple empirical models are not effective as tools for eutrophication management in lakes of this type. Resource managers will require more complex models that take into consideration lake hydrodynamics, with wind as a major external driver.

The emphasis in limnology clearly has shifted from traditional studies of temperate dimictic lakes to a more holistic consideration of the variety of lake types in the world (e.g., see Wetzel, 2001). Large shallow lakes, in particular, have received attention in special symposia in Europe (e.g., Van Liere & Gulati, 1992), and now in Asia (this special issue). Such lakes contain a large percentage of the world's fresh water, provide ecosystem services to hundreds of millions of people, and are located in areas prone to anthropogenic and climactic impacts. An understanding of their main driving factors, especially those related to limiting nutrients such as P, is critical to their protection. A challenge for the future is how to most effectively generalize results of research on lakes such as Okeechobee and Taihu, where funding for research and numerical model development has been favorable, to help address the water quality issues in similar ecosystems at locations where that level of investigation has not been possible.

Acknowledgements The authors are grateful to Peter Doering, Susan Gray, Garth Redfield, and two anonymous reviewers for providing comments on an earlier version of this paper.

References

Aldridge, F. J., E. J. Phlips & C. L. Schelske, 1995. The use of nutrient enrichment bioassays to test for spatial and temporal distribution of limiting factors affecting phytoplankton dynamics in Lake Okeechobee, Florida. Archiv für Hydrobiologie, Advances in Limnology 45: 177–190.
Andersen, J. M., 1974. Nitrogen and phosphorus budgets and the role of sediments in six shallow Danish lakes. Hydrobiologia 74: 528–550.
Aumen, N. G., 1995. The history of human impacts, lake management, and limnological research on Lake Okeechobee, Florida (USA). Archiv für Hydrobiologie, Advances in Limnology 45: 1–16.
Bachmann, R. W., M. V. Hoyer & D. E. Canfield Jr., 1999. The restoration of Lake Apopka in relation to alternative stable states. Hydrobiologia 394: 219–232.
Bachmann, R. W., M. V. Hoyer & D. E. Canfield Jr., 2000. The potential for wave disturbance in shallow Florida lakes. Lake and Reservoir Management 16: 281–291.

Bayley, S., V. Scotts, P. F. Springer & J. Stennis, 1978. Changes in submerged aquatic macrophyte populations at the head of Chesapeake Bay, 1958–1975. Estuaries 1: 171–182.

Brezonik, P. L. & D. R. Engstrom, 1998. Modern and historic accumulation rates of phosphorus in Lake Okeechobee, Florida. Journal of Paleolimnology 20: 31–46.

Burkholder, J. M., R. G. Wetzel & K. L. Klomparens, 1990. Direct comparison of phosphate uptake by adnate and loosely attached microalgae within an intact biofilm matrix. Applied and Environmental Microbiology 56: 2882–2890.

Canfield, D. E. Jr. & M. V. Hoyer, 1988. The eutrophication of Lake Okeechobee. Lake and Reservoir Management 4: 91–99.

Cerco C. F. & T. Cole, 1994. User's guide to the CE-QUAL-ICM three-dimensional eutrophication model. Technical Report EL-95-15, Waterways Experiment Station, US Army Corps of Engineers, Vicksburg, Mississippi, USA.

Caper, G. L. & R. W. Bachmann, 1984. Wind resuspension of sediments in a prairie lake. Canadian Journal of Fisheries and Aquatic Sciences 41: 1763–1767.

Cichra, M. F., S. Badylak, N. Henderson, B. H. Reuter & E. J. Phlips, 1995. Phytoplankton community structure in the open water zone of a shallow subtropical lake (Lake Okeechobee, Florida, USA). Archiv für Hydrobiologie, Advances in Limnology 45: 157–175.

Clugston, J. P., 1963. Lake Apopka, Florida, a changing lake and its vegetation. Florida Academy of Science 26: 168–174.

Cooke, G. D., E. B. Welch, S. A. Peterson & P. R. Newroth, 1993. Restoration and management of lakes and reservoirs. Lewis Publishers, Florida, USA.

Dierberg, F. E., 1992. The littoral zone of Lake Okeechobee as a source of phosphorus after drawdown. Environmental Management 6: 371–380.

Douglas, R. W., B. Rippey & C. E. Gibson, 2002. Interpreting sediment trap data in relation to the dominant redistribution process in a lake. Archiv für Hydrobiologie 155: 529–539.

Fisher, M. M., K. R. Reddy & R. T. James, 2001. Long-term changes in the sediment chemistry of a large shallow subtropical lake. Lake and Reservoir Management 17: 217–232.

Flaig, E. J. & K. E. Havens, 1995. Historical trends in the Lake Okeechobee ecosystem I. Land use and nutrient loading. Archiv für Hydrobiologie, Supplement 107: 1–24.

Hakanson, L., 1982. Lake bottom dynamics and morphometry: the dynamic ratio. Water Resources Research 18: 1444–1450.

Hamilton, D. P. & S. F. Mitchell, 1997. Wave-induced shear stresses, plant nutrients, and chlorophyll in seven shallow lakes. Freshwater Biology 38: 159–168.

Hanlon, C. G., R. L. Miller & B. F. McPherson, 1999. Relationships between wind velocity and underwater irradiance in a shallow lake (Lake Okeechobee, Florida, USA). Journal of the American Water Resources Association 34: 951–961.

Hansson, L. A., 1990. Quantifying the impact of periphytic algae on nutrient availability for phytoplankton. Freshwater Biology 24: 265–273.

Harris, T. T., K. A. Williges & P. V. Zimba, 1995. Primary productivity and decomposition of five emergent macrophyte communities in the Lake Okeechobee marsh ecosystem. Archiv für Hydrobiologie, Advances in Limnology 45: 63–78.

Havens, K. E., 1997. Water levels and total phosphorus in Lake Okeechobee. Lake and Reservoir Management 13: 16–25.

Havens, K. E., 2002. Development and application of hydrologic restoration goals for a large subtropical lake. Lake and Reservoir Management 18: 285–292.

Havens, K. E., 2003. Submerged aquatic vegetation correlations with depth and light attenuating materials in a shallow subtropical lake. Hydrobiologia 493: 173–186.

Havens, K. E., N. G. Aumen, R. T. James & V. H. Smith, 1996. Rapid ecological changes in a large subtropical lake undergoing cultural eutrophication. AMBIO 25: 150–155.

Havens, K. E., H. J. Carrick, E. F. Lowe & M. F. Coveney, 1999. Contrasting relationships between nutrients, chlorophyll a, and Secchi transparency in two shallow subtropical lakes: Lakes Okeechobee and Apopka (Florida, USA). Lake and Reservoir Management 15: 298–309.

Havens, K. E., T. Fukushima, P. Xie, T. Iwakuma, R. T. James, N. Takamura, T. Hanazato & T. Yamamoto, 2001b. Nutrient dynamics and the eutrophication of shallow lakes Kasumigaura (Japan), Donghu (P.R. China), and Okeechobee (USA). Environmental Pollution 111: 263–272.

Havens, K. E., R. T. James, T. L. East & V. H. Smith, 2003. N:P ratios, light limitation, and cyanobacterial dominance in a subtropical lake impacted by non-point source nutrient pollution. Environmental Pollution 122: 379–390.

Havens, K. E., M. C. Harewll, M. A. Brady, B. Sharfstein, T. L. East, A. J. Rodusky, D. Anson & R. P. Maki, 2002. Large-scale mapping and predictive modeling of submerged aquatic vegetation in a shallow eutrophic lake. The Scientific World Journal 2: 949–965.

Havens, K. E., K.-R. Jin, A. J. Rodusky, B. Sharfstein, M. A. Brady, T. L. East, N. Iricanin, R. T. James, M. C. Harwell & A. D. Steinman, 2001a. Hurricane effects on a shallow lake ecosystem and its response to a controlled manipulation of water level. The Scientific World Journal 1: 44–70.

Havens, K. E., E. J. Phlips, M. F. Cichra & B. L. Li, 1998. Light availability as a possible regulator of cyanobacteria species composition in a shallow subtropical lake. Freshwater Biology 39: 547–556.

Havens, K. E. & C. L. Schelske, 2001. The importance of considering biological processes when setting total maximum daily loads (TMDL) for phosphorus in shallow lakes and reservoirs. Environmental Pollution 113: 1–9.

Havens, K. E., B. Sharfstein, M. A. Brady, T. L. East, M. C. Harwell, R. P. Maki & A. J. Rodusky, 2004. Recovery of submerged plants from high water stress

in a large subtropical lake in Florida, USA. Aquatic Botany 78: 67–82.
Havens, K. E. & A. D. Steinman, 1995. Aquatic systems. In Rechcigl, J. E. (ed.), Soil Amendments—Impacts on Biotic Systems. Lewis Publishers, Florida, USA.
Havens, K. E. & W. W. Walker Jr., 2002. Development of a total phosphorus concentration goal in the TMDL process for Lake Okeechobee, Florida (USA). Lake and Reservoir Management 18: 227–238.
Herdendorf, C. E., 1984. Inventory of the morphometric and limnological characteristics of the large lakes of the world. Tech. Bull. Ohio Sea Grant, Columbus, Ohio, USA.
Hilton, J., 1985. A conceptual framework for predicting the occurrence of sediment focusing and sediment redistribution in small lakes. Limnology and Oceanography 30: 1131–1143.
Hutchinson, G. E., 1975. A Treatise on Limnology, volume 1, part 2—Chemistry of Lakes. John Wiley and Sons, New York, USA.
Hwang, S.-J., K. E. Havens & A. D. Steinman, 1998. Phosphorus kinetics of planktonic and benthic assemblages in a shallow subtropical lake. Freshwater Biology 40: 729–745.
Ishikawa, T. & M. Tanaka, 1993. Diurnal stratification and its effects on wind-induced currents and water qualities in Lake Kasumigaura, Japan. Journal of Hydraulic Research 31: 307–322.
James, R. T., B. L. Jones & V. H. Smith, 1995a. Historical trends in the Lake Okeechobee ecosystem II. Nutrient budgets. Archiv für Hydrobiologie, Supplement 107: 25–47.
James, R. T., J. Martin, T. Wool & P. F. Wang, 1997. A sediment resuspension and water quality model of Lake Okeechobee. Journal of the American Water Resources Association 33: 661–680.
James, R. T., V. H. Smith & B. L. Jones, 1995b. Historical trends in the Lake Okeechobee ecosystem III. Water quality. Archiv für Hydrobiologie, Supplement 107: 49–69.
Jin, K.-R. & Z.-G. Ji, 2001. Calibration and verification of a spectral wind-wave model for Lake Okeechobee. Ocean Engineering 28: 571–584.
Jin, K.-R., Z.-G. Ji & J. H. Hamrick, 2002. Modeling winter circulation in Lake Okeechobee, Florida. Journal of Waterway, Port, Coastal and Ocean Engineering 27: 114–125.
Jin, K.-R., J. H. Hamrick & T. Tisdale, 2000. Application of three-dimensional hydrodynamic model for Lake Okeechobee. Journal of Hydraulic Engineering 126: 758–771.
Kristensen, P., M. Sondergaard & E. Jeppesen, 1992. Resuspension in a shallow eutrophic lake. Hydrobiologia 228: 101–109.
Kufel, L. & I. Kufel, 2002. *Chara* beds acting as nutrient sinks in shallow lakes—a review. Aquatic Botany 72: 249–260.
Landsea, C., 2001. Testimony before the Committee on Science, Environment, Technology, and Standards Sub-Committee. United States House of Representatives, Washington, D.C.
Lean, D. R. S., 1973a. Phosphorus dynamics in lake water. Science 179: 678–680.
Lean, D. R. S., 1973b. Movements of phosphorus between its biologically important forms in lake water. Journal of the Fisheries Research Board of Canada 30: 1525–1536.
Luettich, R. A. Jr., D. R. F. Harleman & L. Somlyudy, 1990. Dynamic behavior of suspended sediment concentrations in a shallow lake perturbed by episodic wind events. Limnology and Oceanography 35: 1050–1067.
Maceina, M. J., 1993. Summer fluctuations in planktonic chlorophyll *a* concentrations in Lake Okeechobee, Florida: the influence of lake levels. Lake and Reservoir Management 8: 1–11.
Maceina, M. J. & D. M. Soballe, 1990. Wind-related limnological variation in Lake Okeechobee. Lake and Reservoir Management 6: 93–100.
Mazumder, A. & K. E. Havens, 1998. Nutrient-chlorophyll relationships in temperate vs. subtropical lakes. Canadian Journal of Fisheries and Aquatic Sciences 55: 1652–1662.
Moore, P. A. Jr. & K. R. Reddy, 1994. Role of Eh and pH on phosphorus biogeochemistry in sediments of Lake Okeechobee, Florida. Journal of Environmental Quality 23: 955–964.
Mortimer, C. H., 1941. The exchange of dissolved substances between mud and water in lakes (parts I and II). Journal of Ecology 29: 280–329.
Murphy, T., K. Hall & I. Yesaki, 1983. Co-precipitation of phosphate and calcite in a naturally eutrophic lake. Limnology and Oceanography 28: 58–67.
Otsubo, K. & K. Muraoka, 1987. Field studies on physical properties of sediment and sediment resuspension in Lake Kasumigaura. Japanese Journal of Limnology 48: 131–138.
Peters, R. H. & A. Cattaneo, 1984. The effects of turbulence on phosphorus supply in a shallow bay of Lake Memphremagog. Verhandlungen Internationale Vereinigung für Theoretische und Angewandte Limnologie 22: 185–189.
Phlips, E. J., M. Cichra, K. E. Havens, C. Hanlon, S. Badylak, B. Rueter, M. Randall & P. Hansen, 1997. Relationships between phytoplankton dynamics and the availability of light and nutrients in a shallow subtropical lake. Journal of Plankton Research 19: 319–342.
Phlips, E. J., F. J. Aldridge & P. Hansen, 1995. Patterns of water chemistry, physical and biological parameters in a shallow subtropical lake (Lake Okeechobee, Florida, USA). Archiv für Hydrobiologie, Advances in Limnology 45: 117–135.
Phlips, E. J., P. V. Zimba, M. S. Hopson & T. L. Crisman, 1993. Dynamics of the plankton community in submerged plant dominated regions of Lake Okeechobee, Florida, USA. Verhandlungen Internationale Vereinigung für Theoretische und Angewandte Limnologie 25: 423–426.
Qin, B., G. Zhu, L. Zhang, L. Luo, G. Gao, W. Sumin & B. Gu, 2006. Models and estimation methods of nutrient release from sediments of large shallow lakes:

a case study of Lake Taihu. Science in China (in press).

Reddy, K. R., 1993. Lake Okeechobee Phosphorus Dynamics Study. volume III - biogeochemical processes in the sediments. Technical Report to the South Florida Water Management District, West Palm Beach, Florida, USA.

Reddy, K. R., Y. P. Sheng & B. L. Jones, 1995. Lake Okeechobee Phosphorus Dynamics Study. volume I - summary. Technical Report to the South Florida Water Management District, West Palm Beach, Florida, USA.

Rigler, F. H., 1956. A tracer study of the phosphorus cycle in lake water. Ecology 37: 675–683.

Rigler, F. H., 1964. The phosphorus fractions and the turnover time of inorganic phosphorus in the epilimnion of lakes. Canadian Fish Culturist 32: 3–9.

Rigler, G. H., 1973. A dynamic view of the phosphorus cycle in lakes. In Griffith, E. J., A. Beeton, J. M. Spencer & D. T. Mitchell (eds), Environmental Phosphorus Handbook. John Wiley and Sons, New York, USA.

Reynolds C. S., 1994. The Ecology of Freshwater Phytoplankton. Cambridge Univiversity Press.

Sas, H., 1989. Lake restoration by reduction of nutrient loadings: expectations, experiences, extrapolations. Academia Verlag Richarz, Germany.

Scheffer, M., 1998. Ecology of Shallow Lakes. Chapman and Hall, New York, USA.

Sheng, Y. P., 1993. Hydrodynamics, sediment transport and their effects on phosphorus dynamics in Lake Okeechobee. In Mehta, A. J. (ed.), Coastal and Estuarine Studies 42: Near-shore and Estuarine Cohesive Sediment Transport. American Geophysical Union, Washington, D.C.

Sheng, Y. P., 1999. Effects of hydrodynamic processes on phosphorus distribution in aquatic ecosystems. In Reddy, K. R., G. A. O'Conner & C. L. Schelske (eds), Phosphorus Biogeochemistry in Subtropical Ecosystems. Lewis Publishers, Florida, USA.

Sondergaard, M., J. P. Jensen & E. Jeppesen, 2001. Retention and internal loading of phosphorus in shallow, eutrophic lakes. The ScientificWorld Journal 1: 427–442.

Sondergaard, M., P. Kristensen & E. Jeppesen, 1992. Phosphorus release from resuspended sediment in the shallow and wind-exposed Lake Arreso, Denmark. Hydrobiologia 228: 91–99.

Steinman, A. D., K. E. Havens, N. G. Aumen, R. T. James, K-R. Jin, J. Zhang & B. H. Rosen, 1999. Phosphorus in Lake Okeechobee: sources, sinks, and strategies. In Reddy, K. R., G. A. O'Conner & C. L. Schelske (eds), Phosphorus Biogeochemistry in Subtropical Ecosystems. Lewis Publishers, Florida, USA.

Steinman, A. D., K. E. Havens, H. J. Carrick & R. VanZee, 2002a. The past, present, and future hydrology and ecology of Lake Okeechobee ant its watersheds. In Porter, J. W. & K. G. Porter (eds), The Everglades, Florida Bay, and Coastal Reefs of the Florida Keys, an Ecosystem Restoration Sourcebook. CRC Press, Florida, USA.

Steinman, A. D., K. E. Havens, A. J. Rodusky, B. Sharfstein, R. T. James & M. C. Harwell, 2002b. The influence of environmental variables and a managed water recession on the growth of charophytes in a large, subtropical lake. Aquatic Botany 72: 297–313.

Van Liere, L. & R. D. Gulati, 1992. Restoration and Recovery of Shallow Eutrophic Lake Ecosystems in the Netherlands. Kluwer Academic Publishers, The Netherlands.

Van Rees, K. C. J., K. R. R. Reddy & P. S. C. Rao, 1996. Influence of benthic organisms on solute transport in lake sediments. Hydrobiologia 317: 31–40.

Vermaat, J. E., L. Santamaria & P. J. Roos, 2000. Water flow across and sediment trapping in submerged macrophyte beds of contrasting growth form. Arch. Hydrobiologia 148: 549–562.

Wetzel, R. G., 2001. Limnology—Lake and River Ecosystems, 3rd edn. Academic Press, New York, USA.

Yassuda, E. A. & Y. P. Sheng, 1997. Integrated modeling of Tampa Bay estuarine system. Technical Report, Coastal and Ocean Engineering Department, University of Florida, Gainesville, Florida, USA.

Zimba, P. V., M. S. Hopson, J. P. Smith, D. E. Colle & J. V. Shireman, 1995. Chemical composition and distribution of submersed aquatic vegetation in Lake Okeechobee, Florida (1989–1991). Archiv für Hydrobiologie, Advances in Limnology 45: 241–246.

Hydrobiologia (2007) 581:43–52
DOI 10.1007/s10750-006-0520-6

EUTROPHICATION IN LAKES

Chromophoric dissolved organic matter (CDOM) absorption characteristics in relation to fluorescence in Lake Taihu, China, a large shallow subtropical lake

Yunlin Zhang · Boqiang Qin · Guangwei Zhu · Lu Zhang · Longyuan Yang

Abstract Absorption measurements from chromophoric dissolved organic matter (CDOM) and their relationships with dissolved organic carbon (DOC) and fluorescence were studied in Lake Taihu, a large, shallow, subtropical lake in China. Absorption spectra of lake water samples were measured from 240 nm to 800 nm. Highest values of $a(\lambda)$, DOC and $F_n(355)$ occurred near the river inflow to Meiliang Bay and decreased towards the central lake basin. A significant spatial difference was found between Meiliang Bay and the central lake basin in absorption coefficient, DOC-specific absorption coefficient, exponential slope coefficient, DOC concentration and fluorescence value. The spatial distribution of CDOM suggested that a major part of CDOM in the lake was from river input. CDOM absorption coefficients were correlated with DOC over the wavelength range 280–500 nm, and $a(355)$ was also correlated with $F_n(355)$, which showed that CDOM absorption could be inferred from DOC and fluorescence measurement. The coefficient of variation between $a(\lambda)$ and DOC concentration decreased with increase in wavelength from 240 nm to 800 nm. Furthermore, a significant negative linear relationship was recorded between S value and CDOM absorption coefficient, as well as DOC-specific absorption coefficient. S value and DOC-specific absorption coefficient were used as a proxy for CDOM composition and source. Accurate CDOM absorption measurements are very useful in explaining UV attenuation and in developing, validating remote sensing model of water quality in Lake Taihu.

Keywords Chromophoric dissolved organic matter (CDOM) · Fluorescence · Dissolved organic carbon (DOC) · DOC-specific absorption coefficient · Spectral slope coefficient · Lake Taihu

Guest editors: B. Qin, Z. Liu & K. Havens
Eutrophication of shallow lakes with special reference to Lake Taihu, China

Y. Zhang · B. Qin (✉) · G. Zhu · L. Zhang · L. Yang
Nanjing Institute of Geography & Limnology, Chinese Academy of Sciences, Nanjing 210008, P. R. China
e-mail: qinbq@niglas.ac.cn

Y. Zhang · L. Zhang
Graduate school of Chinese Academy of Sciences, Beijing 100039, P. R. China

Introduction

Chromophoric dissolved organic matter (CDOM) is the light absorbing (ultraviolet and visible) fraction of dissolved organic carbon (DOC) (Kirk, 1994). Spatial and temporal distribution of CDOM in aquatic ecosystems can affect ecosystem productivity; negatively impacting primary productivity since CDOM absorbs light, while positively impacting secondary productivity by providing a substitute for microbial respiration via photo-degraded CDOM (Benner & Biddanda, 1998; Reche et al., 1998; Mazzuoli et al., 2003). Overlaps of pigment absorption spectra with CDOM absorption at blue wavelengths complicate the use of chlorophyll *a* retrieval algorithms based on remotely sensed ocean color (Carder et al., 1991) and lead to chlorophyll *a* concentration overestimation (Nelson et al., 1998; Rochelle-Newall & Fisher, 2002a). In addition, high CDOM concentrations can act as a photoprotectant against UV damage to aquatic organisms (Williamson et al., 1996). Several studies have identified CDOM as the most important factor controlling UV attenuation in natural freshwaters (Morris et al., 1995; Laurion et al., 2000). However, protection of aquatic biota against UV radiation provided by dissolved humic material may be diminished with CDOM photodegradation by UV radiation (Morris & Hargreaves, 1997) and acidification (Schindler et al., 1996; Yan et al., 1996).

In shallow, inland lakes and coastal waters, absorption by non-chlorophyllous substances often exceeds that by phytoplankton (Carder et al., 1991; Frenette et al., 2003). CDOM originates from phytoplankton degradation and river inputs and land drainage, especially in inland waters. Although CDOM influences lake optical properties, remotely sensed spectra, and biogeochemical processes, the CDOM dynamics in diverse aquatic environments are not well known. A thorough inventory of regional CDOM distribution and its light absorption in wavebands covering UV and visible radiation spectra have not been conducted in Lake Taihu. This information is needed to interpret satellite and airplane remote sensing imagery to be used to record lake response to changes in environmental conditions, such as water pollution, eutrophication and cyanobacterial blooms. In this paper, we describe the absorption spectrum of CDOM in lake water samples from north Lake Taihu. To examine CDOM, the following aspects are addressed in this paper:

(1) Spatial distribution of CDOM in different lake regions interpreted by absorption spectra;
(2) The relationship between CDOM absorption coefficients and DOC concentrations;
(3) Variations in the absorption spectra with wavelength and fluorescence.

Materials and methods

Study area

Forty sampling sites were chosen in the northern lake. Sampling site 24 was located at the boundary

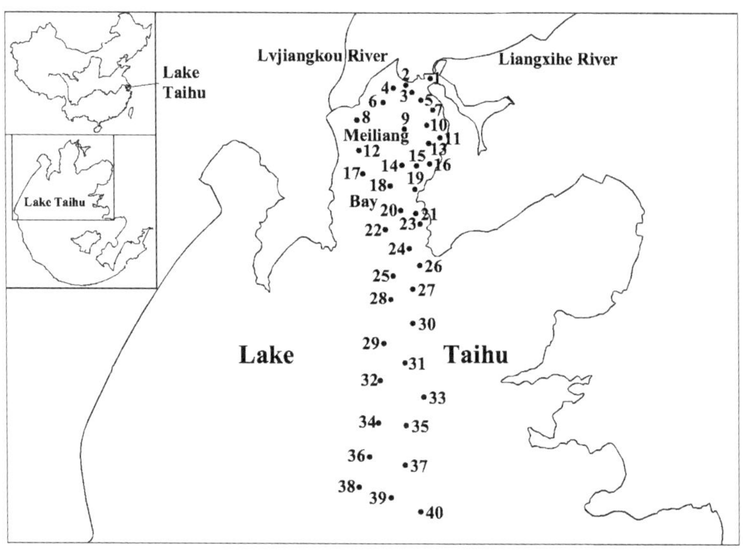

Fig. 1 Position of sampling sites during the cruise

of Meiliang Bay and the central lake. Sites 1–24 were distributed in Meiliang Bay, and other sites were in the central lake (Fig. 1). Meiliang Bay is eutrophic with an area of 132 km^2 and mean depth of 2.0 m. Lujiangkou and Liangxihe Rivers flow into the bay (Fig. 1).

Absorption of CDOM and DOC concentration

Water from 0 m to 0.5 m depth for CDOM absorption and DOC analysis was collected in 250 ml acid-cleaned plastic bottles and held on ice while in the field. In the laboratory, these samples were kept at 4°C until analysis within two days. All samples were filtered at low pressure through pre-combusted Whatman GF/F filters (0.7 μm) into glass bottles pre-combusted at 550°C for 6 h (Del Castillo et al., 1999; Del Castillo & Coble, 2000; Rochelle-Newall & Fisher, 2002b; Yacobi et al., 2003; Frenette et al., 2003; Callahan et al., 2004; Chen et al., 2004). Absorption spectra were obtained between 240 nm and 800 nm at 1-nm intervals using a Shimadzu UV-2401PC UV-Vis recording spectrophotometer with matching 4 cm quartz cells. Milli-Q water was used in the reference cell. Absorption coefficients were obtained by using following expression (Kirk, 1994):

$$a(\lambda') = 2.303D(\lambda)/r \quad (1)$$

where $a(\lambda')$ is uncorrected CDOM absorption coefficient at wavelength λ, $D(\lambda)$ is the optical density at wavelength λ, and r is the cuvette path length in m. Absorption coefficients were corrected for backscattering of small particles and colloids, which pass through filters, using Eq. 2 (Bricaud et al., 1981; Green & Blough, 1994).

$$a(\lambda) = a(\lambda') - a(750) \times \lambda/750 \quad (2)$$

where $a(\lambda)$ = absorption coefficient at wavelength (λ) corrected for scattering, and $a(750)$ = measured absorption coefficient at 750 nm corrected for scattering. Because of the chemical complexity of CDOM, concentration is expressed using the absorption coefficient at some reference wavelength, typically 355 nm, 375 nm or 440 nm (Kirk, 1994; Del Castillo & Coble, 2000; Stedmon et al., 2000). Here, absorption at 355 nm was used to analyze the relationship of CDOM and fluorescence. Molecular size of humic molecules was estimated from the ratio of absorption coefficients at 250 and 365 nm [$a(250)/a(365)$] (De Haan, 1993; Peuravuori & Pihlaja, 1997). The apparent DOC-specific absorption coefficient $a^*(\lambda)$ was calculated by the relationship below (Seritti et al., 1998):

$$a*(\lambda) = a(\lambda)/\text{DOC} \quad (3)$$

where DOC is expressed in mg l^{-1}. DOC in the filtrates was measured on a 1020 TOC analyzer.

Exponential slope coefficient calculation

Exponential slope coefficient was calculated from the absorption spectrum between 280 and 500 nm ($S1$), 280 and 360 nm ($S2$), and 360 and 440 nm ($S3$) using the non-linear regression of Equation (4) (Bricaud et al., 1981). The coefficient of determination (r^2) for the fit of S was consistently higher than 0.99.

$$a(\lambda) = a(\lambda_0)\exp[S(\lambda_0 - \lambda)] \quad (4)$$

where $a(\lambda)$, $a(\lambda_0)$ are the absorption coefficients at wavelengths λ and λ_0, respectively, and S is the exponential slope coefficient.

Fluorescence measurements

Fluorescence was measured using a 1-cm quartz cell in a Shimadzu 5301 spectrofluorometer with an excitation wavelength of 355 nm and an emission wavelength of 450 nm. A Milli-Q water blank was used for comparison (Seritti et al., 1998; Rochelle-Newall & Fisher, 2002b).

Data analysis

Statistical analyses were performed with SPSS 11.0 software (Statistical Program for Social Sciences).

Results and discussion

Spectral character and spatial distribution of CDOM

Figure 2a, b shows representative absorption spectra of samples from northern Meiliang Bay (3), the Bay inlet (24), and the central lake basin (31, 40), which represent a range of CDOM concentrations. Spectral curves for CDOM generally showed near zero absorption at the red end of the visible spectrum (700 nm) and exponentially increase through the ultraviolet (UV) wavelength regimes (280–400 nm). Mean corrected CDOM absorption coefficients at 280 nm [$a(280)$] and 355 nm [$a(355)$] were 18.3 ± 5.6 and 4.6 ± 1.8 m^{-1}, respectively, for the 40 sites. CDOM absorption coefficients at 280, 355 and 440 nm for each site are shown in Fig. 3. The highest value of $a(355)$ was 8.6 m^{-1} at site 15 close to shore in Meiliang Bay, and the lowest value was 2.2 m^{-1} at site 40 near the center of the lake (Figs. 1, 3). In general, $a(355)$ decreased away from the river inflow to Meiliang Bay towards the lake center (Figs. 1–3). CDOM absorption in the Meiliang Bay inlet were around twice those in the central basin. All sites were divided into two groups, one representing Meiliang Bay (sites 1–24) and one representing the central basin (sites 25–40). Significant differences were found between Meiliang Bay and the central basin arising from spatial differences in absorption coefficients, DOC-specific absorption coefficients, exponential slope coefficients, DOC concentrations and fluorescence values using ANOVA ($P \leq 0.005$). Values of $a(280)$ and $a(355)$ in Lake Taihu are higher than those in estuarine and coastal waters (Seritti et al., 1998; Chen et al., 2004), but CDOM absorption values are close to those in large, shallow, fluvial Lake Saint-Pierre (Frenette et al., 2003) and others (Kirk, 1994; Morris et al., 1995). The ratio of $a(250)$ to $a(365)$ ranged from 5.28 to 9.24 with a mean value of 7.10 ± 0.94 (standard deviation), which was close to results from other studies (De Haan, 1993; Huovinen et al., 2003). Higher ratios corresponding to smaller molecules were found in the lake center in agreement with findings from Huovinen et al. (2003) and Yacobi et al. (2003).

Relationship between CDOM absorption and DOC

DOC concentrations ranged from a maximum of 10.9 mg l^{-1} at the Liangxihe mouth (site 1) to a minimum of 6.0 mg l^{-1} in the central basin (Site 37), with a mean value of 8.17 ± 1.38 mg l^{-1}. A significant positive correlation was found between DOC and CDOM absorption coefficients in the ultraviolet and blue band (280–500 nm), especially at shorter wavelengths (Fig. 4a), in agreement with other studies where DOC was dominated by CDOM (Seritti et al., 1998; Stedmon et al., 2000; Laurion et al., 2000). Figure 4b shows that the coefficients of determination (r^2) between CDOM absorption and DOC decrease with increasing wavelength from 280 nm to 500 nm. Similar results were found in shallow lakes in the middle and lower reaches of the Changjiang (Yangtze) River (Zhang et al., 2005).

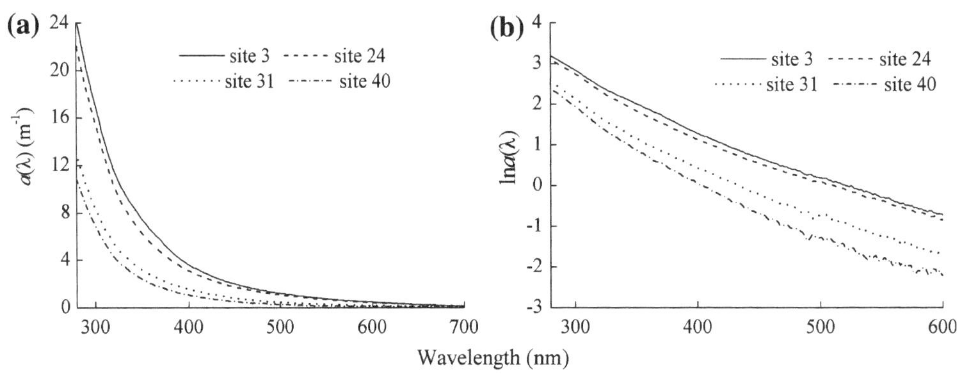

Fig. 2 CDOM absorption spectra of representative sites: (**a**) Wavelength versus $a(\lambda)$ and (**b**) wavelength versus $\ln[a(\lambda)]$

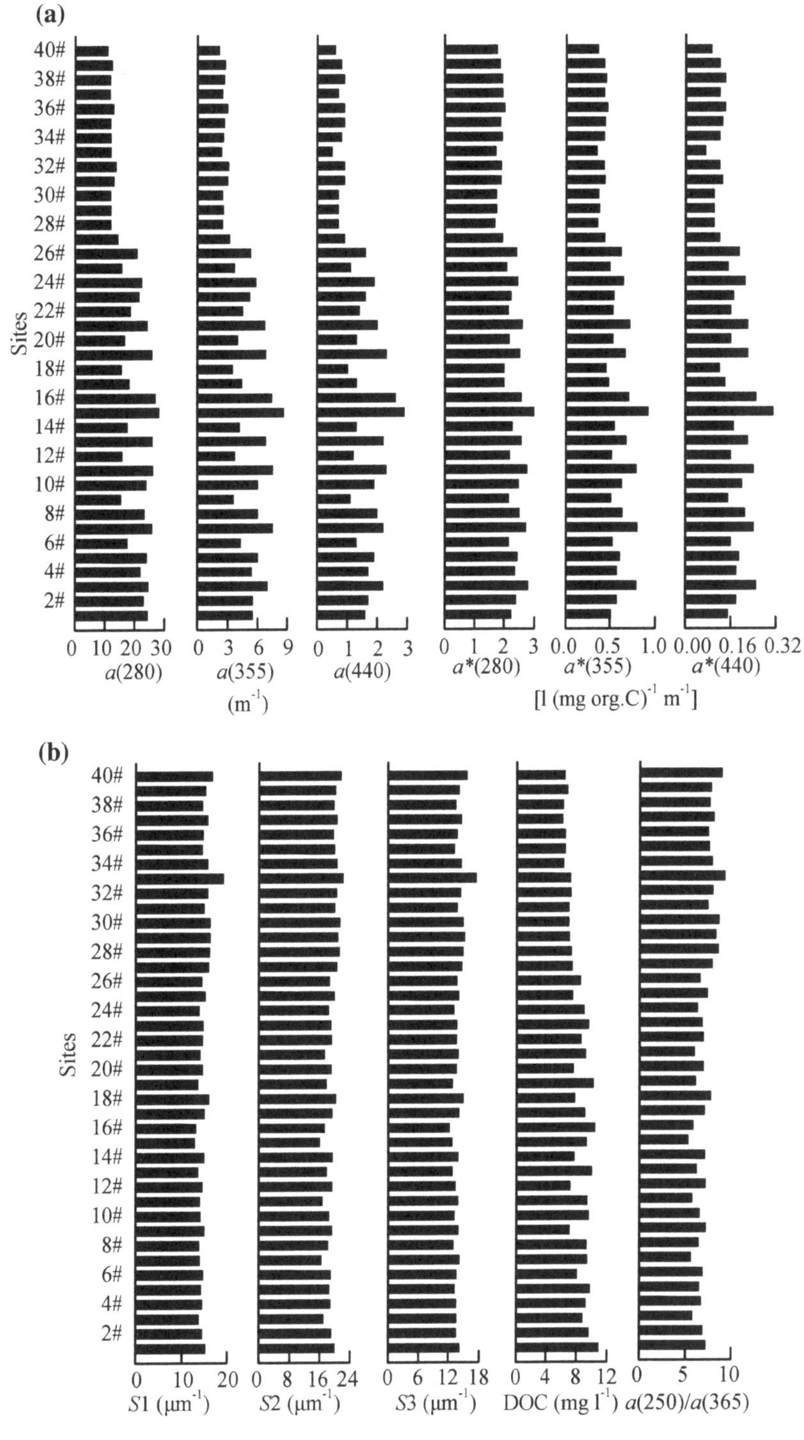

Fig. 3 Absorption coefficients *a*(280), *a*(355), *a*(440), DOC-specific absorption coefficients, *a**(280), *a**(355), *a**(440), spectral slope *S*1 (280–500 nm), *S*2 (280–360 nm), *S*3 (360–440 nm), DOC, fluorescence and ratio of *a*(250) to *a*(365)

Relationships between DOC and CDOM absorption from this study and other studies are presented in Table 1. Published relationships of DOC and *a*(355) are valuable tools to model DOC concentration (Seritti et al., 1998; Rochelle-Newall & Fisher, 2002b). However, differences between models suggest regional and lake-to-lake variations in bio-optical characteristics.

Calculated DOC-specific absorption provides an additional means for distinguishing bio-optical

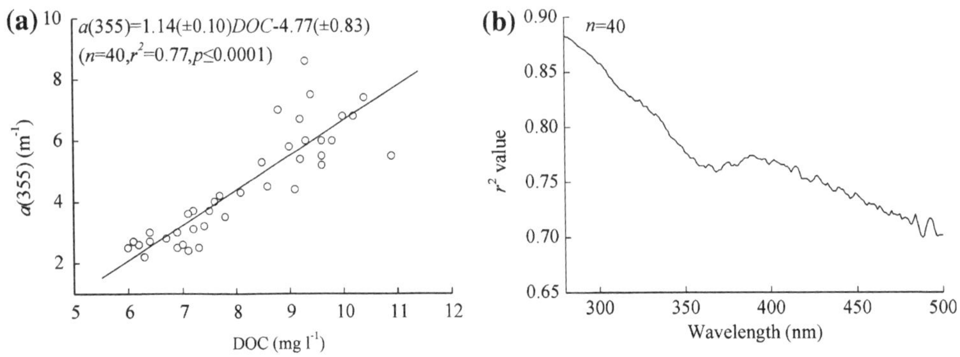

Fig. 4 Plots of CDOM absorption and DOC concentration (**a**) CDOM absorption $a(355)$ versus DOC concentration and (**b**) wavelength versus determination coefficients r^2 value between $a(\lambda)$ and DOC concentration from 280 nm to 500 nm

Table 1 Linear models for the relationship between DOC concentrations and CDOM absorption coefficients

Equations		r^2	n	P	DOC (mg l^{-1})	References
DOC, $a(\lambda)$	$a(355) = 1.465\text{DOC} - 1.315$	0.98	31	≤0.0001	0.89–4.82	Seritti et al. (1998)
	$a(440) = 0.280\text{DOC}$	0.91	6	≤0.0007	4.9–40.7	Yacobi et al. (2003)
	$a(440) = 0.235\text{DOC}$	0.92	6	≤0.0005	7.1–40.5	Yacobi et al. (2003)
	$a(440) = 0.429\text{DOC} - 0.744$	0.73	64	≤0.0001	0.24–23.5	Morris et al. (1995)[a]
	$a(300) = 3.584\text{DOC} - 2.690$	0.71	26	≤0.0001	0.77–3.31	Del Castillo et al. (1999)[a]
	$a(440) = 0.131\text{DOC} + 0.020$	0.58	39	≤0.0001	0.21–3.5	Laurion et al. (2000)[a]
	$a(440) = 0.552\text{DOC} - 1.477$	0.77	12	≤0.0005	4.9–14.9	Huovinen et al. (2003)[a]
	$a(355) = 0.015\text{DOC} - 5.5$	0.90	92	≤0.0001	–	Ferrari et al. (1996)[b]
	$a(355) = 0.00036\text{DOC} + 0.028$	0.66	43	≤0.0001	–	Ferrari (2000)[b]
	$a(355) = 0.744\text{DOC} - 2.176$	0.58	22	≤0.001	2.7–10.1	Zhang et al. (2005)
	$a(355) = 1.14\text{DOC} - 4.77$	0.77	40	≤0.0001	6.0–10.9	Present study
	$a(375) = 0.87\text{DOC} - 3.74$	0.77	40	≤0.0001	6.0–10.9	Present study
	$a(440) = 0.38\text{DOC} - 1.73$	0.73	40	≤0.0001	6.0–10.9	Present study

[a] Fitted with the data in papers ; [b] DOC unit: μM

characteristics of CDOM. Low CDOM absorption was associated with low values of $a^*(\lambda)$. Values of $a^*(355)$ ranged from 0.34 to 0.92 with an average 0.54 ± 0.14 [l (mg org C)$^{-1}$ m^{-1}], which are higher than those observed in estuarine and coastal waters (Seritti et al., 1998). Lowest values of $a^*(320)$ and $a^*(355)$ corresponded with low DOC concentrations and approached those measured in coastal water and mountain lakes (Seritti et al., 1998; Laurion et al., 2000). DOC-specific absorption coefficient represents CDOM composition, which is used to establish relationships with other parameters. Specific absorption was correlated with pH and DOC concentration in three acid lakes (Gallie, 1997). In our study, specific absorption was correlated positively with DOC concentration. The linear regression equation is $a^*(355) = 0.075\text{DOC} - 0.070$ ($r^2 = 0.53$, $n = 40$, $P \le 0.0001$). The linear correlation between DOC concentration and specific absorption is known from other studies (Yacobi et al., 2003). Furthermore, the higher average $a^*(440)$ corresponded to the higher proportion of large and medium fractions in CDOM (Yacobi et al., 2003). In this study, higher $a^*(355)$ and $a^*(440)$ values corresponded to lower $a(250)/a(365)$ values towards increasing molecular size. The regression equation between $a^*(355)$ and $a(250)/a(365)$ values is: $a^*(355) = -0.145a(250)/a(365) + 1.570$ ($r^2 = 0.92$, $n = 40$, $P \le 0.0001$).

Estimation of the spectral slope S

It is widely accepted that spectral slope S can be used as a proxy for CDOM composition (Kowalczuk et al., 2003). The values of the exponential slope coefficient (S) ranged from 12.7 μm^{-1} to 19.0 μm^{-1} with an average value of 14.74 ± 1.13 μm^{-1}, from 15.9 μm^{-1} to 22.0 μm^{-1} with an average value 19.11 ± 1.42 μm^{-1}, and from 12.2 to 17.4 μm^{-1} with an average value 13.86 ± 0.91 μm^{-1} for the wavelength domain from 280 nm to 500 nm ($S1$), 280–360 nm ($S2$) and 360–440 nm ($S3$), respectively. The coefficient of determination (r^2) for the fit of S was higher than 0.99. Although slope values in the present investigation fell within the range of previously reported values (Bricaud et al., 1981; Davies-Colley, 1983; Seritti et al., 1998; Markager & Vincent, 2000), variability of the spectral slope in different wavelength ranges is significant, which illustrates the importance and sensitivity of the wavelength range chosen to calculate the slope coefficient (S). S generally increases with decrease in the wavelength range and may be influenced by CDOM composition (i.e. humic and fulvic acids affecting spectral properties). Samples collected from the Mississippi River plume in the Gulf of Mexico, where CDOM was comprised predominantly from fulvic acids, had a steeper slope (S = 0.0194) than marine samples, where CDOM was mostly humic acids (S = 0.010) (Carder et al., 1989). S values in Lake Taihu were higher than those for oceanic and coastal waters (Markager & Vincent, 2000).

In our study, there are significant negative correlations between spectral slope S and $a(355)$ (Fig. 5a) and $a^*(355)$ (Fig. 5b, Table 2). Determination coefficients between S and $a^*(\lambda)$ were higher than those between S and $a(\lambda)$. It is also interesting that determination coefficients are the highest between $S2$ (280–306 nm) and $a(\lambda)$, $a^*(\lambda)$. Furthermore, significant negative correlations have been obtained between S values and $a^*(440)$ (Yacobi et al., 2003), $a(355)$ (Rochelle-Newall et al., 2004) and $a(375)$ (Del Castillo &

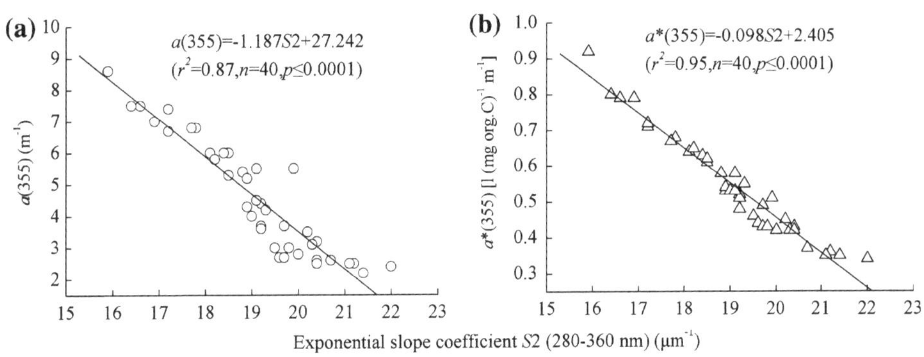

Fig. 5 Plots of exponential slope coefficient versus (**a**) absorption coefficient $a(355)$ and (**b**) specific absorption coefficient $a^*(355)$

Table 2 Linear models for the relationship between $S1$ (280–500 nm), $S2$ (280–360 nm), $S3$ (360–440 nm) and absorption coefficient, DOC-specific absorption coefficient of CDOM (n = 40, $P \leq 0.0001$)

Parameters		Equations	r^2
$a(355)$, S	$S1$ (280–500 nm)	$a(355) = -1.254(\pm 0.160)S1 + 23.045(\pm 2.357)$	0.62
	$S2$ (280–360 nm)	$a(355) = -1.187(\pm 0.075)S2 + 27.242(\pm 1.444)$	0.87
	$S3$ (360–440 nm)	$a(355) = -1.158(\pm 0.262)S3 + 20.613(\pm 3.646)$	0.34
$a(375)$, S	$S1$ (280–500 nm)	$a(375) = -0.976(\pm 0.119)S1 + 17.772(\pm 1.751)$	0.64
	$S2$ (280–360 nm)	$a(375) = -0.910(\pm 0.057)S2 + 20.779(\pm 1.087)$	0.87
	$S3$ (360–440 nm)	$a(375) = -0.919(\pm 0.197)S3 + 16.133(\pm 2.733)$	0.37
$a^*(355)$, S	$S1$ (280–500 nm)	$a^*(355) = -0.103(\pm 0.011)S1 + 2.064(\pm 0.168)$	0.68
	$S2$ (280–360 nm)	$a^*(355) = -0.098(\pm 0.004)S2 + 2.405(\pm 0.067)$	0.95
	$S3$ (360–440 nm)	$a^*(355) = -0.095(\pm 0.020)S3 + 1.858(\pm 0.279)$	0.37
$a^*(375)$, S	$S1$ (280–500 nm)	$a^*(375) = -0.083(\pm 0.008)S1 + 1.627(\pm 0.124)$	0.72
	$S2$ (280–360 nm)	$a^*(375) = -0.077(\pm 0.003)S2 + 1.868(\pm 0.049)$	0.96
	$S3$ (360–440 nm)	$a^*(375) = -0.079(\pm 0.015)S3 + 1.495(\pm 0.211)$	0.42

Coble, 2000; Stedmon et al., 2000). Kowalczuk et al. (2003) presented a trend of increased slope coefficient with a decrease in absorption level.

Relationship between CDOM absorption and fluorescence

CDOM absorption may be measured directly using a spectrophotometer or inferred from fluorescence, which can be measured more rapidly and with greater sensitivity than absorption. In previous studies, a strong relationship was established between CDOM absorption and fluorescence intensity (Vodacek et al., 1995; Seritti et al., 1998; Ferrari & Dowell, 1998; Hoge et al., 1993; Kowalczuk et al., 2003) despite numerous sources of CDOM and different chemical characteristics. CDOM fluorescence $F_n(355)$ and absorption coefficients $a(355)$ were correlated positively in Lake Taihu in this study (Fig. 6). The correlation may enable the use of fluorescence as a proxy for CDOM absorption despite different sources from river inputs and autochthonous humic substances and locations within the lake.

Conclusions

(1) Higher CDOM absorption coefficients were found in the Meiliang Bay inlet of Lake Taihu in association with increased DOC concentrations and fluorescence intensity. There were significant differences between Meiliang Bay and the central lake basin in absorption coefficients, DOC-specific absorption coefficient $a^*(355)$, exponential slope coefficient (S), DOC concentrations and fluorescence values.

(2) There is a significant linear relationship between CDOM absorption and DOC concentrations, especially at shorter wavelengths in Lake Taihu samples.

(3) Exponential slope coefficients fall in the range reported in the literatures. There is a significant negative linear relationship between S values and CDOM absorption coefficients, as well as DOC-specific absorption coefficients.

(4) There is a linear relationship ($r = 0.66$, $P \leq 0.0001$) between CDOM absorption and CDOM fluorescence intensity: $a(355) = 2.12(\pm0.33) + 3.72(\pm0.43)F_n(355)$.

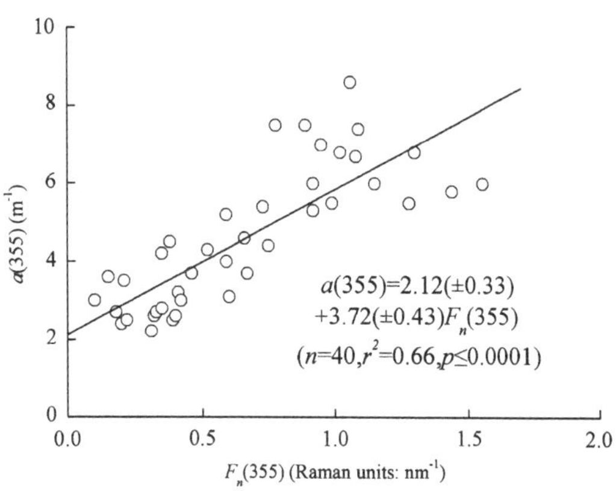

Fig. 6 Plot of $a(355)$ versus $F_n(355)$

Acknowledgements This work was supported by Chinese Academy of Sciences (No. KZCX1-SW-12) and National Natural Science Foundation of China (Nos. 40601099, 40501078), and Director Foundation of Nanjing Institute of Geography and Limnology, CAS (No. NIGLA-S220445). We specially thank Dr. Mark J. McCarthy for his linguistic improvement.

References

Benner, R. & B. Biddanda, 1998. Photochemical transformations of surface and deep marine dissolved organic matter: Effects on bacterial growth. Limnology and Oceanography 43: 1373–1378.

Bricaud, A., A. Morel & L. Prieur, 1981. Absorption by dissolved organic matter of the sea (yellow substance) in the UV and visible domain. Limnology and Oceanography 26: 43–53.

Callahan, J., M. H. Dai, R. F. Chen, X. L. Li, Z. M. Lu & W. Huang, 2004. Distribution of dissolved organic matter in the Pearl River Estuary, China. Marine Chemistry 89: 211–224.

Carder, K. L., S. K. Hawes, R. C. Smith, R. G. Steward & B. G. Mitchell, 1991. Reflectance model for quantifying chlorophyll *a* in the presence of productivity degradation products. Journal of Geophysical Research 9: 20599–20611.

Carder, K. L., R. G. Steward, G. R. Harvey & P. B. Ortner, 1989. Marine humic and fulvic acids: their effects on remote sensing of ocean chlorophyll. Limnology and Oceanography 34: 68–81.

Chen, Z. Q., Y. Li & J. M. Pan, 2004. Distributions of colored dissolved organic matter and dissolved organic carbon in the Pearl River Estuary, China. Continental Shelf Research 24: 1845–1856.

Davies-Colley, R. J., 1983. Absorption of light by yellow substance in freshwater lakes. Limnology and Oceanography 32: 416–425.

De Haan, H., 1993. Solar UV-light penetration and photodegradation of humic substances in peaty lake water. Limnology and Oceanography 38: 1072–1076.

Del Castillo, C. E., P. G. Coble, J. M. Morell, J. M. López & J. E. Corredor, 1999. Analysis of the optical properties of the Orinoco River plume by absorption and fluorescence spectroscopy. Marine Chemistry 66: 35–51.

Del Castillo, C. E. & P. G. Coble, 2000. Seasonal variability of the colored dissolved organic matter during the 1994–95 NE and SW Monsoons in the Arabian Sea. Deep-Sea Research II 47: 1563–1579.

Ferrari, G. M., 2000. The relationship between chromophoric dissolved organic matter and dissolved organic carbon in the European Atlantic coastal area and in the West Mediterranean Sea Gulf of Lions. Marine Chemistry 70: 339–357.

Ferrari, G. M. & M. D. Dowell, 1998. CDOM absorption characteristics with relation to fluorescence and salinity in coastal areas of the Southern Baltic Sea. Estuarine, Coastal and Shelf Science 47: 91–105.

Ferrari, G. M., M. D. Dowell, S. Grossi & C. Targa, 1996. Relationship between the optical properties of chromophoric dissolved organic matter and total concentration of dissolved organic carbon in the southern Baltic Sea region. Marine Chemistry 55: 299–316.

Frenette, J. J., M. T. Arts & J. Morin, 2003. Spectral gradients of downwelling light in a fluvial lake (Lake Saint-Pierre, St-Lawrence River). Aquatic Ecology 37: 77–85.

Gallie, E. A., 1997. Variation in the specific absorption of dissolved organic carbon in Northern Ontario lakes. In Steven, G. A. & F. Robert (eds), Ocean Optics XIII, Proc. SPIE 2963, 417–422.

Green, S. & N. Blough, 1994. Optical absorption and fluorescence properties of chomophoric dissolved organic matter in natural waters. Limnology and Oceanography 39: 1903–1916.

Hoge, F. E., A. Vodacek & N. V. Blough, 1993. Inherent optical properties of the ocean: retrieval of the absorption coefficient of chromophoric dissolved organic matter from fluorescence measurements. Limnology and Oceanography 38: 1394–1402.

Huovinen, P. S., H. Penttolä & M. R. Soimasuo, 2003. Spectral attenuation of solar ultraviolet radiation in humic lakes in Central Finland. Chemosphere 51: 205–214.

Kirk, J. T. O., 1994. Light and Photosynthesis in Aquatic Ecosystem. Cambridge University Press, Cambridge, Britain, 57–71.

Kowalczuk, P., W. J. Cooper, R. F. Whitehead, M. J. Durako & W. Sheldon, 2003. Characterization of CDOM in an organic-rich river and surrounding coastal ocean in the South Atlantic Bight. Aquatic Science 65: 384–401.

Laurion, I., M. Ventura, J. Catalan, R. Psenner & R. Sommaruga, 2000. Attenuation of ultraviolet radiation in mountain lakes: factors controlling the among- and within-lake variability. Limnology and Oceanography 45: 1274–1288.

Markager, W. & W. F. Vincent, 2000. Spectral light attenuation and absorption of UV and blue light in natural waters. Limnology and Oceanography 45: 642–650.

Mazzuoli, S., S. Loiselle, V. Hull, L. Bracchini & C. Rossi, 2003. The analysis of the seasonal, spatial, and compositional distribution of humic substances in a subtropical shallow lake. Acta Hydrochimica et Hydrobiologia 31: 461–468.

Morris, D. P. & B. P. Hargreaves, 1997. The role of photochemical degradation of dissolved organic carbon in regulating the UV transparency of three lakes on the Pocono Plateau. Limnology and Oceanography 42: 239–249.

Morris, D. P., H. Zagarese, C. E. Williamson, E. G. Balseiro, B. R. Hargreaves, B. Modenutti, R. Moeller & C. Queimalinos, 1995. The attenuation of solar UV radiation in lakes and the role of dissolved organic carbon. Limnology and Oceanography 40: 1381–1391.

Nelson, N. B., D. A. Siegel & A. F. Michaels, 1998. Seasonal dynamics of colored dissolved organic material in the Sargasso Sea. Deep-Sea Research I 45: 931–957.

Peuravuori, J. & K. Pihlaja, 1997. Molecular size distribution and spectroscopic properties of aquatic humic substances. Analytica Chimica Acta 337: 133–149.

Reche, I., M. L. Pace & J. J. Cole, 1998. Interactions of photobleaching and inorganic nutrients in determining bacterial growth on colored dissolved organic carbon. Microbial Ecology 6: 270–280.

Rochelle-Newall, E. J. & T. R. Fisher, 2002a. Production of chromophoric dissolved organic matter fluorescence in marine and estuarine environments: an investigation into the role of phytoplankton. Marine Chemistry 77: 7–21.

Rochelle-Newall, E. J. & T. R. Fisher, 2002b. Chromophoric dissolved organic matter and dissolved organic carbon in Chesapeake Bay. Marine Chemistry 77: 23–41.

Rochelle-Newall, E., B. Delille, M. Frankignoulle, J. P. Gattusol, S. Jacquet, U. Riebesell & A. Terbruggen, 2004. Chromophoric dissolved organic matter in experimental mesocosms maintained under different pCO_2 levels. Marine Ecology Progress Series 272: 25–311.

Schindler, D. W., P. J. Curtis, B. R. Parker & M. P. Stainton, 1996. Consequences of climate warming and lake acidification for UV-B penetration in North American boreal lakes. Nature 379: 705–708.

Seritti, A., D. Russo, L. Nannicini & R. Del Vecchio, 1998. DOC, absorption and fluorescence properties of estuarine and coastal waters of the Northern Tyrrhenian Sea. Chemical Speciation and Bioavailability 10: 95–106.

Stedmon, C. A., S. Markager & H. Kaas, 2000. Optical properties and signatures of chromophoric dissolved organic matter (CDOM) in Danish coastal waters. Estuarine, Coastal and Shelf Science 51: 67–278.

Vodacek, A., F. E. Hoge, R. N. Swift, J. K. Yungel, E. T. Peltzer & N. V. Blough, 1995. The use of in situ and airborne fluorescence measurements to determine UV absorption coefficients and DOC concentrations in surface waters. Limnology and Oceanography 40: 411–415.

Williamson, C. E., R. S. Stemberger, D. P. Morris, T. M. Frost & S. G. Paulsen, 1996. Ultraviolet radiation in North American lakes: attenuation estimates from DOC measurements and implications for plankton communities. Limnology and Oceanography 41: 1024–1034.

Yacobi, Y. Z., J. J. Alberts, M. Takács & M. Mcelvaine, 2003. Absorption spectroscopy of colored dissolved organic carbon in Georgia (USA) rivers: the impact of molecular size distribution. Journal of Limnology 62: 41–46.

Yan, N. D., W. Keller, N. M. Scully, D. R. S. Lean & P. J. Dillon, 1996. Increased UV-B penetration in a lake owing to drought-induced acidification. Nature 381: 141–143.

Zhang, Y. L., B. Q. Qin, L. Zhang, G. W. Zhu & W. M. Chen, 2005. Optical properties and fluorescence of chromophoric dissolved organic matter (CDOM) of shallow lakes in the middle and lower reaches of the Yangtze River. Journal of Freshwater Ecology 20(3): 451–459.

Hydrobiologia (2007) 581:53–61
DOI 10.1007/s10750-006-0519-z

EUTROPHICATION IN LAKES

Effects of hydrodynamics on phosphorus concentrations in water of Lake Taihu, a large, shallow, eutrophic lake of China

Guangwei Zhu · Boqiang Qin · Guang Gao · Lu Zhang · Liancong Luo · Yunlin Zhang

Abstract To understand the effect of hydrodynamical process on water phosphorus concentration, wind, wave, and several water quality indices were observed in Meiliang Bay, a shallow and eutrophic bay locates in north of Lake Taihu. During the 7 day observation period, wind speed and significant wave height were recorded more than 3 h per day, and water samples were collected in five water-depth layers once a day. Hydrodynamical disturbance had no significant correlationship with the water quality at the top layer when the significant wave height was smaller than 30 cm, but it significantly increased suspended solids (SS) concentration of the bottom water layer. Concentrations of nutrients showed no positive correlationship with SS concentration in the water body. Intensive sediment resuspension may not have occurred when the hydrodynamic stress on sediment was only a little higher than the critical stress for sediment resuspension. A new method for confirming the critical stress for intensive sediment resuspension and nutrient release still needs to be developed. The range of the water quality indices was quite high during the seven days of observation. High variation seems to be a common character of large shallow lakes like Taihu.

Keywords Shallow lake · Hydrodynamics · Eutrophication · Phosphorus · Algae bloom · Lake Taihu

Guest editors: B. Qin, Z. Liu & K. Havens
Eutrophication of shallow lakes with special reference to Lake Taihu, China

G. Zhu (✉) · B. Qin · G. Gao · L. Zhang · L. Luo · Y. Zhang
Nanjing Institute of Geography and Limnology, Chinese Academy of Sciences, Nanjing 210008, China
e-mail: gwzhu@niglas.ac.cn

Introduction

In shallow lakes, wind-induced waves frequently disturb the water-sediment interface, and often cause intensive sediment resuspension. Strong winds may cause significant increases of suspended solids (SS) concentration in overlying water (Aalderink et al., 1984; Carper & Bachmann, 1984; Luettich et al., 1990; Qin et al., 2000; Zhu et al., 2004; Zhu et al., 2005b). The intensive sediment resuspension may increase the internal release of nutrients from sediment (Søondergåard et al., 1992), but sometimes it may also help deposit soluble nutrients into the bottom sediments by sorption and precipitation (Gunatilaka, 1982). Simulated experiments also indicated that hydrodynamics may have both positive and negative effects on sediment nutrient release (Kristensen et al., 1992; Fan et al., 2001). Therefore, effects of hydrodynamical processes on the exchange of nutrients between sediment and water are quite complicated.

Lake Taihu is the third largest freshwater lake in China (Qin et al., 2007). Water quality of Lake Taihu has been deteriorated quickly since the 1960's due to the drastic effects of anthropogenic activities (Dokulil et al., 2000). The nutrition status of Meiliang Bay (a bay in the northern Lake Taihu) changed from oligo-mesotrophic stage with low algal biomass before 1981 to a eutrophic condition with blooms of *Microcystis* during 1988–1995, and a hyper-eutrophic conditions with the dominance of *Planctonema* and total phosphorus up to 200 mg m^{-3} during 1996–1997 (Chen et al., 2003). Sediment investigations also showed a strong effect of municipal drainage on the water quality of northern bay of Lake Taihu (Qu et al., 2001). At present, the dominant phytoplankton species of Lake Taihu are *Microcystis, Anabaena, Melosira, Cyclotella and Cryptomonas*. *Microcystis* spp. occupies 85% of algae biomass and forms an algal bloom each summer (Chen et al., 1997). Zooplankton grazing is inadequate to control the algae bloom of *Microcystis* (Chen & Nauwerck, 1996; McNaught et al., 1997). Pen fish culture and reconstruction of floating, leave-floating, and submerged plants may benefit water quality improvement (Li & Yang, 1995; Hu et al., 1998; Pu et al., 1998), but are difficult to carry out in a large area like Meiliang Bay.

Wave processes occur frequently in Lake Taihu due to its shallowness and special shape. For example, from 1997 to 1999, daily maximum wind speeds (calculated as hourly average wind speed) higher than 5 ms^{-1} occurred 89.5% of the time, and wind speeds higher than 8 ms^{-1} occurred 34.2% of the time (Fan et al., 2004). A simulation study on Lake Taihu sediment showed that water disturbance caused soluble reactive phosphorus (SRP) release from sediment which was 8–10 times higher than release under undisturbed conditions (Fan et al., 2001). Wave flume experiments showed that the critical shear stress (0.03–0.04 N m^{-2}) that leads to extensive sediment resuspension in Lake Taihu could be induced by a 4–6 ms^{-1} southeast wind event (Qin et al., 2004). Nutrients released from sediment resuspension caused by wind-induced wave may be the main type of internal load of nutrients in Lake Taihu (Qin et al., 2004), but there is little knowledge about the amounts of suspended sediment and released nutrients and also the form of released nutrients in the lake due to the great difficulty in making field observations.

The objectives of this paper are to present field observations on hydrodynamics, wind, water chemistry and algae biomass in Lake Taihu, and to discuss the effect of hydrodynamics on nutrient release from sediments.

Materials and methods

Field observation description

Meiliang Bay locates in the north of Lake Taihu, with an area of 130 km^2. The observation site was selected in the centre of Meiliang Bay to minimize the effect of lake shore on hydrodynamics. Near the observation site, is a regular water quality observation site of Taihu Laboratory for Lake Ecosystem Research (TLLER), which has been monitored every month since 1991. The water depth was 2.4 m at the observation site and did not change significantly during the observation period. On 10th July 2003, a steel-tube platform was set up in the observation site. A wave-meter with three sensors was fixed on the observation platform for wave measurement. An Acoustic Doppler Profile (ADP) was fixed on the platform for lake current measurement. And a Davis meteorological station was fixed on the platform for wind-speed and direction measurement. All of the data were downloaded to computer regularly. A 20 cm long sediment core was collected and sliced to sub-sample each 2 cm depth. Water content (WC, %), bulk density (g cm^{-3}), porosity (%) and loss on ignition (LOI, %) of each sub-sample were measured within 3 h. At the same time, a polymethyl-methacrylate peeper interstitial water sampler with 36 cm long, 1 cm thickness and 0.45 μm membrane was laid at the observation site. After 7 days, the peeper sampler was pulled out and the interstitial water was sampled immediately in each centimeter for Fe^{2+}, PO_4^{3-} and NH_4^+ analysis. Redox potential (Eh) of the interstitial water was immediately measured, too. The meteorological station did not work well the first day, resulting in the loss of some data, so data of wind-speed and direction per hour from the TLLER standard weather station were used for that period.

From 11 July to 17 July, lake water was sampled at 13:00 each day at the five layers (20, 50, 100, 200, and 230 cm). For each water sample, about 100 ml was filtered in the field by 0.45 μm aperture GF/F filter membrane for measuring dissolved total phosphorus (DTP), soluble reactive phosphorus (SRP), ammonia nitrogen (NH_4^+-N), and dissolved organic carbon (DOC). A 250 ml water sample was fixed by alkaline potassium iodide (KI) and manganese sulfate ($MnSO_4$) for dissolved oxygen (DO) determination. Another 1000 ml unfiltered water sample was preserved for total phosphorus (TP), suspended solids (SS) and chlorophyll a (Chl-*a*) analysis. All of the water samples were sent to TLLER for chemical analysis within 2 h. Organic matter in SS was evaluated by loss on ignition (LOI) on the GF/C filter membrane.

Chemical analysis methods

For TP and DTP analysis, water sample was digested by alkaline potassium persulphate in high pressure sterilization pot at 120°C. Then phosphate concentration was determined by molybdenum–antimony–ascorbic acid colorimetry with a Shimadzu UV 2401 spectrophotometer. SRP and NH_4^+-N were determined by colorimetry with Skalar flow-injection analyzer (detection limit for phosphate is 0.001 mg P l^{-1}). DO was determined by sodium hyposulfite titration. DOC was analyzed by I/O 1020A TOC Analyzer. SS was determined by a gravimetric method with Whatman GF/C filter membrane filtration and oven-dried the filtrating residua over 4 h at 105°C. LOI of SS was also determined by gravimetric method with 550°C ignite the SS contained GF/C membrane over 5 h. To determine Chl-*a* concentration, 250 ml water sample was filtered through 0.45 μm pore size synthetic fabric membrane. Chl-*a* concentration was determined by colorimetry after the residue on the membrane was frozen, grinded and extracted by ethanol.

Statistic methods

One way ANOVA analyses of water quality among different wave conditions with Scheffe's method and Tukey's method were carried out by SPSS 11.0 statistical software. Correlation analysis among observed indices was also done by SPSS 11.0.

Results

Physicochemical characteristics of sediment and interstitial water

Eh of interstitial water showed a steep decline in the top 7 cm layer of sediment (Fig. 1), and remained constant in the layer deeper than 7 cm. Concentration of Fe^{2+} showed a continuously increasing trend with depth (Fig. 1), suggesting there is a sharp concentration gradient of Fe^{2+}. This may owe to the release of Fe^{2+} from interstitial water to more oxidation overlying water with increased Eh. Concentration of PO_4^{3-} also showed an increasing trend from overlying water to sediment depth. A peak value at 5–7 cm depth may indicate that internal release of P from sediment by diffusion occurred mainly in the surface 7 cm. The alkaline phosphate activity (APA) and organic phosphorus decomposing bacteria (OPDB) also showed a peak value at about 7 cm and declined to background value at sediment depths over 11 cm in the sediment of Lake Taihu (Wang, 2004). The sediment properties and chemical forms of phosphorus changed drastically in the depth of 5–15 cm in an investigation of vertical distribution of physicochemical characteristics and phosphorus chemical forms of sediment in four sub-lake areas of Lake Taihu (Zhu et al., 2003). Therefore, 7 cm depth of sediment in Lake Taihu may be the peak value of activity of anaerobic bacteria for organic matter and phosphorus regeneration from organic debris. However, no significant increase of NH_4^+ concentration occurred in interstitial water of the top 15 cm except in the surface 1 cm.

Water content, LOI percentage, porocity and bulk density of sediment also indicated that the surface 7 cm sediment is quite different from the

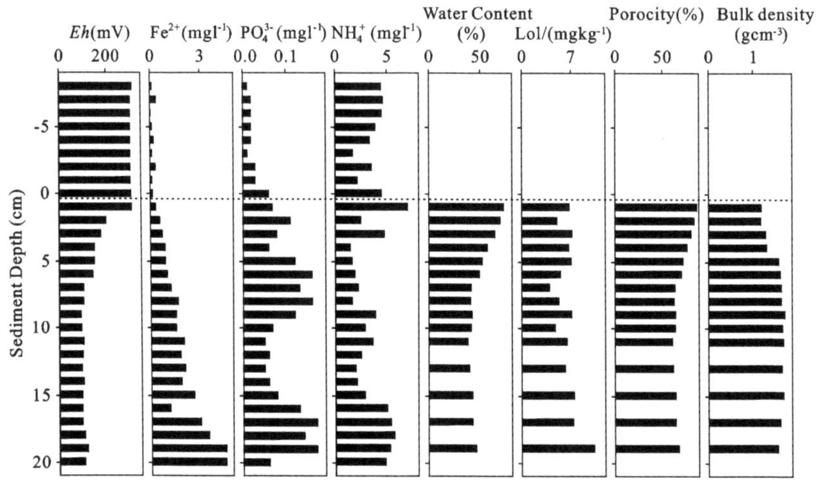

Fig. 1 Vertical distribution of Eh, Fe^{2+} concentration, PO_4^{3-} concentration, NH_4^+ concentration in interstitial water and physicochemical characteristics of sediment of observed site

deeper sediment. Qin et al. (2004) investigated the vertical distribution of physiochemical characteristics of sediment in the whole Lake Taihu and found that the surface 5–15 cm sediment was significantly different from the deeper layer. They concluded that the top 5–15 cm sediment was the active layer participating in the material cycling of Lake Taihu. Thereby, hydrodynamic processes in Lake Taihu may mainly affect the top 5–15 cm sediment that affects nutrient exchange with lake water.

Wind and wave conditions during field observation

During the observation, wind speed varied from 0 ms^{-1} to 7.0 ms^{-1}. Significant wave height varied from 5.85 cm to 28.25 cm. No strong wind-wave process occurred during the observation. The mean wind speed and direction between 12:30–13:30, mean significant wave height per hour of 10:00–11:00, 11:00–12:00, 12:00–13:00 per day, and the weighted mean value of the 3 h are presented in Table 1. Generally, the hydrodynamical disturbance intensity was divided into three kinds of situation according to significant wave height: Over 25 cm at 13th July, 12–20 cm on 16th July, 12th July, and 15th July, and lower than 12 cm on 17th July, 14th July, and 11th July.

The intensity of wind-induced waves was affected significantly by wind direction, though the observation site was located in the central of Meiliang Bay. The wave heights were significantly lower on 12th July, 15th July and 16th July than on other dates because of northeast and southeast winds (Fig. 2). A small hill near the east shore of Meiliang Bay and a relatively short wind-fetch-length explain this observation.

Suspended solids concentration and its LOI percentage

No significant differences of SS concentrations were observed among the 7 days when the SS concentrations of the top four layers were used to denote the condition of water body. However, SS concentrations in the bottom layer were quite different among the different wave conditions. The SS concentration of the bottom layer water was the highest value (154 mg l^{-1}) in 13th July. It is consistent with the strongest wind-induced wave during sampling than the other days (Fig. 2). This result shows that hydrodynamic disturbances did not have significant effects on the top water in Lake Taihu because the significant wave height was lower than 20 cm. This wave intensity may be too common and frequent in Lake Taihu to cause distinguished differences among different wind-induced wave processes.

Concentration of LOI in SS showed a similar trend to that of SS concentration. Generally, there were no significant differences of LOI

Table 1 Wind speed and H_s during the field observation

Date	Wind speed/direction/(ms^{-1})			H_s /(cm)			
	10:00–11:00	11:00–12:00	12:00–13:00	10:00–11:00	11:00–12:00	12:00–13:00	Weighted mean value[a]
11-July	4.1/307	2.1/55	1.2/7	15.14	10.63	6.95	9.54
12-July	3.4/132	2.7/134	2.0/129	19.33	18.52	12.91	15.85
13-July	2.3/235	1.2/239	6.1/278	20.35	26.46	27.05	25.74
14-July	3.3/119	2.1/106	1.8/114	13.82	10.82	8.83	10.32
15-July	3.5/302	4.0/34	4.6/14	12.18	12.57	11.90	12.17
16-July	4.8/71	6.2/60	6.3/70	15.98	17.10	20.28	18.50
17-July	2.3/246	1.9/233	1.1/246	13.48	10.58	8.88	10.21

[a] Weighted mean value of H_s was calculated as following: H_s (mean value) = (H_s (10:00–11:00) + 2 × H_s (11:00–12:00) + 3 × H_s (12:00–13:00))/6

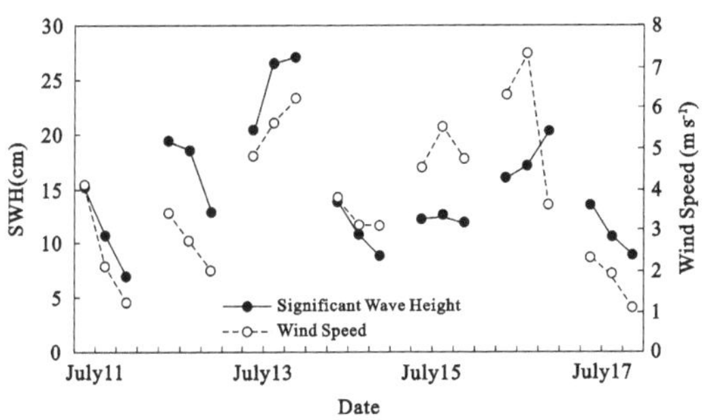

Fig. 2 Wind speed and significant wave height three hours before each days water sampling

concentration in the top four layers among the first 6 days. However, LOI concentration in the top four layers was significantly higher during the last days. This result could be caused by the algae bloom that occurred during the last day. The concentration of Chl-*a* also confirmed this presumption.

Nutrients, DO, and Chl-*a* concentrations

Vertical distributions of TP, DTP and SRP concentrations are presented in Fig. 3. For TP, there was no difference among different days except for 11 July, which was quite low compared to the next 6 days. For DTP, there was no difference among the 7 days. SRP was remarkably high on 15th July. And for NH_4^+, the last 3 days were significantly higher than the first 3 days. For Chl-*a* and DO, the last day is markedly higher than the previous days.

Correlation analysis indicated that there was not any significantly correlation between hydrodynamical disturbance intensive and water quality indices in the top four layers of water. DTP concentration correlated with SRP concentration in the top four layers (r = 0.83, n = 7), and in the bottom layer (r = 0.79, n = 7). LOI concentration correlated with Chl-*a* concentration and DO concentration in the top four layers, which means the water DO and organic suspended solids are determined by algae biomass.

SS and LOI concentrations in bottom layer were significantly affected by the intensive hydrodynamical disturbance (r = 0.83, n = 7). However, there was no significant relationship between SS and the intensive hydrodynamic disturbance in the top four layers. It suggested that wind-induced wave may have caused surface sediment resuspension, but the effect was limited in the bottom layer.

Discussion

Effects of wind-induced wave disturbance on phosphorus release from sediment in shallow lake

Do wind-induced waves cause phosphorus release from sediment to overlying water in shallow lakes? Our observations in Meiliang Bay suggest that they do not. However, we can not conclude that sediment resuspension will not affect nutrient concentrations in overlying water. In fact, field

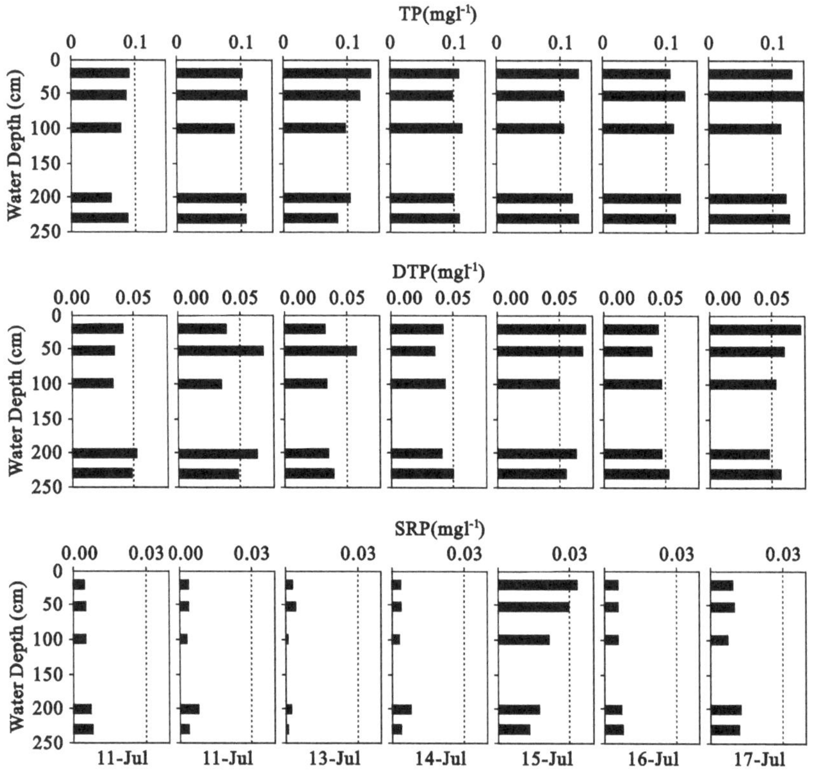

Fig. 3 Vertical distribution of TP, DTP and SRP concentrations in overlying water

observations have indicated that strong wind could increase TP, DTP and SRP concentrations in lake water as the wind speed was over 12 ms^{-1} in Lake Taihu (Zhu et al., 2004, 2005b). However, no significant release of phosphorus from sediment was found in this field observation, perhaps due to the relative low wind speed. In the wave flume experiments, Qin et al. (2004) found that the critical shear stress that leads to extensive sediment resuspension in Lake Taihu was about 0.03–0.04 N m^{-2}, equivalent to a wind speed in situ up to 4 ms^{-1}. The stress of wind-induced wave calculated minute by minute using observed data varied between 0.01 N m^{-2} and 0.11 N m^{-2}, and its peak value occurred between 18:00–18:30, 16th July. But intensive sediment resuspension still was not observed. This result suggests that the critical shear stress of 0.03–0.04 N m^{-2} isn't the correct stress for extensive sediment resuspension, but is simply the stress needed for surface sediment particles to begin to move up. According to hydraulic judging criteria, critical stress for sediment resuspension means the surface sediment particles begin to gradually float upward. In another wave flume experiments, Zhu et al. (2005a) found that SS concentration increased from 2.5 mg l^{-1} without any waves to 13.6 mg l^{-1} when wave height just exceeded the critical wave height (5.93 cm). But SS concentration increased to 224 mg l^{-1} as wave height increased to 13.29 cm. This means that sediment won't suspend extensively as the wave height become a little higher than the critical wave height. The wave height that will cause intensive sediment resuspension by field observation and simulated flume experiment is still not known.

In other field observations of hydrodynamical processes in Meiliang Bay, SS concentration in overlying water increased significantly when the wind-speed over 6.5 ms^{-1} in Meiliang Bay (Qin et al., 2000). SS concentration was about 120 mg l^{-1} with wind-speed over 6.5 ms^{-1}. The SS concentration was over 400 mg l^{-1} in top layer water with wind speed over 12 ms^{-1} (Zhu et al.,

2004, 2005b). At the same time, TP, DTP, SRP concentration in the top layer water increased significantly, too. These results indicate that the wind-speed needs to be between 7 and 12 ms^{-1} for intensive sediment resuspension. However, wind speed seldom exceeded 6.5 ms^{-1} in this field observations. Thus, SS and nutrients concentration in overlying water had not significantly varied with the intensity of hydrodynamical disturbance perhaps due to lower wave height than the critical wave height for intensive sediment resuspension.

Sediment resuspension does not always increase phosphorus concentration in overlying water in shallow lakes. Correlative analysis indicates that concentrations of DTP correlate inversely with SS concentration ($r = 0.83$, $p < 0.05$). The reason is that surface sediments are often oxidized in large shallow lakes. The surface floc is in constant equilibrium with the water column. When the surface floc suspended to overlying water, it absorbs dissolved phosphorus from water phase. This phenomenon was also found in shallow Lake Yangebup (Linge & Oldham, 2004).

Although a positive correlationship between water SS concentration and intensity of hydrodynamic disturbance was not observed in this field observation, it should be noted that a relative strong wave occurred in the third day, with subsequent nutrient concentration increases in the fifth day followed by an algae bloom on the seventh day. However, cause and effect relationships among the three phenomenons are still unclear.

Variation of water quality indices in large shallow lake

Water quality indices of shallow lake are highly variable because of the spatial and vertical inhomogeneity of water quality, as well as the violent influence of hydrodynamical disturbance. Most of the observed water quality indices varied drastically in the 7 days of observation (Fig. 4). For example, DTP concentration in the surface 20 cm layer water varied between 0.34 mg l^{-1} and 0.77 mg l^{-1}, SRP varied between 0.002 mg l^{-1} and 0.033 mg l^{-1}, NH_4^+–N varied between 0.056 mg l^{-1} and 0.832 mg l^{-1}, and Chl-*a* varied between 33 mg l^{-1} and 117 mg l^{-1} during the seven observation days.

Compared to the monthly monitored data from 2000 to 2003, considerable variation of the water quality indices occurred among the observed 7 days (Fig. 4). This observation suggests that monthly monitored data may be inadequate to reflect the water quality variation in large shallow lakes like Lake Taihu. For example, the cycle time of phytoplankton colony is generally a week. Thus, Chl-*a* concentration must be quite different between water sampling events depending on the

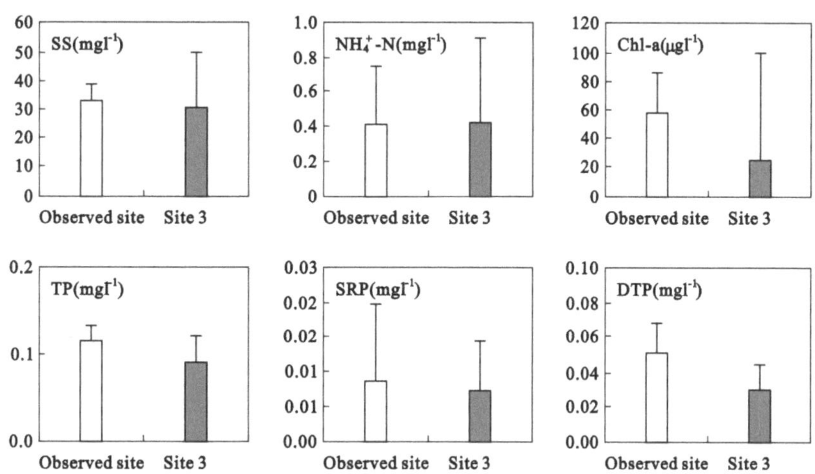

Fig. 4 Variation of the observed surface 20 cm water quality indices during the 7 days and that of the monthly monitored data at No. 3 regular monitoring site of TLLER

presence or absence of an algae bloom. Additionally, wind speed and direction during water sampling can affect water quality indices, such as Chl-*a* concentration, phytoplankton structure (Chen et al., 2003). Therefore, one must be careful in using monthly monitored data to explain water quality change in large shallow lakes.

Acknowledgements This work was supported by The Key Project of Chinese Academy of Sciences (KZCX1-SW-12) and Natural Science Foundation of China (No. 40673078, 40573062). We also would like to thank Chen Weimin, Hu Weiping, Zhang Fabing, Ji Jiang, Qian Rongshu, Song Yuzhi, Yang Liuyan, Zhang Xiaofeng, Yang Longyuan for their help on field observation, sampling and chemical analysis.

References

Aalderink, R. H., L. Lijklema, J. Breukelman, W. Raaphorts & A. G. Brinkman, 1984. Quantification of wind-induced resuspension in a shallow lake. Water Science & Technology 17: 903–914.

Carper, G. L. & R. W. Bachmann, 1984. Wind Resuspension of Sediments in a Prairie Lake. Canadian Journal of Fisheries and Aquatic Sciences 41: 1763–1767.

Chen, W. M. & A. Nauwerck, 1996. A note on composition and feeding of the crustacean zooplankton of Lake Taihu, Jiangsu Province, China. Limnologia 26: 275–279.

Chen, W. M., Y. W. Chen, X. Y. Gao & I. Yoshida, 1997. Eutrophication of Lake Taihu and its control. Agricultural Engineering Journal 6(2): 109–120.

Chen, Y. W., B. Q. Qin, K. Teubner & M. Dokulil, 2003. Long-term dynamics of phytoplankton assemblages: Microcystis- domination in a large shallow lake in China. Journal of Plankton Research 25: 445–453.

Dokulil, M., W. Chen & Q. Cai, 2000. Anthropogenic impacts to large lakes in China: the Tai Hu example. Aquatic Ecosystem Health and Management 3: 81–94.

Fan, C. X., L. Zhang & W. C. Qu, 2001. Lake sediment resuspension and caused phosphate release - a simulation study. Journal of Environmental Sciences 13: 406–410.

Fan, C. X., L. Zhang, B. Q. Qin, S. M. Wang, W. P. Hu & C. Zhang, 2004. Estimation on dynamic release of phosphorus from wind-induced suspended particulate matter in Lake Taihu. Science in China Series D 47: 710–719.

Gunatilaka, A., 1982. Phosphate adsorption kinetics of resuspended sediments in a shallow lake, Neusiedlersee, Austria. Hydrobiologia 91: 293–298.

Hu, W. P, J. Salomonsen, F. L. Xu & P. M. Pu, 1998. A model for the effects of water hyacinths on water quality in an experiment of physico-biological engineering in Lake Taihu, China. Ecological Modelling 107: 171–188.

Kristensen, P. M., M. Søondergåard & E. Jeppesen, 1992. Resuspension in a shallow lake. Hydrobiologia 228: 101–109.

Li, W. C. & Q. X. Yang, 1995. Wetland utilization in Lake Taihu for fish faming and improvement of lake water quality. Ecological Engineering 5: 107–121.

Linge, K. L. & C. E. Oldham, 2004. Control mechanisms for dissolved phosphorus and arsenic in a shallow lake. Applied Geochemistry 19: 1377–1389.

Luettich, R. A., D. R. F. Harleman & L. Somlódy, 1990. Dynamic behaviour of suspended sediment concentrations in a shallow lake perturbed by episodic wind events. Limnology & Oceanography 35: 1050–1067.

McNaught, D. C., W. M. Chen & H. J. Cai, 1997. Grazing on a Microcystis bloom by cladocerans and copepods in Tai Lake. Verhandlungen Internationale Vereinigung fur Theoretische und Angewandte Limnologie 26: 550–552.

Pu, P. M., W. P. Hu, J. S. Yan, G. X. Wang & C. H. Hu, 1998. A physico-ecological engineering experiment for water treatment in a hypertrophic lake in China. Ecological Engineering 10: 179–190.

Qin, B. Q., W. P. Hu, W. M. Chen, C. X. Fan, J. Ji, Y. W. Chen, X. Y. Gao, L. Y. Yang, G. Gao, W. Y. Huang, J. H. Jiang, C. Zhang, Y. B. Liu & Z. Y. Zhou, 2000. Studies on the hydrodynamic processes and related factors in Meiliang Bay, Northern Taihu Lake, China. Journal of Lake Sciences 12: 327–334 (in Chinese).

Qin, B. Q., W. P. Hu, G. Gao, L. C. Luo & J. S. Zhang, 2004. Dynamics of sediment resuspension and the conceptual schema of nutrient release in the large shallow Lake Taihu, China. Chinese Science Bulletin 49: 54–64.

Qin, B. Q., P. Z. Xu, Q. L. Wu & L. C. Luo, 2007. Environmental issues of Lake Taihu, China. Hydrobiologia 581: 3–14.

Qu, W. C., M. Dickman & S. M. Wang, 2001. Multivariate analysis of heavy metal and nutrient concentrations in sediments of Taihu Lake, China. Hydrobiologia 450: 83–89.

Søondergåard, M., P. Kristensen & E. Jeppesen, 1992. Phosphorus release from resuspended sediment in the shallow and wind exposed Lake Arresø, Denmark. Hydrobiologia 228: 91–99.

Wang, J. J., 2004. Effect of biogeochemistry processes of new born surface caused by sludge dredging on the internal loading recovery in a lake. Master thesis of Nanjing Institute of Geography and Limnology, Chinese Academy of Sciences.

Zhu, G. W., G. Gao, B. Q. Qin, L. Zhang & L. C. Luo, 2003. Geochemical characteristics of phosphorus in sediments of a large shallow lake. Advances in Water Science 14: 714–719 (in Chinese).

Zhu, G. W., B. Q. Qin & G. Gao, 2004. Vertical distribution of the concentrations of phosphorus and suspended solid in Lake Taihu affected by wind-induced wave. Advances in Water Science 15: 775–780 (in Chinese).

Zhu, G. W., B. Q. Qin, L. Zhang, L. C. Luo, X. J. Sun, D. L. Hong, Y. J. Gao & R. Xie, 2005a. Study on the effects of wave on nutrients releasing of sediments from Lake Taihu by flume experiments. Journal of Lake Sciences, 17(1): 61–68 (in Chinese).

Zhu, G. W., B. Q. Qin & G. Gao, 2005b. Direct evidence of phosphorus outbreak release from sediment to overlying water in a large shallow lake caused by strong wind. Chinese Science Bulletin 50: 577–582.

Hydrobiologia (2007) 581:63–70
DOI 10.1007/s10750-006-0515-3

EUTROPHICATION IN LAKES

Total inputs of phosphorus and nitrogen by wet deposition into Lake Taihu, China

Liancong Luo · Boqiang Qin · Longyuan Yang · Yuzhi Song

Abstract Lake Taihu suffers from eutrophication caused by riverine nutrient inputs and air deposition. To characterize wet deposition of phosphorus (P) and nitrogen (N) to the lake, precipitation collection and measurements of total phosphorus (TP) and total nitrogen (TN) and other components at five cities around Lake Taihu were made from July 2002 to June 2003. TP and TN concentrations and deposition rates exhibited strong spatial variation in the whole catchment. An inverse correlation between station-averaged TP and TN concentrations and precipitation amount was found. Maximal TP concentration in rainfall was found in Suzhou, and maximal TN in Wuxi. However, highest wet deposition rates of TP and TN were found in Suzhou, which suggests that atmospheric nutrients are mostly from the east and northwest area of Lake Taihu. Mean TP and TN deposition rates were 0.03 and 2.0 t km^{-2} year^{-1} respectively in Lake Taihu, which are greater than reported values in other areas by comparision. Total N and P contributed to the lake by wet deposition were 75 and 4720 t per year, respectively, which represent about 7.3% and 16.5% of total annual N and P inputs via inflow rivers. Wet deposition, especially N, could have significant effects on eutrophication in the lake, which shows that air deposition should be taken into account while reducing the external nutrients in the lake.

Keywords Nitrogen · Phosphorus · Wet deposition · Eutrophication · Riverine inputs · Lake Taihu

Guest editors: B. Qin, Z. Liu & K. Havens
Eutrophication of shallow lakes with special reference to Lake Taihu, China

L. Luo · B. Qin (✉) · L. Yang · Y. Song
Nanjing Institute of Geography & Limnology,
Chinese Academy of Sciences, Nanjing 210008, China
e-mail: qinbq@niglas.ac.cn

Introduction

Wet deposition may impact freshwater ecosystems, terrestrial environments (Buijsman & Erisman, 1988; Likens, 1989; Khemani et al., 1989; Lovett & Kinsman, 1990; Owens et al., 1992; Spokes et al., 1993; Helmers & Schrems, 1995; Raper & Lee, 1996) and coastal seas (De Leeuw et al., 2003). Input of nutrients by rainfall (e.g. P and N) and acidic rain characterized by low pH (<5.6) can affect the chemistry of lakes, soils, forests and wetlands (Gorham, 1998) and make ecosystems nutrient-limited or susceptible to acid deposition. For instance, it was noted (Gordon & Gorham, 1963) that acid deposition close to the iron-sintering plant at Lakes Wawa and Ontario was causing substantial leaching of calcium from local soils into adjacent ponds and lakes. It then was observed that acid deposition led to dissolution of aluminum in soil minerals and strong enrichment of aluminum ions in lake water (Wright et al., 1976). Many other researchers have addressed acidic rain in different areas (Cowling & Nilsson, 1995; Lin et al., 1998; Okochi et al., 2000; Peart, 2000; Vasjari et al., 2000).

Polluted rain also may deposit nutrients and trace elements, resulting in eutrophication and iron-concentration changes. Input of nutrients by wet deposition can contribute to total nutrients in water, especially in seepage lakes, regulate primary productivity and affect algal community structure (Pollman et al., 2002). Atmospheric N can be available for algae growth, which makes this contribution more significant because only a fraction of river runoff contributes N that is incorporated into biological material (De Leeuw et al., 2003). Heavy metals from rainfall (e.g. Zn, Cu, Fe, Ca and Mg) can affect surface water production (Morel et al., 1991) and influence biological activity in marine waters (Coale & Bruland, 1990; Sunda & Huntsman, 1992).

Lake Taihu is located in the lower Changjiang (Yangtze) River delta and is important for water quantity regulation, water supply, cultivation, irrigation, navigation, culture fisheries and tourism. In the 1950s, lake water was oligotrophic (Chang, 1995). With rapid industrial and agricultural development and population growth, increasing amounts of nutrients, heavy metals (Shen et al., 2001) and other pollutants (Feng et al., 2003) have been delivered to the lake via river runoff (Zhou & Zhu, 2003; Chen et al., 2003; Qu et al., 2001) or bulk deposition (Qu et al., 2002; Pu et al., 1998). Pollutants can be absorbed onto or associated with fine-grained sediment particles, resulting in limnic ecosystem recovery prevention and long-term eutrophication problems by sediment-water exchange. The lake is hypertrophic in three northern bays, with seasonal algae blooms increasing the cost of water purification. Consequently, reasons for deterioration and improvement of lake water-quality have received much attention in recent years (Wang et al., 1998; Qin, 1998; Yang et al., 2001; Chen, 2002; Yang et al., 2003; Qin et al., 2004). The largest project, "Zero-point Action", initiated by the State Environmental Protection Administration (SEPA) in 1998, required "Standard Discharge" of all industrial units. However, water quality has not improved remarkably despite reduction of external loading from rivers, which raises the possibility that net air-deposition and sediment resuspension may have considerable impacts.

Many reports have focused on effects of sediment and river inputs (Zheng et al., 2001) on water quality degradation in Lake Taihu, including acid rain impacts (Chen et al., 2000, 2001; Yang et al., 2001; Yang, 2001). Little is known about amounts of N & P contributed by wet deposition. The aim of this study is to estimate wet-deposition of P and N into the lake based on data from samples collected from July 2002 to June 2003.

Methodology

Site description

Fifteen sampling sites, located in five cities (Wuxi, Suzhou, Huzhou, Jintan and Changzhou; Fig. 1) around Lake Taihu, were selected for rain collection. Four were at Changzhou situated northwest of the lake, four in Suzhou in the east, three in Huzhou in the south, two in Wuxi in the north and two in Jintan in the northwest (Table 1). Several sites in each city were sampled

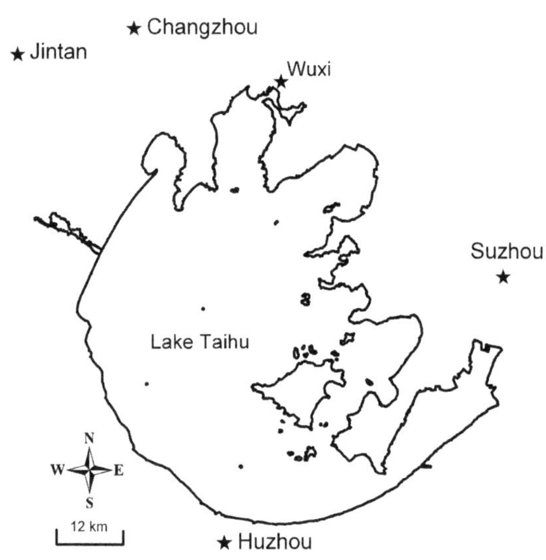

Fig. 1 Map of the study area

Table 1 Sample stations and their locations in Taihu Basin

Station	Longitude (°E)	Latitude (°N)	Site number	Location
Wuxi	120.25	31.58	2	North
Suzhou	120.61	31.30	4	East
Changzhou	119.97	31.77	4	Northwest
Jintan	119.58	31.73	2	Northwest
Huzhou	120.10	30.89	3	South

to evaluate different land-use types (e.g. rural, urban, industrial, forest, etc.).

Sample collection and analysis

Precipitation samples were collected automatically by a tipping-bucket rain gauge at all sites from July 2002–June 2003. Analytical parameters included pH, conductivity, K^+, Na^+, Ca^{2+}, Mg^{2+}, Cl^-, SO_4^{2-}, NO_3^-, NH_4^+, TP and TN (no TP & TN in Huzhou due to technical limits). All chemical components were analyzed monthly in the local Environmental Monitoring Center (EMC). pH was measured immediately without stirring by a digital pH meter, and conductivity was measured using an ion chromatography detector. All ions were determined using suppressed ion chromatography (Chen et al., 2001). All data from different EMCs were assumed to have similar accuracy since ion concentration measurements were done using standard SEPA methods.

Calculation of annual amount of deposited nutrients

Annual amounts (in: t year^{-1}) of TP and TN can be calculated from the following equation,

$$AA = \frac{1}{m}\sum_{j=1}^{m}\sum_{i=1}^{n} S \times RR_{ij} \times C_{ij} \times 10^{-3} \quad (1)$$

where S is the lake surface area (km^2), RR_{ij} (i = 1,2,...12; j = 1,2,3,4) is the precipitation rate (mm month^{-1}) at station j and month i, C_{ij} is the concentration of a substance in precipitation (mg l^{-1}) with the same subscript as RR_{ij}, m is the number of stations (4 in this case due to lack of data in Huzhou). The wet deposition rate (in: t km^{-2} year^{-1}) of chemical components can be derived from Eq. 1 without consideration of S.

Results

Seasonality

Monthly precipitation (RR) and TN and TP concentrations are shown in Fig. 2. RR ranged from 24–130 mm per month. Minimum RR was in January, with a value of 24.3 mm, and November, with a value of 50.7 mm. There was more rainfall in summer (128 mm in August and 117 mm in July). RR exhibited a summer (June–August) maximum and a fall (September–November) minimum, demonstrating strong seasonality.

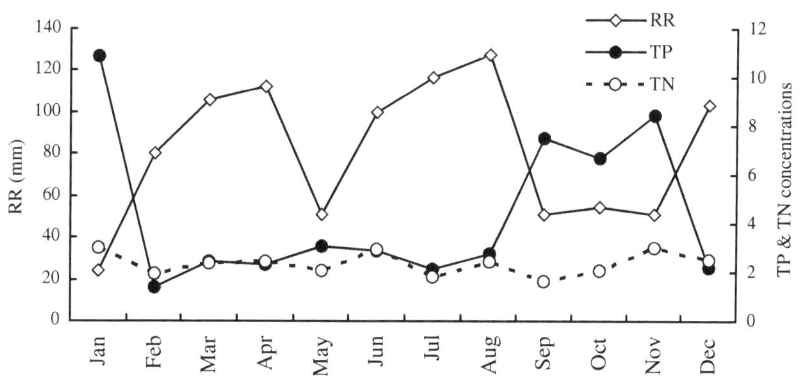

Fig. 2 Monthly station-averaged precipitation amount (RR, in mm), precipitation-weighted concentrations of TP (in: 100 mg l^{-1}) and TN (in: mg l^{-1})

TP concentration varied from 0.01 to 0.11 mg l^{-1} with the peak value in January, which is, in magnitude, similar with those reported by Brown et al. (1984, 0.015–0.15 mg l^{-1}) and Migon and Sandroni (1999, 0.002–0.14 mg l^{-1},), but significantly greater than those reported by Chen et al. (1985) and Filoso et al. (1999). Although no significant seasonal variation in TN was found (Fig. 2), there was a seasonality of TP and TN despite slight differences among the 12 months. TN and TP followed an inverse pattern to RR.

TN contributed by rain followed a similar seasonal pattern to RR and peaked in summer and fall, with minima in winter and spring. However, TP had a maximum in fall and winter and minimum in spring and summer (Fig. 3). These seasonal differences varied from those of TP and TN concentrations throughout the year due to differences in precipitation amounts.

Total annual TP and TN deposition rates

Deposition rates of NO_3^-, NH_4^+, TP and TN were estimated from Eq. (1) with lake surface area excluded. In each city, monthly precipitation amounts were used to calculate average precipitation rates. Table 2 shows month-averaged wet deposition rates of the five cities. There was no data on TN and TP in Huzhou, but concentrations of NO_3^- and NH_4^+ suggest a moderate contribution. Spatial differences suggest that the east region of Lake Taihu (Suzhou), and then the northwest (Changzhou and Jintan) have the largest N and P contribution from rainfall. Wuxi and Huzhou have moderate influences. Wet deposition of TP within the lake catchment was 0.03 t km^{-2} year^{-1}, TN was 2.0 t km^{-2} year^{-1}, NO_3^- was 3.70 t km^{-2} year^{-1}, and NH_4^+ was 1.56 t km^{-2} year^{-1}. Based on individual deposition rates, total annual contribution from precipitation was 75 and 4720 t per year for TP and TN, respectively.

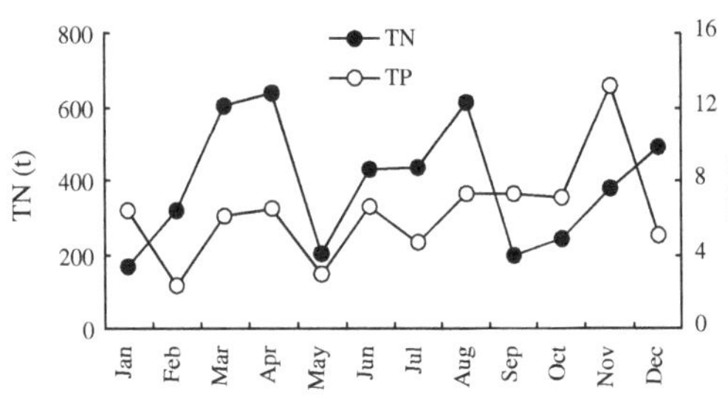

Fig. 3 Monthly station-averaged amounts of TP & TN (in: t)

Discussion

The simple model based on Eq.1 may overestimate total input of N & P into Lake Taihu because precipitation chemistry over the lake may differ from cities. Raindrops are formed when small water droplets adhere to particles, including chemical components in air, and, consequently, depend on local air movement and concentrations of vapor and particles in the atmosphere. Due to low nutrient concentrations in air over the lake, especially the center, precipitation inputs of P and N in these regions may be lower than those in the shore areas. However, this uncertainty for regional differences is difficult to quantify unless many observation sites are established over the water surface of Lake Taihu.

Despite the spatial variability of nutrient concentrations in precipitation, results obtained from observation data and calculations should indicate

Table 2 Wet-deposition rates (t km^{-2} year^{-1}) and concentrations (mg l^{-1}) of TP and TN at different stations

City	Wet deposition rate (t km^{-2} year^{-1})				Averaged concentration (mg l^{-1})	
	TP	TN	NO_3^-	NH_4^+	TP	TN
Wuxi	0.016	2.25	2.76	1.44	0.022	3.28
Suzhou	0.06	2.81	4.36	2.26	0.075	2.58
Changzhou	0.018	1.81	0.82	1.52	0.023	2.22
Jintan	0.033	1.20	6.70	0.95	0.057	1.33
Huzhou			3.87	1.65		
Average	0.03	2.0	3.70	1.56	0.044	2.35

effects of wet deposition on lake eutrophication. In Lake Taihu, the primary source of N and P is from Suzhou, where the highest TP concentration and deposition rates were found. Changzhou and Jintan had the second greatest contribution of TN and TP. Thus, the main N and P inputs via rainfall are from the east and northwest. The smallest TP deposition rate and the greatest TN concentration were found in Wuxi. Therefore, Wuxi, with a greater annual precipitation, may make the greatest contribution to total N input into the lake.

Deposition rates of NO_3^- and NH_4^+ are high relative to TN inputs, with values of 3.70 and 1.56 t km^{-2} year^{-1}, respectively, which are 2 and 5 times higher than those for Barnegat Bay (1.81 and 0.34 t km^{-2} year^{-1} for NO_3^- and NH_4^+; Gao, 2002). Estimated TP deposition rate (0.03 t km^{-2} year^{-1}) near Lake Taihu also is high compared to that in the Amazon (0.004 t km^{-2} year^{-1}; Swap et al., 1992), Connecticut (0.004 t km^{-2} year^{-1}; Yang et al., 1996), New Zealand (0.01 t km^{-2} year^{-1}; Chen et al., 1985), Corsica (0.01–0.02 t km^{-2} year^{-1}; Bergametti et al., 1992), Apalachicola, FL (0.012 t km^{-2} year^{-1}; Fu & Winchester, 1994) and Ireland (0.021 t km^{-2} year^{-1}; Gibson et al., 1995). Also, TN deposition rate is about 10 times that in Colorado and southern Wyoming (<0.2 t km^{-2} year^{-1}; Burn, 2003). High N and P deposition rates result from livestock and fertilization of croplands (Vitousek et al., 1997), crop senescence, fossil fuel combustion, automobile emissions and factories. In Taihu Basin, agriculture and coal combustion are the main contributors to high rates of N and P deposition.

Wet deposition is greater than dry deposition, especially in ocean areas (Owens et al., 1992; Hertel et al., 2002), the Mississippi River Basin (Lawrence et al., 2000), and Barnegat Bay (Gao, 2002). With this hypothesis, annual net atmospheric inputs of TP and TN into Lake Taihu should be slightly higher than estimated values. However, deposition effects on lake eutrophication are significant even excluding dry deposition. Based on nutrient concentrations observed at inflow rivers connected to Lake Taihu, annual riverine input of TP was estimated to be 1,030 t and TN about 28,650 t (Xu & Qin, 2005). Therefore, atmospheric inputs of TP and TN represent about 7.3% and 16.5% of annual total inputs of P and N via inflow rivers, which suggests that air deposition should be considered in the evaluation nutrient inputs to Lake Taihu. Nitrogen from air deposition may promote primary productivity. For example, 5.3 mmol C m^{-2} day^{-1} could result from average deposition of 0.8 mmol N m^{-2} day^{-1} (De Leeuw et al., 2003). Also, atmospheric deposition can affect nitrogen-saturated forests and acidic freshwater lakes and streams (Stoddard, 1994; Aber et al., 1995; Fenn et al., 1998). However, Lake Taihu generally is P-limited and inherently vulnerable to N deposition (Yin et al., 1996). As a result, deposited P could promote primary productivity. Moreover, some deposited N and P may be absorbed onto or associated with sediment particles, which could make bottom sediment a source of nutrients. In shallow lakes, perturbation has substantial effects on sediment-water interactions (Sheng & Lick, 1979; Furumai et al., 1989; Shrestha, 1996; James et al., 1997; Bailey and Hamilton, 1997; Blom & Winkels, 1998; Poelma & Ooms, 2002), and therefore, waves may resuspend sediment and release nutrients resulting in increased N & P. For example, 1.64×10^4 t P can be suspended from sediment into water of Lake Taihu with a 15.0 m s^{-1} SE wind event, resulting in an average TP increase of 4.70 mg l^{-1} (Luo, 2004). Some suspended P may be available for aquatic plants, and some may be redeposited.

Atmospheric N originates mainly from ammonium nitrate aerosol (NH_4NO_3), ammonia gas (NH_3) and nitric acid gas (HNO_3) (Lindberg et al., 1990). NO_x exists primarily as NH_4NO_3 and HNO_3, and can be transformed to HNO_3 after photochemical processes. HNO_3 and NH_3 (primary pollutant) can form an equilibrium with NH_4NO_3 (Tarnay et al., 2001). These compounds are mainly derived from fossil fuel combustion and agricultural production (e.g. fertilizer and animal waste). The origin of atmospheric deposition can reflect nutrient-source patterns of riverine inputs. Consequently, external N loading control in Lake Taihu Basin should include production of agricultural and fuel-combustion emissions. The "Zero-point Action" initiated by SEPA in 1998 has resulted in significant reduction of TP and TN inputs in Lake Taihu, but agricultural inputs of TP and TN still constitute a large proportion. Thus, control of only industrial discharge is not an effective method to control eutrophication. Municipal policy should include control of agricultural emission and coal combustion inputs in Lake Taihu Basin.

Conclusion

A study of wet deposition in Lake Taihu Basin examined the effects of N and P contributed by rainfall on lake eutrophication. Fifteen sites were selected in five cities around the lake. Results lead to the following conclusions:

(1) Deposition rates and concentrations of TP and TN in precipitation over Lake Taihu exhibited strong spatial variation. N and P in precipitation come mainly from the east (Suzhou) and northwest (Changzhou and Jintan). Wuxi and Huzhou make only moderate contributions to atmospheric inputs into Lake Taihu.

(2) Precipitation had a summer maximum and a fall minimum and followed an inverse seasonal cycle compared to TP and TN concentrations.

(3) Deposition rates of NO_3^- and NH_4^+ were 3.70 and 1.56 t km^{-2} $year^{-1}$, respectively. TP and TN deposition rates were 0.03 and 2.0 t km^{-2} $year^{-1}$, respectively.

(4) Total amounts of N and P contributed to the lake by wet deposition were 75 and 4720 t per year, respectively, which may contribute to lake eutrophication.

Acknowledgements We would like to gratefully acknowledge financial support from Chinese Academy of Sciences (No. KZCX1-SW-12), National Natural Science Foundation of China (No. 40501078, 40131160734), Ministry of Science and Technology of China (No. 2002AA601011) and Director Foundation of Nanjing Institute of Geography & Limnology, Chinese Academy of Sciences. Thanks to the Environmental Monitoring Centers of Wuxi, Suzhou, Changzhou, Jintan and Huzhou for their sample collection and analysis. The authors are indebted to Professor Guang Gao, Dr. Guangwei Zhu, Dr. Xiaoli Shi and Dr. Lu Zhang for their help with mathematical analysis on chemical data. This manuscript benefited from the comments of two anonymous reviewers.

References

Aber, J. D., A. Magill, S. G. McNulty, R. D. Boone, K. J. Nadelhoffer, M. Downs & R. Hallett, 1995. Forest biogeochemistry and primary production altered by nitrogen saturation. Water, Air and Soil Pollution 85: 1665–1670.

Bailey, M. C. & D. P. Hamilton, 1997. Wind induced sediment resuspension: a lake-wide model. Ecological Modeling 99: 217–228.

Bergametti, G., E. Remoudaki, R. Losno & E. Steiner, 1992. Source, transport and deposition of atmospheric phosphorus over the northwestern Mediterranean. Journal of Atmospheric Chemistry 14: 501–513.

Blom, G. & H. J. Winkels, 1998. Modeling sediment accumulation and dispersion of contaminants in Lake Ijsselmeer (the Netherlands). Journal of Environmental Management 37: 17–24.

Brown, G., D. T. Mitchell & W. D. Stock, 1984. Atmospheric deposition of phosphorus in a Coastal Fynbos Ecosystem of the southwestern cape, South-Africa. Journal of Ecology 72(2): 547–551.

Buijsman, E. & J. W. Erisman, 1988. Wet deposition of ammonium in Europe. Journal of Atmospheric Chemistry 6: 265–280.

Burn, D. A., 2003. Atmospheric nitrogen deposition in the Rocky Mountains of Colorado and southern Wyoming – a review and new analysis of past study results. Atmospheric Environment 37: 921–932.

Chang, Y. B., 1995. Major environmental changes since 1950 and the onset of accelerated eutrophication in Taihu Lake, China. Acta Palaeontologica Sinica 35: 155–174 (in Chinese).

Chen, H. S., 2002. Analysis of water environmental changes in Taihu Basin. Journal of Lake Sciences 14: 111–116 (in Chinese).

Chen, J. M., G. X. Pan, L. Cang, J. J. Yang & J. H. Wang, 2000. Effect of simulated acid rain on plant growth and paddy soils pH in Taihu area. Journal of Nanjing Agricultural University 23: 116–118 (in Chinese).

Chen, J. M., G. X. Pan & J. W. Zheng, 2001. Effect of simulated acid rain on adsorption and desorption of copper by paddy soils in Lake Taihu area. Acta Pedologica Sinica 38: 333–340 (in Chinese).

Chen, L., R. Arimoto & R. A. Duce, 1985. The sources and forms of phosphorus in marine aerosol – particles and rain from Northern New Zealand. Atmospheric Environment 19(5): 779–787.

Chen, X. M., Q. R. Shen & G. X. Pan, 2003. Characteristics of nitrate horizontal transport in a paddy field of the Tai Lake region, China. Chemosphere 50: 703–706.

Coale, K. H. & K. W. Bruland, 1990. Spatial and temporal variability in copper complexation in the North Pacific. Deep-sea Research 34: 317–336.

Cowling, E. B. & J. Nilsson, 1995. Acidification research: lessons from history and visions of environmental futures. Water, Air and Soil Pollution 85: 279–292.

De Leeuw, G., L. Spokes, T. Jickells, C. A. Skjoth, O. Hertel, E. Vigniti, S. Tamm, M. Schulz, L. L. Sorensen, B. Pedersen, L. Klein & K. H. Schlunzen,

2003. Atmospheric nitrogen inputs into the North Sea: effect on productivity. Continental Shelf Research 23: 1743–1755.

Feng, K., B. Y. Yu, D. M. Ge, M. H. Wong, X. C. Wang & Z. H. Cao, 2003. Organo-chlorine pesticide (DDT and HCH) residues in the Taihu Lake Region and its movement in soil – water system I. Field survey of DDT and HCH residues in ecosystem of the region. Chemosphere 50: 683–687.

Fenn, M. E., M. A. Poth, J. D. Aber, J. S. Baron, B. T. Bormann, D. W. Johnson, A. D. Lemly, S. D. McNulty, D. F. Ryan & R. Stottlemyer, 1998. Nitrogen excess in North American ecosystems: a review of predisposing factors, geographic extent, ecosystem responses and management strategies. Ecological Applications 8: 706–733.

Filoso, S., M. R. Williams & J. M. Melack, 1999. Composition and deposition of throughfall in a flooded forest archipelago (Negro River, Brazil). Biogeochemistry 45(2): 169–195.

Fu, J. M. & J. W. Winchester, 1994. Sources of nitrogen in 3 watersheds of Northern Florida, USA – mainly atmospheric deposition. Geochimica et Cosmochimica Acta 58: 1581–1590.

Furumai, H., T. Kondo & S. Ohgaki, 1989. Phosphorus exchange kinetics and exchangeable phosphorus forms in sediments. Water Research 23: 685–691.

Gao, Y., 2002. Atmospheric nitrogen deposition to Barnegat Bay. Atmospheric Environment 36: 5783–5794.

Gibson, C. E., Y. Wu & D. Pinkerton, 1995. Substance budgets of an upland catchment-the signi.cance of atmospheric phosphorus inputs. Freshwater Biology 33: 385–392.

Gordon, A. G. & E. Gorham, 1963. Ecological aspects of air pollution from an iron-sintering plant at Wawa, Ontario. Canadian Journal of Botany 41: 1063–1078.

Gorham, E., 1998. Acid deposition and its ecological effects: a brief history of research. Environmental Science & Policy 1: 153–166.

Helmers, E. & O. Schrems, 1995. Wet deposition of metals to the tropical north and the south Atlantic Ocean. Atmospheric Environment 29: 2475–2484.

Hertel, O., C. Ambelas Skjoth, L. M. Frohn, E. Vignati, J. Frydendall, G. De Leeuw, U. Schwarz & S. Reis, 2002. Assessment of the atmospheric nitrogen inputs into the North Sea using a Lagrangian model. Physics and Chemistry of the Earth 27: 1507–1515.

James, R. T., J. Martin, T. Wool & P. F. Wang, 1997. A sediment resuspension and water quality model of Lake Okeechobee. Journal of the American Water Resources Association 33: 661–680.

Khemani, L. T., G. A. Momin, P. S. P. Rao, P. D. Safai, G. Singh & R. K. Kapoor, 1989. Spread of acid rain over India. Atmospheric Environment 23: 757–762.

Lawrence, G. B., D. A. Goolsby, W. A. Battaglin & G. J. Stensland, 2000. Atmospheric nitrogen in the Mississippi River Basin – emissions, deposition and transport. The Science of the Total Environment 248: 87–99.

Likens, G. E., 1989. Some aspects of air pollutant effects on terrestrial ecosystems and prospects for the future. AMBIO 14: 9–14.

Lin, C., M. D. Melville, M. M. Islam, B. P. Wilson, X. Yang & P. Van Oploo, 1998. Chemical controls on acid discharges from acid sulfate soils under sugarcane cropping in an eastern Australian estuarine floodplain. Environmental Pollution 103: 269–276.

Lindberg, S. E., M. Bredemeier, D. A. Schafer & L. Qi, 1990. Atmospheric concentrations and deposition of nitrogen and major ions in conifer forests in the United States and Federal Republic of Germany. Atmospheric Environment 24: 2207–2220.

Lovett, G. M. & J. D. Kinsman, 1990. Atmospheric pollutant deposition to high-elevation ecosystems. Atmospheric Environment 24: 2767–2786.

Luo, L. C., 2004. Hydrodynamics and its effects on aquatic environment in Lake Taihu, Ph.D thesis, Nanjing Institute of Geography and Limnology, Chinese Academy of Sciences.

Migon, C. & V. Sandroni, 1999. Phosphorus in rainwater: partitioning inputs and impact on the surface coastal ocean. Limnology and Oceanography 44(4): 1160–1165.

Morel, F. M. M., R. J. M. Hudson & N. M. Price, 1991. Limitation of productivity by trace metals in the sea. Limnology & Oceanography. 36: 1742–1755.

Okochi, H., H. Kameda, S. Hasegawa, N. Saito, K. Kubota & M. Igawa, 2000. Deterioration of concrete structures by acid deposition an assessment of the role of rainwater on deterioration by laboratory and field exposure experiments using mortar specimens. Atmospheric Environment 34: 2937–2945.

Owens, N. J. P., N. J. Galloway & R. A. Duce, 1992. Episodic atmospheric nitrogen deposition on oligotrophic oceans. Nature 357: 397–399.

Peart, M. R., 2000. Acid rain, storm period chemistry and their potential impact on stream communities in Hong Kong. Chemosphere 41: 25–31.

Poelma, C. & G. Ooms, 2002. Influence of hydrodynamic interactions between particles on the turbulent flow in a suspension. Experimental Thermal and Fluid Sciences 26: 653–659.

Pollman, C. D., W. M. Landing, J. J. Perry, Jr & T. Fitzpatrick, 2002. Wet deposition of phosphorus in Florida. Atmospheric Environment 36: 2309–2318.

Pu, P. M., W. P. Hu, J. S. Yan, G. X. Wang & C. H. Hu, 1998. A physico-ecological engineering experiment for water treatment in a hypertrophic lake in China. Ecological Engineering 10: 179–190.

Qin, B. Q., W. P. Hu, G. Gao, L. C. Luo & J. S. Zhang, 2004. The Dynamics of resuspension and conceptual mode of nutrient releases from sediments in large shallow Lake Taihu, China. In Chinese Sciences Bulletin 49: 54–64.

Qin, B. Q., 1998. A review and prospect about the aquatic environment studies in Lake Taihu. Journal of Lake Sciences 10: 1–9 (in Chinese).

Qu, W. C., M. Dickman, C. X. Fan, S. M. Wang, W. C. Su, L. Zhang & H. X. Zou, 2002. Distribution, sources and potential toxicological significance of polycyclic aromatic hydrocarbons (PAHs) in Lake Taihu sediments, China. Hydrobiologia 485: 163–171.

Qu, W. C., M. Dickman & S. M. Wang, 2001. Multivariate analysis of heavy metal and nutrient concentrations in

sediments of Lake Taihu, China. Hydrobiologia 450: 83–89.
Raper, D. W. & D. S. Lee, 1996. Wet deposition at the sub–20 km scale in a rural upland area of England. Atmospheric Environment 30: 1193–1207.
Shen, J. H., B. Gutendorf, H. H. Vahl, L. Shen & J. Westendorf, 2001. Toxicological profile of pollutants in surface water from an area in Taihu Lake, Yangtze Delta. Toxicology 166: 71–78.
Sheng, Y. P. & W. Lick, 1979. The transport and resuspension of sediments in a shallow lake. Journal of Geophysical Research 84: 1809–1826.
Shrestha, P. L., 1996. An integrated model suite for sediment and pollutant transport in shallow lake. Advances in Engineering Software 27: 201–212.
Spokes, L., T. Jickells, A. Rendell, M. Schulz, A. Rebers, W. Dannecker, O. Kruger, M. Leermakers & W. Baeyens, 1993. High atmospheric nitrogen deposition events over the North Sea. Marine Pollution Bulletin 26: 689–703.
Stoddard, J. L., 1994. Long-term changes in watershed retention of nitrogen. In Baker, L. A. (ed.), Environmental Chemistry of Lakes and Reservoirs. American Chemical Society Advances in Chemistry 227: 223–284.
Sunda, W. G. & S. A. Huntsman, 1992. Feedback interactions between zinc and phytoplankton in seawater. Limnology & Oceanography 37: 9–40.
Swap, R., M. Garstang, S. Greco, R. Talbot & P. Kallberg, 1992. Saharan dust in the Amazon Basin. Tellus, Series B. Chemical and Physical Meteorology 44: 133–149.
Tarnay, L., A. W. Gertler, R. R. Blank & G. E. Taylor, Jr, 2001. Preliminary measurements of summer nitric acid and ammonia concentrations in the Lake Tahoe Basin air-shed: implications for dry deposition of atmospheric nitrogen. Environmental Pollution 113: 145–153.
Vasjari, M., A. Merkoci & S. Alegret, 2000. Potentiometric characterisation of acid rains using corrected linear plots. Analytica Chimica Acta 405: 173–178.
Vitousek, P. M., J. D. Aber, R. W. Howarth, G. E. Likens, P. A. Matson, D. W. Schindler, W. H. Schlesinger & D. G. Tilman, 1997. Human alternation of the global nitrogen cycle: sources and consequences. Ecological Applications 7: 737–750.
Wang, G. X., P. M. Pu, S. Z. Zhang, W. C. Li, W. P. Hu & C. H. Hu, 1998. The purification of artificial complex ecosystem for local water in Lake Taihu. China Environmental Science 18: 410–414 (in Chinese).
Wright, R. F., T. Dahl, E. T. Gjessing, G. R. Hendrey, A. Henriksen, M. Johannessen & I. P. Muniz, 1976. Impact of acid precipitation on freshwater ecosystems in Norway. Water, Air and Soil Pollution 6: 483–499.
Xu, P. Z & B. Q. Qin, 2005. Water quantity and pollutant fluxes of the surrounding rivers of Lake Taihu during the hydrological year of 2001–2002. Journal of Lake Sciences 17(3): 213–218 (in Chinese).
Yang, L. Y., B. Q. Qin & R. J. Wu, 2001. Preliminary study for potential impacts on the aquatic environment of Lake Taihu by acid rain. Journal of Lake Sciences, 13: 135–142 (in Chinese).
Yang, X. S., D. R. Miller & X. Xu, 1996. Spatial and temporal variations of atmospheric deposition in interior and coastal Connecticut. Atmospheric Environment 30: 3801–3810.
Yang, X. H., 2001. Acid rain and countermeasures in southern Lake Taihu. Journal of Huzhou Teachers College 23: 68–72 (in Chinese).
Yang, Z. F., W. G. Shi, L. Q. Chen, Y. Chen & Z. L. Zhou, 2003. Ecological environmental succession and countermeasure of East Taihu. China Environmental Science 23: 64–68 (in Chinese).
Yin, D. Q., Z. H. Wu, X. R. Wang & S. Yan, 1996. Algal growth potential as affected by lake water and phosphorus released from sediments in Lake Taihu. Journal of Nanjing University (Natural Sciences) 32(2): 253–260 (in Chinese).
Zheng, Y., X. J. Wang & Y. C. Jiang, 2001. Water quality of the inflow rivers connected to Lake Taihu and totally inputted nutrients. Geography and Territorial Research 17: 40–44 (in Chinese).
Zhou, Q. X. & Y. M. Zhu, 2003. Potential pollution and recommended critical levels of phosphorus in paddy soils of the southern Lake Tai area, China. Geoderma 115: 45–54.

Hydrobiologia (2007) 581:71–77
DOI 10.1007/s10750-006-0499-z

EUTROPHICATION IN LAKES

The effects of different electron donors on anaerobic nitrogen transformations and denitrification processes in Lake Taihu sediments

Donghong Wang · Qinghui Huang · Chunxia Wang · Mei Ma · Zijian Wang

Abstract Nitrogen transformations in anaerobic sediments and leachate in Lake Taihu were simulated in the laboratory. Ammonium, nitrate and nitrite were analyzed after incubation under anaerobic conditions. Different reductive states and pH values were obtained by using different electron donors, such as glucose, sucrose, potato starch and sodium acetate. Chemical nitrogen transformation mechanisms were discussed relative to physico-chemical properties of lake sediment. Results demonstrated that nitrogen transformations in anaerobic conditions supplemented with different electron donors varied, and supplementation with certain electron donors may enhance nitrogen removal from anaerobic sediments. Among the four electron donors studied, higher nitrogen removal efficiencies were observed with acetate and starch. Saccharides, such as glucose, sucrose and starch, stimulate nitrate reduction to nitrite, while acetate stimulates nitrate reduction to ammonium.

Guest editors: B. Qin, Z. Liu & K. Havens
Eutrophication of shallow lakes with special reference to Lake Taihu, China

D. Wang · Q. Huang · C. Wang · M. Ma ·
Z. Wang (✉)
State Key Laboratory of Environmental Aquatic Chemistry, Research Center for Eco-Environmental Sciences, Chinese Academy of Sciences, Shuangqing Road, Haidian District, Beijing 100085, China
e-mail: zjwang@mail.rcees.ac.cn

Keywords Electron donor · Anaerobic · Nitrogen · Transformation · Sediment · Lake Taihu

Introduction

Denitrification is a microbial process that converts nitrate to nitrogen gas via several intermediates, including nitrite and nitrous oxide (Tiedje, 1982). It is recognized as the most important process removing fixed nitrogen in most natural environments (Boicourt et al., 1996). Transformations involve nitrogen reduction and therefore, require electron donors (Devlin et al., 2000). Denitrification is a respiratory process which also requires an electron donor for energy (Islas-Limaa et al., 2004). Denitrifying bacteria can use nitrate and/or nitrite as an alternative electron acceptor for metabolic activities when molecular oxygen is unavailable (Bae et al., 2004). However, in natural anaerobic systems including sediments, denitrification often is limited by readily biodegradable electron donors.

Studies have demonstrated enhancement of nitro-reduction rates following addition of organic electron donors, such as glucose, pyruvate, formate, starch, acetate, hydrogen gas, elemental sulphur, thiosulphate, aqueous ferrous iron and pyrite (Preuss et al., 1993; Boopathy, 1994; Roberts et al., 1995; Devlin et al., 2000).

Saccharide is an abundant biomolecule in the natural environment. Glucose is a monosaccharide, and sucrose is a disaccharide, which yields two monosaccharides upon hydrolysis. Starch contains two polysaccharides: amylose and amylopectin (Yu et al., 2004). Acetate is an organic acid and has been used in waste water treatment for nitrogen removal (Graaf et al., 1996; Cervantes et al., 1999).

Addition of electron donors to enhance reduction processes is, therefore, a potential method for in-place treatment of contaminated sediments (Gerlach et al., 1999). This may be an important treatment for eutrophic lakes, such as Lake Taihu, which is the third largest freshwater lake in China (Qin et al., 2007). Water quality in Lake Taihu has deteriorated due to rapid economic development and intensive use of water resources in recent years. As a result, cyanobacteria blooms occur over large areas in summer, and these blooms impact the aquatic ecosystem detrimentally (Pu et al., 1998a, b).

Most studies have focused on nitrate and ammonium removal in waste water systems (Cervantes et al., 2001; Schmidt et al., 2003), and comparative studies on denitrification and electron donor supplements in fresh water sediments are scarce.

Objectives of the present study were to compare total nitrogen (TN) reduction effects in Lake Taihu sediments stimulated by saccharide electron donors, such as glucose, sucrose, potato starch and organic acid acetate; and to evaluate effects of nitrogen species transformation under different electron donor additions.

Materials and methods

Sediment samples collection

Six surface sediment samples were collected using a Peterson grab bucket on 13–15 July 2003 from Lake Taihu. Locations of sampling sites are shown in Fig. 1. Samples were stored in a cooler until return to the laboratory, freeze-dried, sieved to ~0.28 mm aggregate size, mixed together, homogenized by gentle shaking, and stored at 4°C in a sealed polyethylene bag.

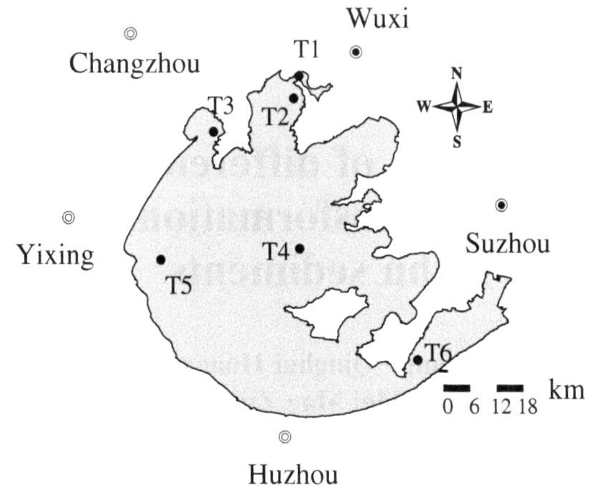

Fig. 1 Geographic locations of sampling sites

Anaerobic incubation experiment

Microcosms were constructed using 20 g of sediment and 150 ml of artificial freshwater (1 mg l^{-1} ammonium, 1 mg l^{-1} nitrate and 0.1 mg l^{-1} nitrite), which was purged with high purity nitrogen (99.99% N_2) for 10 min to remove oxygen before use in 250-ml serum bottles. Microcosms were prepared in triplicate, and electron donors glucose, sucrose, potato starch (Sigma Co.) and sodium acetate added 2 g in stoichiometric excess. Two electron donor-free control groups, one sterilized with 0.1 g $HgCl_2$, were established. Table 1 lists the treatment conditions and microcosm abbreviations. Test bottles were aerated with high purity nitrogen for 3–5 additional minutes, sealed with butyl rubber stoppers and continuously mixed with a planar shaker rotating at 200 rpm at 25°C.

Table 1 Treatment conditions and abbreviations of microcosms

Abbreviation	Treatment
A	Sterilized control
B	Electron donor free control
C	Microcosm of glucose as electron donor
D	Microcosm of sucrose as electron donor
E	Microcosm of potato starch as electron donor
F	Microcosm of sodium acetate as electron donor

Sampling was done in a glove bag (I^{2}R Glove Bag^{TM}, Model X-27-27, USA) purged with nitrogen before use. Bottles were vigorously shaken; pH and Eh were determined using a portable meter (Thermo Orion Model 250, USA). Ten milliliter of sediment/water slurry was poured into a 10 ml centrifuge tube, and the same quantity was added (10 ml artificial freshwater contained 1.33 g sediment). After sampling, sediment samples were centrifuged at 4000 rpm for 20 min and filtered through a 0.45 μm pore-size cellulose acetate filter to separate sediment and leachate at once.

Chemical analysis

For leachate samples, ammonium, nitrate and nitrite were determined within 24 h after sampling. Nitrate was determined by ion chromatography (Dionex 4500i using an AS-14 4-mm anion column), nitrite with *N*-(1-naphthyl) ethylene diamine dihydrochloride (APHA, 1998a), and ammonium with salicylic-hypochlorite colorimetry (Standard method of People's Republic of China, 1987).

Total nitrogen in sediments was analyzed using persulfate according to standard methods for examination of water and wastewater (APHA, 1998b). Weights of sediments reported are on an oven-dry basis.

Results and discussion

Formal potentials and pH conditions in the microcosms

Formal potential consists of measurement of the electromotive force of an electrochemical cell in which, under the specified conditions, the analytical concentration of the two oxidation states is varied. The formal potential includes correction factors for activity coefficients, acid-base phenomena, complex formation, and the liquid junction potential used between the reference electrode and the half-cell in question. It may often leads to better predictions than standard potential because it represents quantities subject to direct experimental measurement. In this paper, it is expressed by pε, which is calculated from the determined Eh values using Nernst Equation. The pε can be defined as the negative logarithm of electron activity in a solution. It is a parameter for the redox intensity gives the electron activity at equilibrium and measures the relative tendency of a solution to accept or transfer electrons (Stumm & Morgan, 1981).

Values of pH and pε are shown in Fig. 2. In all microcosms, pH declined with time. Trends are divided into three groups. Microcosms treated with glucose and sucrose had similar trends. pH declined sharply in the first 96 h and dropped to 4.3 by completion. pH in starch additions also declined, but more slowly than glucose and sucrose, dropping to 4.8 at the end of the incubation. pH in the sterilized control, electron donor-free control and sodium acetate addition declined in the first 48 h, then remained around 7.4 to 7.5.

Formal potentials in electron donor systems reached a strong reductive environment after 2 or 3 days. pε dropped in the first 48 h in glucose and sucrose microcosms and in the first 96 h in starch and sodium acetate microcosms. Afterwards, the

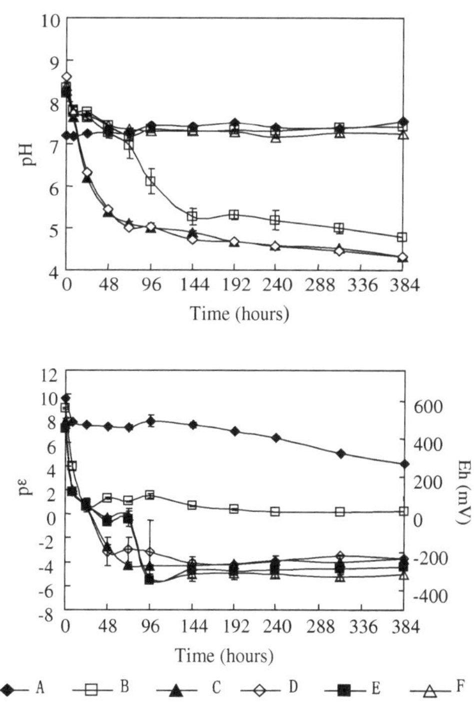

Fig. 2 pH and pε values of microcosms

reductive environment in these microcosms reached steady-state and remained constant at –4 to –6. The sterilized control did not attain reductive conditions, and the electron donor-free control barely reached a mild reductive environment.

Systems supplemented with electron donors achieved more reduction potential than controls. It is believed that Eh values is featured by reduction when they below 200 mV. (Song et al., 1990) Sediments from Lake Taihu in our experiment were strongly anaerobic except two control groups, and additional electronic donors could accelerate reducing conditions in microcosms.

pH and reductive potential are important factors controlling the fate of pollutants, especially those at the soil-water interface (Cao et al., 2001). pH and reductive potential also affect nitrogen composition (Stumm & Morgan, 1981). pH in all microcosms treated with electron donors declined, except the system treated with sodium acetate, likely due to acetate hydrolysis. In our experiment, the acetate microcosm may have maintained pH of 7.2–7.4 via its strong buffering capacity (Terra & Regel, 1995). However, in other three electron donor microcosms, pH values declined and remained below 5. At this pH, acid producing microorganisms may not survive, leading to stable pH.

Transformation of nitrogen species

Nitrogen transformations in control and electron donor microcosms are shown in Figs. 3 and 4. Results show that, in all microcosms, nitrate in leachate increased early in the incubation, then decreased. Nitrate, which is formed during organic matter degradation under oxic conditions (ammonification/nitrification), may be transported (mixing, currents, upwelling etc.) to anoxic environments where it can undergo denitrification (Boyd, 2001). In the present study, early nitrate increases may be caused by degradation of organic matter using residual oxygen in the sediment. For in the first 24 h, formal potential in all microcosms was positive, and pH was about 7, indicating characteristics of sediments were weak oxic or weak reduction in this condition.

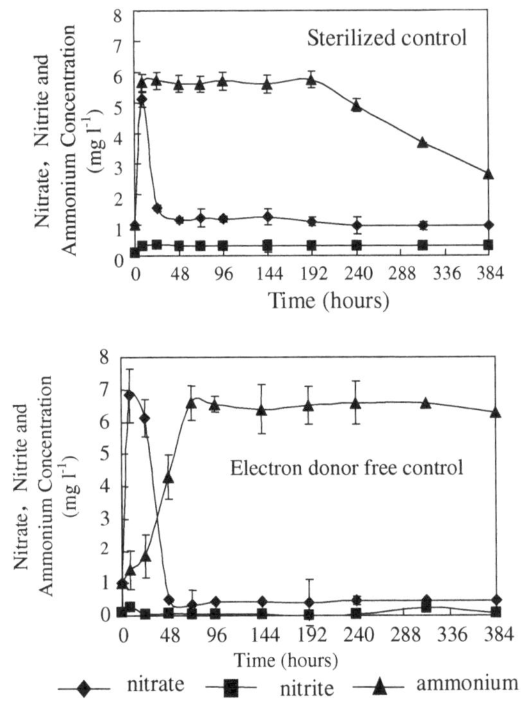

Fig. 3 Transformation of nitrogen species in two control microcosms

Nitrogen transformations in glucose and sucrose microcosms were similar. Nitrate in glucose and sucrose systems were lower than controls. Highest nitrate concentrations in glucose and sucrose microcosms were 5.2 and 4.7 mg l^{-1}, respectively, in the first 8 h and then decreased sharply. Highest nitrite concentrations in leachate occurred simultaneously with nitrate maxima. Ammonium decreased to below detection limits in 48 h, which indicates that monosaccharide and disaccharide electron donors may stimulate anerobic nitrogen transformations.

Nitrate concentration in the starch addition reached 7.4 mg l^{-1} in 24 h, and ammonium concentration increased to a similar value as the controls, then decreased to below detection limits after 96 h.

In anoxic, nitrate-containing environments, two pathways of dissimilatory nitrate reduction have been identified: denitrification, by which nitrogen oxides (mainly NO_3^-) are reduced to nitrite (NO_2^-) and then to dinitrogen gases (N_2O and N_2), and dissimilatory nitrate reduction to ammonia (DNRA), producing ammonia by

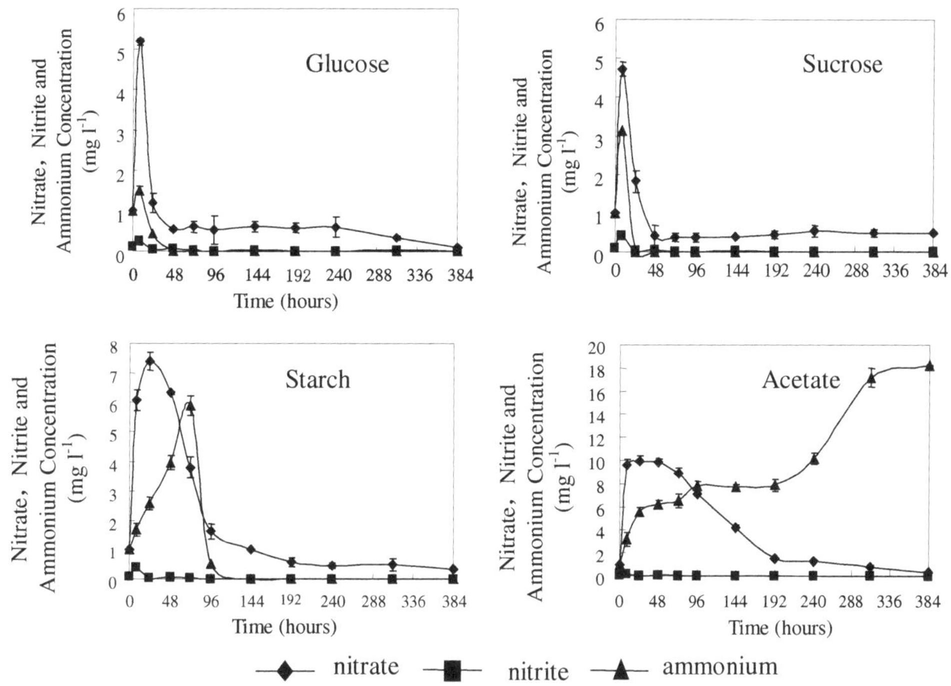

Fig. 4 Transformation of nitrogen species in different electron donor treated microcosms

reduction of nitrate and/or nitrite (Brunet & Garcia-Gil, 1996).

Apparently, nitrate reduced in microcosms supplemented with glucose, sucrose and starch as electronic donors were denitrified. Ammonium could not be detected after 48 h in glucose and sucrose systems and 96 h in the starch system. This indicates that saccharide, including monosaccharide (glucose), disaccharide (sucrose) and polysaccharides (starch), may stimulate nitrate reduction to nitrogen gas, although the dinitrogen gases (N_2O and N_2) were not determined in our experiment.

Nitrate and nitrite in the sodium acetate system was different from other electron donor systems. Nitrate in leachate reached 10 mg l^{-1} and maintained this value for 96 h. Nitrite concentration was half that of glucose and sucrose microcosms. However, ammonium increased to about 8 mg l^{-1} within 96 h and remained steady to 240 h. Afterwards, ammonium concentration increased to 18.3 mg l^{-1} by the end of incubation. Ammonium increased early maybe caused by ammonification of organic matter, for nitrate did not decrease at the first 96 h. But latter the high ammonium value indicated that nitrate was reduced to ammonium in this microcosm. The removal of eight electrons during NO_3^- reduction to NH_4^+ makes this process potentially advantageous in reduced environments (Tiedje et al., 1982). DNRA may be quantitatively important nitrogen cycling of marine sediments (Sorensen, 1978).

Ammonium concentrations in the control systems were higher than electron donor addition systems except for acetate. Ammonium in the sterilized control increased for 8 h, then remained constant at approximately 6 mg l^{-1} from 8 h to 192 h. In the electron donor-free control system, ammonium increased slowly for 192 h. These systems may be controlled by pH and redox potential. According to pε~pH diagrams (Wang, 2001), at pH about 7.5 and pε about 0–8, ammonium is the predominant nitrogen species. This may explain why ammonium in the starch microcosm increased for 96 h and then decreased to detection limits. pH and pε conditions in the starch microcosm up to 96 h were within the ammonium predominance area. And, nitrate reduction seems follow denitrification for 96 h in the starch system since pH was lower than 6 and pε decreased sharply to about –6. This condition overstepped the ammonium predominance area. So, nitrate reduction to ammonium stimulated by sodium acetate may be due to comparatively high pH caused by acetate hydrolysis. In soil, similar results were reported that DNRA was favored at high pH (Stevens et al., 1998).

TN removal by denitrification

The ability to enhance sediment denitrification via addition of various electron donors was tested in sediment–water microcosms using TN reduction as a gross indicator of denitrification activity. TN removal from sediments in microcosms over 384 h are shown in Fig. 5. TN removal was observed in all systems but was not significant in the sterilized control, which increased for 48 h, then stayed steady for 384 h. Denitrification in the electron donor-free control increased for 144 h, and TN reduction percentage was 21.8%. Compared to control microcosms, potato starch and sodium acetate systems obtained higher reductive efficiency (45.5% and 42.5%, respectively). At the beginning of the incubation, denitrification in glucose and sucrose additions was lower than controls. However, reductive efficiency increased sharply at the end of the incubation (31.4% and 34.7%, respectively).

These results demonstrate that additional electronic donors may accelerate. Among the four electron donors, starch and acetate have largest effects. TN reduction ratios of starch and acetate were 10% higher than glucose and sucrose. Since the starch used had low solubility, starch may have been more associated with sediments than the other electron donors (Gerlach et al., 1999) However, TN reduction in the sterilized control increased slightly throughout the experiment, indicating that $HgCl_2$ addition may not have been sufficient to eliminate microbial activity; but the increase in TN reduction was significantly less than the non-inhibited microcosms. However, this TN reduction was not observed in the electronic donor-free controls, indicating that either the sterilization agent affected abiotic redox activity, or microorganisms could reduce TN without electron donor addition.

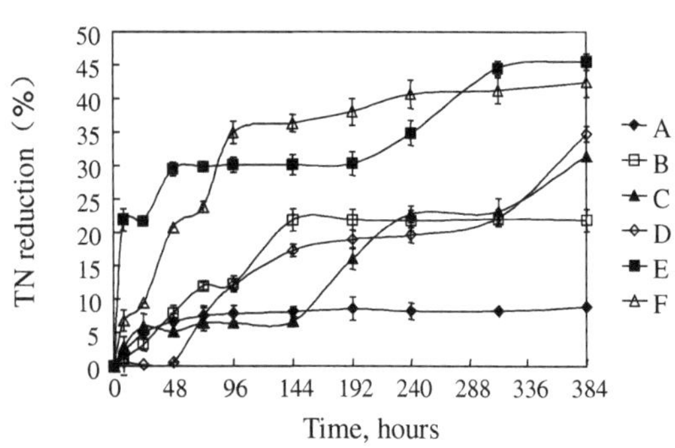

Fig. 5 TN removal ratio of sediments in microcosms

Conclusions

The results presented in this paper show certain electron donors may be used to enhance TN removal in anaerobic sediments. Higher nitrogen removal efficiencies were observed when acetate and starch were added. However, nitrogen transformations in anaerobic conditions supplemented with different electron donors varied. Saccharides, such as glucose, sucrose and starch, stimulated nitrate reduction to nitrite, and acetate stimulated nitrate reduction to ammonium. Nitrogen transformations in these microcosms were controlled by pH and formal potential.

Acknowledgements This work was supported by the Key Project of Chinese Academy of Sciences, (Grant No.KZCX1-SW-12-II-32) and National Key Basic Research Support Foundation of China (Grant No.2002CB412304).

References

APHA, AWWA & WEF, 1998a. 4500- NO_2^--B Colorimetric Method. Standard Methods for the Examination of Water and Wastewater, 20th edn.

APHA, AWWA & WEF, 1998b. Persulfate Method. Standard Methods for the Examination of Water and Wastewater, 20th edn, 4500-N C: 4–102.

Bae, H.-S., T. Yamagishi & Y. Suwa, 2004. An anaerobic continuous-flow fixed-bed reactor sustaining a 3-chlorobenzoate-degrading denitrifying population utilizing versatile electron donors and acceptors. Chemosphere 55: 93–100.

Boicourt, W. C., W. R. Boynton, T. M. Church, D. M. Ditoro, R. Elmgren, J. H. Garber, A. E. Giblin, R. A. Jahnke, N. J. P. Owens, M. E. Q. Pilson & S. P Seitzinger, 1996. The fate of nitrogen and phosphorus at the land-sea margin of the North Atlantic Ocean. Biogeochemistry 35: 141–180.

Boopathy, R., 1994. Transformation of nitroaromatic compounds by a methanogenic bacterium, *Methanococcus* sp. strain B. Archives of Microbiology 162: 167–172.

Boyd, S. R., 2001. Nitrogen in future biosphere studies. Chemical Geology 176: 1–30.

Brunet, R. C. & L. J. Garcia-Gil, 1996. Sulfide-induced dissimilatory nitrate reduction to ammonia in anaerobic freshwater sediments. FEMS Microbiology Ecology 21: 131–138.

Cao, X., Y. Chen, X. R. Wang & X. H. Deng, 2001. Effects of redox potential and pH value on the release of rare earth elements from soil. Chemosphere 44: 655–661.

Cervantes, F., O. Monroy & J. Gómez, 1999. Influence of ammonium on the performance of a denitrifying culture under heterotrophic conditions. Applied Biochemistry and Biotechnology 81: 13–23.

Cervantes, F. J., A. David De la Rose & J. Gómez, 2001. Nitrogen removal from wastewater at low C/N ratios with ammonium and acetate as electron donors. Bioresource Technology 79: 165–170.

Devlin, J. F., R. Eedy & B. J. Butler, 2000. The effects of electron donor and granular iron on nitrate transformation rates in sediments from a municipal water supply aquifer. Journal of Contaminant Hydrology 46: 81–97.

Gerlach, R., M. Steiof, C. L. Zhang & J. B. Hughes, 1999. Low aqueous solubility electron donors for the reduction of nitroaromatics in anaerobic sediments. Journal of Contaminant Hydrology 36: 91–104.

Graaf, A., P. Bruijin, L. A. Robertson, M. S. M. Jetten & J. G. Kuenen, 1996. Autotrophic growth of anaerobic, ammonium-oxidizing microorganisms in a fluidized bed reactor. Microbiology 142: 2187–2196.

Islas-Limaa, S., F. Thalassoa & J. Gómez Hernandezb, 2004. Evidence of anoxic methane oxidation coupled to denitrification. Water Research 38: 13–16.

Pu, P. M., W. P. Hu, G. X. Wang, S. Z. Zhang, C. G. Hu & J. S. Yan, 1998a. The new strategy for improving the aqua-ecological environmental in Taihu Lake Basin, China. How can we solve the problem of lack of qualified water and deterioration of environment and natural resources in Taihu Lake basin. Journal of Lake Science 10: 47–58 [Suppl].

Pu, P. M., W. P. Hu, J. S. Yan, G. X. Wang & C. G. Hu, 1998b. A physico-ecological engineering experiment for water treatment in a hypertrophic lake in China. Ecologied Engineering 10: 179–190.

Preuss, A., J. Fimpel & G. Diekert, 1993. Anaerobic transformation of 2,4,6-trinitrotoluene TNT. Archives of Microbiology 159: 345–353.

Qin, B. Q, P. Z. Xu, Q. L. Wu, L. C. Luo & Y. L. Zhang, 2007. Environmental Issues of Lake Taihu, China. Hydrobiologia 581: 3–14.

Roberts, D. J., S. Pendharkar & F. Ahmad, 1995. Factors Affecting TNT Degradation by Anaerobic Consortia. Platform-Presentation, 3rd International Symposium on In Situ and On-site Bioreclamation. San Diego, CA, April 24–27.

Schmidt, I., O. Sliekers, M. Schmid, E. Bock, J. Fuerst, J. G. Kuenen, S. M. Jetten & M. Strous, 2003. New concepts of microbial treatment processes for the nitrogen removal in wastewater. FEMS Microbiology Reviews 27: 481–492.

Song, J. M., Y. Li & Z. Zhu, 1990. Relationship between Eh value and redox environment in marine sediments. Marine Science Bulletin 9(4): 33–39.

Sorensen, J., 1978. Capacity for denitrification and reduction of nitrate to ammonia in a coastal marine sediment. Applied & Environmental Microbiology 35: 301–305.

Standard method of People's Republic of China, 1987. Water quality-determination of ammonium-spectrophotometric method with salicylic acid, GB 7481–87:211.

Stevens, R. J., R. J. Laughlin & J. P. Malone, 1998. Soil pH affects the processes reducing nitrate to nitrous oxide and di-nitrogen. Soil Biology and Biochemistry 30: 1119–1126.

Stumm, W., J. J. Morgan, 1981. Aquatic Chemistry: An Introduction Emphasizing Chemical Equilibria in Natural Waters, 2nd edn. John Wiley, Sons, Inc.

Terra, W. R. & R. Regel, 1995. pH buffering in *Musca domestica* midguts, comparative biochemistry and physiology Part A. Physiology 112: 559–564.

Tiedje, J. M., 1982. Denitrification. In Page, A. L. (ed.), Methods of Soil Analysis: Part 2. Chemical and Microbiological Properties. 2nd edn. ASA-ASSA, Madison, WI, 1011–1026.

Tiedje, J. M., A. J. Sexstone, D. D. Myrold & J. A. Robinson, 1982. Denitrification: ecological niches, competition and survival. Antonie Van Leeuwenhoek Journal of Microbiology 48: 569–583.

Wang, K. X., 2001. Aquatic chemistry, Beijing: Chemical Industry Press, Ch.6, Oxidation and Reduction Chemistry, 207 pp.

Yu, H., M. Amann, T. Hansson, J. Köhler, G. Wich & W. F. van Gunsterena, 2004. Effect of methylation on the stability and solvation free energy of amylose and cellulose fragments: a molecular dynamics study. Carbohydrate Research 339: 1697–1709.

Hydrobiologia (2007) 581:79–87
DOI 10.1007/s10750-006-0514-4

EUTROPHICATION IN LAKES

A 200-year historical modeling of catchment nutrient changes in Taihu basin, China

Ge Yu · Bin Xue · Geying Lai · Feng Gui · Xiaomei Liu

Abstract Sedimentary records provide important information for understanding changes in the history of eutrophication in Lake Taihu. In addition, the catchment nutrient model SWAT provides a powerful tool to examine eutrophic changes in a long-term context. Since it is difficult to evaluate impacts of natural eutrophic development and anthropogenic changes in catchment discharge and land use, simulation of past changes provides a mirror on processes and dynamics. Boundaries in the simulations are set to a pre-industrial time to evaluate natural-agricultural nutrient changes in Taihu Basin a 100 years ago. Total nitrogen (TN) and total phosphorus (TP) in the main channel flowing into the lake are simulated in four sub-basins for 200 model years. Results show that modeling can capture basic features of basin nutrient development, where mean TN concentration (0.12 mg l^{-1}) can be compared in broad scale to mean TN concentration (0.17 mg kg^{-1}) from Lake Taihu sedimentary cores dating back about 100 years. Spatial nutrient simulations suggest that the two major nutrient sources are from the southwestern sub-basin (48% TN and 68% TP of the basin total) and the northwestern sub-basin (18% TN and 17% TP). There are differences of $+7.3 \times 10^4$ kg TN and $+2.0 \times 10^5$ kg TP between total input and output values, simulating mean annual amounts of nutrient deposited into the lake. TN and TP concentration differences between input and output sub-basins become smaller in the second 100 years than the first 100 years, suggesting a 100 year period to reach a balance of net nutrients. Catchment nutrient modeling provides a basis to evaluate how nutrient production and balance responded to environmental changes over 200 years in Taihu Basin.

Keywords Natural-agricultural environment · TN and TP simulations · Nutrient change · Decadal-centurial time scale · Taihu Basin

Introduction

Eutrophication in Lake Taihu became a problem in the last 20 years, and anthropogenic change is a major factor (Jin et al., 1990; Huang et al., 2001; Qin et al., 2004). Nutrient release from lake bottom sediments (Qin et al., 2004) and nutrient discharge from the catchment (Xia & Yang, 2003) are two important sources leading to eutrophication, whereas nutrient source changes from

Guest editors: B. Qin, Z. Liu & K. Havens
Eutrophication of shallow lakes with special reference to Lake Taihu, China

G. Yu (✉) · B. Xue · G. Lai · F. Gui · X. Liu
Nanjing Institute of Geography & Limnology,
Chinese Academy of Sciences, Nanjing 210008, China
e-mail: geyu@niglas.ac.cn

bottom sediments or the catchment are not only controlled by human activities on an annual-interannual timescale, but also are related to natural system changes in a decadal-century time scale (Last & Smol, 2001; Smol, 2002). Paleolimnological techniques can be used to reconstruct past water quality changes on a number of time scales using information preserved in sedimentary profiles (e.g. Ramstack et al., 2003; Rose et al., 2004; Bradbury et al., 2004). When tracking eutrophic changes using lake sediments, a number of rich-organic layers with high nitrogen content have been found in lacustrine sediments from Lake Taihu. At depths of 22, 30, 35, and 42 cm in cores from Lake Taihu (Liu et al., 2004a; b), total nitrogen contents are similar to or higher than values at 1–5 cm (1990s; Fig. 1). While examining longer sediment cores from Lake Taihu (Xue et al., 1998), nitrogen contents at two layers ^{14}C-dated to 5,930 year B.P. (i.e. Before Present) and 7,899 year B.P. were higher than those in surface sediments. The lake may have experienced natural eutrophication before industrial and urban modernization in Taihu Basin. Examining and modeling long-term changes in lake eutrophication will provide a better understanding of the trajectories of lake environmental degradation and help set realistic mitigation goals for recovery.

Taihu Basin (Fig. 2) is located in a lowland area of the Changjiang River Plains in China approximately between 29–32° N and 118–121° E and about 50–100 km from the East China Sea (Sun & Huang, 1993). Climate is warm and wet, typical of subtropical monsoon climate. The basin is covered naturally by subtropical evergreen broadleaf forest, and the lake has a high aquatic production (Huang et al., 2001). Natural soil horizons developed on parent rock are composed of loose and organic-rich fluvial-lacustrine sediments deposited in the Holocene (Sun & Huang, 1993; Wang et al., 1996). The basin is in a shallow depression in the Changjiang River Plains, and there is a seasonal water exchange between the Changjiang River and Lake Taihu. Thus these natural conditions are potentials to develop the lake eutrophication.

To understand eutrophic changes and factors controlling nutrients in Taihu Basin (e.g. Zhang et al., 2002; Qin, 2002), it must be determined if natural nutrient sources exist that can result in eutrophication. If so, what are the rates and trends of these natural changes? Long-term catchment nutrient modeling of a natural system is needed to address this question. Other issues, such as how these natural changes were different from anthropogenic effects, need to be addressed. Hydrological-based catchment models provide a powerful tool to understand these issues in processes and dynamics (Schulze, 2000; Wang et al., 2003).

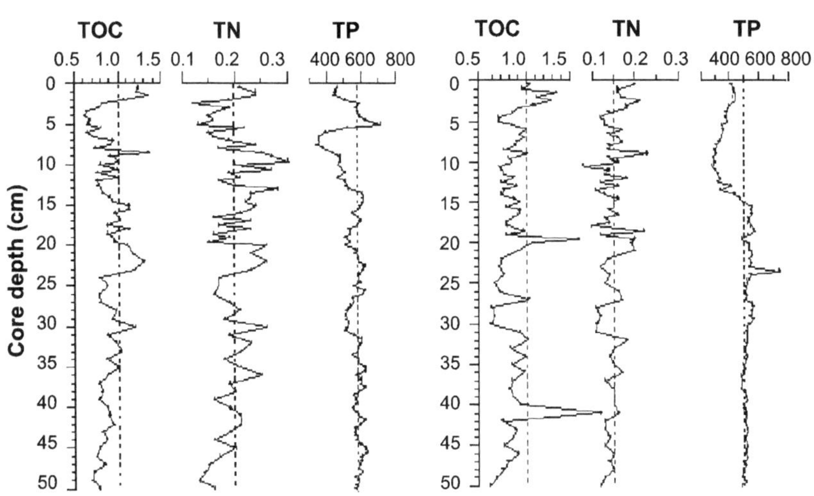

Fig. 1 Nutrient records from sedimentary cores Dls (left) and Ms (right) in Lake Taihu, which reflect higher total nitrogen concentrations (TN,%) than that in the 1990s on the surface sediments

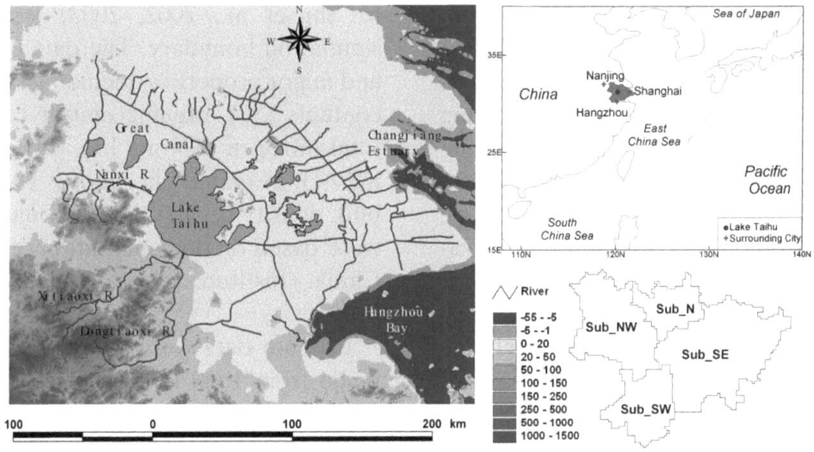

Fig. 2 Geographical location (Right upper), physical and hydrological settings (Left) and four sub-basin divisions (Right lower) of Taihu Basin

The SWAT (Soil and Water Assessment Tool) model is a river basin-scale model developed to predict the impact of land management practices on water, sediment, and agricultural chemical yields in large, complex basins with varying soils, land use, and management conditions over long periods of time (Arnold et al., 1998; Di Luzio et al., 2002a). The model is physically based and computationally efficient, uses readily available inputs, and enables users to study long-term impacts (Di Luzio et al., 2002b). SWAT is being used extensively to assess the impact of global climate on water supply and quality (e.g. Rosenberg et al., 1999). In this study, SWAT was applied to simulate nutrient changes in Taihu Basin dating back to pre-industrial time. The model highlights the whole basin and can simulate catchment nutrients as sources and feedback. Second, climatic and geographic conditions are major controls in natural nutrient development, and the model can be driven by climate-based hydrology, topographical-based Hydrological Response Units (HRUs), and nutrients sourced from vegetation and soil coverage. Finally, natural eutrophication in a lake is a long-term process on a decade and century timescale (Smol, 2002), and SWAT is a continuous integration model capable of simulation 10^1–10^2 year timescales.

This study examines major nutrient changes in a pre-industrial time mode for Taihu Basin to identify nutrient sources from the sub-basins, estimate nutrient balance in the lake, and assess trends on a timescale of 10^1–10^2 years.

Model and experimental design

The model used in this study is the AVSWAT-2000 model (Di et al., 2002a, b), which is an ArcView graphical user interface for the SWAT model (Arnold et al., 1998). Basin topography is a basic boundary in the SWAT model. Vector-based 1:250,000 topographical maps in electronic files (National GIS Center of China, 1999) are converted using ARCINFO into a grid-based Digital Elevation Model (DEM). Grids within 28–3° N and 118–121° E were taken as the boundary of Taihu Basin, and total area was 30,246 km^2 with a maximum elevation difference of 1,266 m from the southwestern mountains to the eastern coastal plains. Four river systems in the basin constitute four sub-basins (Fig. 2). The Nanxi River and tributaries flows to the lake from the northwestern part of the basin (Sub_NW). The Tiaoxi River, with two eastern and western tributaries, discharges to the lake from the southwestern part of the basin (Sub_SW). The southeastern part of the basin is a watershed area from Lake Taihu towards the Changjiang Estuary – Hangzhou Bay (Sub_SE). The northern sub-basin connects Changjiang River and Lake Taihu by the Great Canal of China system (Sub_N).

Thus, Lake Taihu receives stream flows from Sub_SW and Sub_NW and drains to Sub_SE. Different from the other subbasins, there is seasonal water exchange between the Changjiang River and Lake Taihu in Sub_N; i.e. during Changjiang flooding, discharge is orientated from the River to the Lake, while in other seasons, water flows toward the River. Based on observations in extreme years during the 1960–1980s at Liangxihe River Station (Sun and Huang, 1993; Huang et al., 2001), the ratio of river discharge volumes to lake discharge volumes was 0.55:0.45, and the ratio of river discharge days to lake discharge days was 0.46:0.54. Thus, the ratios are approximately 1:1. Based on the estimates above, water input is from the 2.5 sub-basins, and outflow is from 1.5 sub-basins (Table 1).

Land-use in the SWAT contains vegetation coverage and agricultural, industrial, and urban land. In this study, natural and agricultural plants were set at a pre-industrial time mode. Vegetation data were sourced from a vector-based 1:1,000,000 China Vegetation Map (Hou et al., 1982), which summarizes Chinese vegetation investigations since the early 1960s. A digital map from Hou et al. (1982) at a resolution of 10 × 10 min was interpolated into the same resolution as the DEM, consistent with applications in SWAT modeling. According to the dominant plants and relative physical and biological properties (Hou et al., 1982; China Academy of Sciences, 1988) and the agricultural crop atlas (Zui et al., 1984; Bi, 1995), vegetation types in Taihu Basin consistent with the land-use types in the SWAT model setting were assigned (Table 2).

Soil data were obtained from Chinese soil investigations and mapping post-1980s (Institute of Soil Science, 1986; National Soil Survey Office of China, 1993–1997). Vector-based soil data at a resolution of 1:1,000,000 (Unpublished data; ref. to Shi et al., 2002, 2003) were used in the simulation boundary. The data contain soil types and major properties, including sand, silt, and clay contents (%), organic matter content (%), pH, total nitrogen (TN mg l^{-1}), total phosphorus (TP mg l^{-1}), available phosphorus (AP mg kg^{-1}), and bulk density (g cm^{-3}). Additional soil parameters were based on documentation of regional soils in China (Institute of Soil Science, 1986; Zhang, 1998).

Climate data were sourced from observations made from 1960 to 1990 (National Meteorology Center of China, 2001). The 30-year mean monthly temperature and precipitation data were the two major climate variables for the long-term simulation. Relative climate parameters were based on the Atlas of China Climate (Lu & Gao, 1984).

The pre-industrial model was designed to simulate natural-agricultural impacts on water, sediments, and nutrients in Taihu Basin. Model results were compared to geological records from lake cores, which preserve information on decadal-centurial time scales.

Results and analysis

The experiment runs 200 model years in four sub-basins and simulates the flow of major nutrients through the main channel in the catchment, nutrients transported with water during each time step in concentrations in HRUs, and production in each sub-basin. The simulation outputs contain organic nitrogen, NO_3^-, NO_2^- and NH_4^+, organic phosphorus and mineral phosphorus. Total nitrogen (TN) is calculated as a sum of organic nitrogen, NO_3^-, NO_2^- and NH_4^+ and total phosphorus (TP) as the sum of mineral phosphorus and organic phosphorus.

Table 1 Sub-basin topography and discharge features

Subbasin	Sub_SW	Sub_NW	Sub_N	Sub_SE	Total
Area (km^2)	5254	8298	3710	12984	30246
Elevation range (m a s l.)	21–1266	2–453	1–283	0–178	–
Mean elevation (m a s l.)	180	31	11	7.8	–
Input basin	1	1	0.5	0	2.5
Output basin	0	0	0.5	1	1.5

Table 2 Dominant plants in Taihu Basin and assigned land use types in the SWAT modeling

Land use type in SWAT model	Dominant plants in Taihu Basin
Forest - Mixed	Pinus massoniana forest: Shrubby layer with *Rhododendron simsii*, *Vaccinium bracteatum*, *Loropetalum chinense* and *Eurya nitida*; *Cunninghamia lanceolata* forest
Forest - Deciduous	Deciduous oak forest: *Quercus Mongolic*, *Quercus variabilis*, *Q. Acutedentata* and *Q. glandulifera*
Forest - Evergreen	Mixed forest containing *Cyclobalanopsis*, *Castanopsis* and *Fagus*; *Phyllostachys pubescens* forest
Wetlands - Mixed	Halophytic grass and forb meadow *Aeluropus littoralis* var. *sinensis* and *Suaeda salsa*
Winter Wheat	Winter wheat, coarse grains (corn, millet, sweet potatoes), two crops annually
Rice	Summer rice, winter wheat (rapeseed), two crops annually (double cropping rice locally); Double-cropping rice followed by winter wheat (rapeseed) or green manure; Single or double-cropping rice followed by winter wheat, rapeseed, green manure annually
Potato	Potatoes, sweet potatoes, grains, soybean green upland crops annually
Water	Water

Simulation of TN and TP production from four sub-basins

The model simulates 200-year processes for nitrogen and phosphorus production for each sub-basin (Fig. 3). When annual means for each sub-basin (column diagrams in Fig. 4) are divided by sub-basin areas (pie diagrams in Fig. 4), the largest portions of TN and TP are produced in the southwestern mountain areas of Sub_SW, in which TN is 48% and TP is 68% of total production in the basin. Smaller portions of nutrients are produced in Sub_NW (18% TN and 17% TP) and Sub_N (18% TN and 10% TP). The smallest portion is generated in Sub_SE (16% TN and 5% TP).

About half of the nitrogen production comes from Sub_SW, which relates to high biome productions of evergreen forests over the areas (Huang et al., 2001). The model also simulated a

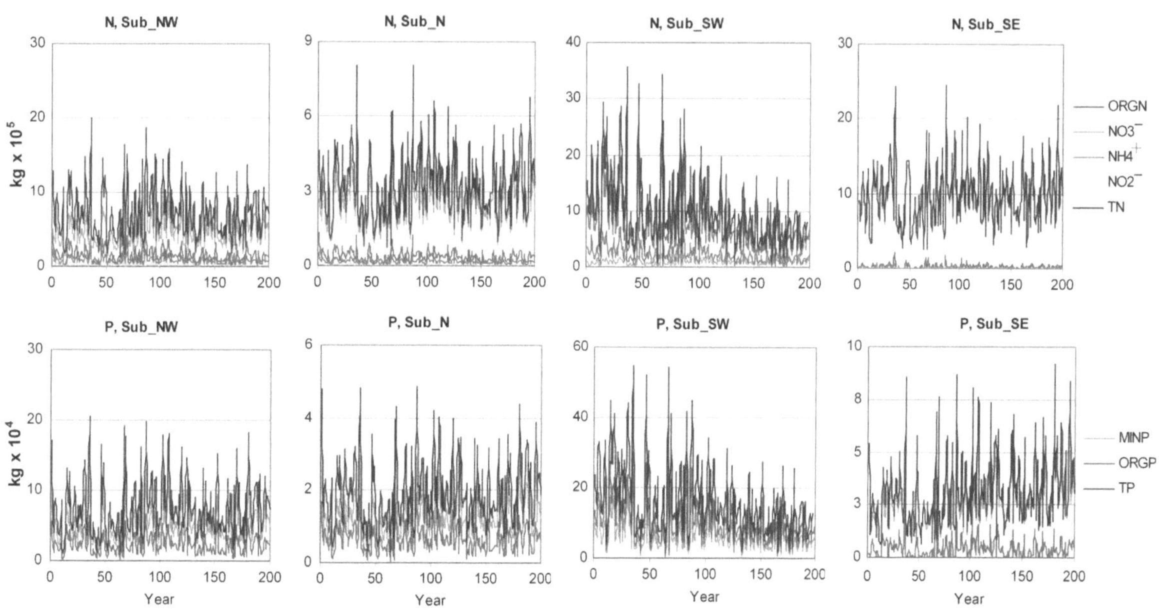

Fig. 3 Annual nitrogen and phosphorus production (kg) simulated from four sub-basins in Taihu Basin, where total nitrogen (TN) as a sum of organic nitrogen (ORGN), NO_3^-, NO_2^- and NH_4^+, total phosphorus (TP) a sum of mineral phosphorus (MINP) and organic phosphorus (ORGP)

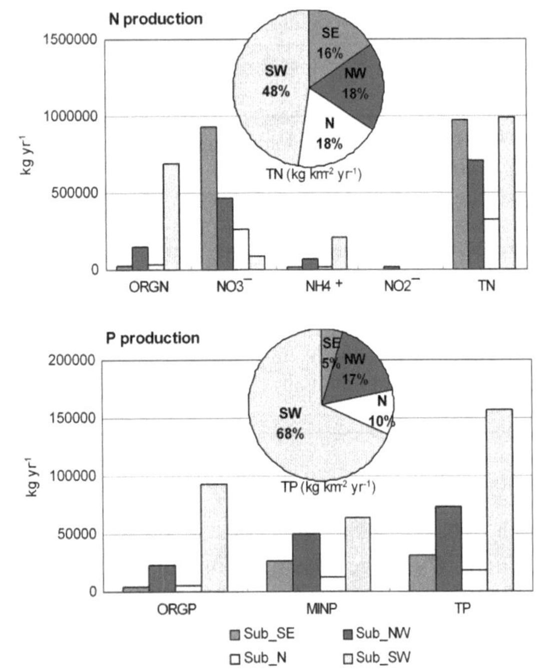

Fig. 4 Distribution simulations of TN and TP in the Lake Taihu sub-basins. Nutrient codes as in Fig. 3

high phosphorus production (~67%) for this area, reflecting strong weathering and erosion of limestone and metamorphic bedrock/outcroppings in the sub-basin. Additionally, relatively steep slopes in this mountain area (mean 2.0%) result in a high capability of discharge and transportation. Lower nutrient production was simulated in Sub_SE, which is an output area for Lake Taihu. Although nutrients discharge to coastal areas, there is small slope (mean < 0.1%), which often slacks water flows and favors nutrient deposition into the lake.

Annual and areal means of TN and TP production are plotted in Fig. 5. Total TN input from the 2.5 sub-basins to the lake is 1.86×10^5 kg, and total TN from 1.5 output sub-basins is 1.14×10^5 kg. This indicates that there is net deposition of 7.3×10^4 kg TN to the lake. Annual TP mean is 2.39×10^5 kg in 2.5 input sub-basins and 0.41×10^5 kg in 1.5 output sub-basins. TP difference between input and output is 1.99×10^5 kg, suggesting that there is TP deposition in the lake each year.

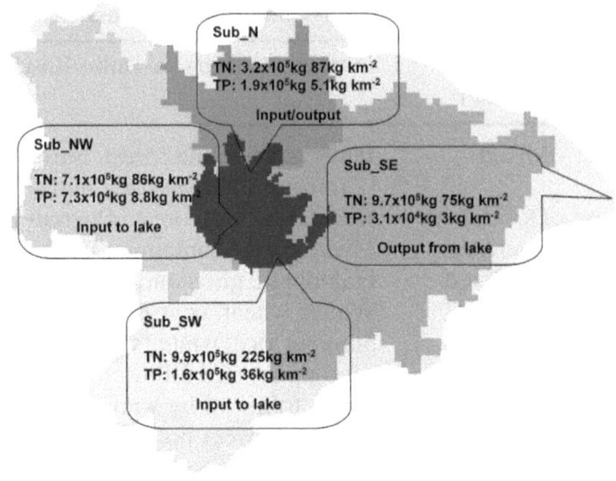

Fig. 5 Simulations of TN and TP annual means (kg) and the annual area means (kg km^{-2}) in four sub-basins

Analysis of catchment-sourced lake eutrophication

To examine lake eutrophication changes, the nutrient balance was evaluated by calculating nutrient differences between input sub-basins and output sub-basins. TN and TP concentrations used are weighted averages of all HRUs within the sub-basins.

The 200-year simulations show that annual TN concentration changes are characteristic of 12-month cycles (Fig. 6a). In the first 100 years, mean TN concentration difference was 0.044 mg l^{-1}, suggesting a positive deposition in the lake. In the second 100 years, the mean difference decreases to 0.017 mg l^{-1}, reflecting positive nitrogen deposition but at a lower rate than the first 100 years. Slopes for the trends are -3.3×10^{-5} in the first 100 years and -1.4×10^{-5} in the second 100 years (Fig. 6a), indicating that TN change stabilized after the first 100 years.

TP concentration simulations show phosphorus deposition in the lake (Fig. 6b). The TP difference in the first 100 years (+0.012 mg l^{-1}) is higher than the second 100 years (+0.008 mg l^{-1}), indicating a declining trend in phosphorus deposition. The slope for the first 100 years (-5.0×10^{-6}) was steeper than the second 100 years (slope -2.4×10^{-6}), suggesting that phosphorus concentration changes are small in the lake.

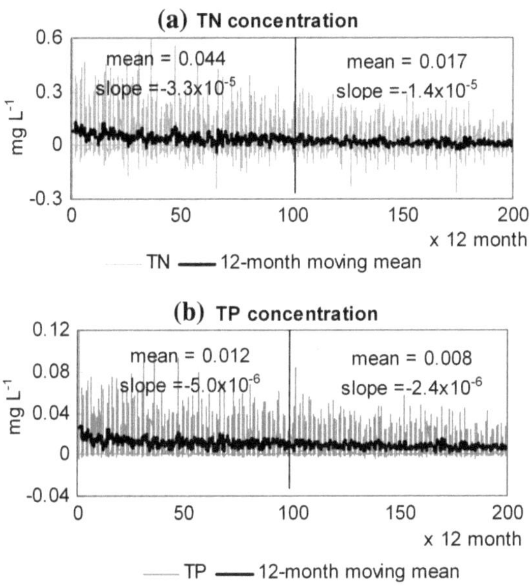

Fig. 6 Monthly differences of TN and TP concentration with the 12-month moving means between input and output sub-basins in Taihu Basin

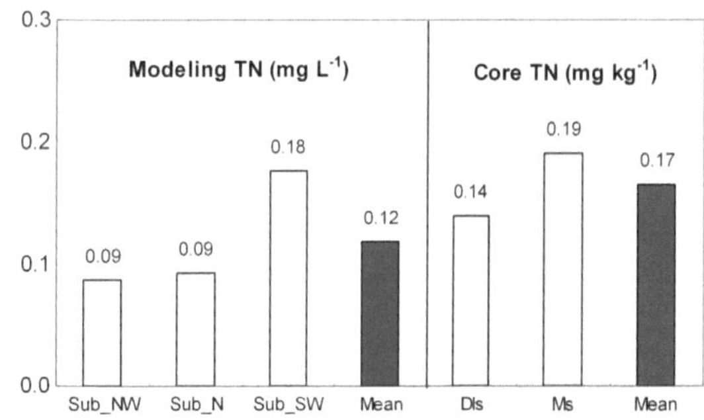

Fig. 7 Modeling TN means for a 100 year ago and measured TN from sediment records in Lake Taihu. Modeling TN concentrations (mg l^{-1}) for the lake water are tentatively compared to concentrations from sediment deposits (mg kg^{-1})

Discussion and conclusions

Comparison with nutrient records from lake sediment cores allows evaluation of the simulations when the area was agricultural before modern industrial development in Taihu Basin. Two sediment cores provided TN concentrations dating back 100 years (Liu et al., 2004a; Fig. 1). Core Dls was taken from the southern part and Core Ms from the northern part of Lake Taihu. TN concentrations at core depth of 20–40 cm are 0.14–0.19 mg kg^{-1} (Liu et al., 2004b), where sediment samples were ^{210}Pb- and ^{137}Cs-dated to before 1900 AD (Liu et al., 2004a). 200-year means show TN concentrations between 0.09 and 0.18 mg l^{-1} (Fig. 7). Geochemical nitrogen values from lake sediments may be subjected to post-depositional mobility (Smol, 2000). Diatom-inferred nutrients, together with geochemical data, may be used for historical eutrophication studies in the future.

The simulation mean (0.12 mg l^{-1}) and sub-basin means (0.09–0.18 mg l^{-1}) are significantly lower than the data mean (0.17 mg kg^{-1}) and two cores (0.14–0.19 mg kg^{-1}) with 95% confidence interval in T-test. Nitrogen from sediment cores would respond to both catchment sources and lake water production. However, the present modeling is focused on nutrient production in the catchment and transport to the lake. Thus, the nitrogen concentration simulation should be lower than in the cores. Modeling may adequately simulate basin nutrients for pre-industrial times.

In summary, the present simulation may capture major factors of lake eutrophication over a 200-year history. Analysis of spatial nutrient distributions suggests that the major source (48% TN and 68% TP) comes from the southwestern sub-basin. Due to a gentle slope in the eastern plains of Taihu Basin, the smallest portion of nutrient input (16% TN and 5% TP), from the southeastern sub-basin, cannot be fully discharged to the lake. There is a net annual nutrient deposition of 7.3×10^4 kg TN and 1.99×10^5 kg TP to the lake. Time series analysis for the 200-year simulation suggests that nitrogen and phosphorus deposition has decreased over the last 100 years.

Catchment nutrient modeling provides a basis to evaluate nutrient production and balance response to environmental changes for the last 200 years in Taihu Basin. There is considerable uncertainty about the direction of future nutrient changes in Lake Taihu due to anthropogenic changes. The accuracy of model-based predictions of future eutrophication cannot be evaluated, but the models used to generate such predictions can be tested by simulating the past and comparing these simulations with sediment records. This effort, along with future investigations, e.g. using regional climate modeling (Zheng et al., 2003) and fossil pollen-based vegetation reconstructions (Yu et al., 2000), would provide more credibility for modeling the future.

Acknowledgements Financial support was provided by Chinese Academy of Sciences (No. KZCX1-SW-12-1) and National Basic Research Program of China (No. 2002CB412301). We thank Jinfang Dai (Nanjing Institute of Geography & Limnology, CAS) for providing land use data, Haiyu Li (Department of Geography, Nanjing University) for assistance in DEM transforming, and Xuesheng Shi (Institute of Soil Sciences, CAS) for providing soil data. We thank Sumin Wang for constructive suggestions on the manuscript.

References

Arnold, J. G., R. Srinivasan, R. S. Muttiah & J. R. Williams, 1998. Large area hydrologic modeling and assessment part I: model development. Journal of American Water Resources Association 34:73–89.

Bi, Y. Y., 1995. Farmland in China. Agricultural Science and Technology Press, Beijing, 231 (in Chinese).

Bradbury, J. P., S. M. Colman & R. L. Reynolds, 2004. The history of recent limnologic changes and human impact on Upper Klamath Lake, Oregon. Journal of Paleolimnology 31: 151–165.

Chinese Academy of Sciences, 1988. Physical Geography of China. Biogeography Science Press, Beijing, 318 (in Chinese).

Di Luzio, M., R. Srinivasan, J. G. Arnold & S. L. Neitsch, 2002a. Soil and Water Assessment Tool. ArcView GIS Interface Manual: Version 2000. GSWRL Report 02–03, BRC Report 02–07, Published by Texas Water Resources Institute TR-193, College Station, TX, 346.

DiLuzio, M., R. Srinivasan & J. G. Arnold, 2002b. Integration of watershed tools and SWAT model into basins. Journal of the American Water Resources Association 38: 1127–1141.

Hou, X. Y., S. Z. Sun, J. W. Zhang, M. G. He, Y. F. Wang, D. Z. Kong & S. Q. Wang, 1982. Vegetation Map of the People's Republic of China. China Cartography, Beijing (in Chinese).

Huang, Y. P., C. F. Fan, W. M. Pu, J. H. Jiang & D. Y. Dai, 2001. Water Environments of Lake Taihu and Controlling the Pollution. Science Press, Beijing, 298 (in Chinese).

Institute of Soil Science of Chinese Academy of Sciences, 1986. The Soil Atlas of China (1:14,000,000). Cartographic Publishing House, Beijing (in Chinese).

Jin, X. C., H. L. Liu, Q. Y. Tu, Z. S. Zhang & X. Zhu, 1990. Lake Eutrophications of China. Chinese Environment Press, Beijing, 614 (in Chinese).

Last, W. M. & J. P. Smol, 2001. Tracking Environmental Change Using Lake Sediments, Volume 1, Basin Analysis, Coring, and Chronological Techniques. Kluwer Academic Publishers, Dordrecht, 548.

Liu, E. F., J. Shen, S. C. Yao & B. Xue, 2004a. Analysis of heavy metal element sources in surface sediment of Lake Taihu. Journal of Lake Sciences 16: 113–119 (in Chinese).

Liu, E. F., J. Shen, Y. X. Zhu, W. L. Xia, H. X. Pan & Z. D. Jin, 2004b. Heavy metals and nutrients pollution in sediments of Taihu Lake. Acta Sedimentologica Sinica 22: 507–512 (in Chinese).

Lu, Y. R. & G. D. Gao, 1984. Atlas of Atmosphere-Hydrology Climate of China. Meteorology Press, Beijing, 183 (in Chinese).

National GIS Center of China, 1999. Data Base of 1:2,500,000 topography of China, Beijing.

National Meteorology Center of China, 2001. Annual Meteorological Observation Report in land surface of China, Beijing (in Chinese).

National Soil Survey Office of China, 1993–1997. Soils in China. Agricultural Publishing House: Beijing, 1–6 (in Chinese).

Qin, B. Q, 2002. Eutrophication of shallow water lakes in the mid-lower reaches of Changjiang and their controlling strategy. Journal of Lake Sciences 14: 193–202 (in Chinese).

Qin, B. Q., W. P. Hu & G. Gao, 2004. Dynamics of sediment resuspension and the conceptual schema of nutrient release in the large shallow Lake Taihu, China. Chinese Science Bulletin 49: 54–64.

Ramstack, J. M., S. C. Fritz, D. R. Engstrom & S. A. Heiskary, 2003. The application of a diatom -based transfer function to evaluate regional water-quality trends in Minnesota since 1970. Journal of Paleolimnology 29: 79–94.

Rose, N. L., J. F. Boyle, Y. Du, C. Yi, X. Dai, P. G. Appleby, H. Bennion, S. Cai & L. Yu, 2004. Sedimentary evidence for changes in the pollution status of Taihu in the Jiangsu region of eastern China. Journal of Paleolimnology 32: 41–51.

Rosenberg, N. J., D. L. Epstein, D. Wang, L. Vail, R. Srinivasan & J. G. Arnold, 1999. Possible impacts of global warming on the hydrology of the Ogallala aquifer region. Journal of Climate 42: 677–692.

Schulze, R. E., 2000. Modelling hydrological responses to land use and climate change: a southern african perspective. AMBIO 29: 12–22.

Shi, X. Z. & D. S. Yu, 2002. A framework for the 1:1,000,000 soil database of China. In Warner, E. D. & G. W. Peterson (eds), Proceedings of the 17th World Congress of Soil Science, Bangkok, 1757: 1–5.

Shi, X. Z., D. S. Yu, X. Z. Pan, W. X. Sun, H. J. Wang & Z. T. Gong, 2003. The 1:1,000,000 China Soil Database and the Application. The Proceedings of China Soil Science Congress (in Chinese).

Smol, J. P., 2002. Pollution of Lakes and Rivers: A Paleoenvironmental Perspective. London, Arnold Publishers, New York, 280.

Sun, S. C. & Y. P. Huang, 1993. Lake Taihu. Ocean Press, Beijing, 273 (in Chinese).

Wang, J., Y. J. Wang & J. L. Liu, 1996. Evolution of sedimentary environment in Lake Taihu during the last 16,000 years. Acta Palaeontologica Sinica 35: 213–223 (in Chinese).

Wang, Z. G., C. M Liu & X. F. Wu, 2003. A perspective for the DEM-based hydrological catchment model. Journal of Natural Resources 18: 168–173 (in Chinese).

Xia, L. Z. & L. Z. Yang, 2003. Studying and controlling for catchment – sourced pollutions in Lake Taihu. Resources and Environments in the Changjiang Catchment 12: 45–49 (in Chinese).

Xue, B., W. C. Qu, Y. H. Wu, S. M. Wang & R. J. Wu, 1998. Sedimentological record of paleoenvironment of Lake Taihu in Late-Glacial to Holocene. Journal of Lake Sciences 10: 30–36 (in Chinese).

Yu, G., X. Chen, J. Ni, R. Cheddadi, J. Guiot, H. Han, S.P. Harrison, C. Huang, M. Ke, Z. Kong, S. Li, W. Li, P. Liew, G. Liu, J. Liu, Q. Liu, K. B. Liu, I. C. Prentice, W. Qui, G. Ren, C. Song, S. Sugita, X. Sun, L. Tang, E. Van Campo, Y. Xia, Q. Xu, S. Yan, X. Yang, J. Zhao & Z. Zheng, 2000. Palaeovegetation of China: a pollen data-based synthesis for the mid-Holocene and last glacial maximum. Journal of Biogeography 27: 635–664.

Zhang, W. R., 1998. Soil Conditions of the Dominant Plants in the Forest of China. China Science and Technique Press, Beijing, 432 (in Chinese).

Zhang, Y. L., B. Q. Qin & F. Q. Huang, 2002. Evolution and zonal analysis on lake eutrophication in East Plain Region. Shanghai Environmental Sciences 21: 549–553 (in Chinese).

Zheng, Y. Q., G. Yu & S. M. Wang, 2003. Numerical climate modeling since the last 140 years by RegMC2 for East Asia. Progresses of Natural Sciences 13: 951–956 (in Chinese).

Zui, D. C., H. S. Liu, J. R. Min & J. M. He, 1984. Atlas of Climatic Resources for Major Crops in China (1:17,000,000). Meteorological Press, Beijing (in Chinese).

Hydrobiologia (2007) 581:89–95
DOI 10.1007/s10750-006-0510-8

EUTROPHICATION IN LAKES

Evaluation of biogenic and anthropogenic inputs of aliphatic hydrocarbons to Lake Taihu sediments using biomarkers

Wenchuan Qu · Bin Xue · Chenwei Su · Sumin Wang

Abstract Surficial sediments from 13 sites throughout Lake Taihu, one of the largest urbanized freshwater lake systems in China, were analyzed for biomarkers (e.g., *n*-alkanes and hopanes) to track the origin of organic inputs (biogenic or anthropogenic), and, thus, to identify any 'hot spots' of hydrocarbon contamination. A distinct spatial distribution of aliphatic hydrocarbons in sediments was observed in Lake Taihu. At the northern tip of the lake (i.e., Meiliang and Wuli Bays), the highest mean aliphatic hydrocarbon concentration, with a significant contribution of an unresolved complex mixture (UCM), was found, indicating possible anthropogenic petroleum contamination (mainly from untreated and partially treated industrial and domestic sewage from Wuxi, Changzhou and other cities). This was supported by the n-alkane indices (e.g., small Carbon Preference Index and odd-to-even values) and a high degree of maturity of the hopane biomarkers. However, hydrocarbons from East Taihu were mainly biogenic, with the lowest mean concentrations, negligible or no contribution of UCM, abundance of vascular plant C_{23}–C_{33} *n*-alkanes with a high odd-to-even predominance, and the presence of biogenic hopanes (e.g., 17β(H), 21β(H)-hopanes and hopenes). In the other areas of the lake, however, the predominance of biogenic in combination with petrogenic hydrocarbons was indicated by the biomarkers.

Keywords Lake Taihu · Sediment · Biomarkers · Sources · Biogenic · Anthropogenic

Guest editors: B. Qin, Z. Liu & K. Havens
Eutrophication of shallow lakes with special reference to Lake Taihu, China

W. Qu (✉) · B. Xue · S. Wang
Key Laboratory of Lake Sediment & Environment, Chinese Academy of Sciences, Nanjing 210008, China
e-mail: wenchuan.qu@gmail.com

W. Qu
Australian Bureau of Statistics, Canberra, ACT 2615, Australia

C. Su
Research School of Biological Sciences, The Australian National University, Canberra, ACT 0200, Australia

Introduction

Aliphatic hydrocarbons (AHCs) are ubiquitous in aquatic sediments, and they derive from anthropogenic and natural sources (Zanardi et al., 1999). Potential long-term deleterious effects of hydrocarbon pollutants to organisms and human health, through trophic transfer, has driven intensive research to study hydrocarbon pollutant sourcing in aquatic environments using biomarkers and related indices (Albaiges & Albrecht,

1979; Qu et al., 1998; Commendatore et al., 2000; Barakat et al., 2002).

Lake Taihu is the third largest freshwater lake in China (Sun & Huang, 1993; Qu et al., 1999; Qin et al., 2007). Pollution is one of the key environmental problems in Lake Taihu due to recent urbanization and industrial development, with associated increases in urban, industrial and rural pollution (TBA, 1998). For example, Meiliang Bay (at the northern tip) receives large amounts of untreated effluent from industries in Changzhou and Wujin, mainly through Zhihugang. Just beside Meiliang Bay, there is a small lagoon (area ~10 km^2) named Wuli Bay (Fig. 1), which receives millions of tons of domestic and industrial sewage each day, mainly via Liangxihe from Wuxi, a large city located ~2 km northeast of Wuli Bay (Zou et al., 1996). The southeast part of Lake Taihu (East Taihu; area 131 km^2) has macrophytes covering ~97% of its total area and is referred to as a macrophyte-type basin.

In this article, surficial sediments from 13 sites in Lake Taihu were analyzed for biomarkers (e.g., *n*-alkanes and hopanes) to track origins of organic inputs (biogenic or anthropogenic) and identify 'hot spots' of hydrocarbon contamination.

Materials and methods

Sample collection

Surface sediments were collected using a metallic bucket grab sampler in February 2000 at stations shown in Fig. 1. Sites 1, 2 and 3 are located in Wuli Bay. Site 1 is close to the mouth of Liangxihe. Site 4 is located in west Meiliang Bay near the mouth of Zhihugang. Site 5 is located in central Meiliang Bay. Sites 11, 12, and 13 are located in East Taihu, and the other sites (6–10) are distributed in the West Basin. Sediment samples were returned to the laboratory and kept frozen until analyzed.

Extraction and fractionation

Sediment samples were dried in the air and sieved (100 mesh or 149 μm) to remove large particles of plant debris. About 50–100 g of sediment were extracted in a Soxhlet (500 ml; (2:1) CH_2Cl_2/MeOH) solution for 72 h. Sulphur was removed by activated copper. Extracts were fractionated into aliphatic hydrocarbons and aromatic hydrocarbons on an activated silica/alumina column using hexane and benzene as eluents, respectively.

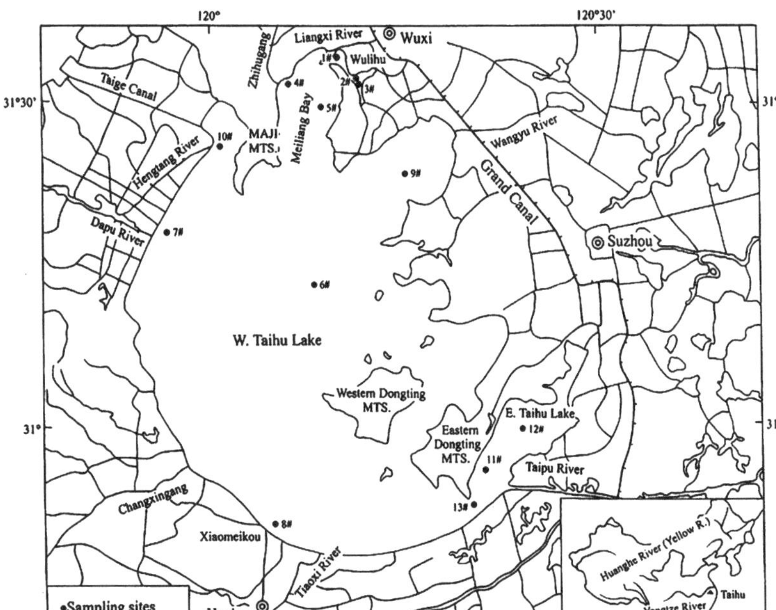

Fig. 1 Sampling sites in Lake Taihu

Gas chromatography and mass spectrometry

Aliphatic hydrocarbons were analyzed on a gas chromatography-mass spectrometry (GC-MS) instrument (HP5890 capillary GC-HP5989A quadrupole MS). Operating conditions were: a HP-5 fused silica capillary column (30 m × 0.25 mm), high-purity He carrier gas (>4.6); gradient temperature: 70 → 300°C, at 4°C/60 s; ion source temperature: 250°C and electron energy: 1.12×10^{-17} J (Qu et al., 1998). Biomarkers were analyzed by monitoring m/z 85 (alkanes) and m/z 191 (hopanes) using the GC/MS in Selected Ion Monitoring (SIM) mode (Qu et al., 1998).

Total organic carbon analysis

Total organic carbon (Org-C) was determined on dried sediments by $K_2Cr_2O_7$ volumetry (Marr & Cresser, 1983).

Results and discussion

Gas Chromatography—total aliphatic hydrocarbon fraction

Concentration of total aliphatic hydrocarbons (AHCs, defined as the sum of resolved peaks plus the unresolved complex mixture, UCM) and organic carbon for 13 surface sediments in Lake Taihu are summarized in Table 1.

A distinct spatial distribution of aliphatic hydrocarbons in surface sediments was observed. High AHC concentrations (>300 $\mu g\ g^{-1}$ dw) were found in sediments from Wuli Bay (Sites 1–3), with the highest concentration (1,630 $\mu g\ g^{-1}$ dw) occurring at Site 1 (Table 1). This may be because Site 1 is close to the mouth of the Liangxihe, which carries large amounts of untreated domestic and industrial sewage from Wuxi and other small cities near Wuli Bay (Fig. 1). AHC concentration at Site 4 also was high (342 $\mu g\ g^{-1}$ dw). This may be explained by its proximity to the mouth of Zhihugang, which carries about 20 million tons of industrial effluents annually from industrialized cities and towns (e.g., Changzhou and Wujin) (Sun & Huang, 1993). AHC concentrations in sediments from other sites were lower (<100 $\mu g\ g^{-1}$ dw), with the lowest occurring in East Taihu (36, 33 and 54 $\mu g\ g^{-1}$ dw for Sites 11, 12 and 13, respectively) (Table 1), which is likely due to the greater distance from major urban and industrial areas and, thus, anthropogenic hydrocarbon sources. The same spatial distribution pattern was found for other micropollutants, such as polycyclic aromatic hydrocarbons (PAHs) (Qu et al., 2002).

To reduce variability from organic matter dilution by inorganics and, thus, better evaluate sediment hydrocarbon pollution, AHC concentrations (in $\mu g\ g^{-1}$) were normalized to organic carbon content (Org-C%). The ratio of AHC concentration to Org-C% has been used as a diagnostic tool, proposing that hydrocarbon (petroleum) pollution is present when the ratio is >50 (Boehm, 1984; Hong et al., 1995). In this study, Sites 1–4 showed the highest Org-C-normalized AHC concentrations (ranging from 144 to 426), indicating the presence of substantial hydrocarbon contamination at the northern tip of the lake. Sediments from East Taihu (Sites 11,12 and 13) had the lowest ratios (ranging from 14 to 52) and were considered unpolluted. AHCs in the remaining sediment samples (mainly from the West Basin) were intermediate relative to Org-C and appeared to be slightly contaminated

GC/MS SIM analysis of alkane biomarkers for AHC source tracing

To distinguish probable sources of AHCs (anthropogenic or biogenic) in surface sediments, different n-alkane evaluation parameters, such as UCM index, carbon preference index (CPI) and even-to-odd carbon number predominance (OEP) ratios, were analysed in addition to absolute concentrations.

UCM is a mixture of chromatographically unresolved components (e.g., many structurally complex isomers and homologues of branched and cyclic hydrocarbons) seen as a 'hump' in the chromatogram (Hostettler et al., 1989; Hong et al., 1995). UCM magnitude is related to the degree of anthropogenic (petroleum) contribution (Mazurek & Simoneit, 1984; Commendatore et al., 2000). UCM is shown clearly in gas chromatograms of aliphatic hydrocarbons in heavily

Table 1 Concentrations of total aliphatic hydrocarbons (AHCs), organic carbon (Org-C), alkane and hopane biomarker parameters for 13 sediment samples in Lake Taihu

Locations	Org-C (%)	AHC μg g^{-1} dw	UCM[a](%)	AHC/ Org-C	Odd/ Even (OEP)	CPI[b]$_{24-34}$	%C_{30}-hopenes/ total hopanes	Tm/ Ts[c]	$C_{30}\beta\alpha$/ $(\alpha\beta+\beta\alpha)$	$C_{30}\beta\beta$/ $\alpha\beta$	C_{31}S/ (S + R)
1	3.83	1630	80	426	1.6	1.8	1.1	0.90	0.08	0.00	0.53
2	2.05	337	60	164	1.9	2.4	4.9	0.91	0.10	0.00	0.54
3	2.16	312	65	144	2.0	2.0	4.7	0.92	0.10	0.00	0.55
4	1.86	342	60	184	1.6	2.1	1.6	0.80	0.06	0.00	0.58
5	1.06	93	45	87	2.1	2.6	6.0	0.94	0.18	0.00	0.53
6	0.87	66	15	76	2.3	3.7	6.9	1.50	0.18	0.00	0.48
7	0.82	73	40	89	2.8	3.7	6.3	1.00	0.13	0.00	0.52
8	0.81	74	35	91	2.1	2.8	5.1	1.00	0.11	0.00	0.53
9	0.94	52	15	55	2.3	3.3	9.6	1.70	0.15	0.00	0.52
10	0.81	112	45	138	2.5	3.8	9.7	1.00	0.11	0.00	0.50
11	1.99	36	6	18	3.1	5.8	36.1	2.00	0.20	0.50	0.41
12	2.31	33	8	14	3.4	8.0	13.1	3.28	0.19	0.12	0.46
13	1.04	54	7	52	2.9	4.4	10.0	1.80	0.18	0.10	0.48

[a] UCM = Unresolved complex mixture

[b] $CPI_{24-34} = 1/2([(C_{25}+C_{27}+C_{29}+C_{31}+C_{33})/(C_{24}+C_{26}+C_{28}+C_{30}+C_{32})]+[(C_{25}+C_{27}+C_{29}+C_{31}+C_{33})/(C_{26}+C_{28}+C_{30}+C_{32}+C_{34})])$

[c] Tm/Ts = 17α(H)-22,29,30-trisnorneohopane /18α(H)-22,29,30-trisnorneohopane (Please refer to the text for more explanations)

AHC polluted sediments (e.g., Site 4, Fig. 2a). Table 1 suggests that UCM accounted for >60% of total AHCs in sediments from Sites 1 to 4, indicating possible petroleum contamination. In contrast, sediments from East Taihu (Sites 11, 12 and 13) showed negligible UCM contribution (<10%) (Fig. 2b), indicating absence of petrogenic inputs. UCM in other samples contributed about 10–50% to total AHCs, suggesting predominance of biogenic in combination with petrogenic hydrocarbons.

The CPI and OEP indices also are used as diagnostic criteria of hydrocarbon sources (anthropogenic or biogenic) in sediments (Hostettler et al., 1989). In most plant waxes, odd-chain alkanes are 8–10 times more abundant than even-chain *n*-alkanes. Thus, vascular plants and unpolluted sediments usually show high OEP and CPI_{24-34} values (ranging from 3 to 6). However, petrogenic hydrocarbons often show a wide distribution range of *n*-alkanes, and the even/odd ratio shows no predominance, with CPI values around 1 (Colombo et al., 1989). As expected, samples from Wuli and Meiliang Bays displayed low CPI (ranging from 1.8 at Site 1 to 2.6 at Site 5) and OEP (ranging from 1.6 at Site 1 to 2.1 at Site 5) ratios (Table 1, Fig. 2c), which evidenced the presence of (riverine) petroleum residues in surface sediments at the northern tip of the lake. Lower CPI and OEP values occurred in samples from the West Basin (Sites 6–10), which indicates decreased contribution of petrogenic hydrocarbons. Samples from East Taihu (Sites 11, 12 and 13) showed the highest CPI (5.8, 8 and 4.4, respectively) and OEP ratios (3.1, 3.4 and 2.9, respectively) (Table 1, Fig. 2d), confirming that these sediments contain mainly contributions from biogenic sources (i.e., vascular plants).

GC/MS SIM analysis of hopane biomarkers for source tracing

Chromatograms of the hexane fraction (Fig. 2) also showed some hopane biomarker peaks. Therefore, GC/MS SIM analysis of hopanes (m/z 191) was conducted to provide more information on AHC sources and spatial patterns, since these biomarkers are more resistant to degradation than alkanes and isoprenoids and retain their composition and distribution during various processes (Barakat et al., 2002).

In the AHC polluted sediments (Sites 1–4), the thermodynamically stable 17α(H), 21β(H)-hopane (that is, αβ-hopane) series were more

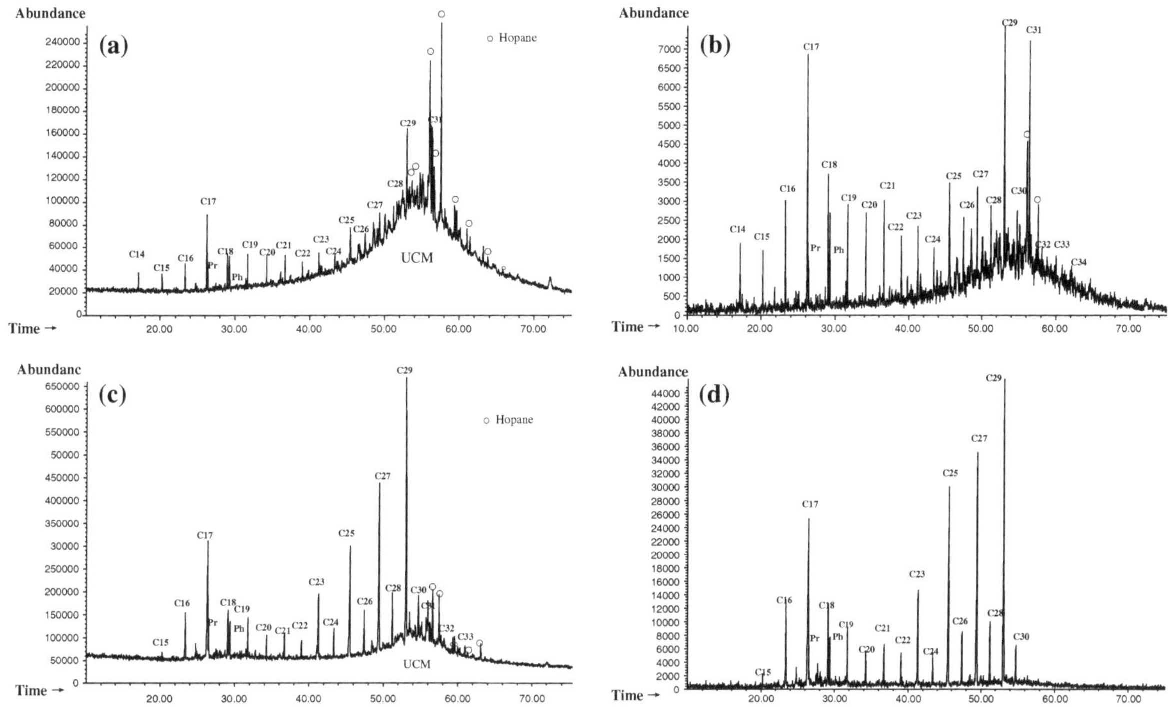

Fig. 2 Capillary column gas chromatograms of the hexane fraction of sediment samples from (**a**) Site 4 (representative of locations with mainly anthropogenic contributions) and (**b**) Site 13 (representative of locations with mainly biogenic contributions); the m/z 85 ion chromatograms of sediment samples from (**c**) Site 4 and (**d**) Site 13 in Lake Taihu. Peaks labeled are the *n*-alkanes from C_{14} to C_{33} (14–33), pristine (Pr), phytane (Ph), and hopanes (*o*)

abundant than the less stable $\beta\alpha$-hopane series, natural biogenic $\beta\beta$-hopane series or mono-unsaturated hopanes (e.g., C_{30}-hopenes) (Fig. 3a). These samples thus had a high degree of hopane biomarker maturity (e.g., lower ratios of $C_{30}\beta\alpha/(\alpha\beta + \beta\alpha)$, $C_{30}\beta\beta/\alpha\beta$ and % C_{30}-hopenes) (Table 1), suggesting a contribution of petrogenic hydrocarbons. Further confirmation of petrogenic pollution was obtained from other geochemical maturity and source indicators. Mature sediments often have low 17α(H)-/18α(H)-22,29,30-trisnorneohopane (Tm/Ts) ratios; extended $\alpha\beta$-hopanes (i.e., series with more than 30 carbon atoms) are resolved into two diastereomers (22S and R) with high S/(S+R) ratios near values for full maturity (equilibrium distribution) (e.g., approximately 0.6 for $C_{31}\alpha\beta$-hopanes) (Hostettler et al., 1999). This is the case in samples from Sites 1 to 4 (Fig. 3a, Table 1).

However, hopane in East Taihu sediments were biogenic (e.g., $\beta\beta$-hopane series and mono-unsaturated hopanes) (Fig. 3b). These samples had the highest ratios of $C_{30}\beta\alpha/(\alpha\beta + \beta\alpha)$ (ranging from 0.18 to 0.20), $C_{30}\beta\beta/\alpha\beta$ (0.10 to 0.50) and % C_{30}-hopenes (10.0–36.1%) (Table 1). Mass chromatograms (m/z 191) of these samples (e.g., Site 13, Fig. 3b) also showed extended $\alpha\beta$-hopane epimers and trisnorhopanes, Ts and Tm, indicating lower maturity (i.e., higher Tm/Ts and lower C_{31} S/(S+R) epimer ratios) (Table 1). All these features are indications of low geochemical maturity and biogenic hydrocarbon sources. In addition, the rest of the samples, mainly from the Western Basin, showed a mixture of mature and immature constituents (Table 1), supporting the conclusion that these sediments contain both biogenic (i.e., vascular plants) and anthropogenic (i.e., petrogenic) sources.

Conclusions

In summary, Lake Taihu surficial sediments showed different levels and sources of aliphatic

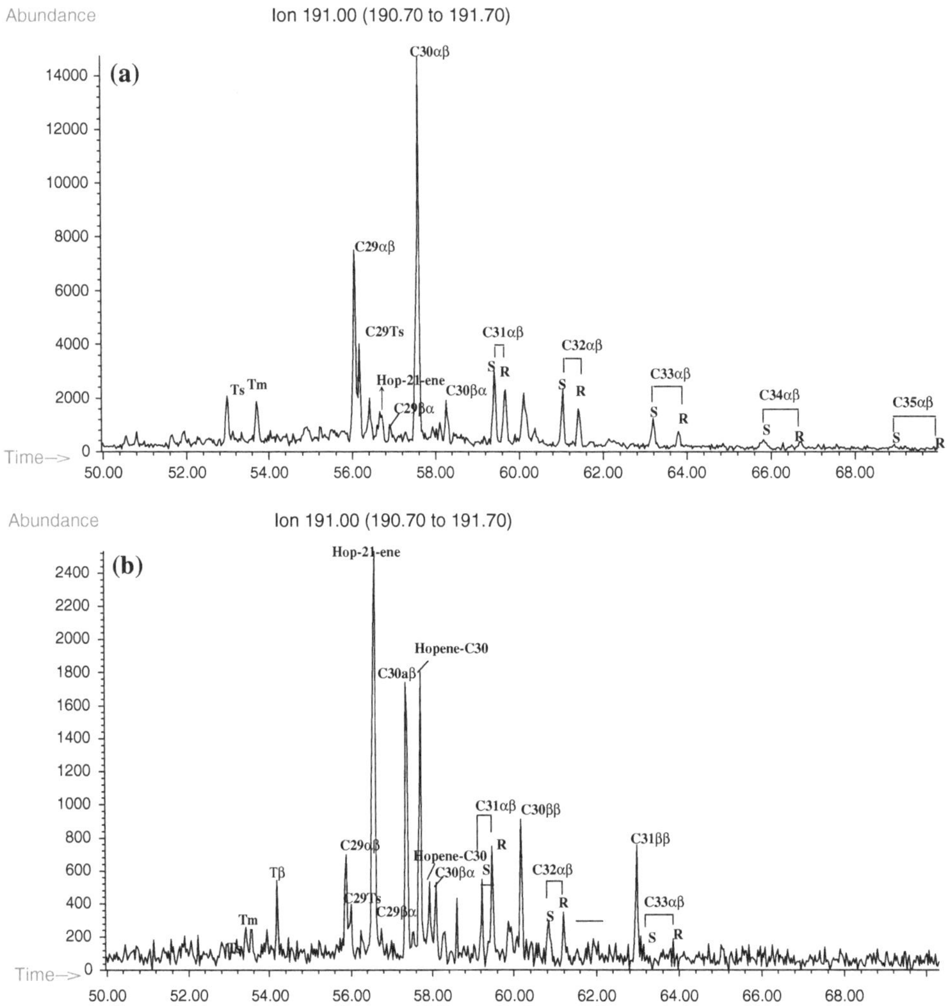

Fig. 3 Mass chromatograms of m/z 191 (hopanes) of surface sediment samples from (**a**) Site 4 and (**b**) Site 13 in Lake Taihu. Peaks identified are hopanes: Ts = 18α(H)-22,29,30-trisnorneohopane; Tm = 17α(H)-22,29,30-trisnorneohopane; T_β = 17β(H)-22,29,30-trisnorneohopane; $C_{29}\alpha\beta$ = 17α(H), 21β(H)-30-norhopane; 29Ts = 18α(H)-norneohopane; $C_{29}\beta\alpha$ = 17β(H), 21α(H)-30-normoretane; $C_{30}\alpha\beta$ = 17α(H), 21β(H)-hopane; $C_{30}\beta\alpha$ = 17β(H), 21α(H)-moretane; S- and R-$C_{31}\alpha\beta$ = S and R isomers of 17α(H), 21β(H)-homohopanes; $C_{30}\beta\beta$=17β(H), 21β(H)-hopane; S- and R- $C_{32}\alpha\beta$ = S and R isomers of 17α(H), 21β(H)-bishomohopanes; $C_{31}\beta\beta$=17β(H), 21β(H)-homohopane; S- and R- $C_{33}\alpha\beta$ = S and R isomers of 17α(H), 21β(H)-trishomohopanes

hydrocarbons, and biomarkers and related indices can provide information on anthropogenic contamination, especially related to petrogenic sources, as well as recent biogenic materials.

Near urban (e.g., Wuxi and Changzhou) and riverine (e.g., Zhihugang and Liangxihe) discharges, anthropogenic petrogenic AHC inputs were predominant. This was supported by: (1) high Org-C-normalized AHC concentrations; (2) significant contributions of UCM; (3) low n-alkane CPI and OEP; and (4) the high degree of maturity of hopane biomarkers. On the contrary, East Taihu sediments exhibited: (1) the lowest level of AHCs; (2) negligible or no contributions of UCM; (3) high n-alkane CPI and OEP; and (4) low degree of maturity of hopane biomarkers. These features suggested that AHCs in these sediments have a dominant biogenic (vascular plants) origin. Biomarker parameters for Western Basin sediment samples, however, indicated the predominance of biogenic in combination with petrogenic hydrocarbons.

Information on sources and spatial variations of AHCs, together with other micropollutants, such as PAHs and heavy metals, may be useful in designing future strategies for protection and management of Lake Taihu.

Acknowledgements The authors thank Jiang Ji of Nanjing Institute of Geography & Limnology, Chinese Academy of Sciences (CAS), for his help in the field. They are grateful to Minzhong Cui and Wanren Ding of Lanzhou Institute of Geology, CAS, for their assistance in the experimental analysis. Thanks are also due to the National Key Project for Basic Research on the Processes of Lake Eutrophication and the Mechanism of Cyanobacterial Blooming (No. 2002CB412301), Chinese Academy of Sciences; Knowledge Innovation Project "Environmental Effects Studies of Lake Sediment Dredging" (No. KZCX3-SW-348) and the 2005 Endeavour Australia Cheung Kong Awards for their financial aid.

References

Albaiges, J. & P. Albrecht, 1979. Fingerprinting marine pollutant hydrocarbons by computerized gas chromatography-mass spectrometry. International Journal of Environmental Analytical Chemistry 6: 171–190.

Barakat, A. O., A. Mostafa, Y. Qian & M. C. Kennicutt, 2002. Application of petroleum hydrocarbon chemical fingerprinting in oil spill investigation - Gulf of Suez, Egypt. Spill Science & Technology Bulletin 7: 229–239.

Boehm, P. D., 1984. Aspects of the saturated hydrocarbon geochemistry of recent sediments in the Georges Bank region. Organic Geochemistry 7: 11–23.

Colombo, J. C., E. Pelletier, C. Brochu & M. Khalil, 1989. Determination of hydrocarbons sources using n-Alkane and polyaromatic hydrocarbon distribution indexes. Case study: Rio de La Plata Estuary, Argentina. Environmental Science & Technology 23: 888–894.

Commendatore, M. G., J. L. Esteves & J. C. Colombo, 2000. Hydrocarbons in coastal sediments of Patagonia, Argentina: levels and probable sources. Marine Pollution Bulletin 40(11): 989–998.

Hong, H., L. Xu, L. Zhang, J. C. Chen, Y. S. Wong & T. S. M. Wan, 1995. Environmental fate and chemistry of organic pollutants in the sediment of Xiamen and Victoria Harbours. Marine Pollution Bulletin 31: 229–236.

Hostettler, F. D., W. E. Pereira, K. A. Kvenvolden, A. Geen, S. N. Luoma, C. C. Fuller & R. Anima, 1999. A record of hydrocarbon input to San Francisco Bay as traced by biomarker profiles in surface sediment and sediment cores. Marine Chemistry 64: 115–127.

Hostettler, F. D., J. B. Rapp, K. A. Kvenvolden & S. N. Luoma, 1989. Organic markers as source discriminants and sediment transport indicators in south San Francisco Bay, California. Geochimica et Cosmochimica Acta 53: 1563–1576.

Marr, I. L. & M. S. Cresser, 1983. Environmental Chemical Analysis. Blackie & Son Ltd, London: 258.

Mazurek, M. A. & B. R. T. Simoneit, 1984. Characterization of biogenic and petroleum-derived organic matter in aerosols over remote, rural and urban areas. Ann Arbor Science. Boston: 353–378.

Qin, B. Q, P. Z. Xu, Q. L. Wu, L. C. Luo & Y. L. Zhang, 2007. Environmental issues of Lake Taihu, China. Hydrobiologia 581: 3–14.

Qu, W. C., M. Dickman, C. X. Fan, S. M. Wang, C. W. Su, L. Zhang & H. X. Zou, 2002. Distribution, sources and potential toxicological significance of polycyclic aromatic hydrocarbons (PAHs) in Lake Taihu sediments, China. Hydrobiologia 485: 163–171.

Qu, W. C., M. Dickman, S. M. Wang & R. J. Wu, 1999. Evidence for an aquatic origin of ketones found in Lake Taihu sediments. Hydrobiologia 397: 149–154.

Qu, W. C., M. Dickman, S. M. Wang, R. J. Wu, P. Z. Zhang, J. F. Chen & H. J. He, 1998. Lake typology based on the use of sediment alkanes in the east and west basins of Lake Taihu, China. Hydrobiologia 364: 219–223.

Sun, S. C. & Y. P. Huang, 1993. Taihu. The Ocean Press, Beijing: 271.

TBA, 1998. Eutrophication Assessment of Lake Taihu - Report of China-Japan Co-operation Project in Lake Taihu. Taihu Basin Authority, Shanghai, 200 pp.

Zanardi, E., M. C. Bicego, L. B. Miranda & R. R. Weber, 1999. Distribution and origin of hydrocarbons in water and sediment in Sao Sebastiao, SP, Brazil. Marine Pollution Bulletin 38(4): 261–267.

Zou, H. X., G. Y. Sheng, C. Sun & O. Y. Xu, 1996. Distribution of organic contaminants in Lake Taihu. Water Research 30: 2003–2008.

Hydrobiologia (2007) 581:97–108
DOI 10.1007/s10750-006-0494-4

EUTROPHICATION IN LAKES

Spatio-temporal distribution of nitrogen in the undulating littoral zone of Lake Taihu, China

Hongjun Wang · Jinwei Lu · Weidong Wang · Peisheng Huang · Chengqing Yin

Abstract Spatio-temporal distribution of nitrogen was examined along the gradient from open water to lakeshore in Lake Taihu, China. Two types of undulating littoral zones were selected: natural reed belt and bare lakeshore. The reed belt affected nitrogen transformation and was sink for internal-lake nitrogen, whereas the bare lakeshore showed little effect. During the growing season, NO_3^--N concentration increased by up to 3–5 times from open water to reed belt, while NH_4^+-N concentration decreased. It suggested that nitrification was the main nitrogen process in reed belt. Total dissolved nitrogen (TDN) showed little spatial variation, indicating that most of nitrogen released from sediment did not move into open water again. Significant temporal variation of dissolved nitrogen occurred and was similar in both the littoral zone and the open water. Maximum TDN and NH_4^+-N concentrations occurred in January, and NO_3^--N in March. Minimum NH_4^+-N and NO_3^--N concentrations occurred in July and August, respectively. An increasing total soil nitrogen was found in the surface sediments from reed belt to open water. This further suggested that the reed-covered littoral zone had strong nitrogen transformation potential.

Keywords Nitrogen · Distribution · Spatial · Temporal · Littoral zone

Guest editors: B. Qin, Z. Liu & K. Havens
Eutrophication of shallow lakes with special reference to Lake Taihu, China

H. Wang · J. Lu · W. Wang · C. Yin (✉)
SKLEAC, Research Center for Eco-Environmental Sciences, CAS, Beijing 100085, P.R. China
e-mail: cqyin@263.net

H. Wang · J. Lu
Graduate University of the Chinese Academy of Sciences, Beijing 100039, P.R. China

P. Huang
Institute of Hydrobiology, Ji'nan University, Guangzhou 510632, P.R. China

Introduction

Riparian zone, the transitional boundary between terrestrial and aquatic ecosystems, influences water movement and waterborne contaminants (Fennessy & Cronk, 1997) and is the principal route for transport of allochthonous matter from terrestrial to aquatic systems (Naiman et al., 1989). Many studies have shown that riparian buffer strips with small dimensions but playing a disproportionately large role, can attenuate terrestrial non-point source pollutants, including sediment, phosphorus and nitrogen (Cooper et al., 1987; Yin & Lan, 1995; Bratli et al., 1999). Recent industrial growth and intensive agriculture have resulted in many riparian zones being disturbed or destroyed (Mitsch, 1995). If 70–95% of natural riparian systems are disturbed or destroyed, the landscape would lose much of its water purification and protection capacity (Fennessy & Cronk, 1997). In China, many lakes, such as Lake Taihu (Hu et al., 1998) and Lake Chaohu (Xu et al., 2003), have deteriorated with nutrient loading sufficiently high to maintain a eutrophic state, even after external nutrient source reduction (Xu et al., 2003). Removing internal nutrients from aquatic systems may help control eutrophication. Dredging sediment from lakes is a direct, physical method (Xu et al., 2003), but it is expensive and difficult to apply to the whole lake. Instead, it is urgent to find alternative biological techniques based on natural processes.

Lake Taihu is an important drinking water source and supports local agriculture, tourism and fisheries. Since the 1980s, rapid industrial development, heavy application of chemical fertilizers and intensive use of water resources have resulted in nutrient-rich water discharge into the lake, causing hyper-eutrophication. According to the Chinese environmental quality standards for surface water (GB 3838-2002), Lake Taihu water quality has deteriorated from grade I–II to grade IV–V, affecting life and economic development. Thus, studies on improving water quality, via non-point and point pollutant attenuation schemes, have been carried out in Lake Taihu (Dou et al., 1995; Pu & Hu, 1996). While internal pollution and sediment/water interactions are stronger in shallow versus deep lakes (Threlkeld, 1994), internal loading may result from accumulation of pore water nutrients and subsequent release across the sediment–water interface in shallow lakes (Kleebery & Kozerski, 1997). Internal nutrient loading from sediments may have an equal or greater contribution than external loading from the watershed (Detenbeck et al., 1993). Reducing internal nutrient loading is difficult, especially with respect to nitrogen, which is comparatively active.

Riparian zones can decrease the energy of flowing water, often resulting in increased sedimentation (Naiman et al., 1989). Finer suspended or resuspended sediment with abundant labile nutrient and absorbed pollutants (Naiman & Décamps, 1997) carried by flowing water enrich riparian sediments proportionally (Johnston et al., 1984). Whether riparian zones will be a sink or source to open water is not clear. A review of nutrient budgets for lacustrine wetlands revealed that the lakeshore could be a net sink for nitrogen and phosphorus either seasonally or annually (Van der Valk et al., 1978). Prentki et al. (1978) suggested that wetland plants and macrophytes act as nutrient pumps from the sediment, while Sager et al. (1985) implied that lakeshore marshes act as nutrient transformers rather than nutrient sinks. If plant-covered littoral zones can act as nutrient sinks, with nutrients mostly converted to inactive forms with low mobility and bioavailability or removed completely, then the riparian zones could mitigate internal pollution. However, little information is available, since most researchers have focused on non-point source or external pollutant removal (Mitsch, 1995). To explore this problem, two types of littoral zones with different plant coverage and lacking terrestrial inputs were selected. In the first year, we mainly developed some fundamental work for our further long-term research in the two littoral zones, involving water, sediment, plant and phytoplankton, and focused on nitrogen in this paper.

Riparian zones are biogeochemical hot-spots for nitrogen (Cirmo & McDonnell, 1997), where complicated nitrogen transformations take place. Nitrogen distribution is a function of physical, chemical and biological processes in the water column and sediments (Watts et al., 1998). Spatio-temporal distribution of nitrogen in mesocosmic environments may be representative of such processes. This paper presents results of a 13-month field study on spatio-temporal nitrogen distribution in the littoral zone of Meiliang Bay, Lake Taihu.

Materials and methods

Lake Taihu is a shallow, eutrophic lake in Jiangsu Province, East China. Monsoon climate dominates this region with annual mean air temperature of 14.9–16.2°C and a growing season lasting from April to November. Meiliang Bay, in the

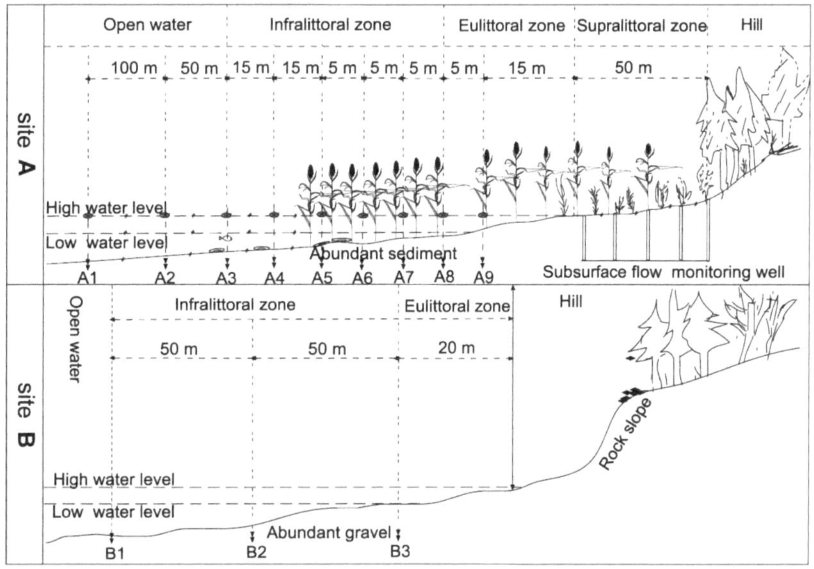

Fig. 1 Sampling station distribution in Sites A and B

northern part of the lake, is hyper-eutrophic. Two types of littoral zones with differing plant coverage and one open water site (control) were chosen in Meiliang Bay (Fig. 1). Site A is a reed-covered littoral zone with an above-ground dry biomass density of 3–4 kg m^{-2}, and Site B is a bare zone without plants. Site A is covered by a well-developed reed population with few other macrophytes. From open water, the first 25 m is inundated continuously, and the next 20 m is inundated intermittently. Toward land, the first 50 m are dominated by a reed (*Phragmites australis*) community and *Gramineae*, then a change from herbs to shrubs toward hills. All shrub stands are at least 30 years old, dominated by *Myrica rubra*, *Symplocos setchensis*, *Loropetalum chinense* and *Eurya murica*, and without visual evidence of recent natural or human disturbance, such as fire, wind-throw, logging or widespread insect infestation. In the lake-ward zone, there are sparse macrophytes with *Myriophyllum verticillatum* predominant in spring. The lake bottom is overlain by 30–40 cm soft sediment. Site B is a bare zone with little bottom sediment, and the interface between water and hills is cliff rock about 10 m wide. Shore vegetation resembles Site A with abundant shrubs. The open water site (C) is regarded as a control point. Initially, Site C was about 500 m from shore. After three preliminary surveys, due to little variation from 100 to 500 m, another station about 200 m from Site A was chosen as the control point.

To collect water and sediment and reduce sampling disturbance, a sampling bridge (50 cm width, 60 m long) was built from dry shore to open water at Site A. To investigate subsurface water in the supralittoral zone, a soil auger (Ø = 11.5 cm) was used to dig 5 vertical holes about 2 m deep as monitoring wells, and polypropylene pipes (Ø = 11 cm) equipped with nylon netting (< 300 μm) at the bottom were inserted into the wells.

Nine sampling stations at Site A and three at Site B were established for monthly sampling along open water to lakeshore (Fig. 1), named A1–A9 and B1–B3, respectively. The interval distances of stations A1–A9 were 100, 50, 15, 15, 5, 5, 5 and 5 m, respectively, with A9 10–15 m from the lake edge at high water, A5 at the fringe of the reed belt, and A1–A4 in the water without macrophytes. On occasion, low water levels left A9 out of water. At Site B, the interval was 50 m. Sampling stations were selected after an extensive survey of hydrochemistry reflecting the fringe effect of reed belt.

Monthly water sampling was carried out from Sep. 2003 to Sep. 2004. At each sampling, after 4–5 days of calm water, triplicate samples were taken using a 3-m long tube providing a vertical sample from the whole water column (Présing et al., 2001). Samples were placed in acid-washed polyethylene bottles and stored in a heat-preservation box on site. Dissolved oxygen (DO), pH, and Eh were measured in situ using an oxygen meter (YSI Model 57) and a pH/Eh meter (HI 8424 Microcomputer HANNA), respectively. In the laboratory, 250 ml from each water sample was filtered within 24 h for analyses of nutrient and chlorophyll-*a* concentrations. Ammonium nitrogen (NH_4^+-N), nitrate nitrogen (NO_3^--N), nitrite nitrogen (NO_2^--N), and chlorophyll-*a* (chl-*a*) were determined using APHA standard methods (APHA, 1998). Total nitrogen (TN), total phosphorus (TP), total dissolved nitrogen (TDN), and total dissolved phosphorus (TDP) were determined by peroxodisulfate oxidation (Ebina et al., 1983) in original water samples and filtrates, respectively. Total particulate nitrogen (TPN) and total particulate phosphorus (TPP) were calculated as the difference between TN and TDN, and TP and TDP, respectively.

About 300–400 g wet weight of mixed sediment (0–10 cm) and surface sediment (0–1 cm) was collected in duplicate using an Ekman-bottom sampler (Hydro-Bios, Kiel, Germany). At Site B, no sediment was present or collected. Sediment samples were stored in polyethylene bags and air-dried in the laboratory. Dried sediments were sieved (≤150 μm) for analyses of soil organic matter (OM) and total soil nitrogen (TSN) contents. OM was obtained from the loss on ignition (500°C, 2 h; Zhu & Carreiro, 2004), and TSN was determined by the semi-microdistillation method using a K_2SO_4–$CuSO_4$–Se catalyst (Lao, 1996).

From Apr. to Sep. 2004, particle sedimentation rates were measured with three sediment traps (Wahbah & Zughul, 2001) placed in different locations in the two littoral zones. Each trap consisted of a cylindrical plastic jar (19 cm height, 7 cm diameter) attached to a rope stretched by an anchor and subsurface buoy marked by a floating buoy. Traps were retrieved monthly, sampled and then redeployed. Material collected in the trap was oven-dried at about 60°C for measuring total weight, and TSN and OM contents were measured as described above.

For subsurface water, samples were collected twice in the growing (May and June 2004) and non-growing season (Jan. and Feb. 2004). Two days prior to sampling, stagnant water in each well was removed three times, allowing the water level to recover to its original level on the sampling day. Nutrient analyses were carried out as above.

Storm runoff may be an important nutrient source in littoral zone. During the study, storms with duration more than 1 h were few. Surface runoff samples in the littoral zone could not be collected during any of the storms, so overland runoff effects remain untested.

One-way ANOVA was used to test temporal and spatial differences by means of parameters. Unless otherwise noted, all significant differences were at a probability of 0.05 or less. All analyses were performed using SAS 8.01 software (SAS, Inc., Cary, NC, USA).

Results

Spatial distribution of inorganic and total nitrogen

At Site A, water column NH_4^+-N and NO_3^--N concentrations exhibited opposite trends from open water to reed belt in the growing season (Fig. 2). NH_4^+-N concentrations decreased from 3.3 to 1.5 mg l^{-1} in April and 1.5 to 0.7 mg l^{-1} in August, while NO_3^--N increased from 2.4 to 4.4 mg l^{-1} in April and 0.4 to 2.3 mg l^{-1} in August. Moreover, spatial variation was highly significant ($P < 0.01$) between the sampling points in reed belt. Figure 2 shows that reed belt fringe effects on open water extend to 20 m, while between 50 and 100 m offshore, inorganic nitrogen concentrations change little. In the non-growing season, the spatial variation trend was similar to that in the growing season, but with a gentler gradient. From surface water in reed belt to subsurface water in supralittoral zone, each nitrogen form decreases in concentrations, except NH_4^+-N in the growing season (Table 1). One-way ANOVA showed significant difference between reed belt and open water in the growing season ($P < 0.01$) but not in the non-growing season. At Site B, inorganic nitrogen concentrations in the shore were similar to those in open water and no significant variation was detected.

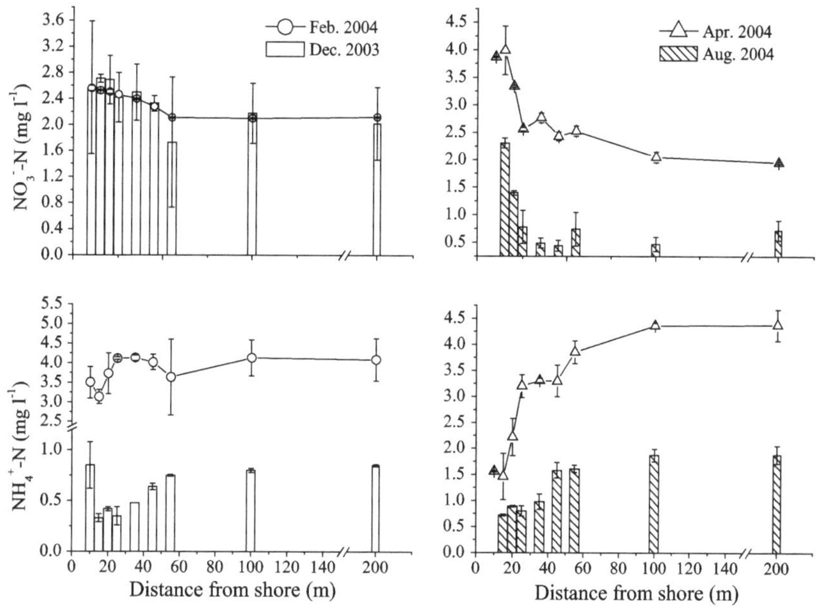

Fig. 2 Spatial variation of nitrate (NO_3^--N, top) and ammonium nitrogen (NH_4^+-N, bottom) concentrations (mean ± SD) in the reed-covered littoral zone (A) on two occasions

Table 1 The nitrogen forms in surface water of reed belt and subsurface water of supralittoral zone, n = 36

Parameters (mg l^{-1})	Growing season (mean ± SD)		Non-growing season (mean ± SD)	
	Surface water	Subsurface water	Surface water	Subsurface water
NH_4^+-N	0.37 ± 0.26	0.40 ± 0.24	3.37 ± 0.37	1.23 ± 0.81
NO_3^--N	2.32 ± 0.24	0.52 ± 0.59	2.83 ± 0.21	1.28 ± 0.59
NO_2^--N	0.12 ± 0.14	0.04 ± 0.03	0.22 ± 0.13	0.11 ± 0.15
TN	4.17 ± 1.15	2.62 ± 1.97	17.72 ± 7.82	5.81 ± 2.24

Spatial distribution of TN and TDN at Site A is shown in Fig. 3. TDN in reed belt varied little, and concentrations were similar to those in open water. Only at Station A9 at 10 cm water depth, TN concentrations increased from 12 to 28 mg l^{-1} and 4 to 54 mg l^{-1} in February and August, respectively. TN and TDN in the shore and open water was homogeneous at Site B.

Temporal variation of inorganic and total nitrogen

Temporal variations of NH_4^+-N, NO_3^--N and NO_2^--N in various zones are given in Fig. 4. Every form of nitrogen showed noticeable variation trend. NH_4^+-N concentration maxima (3.2–5.2 mg l^{-1}) occurred in January, with temperature rise and increasing biological activities and NH_4^+-N concentrations decreased to 0.3–0.9 mg l^{-1} in September. NO_3^--N concentrations reached highest values (2.9–4.8 mg l^{-1}) in March, with a "lag" of about 2 months, and the minimum value occurred in August. Temporal variation of NO_2^--N concentrations showed peak values (1.4 and 0.7 mg l^{-1}) in May and September, respectively.

TN can be categorized as two components, TDN and TPN. There was a similar temporal trend between open water and bare littoral zones (Fig. 5). Maxima for TN (14.9–18.8 mg l^{-1}) and TDN (9.8–11.6 mg l^{-1}) occurred in January while minima in summer. TPN was highest (4.4–11.1 mg l^{-1}) in winter and lowest (0.5–1.6 mg l^{-1}) in spring. However, irregular TN and TPN patterns were observed in the reed-covered littoral

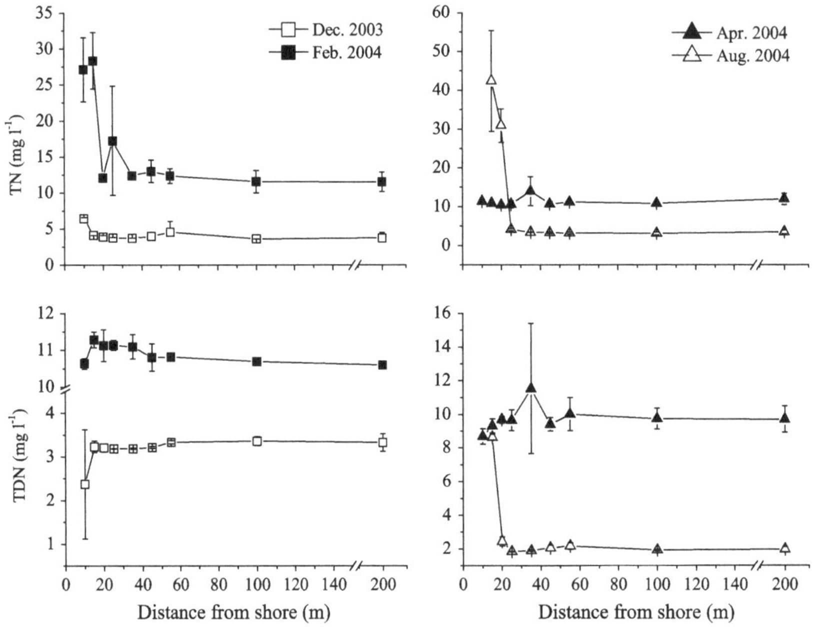

Fig. 3 Spatial distribution of total nitrogen (TN, top) and total dissolved nitrogen (TDN, bottom) concentrations (mean ± SD) in the reed-covered littoral zone (A)

zone, with peaks in summer and winter, respectively. From July to August, an algal bloom developed, and algae accumulated in reed belt (100,000–300,000 μg l^{-1} chl-*a*), so there was an extraordinary peak value (about 30 mg l^{-1}) of TPN with a positive correlation between TN and chl-*a* in summer ($r^2 = 0.925$, $P < 0.0001$).

Spatial variation of TSN and OM in sediment

In the reed-covered littoral zone, there were plentiful soft sediments about 30–40 cm in thickness. Most were freshly settled from suspension in open water where mechanical energy is dampened by littoral reed stands. These sediments were dominated by decayed algae rich in nutrients, and TSN and OM contents in vertical mixed sediment (0–10 cm) were up to 0.5–0.6% and 7–8%, respectively. TSN and OM contents in vertical mixed sediment had little spatial variation, however, they decreased from 0.5–0.6% to 0.1–0.2% and 7–8% to 2–3% in surface sediment (0–1 cm), respectively from open water to reed belt (Fig. 6).

Discussion

External sources of nitrogen in the littoral zone

Littoral zone is the interface between terrestrial and aquatic systems, and its dynamics are influenced by both. External nitrogen sources mainly come from land and open water, and the contribution from wet precipitation is negligible (0.18 g m^{-2} a^{-1} in the case of Lake Taihu: Qin et al., 2004). Nitrogen enters littoral zone via surface runoff and subsurface flow from land. There were no conspicuous point sources, such as a factory or village, or non-point pollution sources in this study. Vegetation covered the littoral zone, and the rainy season coincided with vegetation density, where plants dampened the energy of flowing water and retained most waterborne materials with little subsequent erosion. However, there were few long duration rainstorms. Five surface runoff events were observed, but overland transport to littoral zone was not detected. Although some heavy rainstorms may

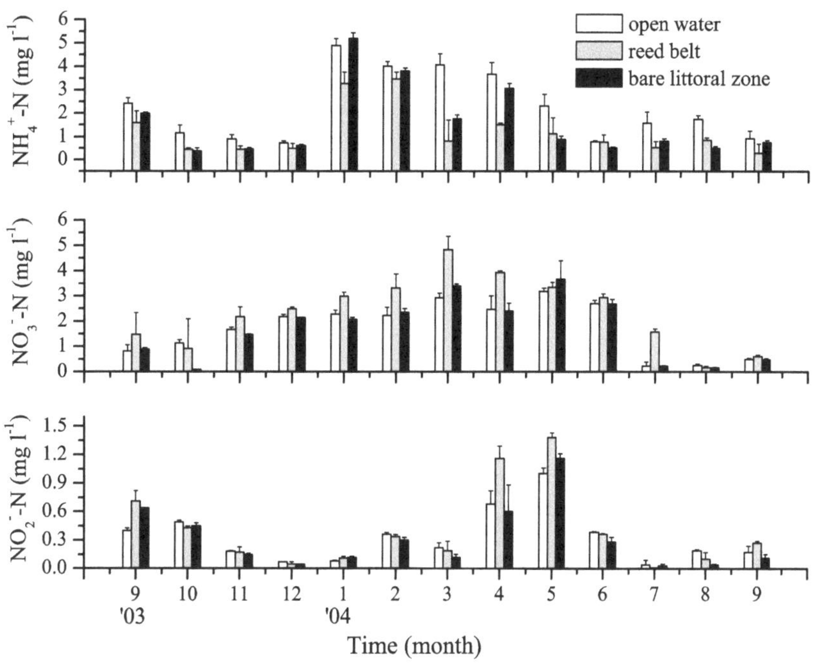

Fig. 4 Temporal variation of ammonium nitrogen (NH_4^+-N), nitrate nitrogen (NO_3^--N), and nitrite nitrogen (NO_2^--N) concentrations (mean ± SD) in open water, reed belt and bare littoral zone

have been missed, we believe they were few. In this study, overland runoff effects have been disregarded. Subsurface flow is mostly unidirectional toward aquatic systems across the terrestrial boundary (Naiman & Décamps, 1997), and our investigation suggested that few nutrients came from subsurface flow. TN and NO_3^--N concentrations in subsurface water were less than those in the reed belt water column in the growing and non-growing seasons. Moreover, NO_3^--N concentrations decreased from the lakeside to the hill and was lowest in the saturated area, which indicates (1) subsurface flow can carry little NO_3^--N to the inundated littoral zone; and (2) water column NO_3^--N may infiltrate into the supralittoral zone; and (3) the littoral reed belt may remove NO_3^--N efficiently from infiltrated lake water, as found in Lake Baiyangdian (Wang et al., 2002).

Material retention was high in ecotones or riparian zones with a sharp decrease in the kinetic energy of wind or water (Naiman et al., 1989). In a eutrophic lake, the retentate contained abundant nutrients, which would enrich the riparian zone (Johnston et al., 1984). In the littoral, especially where there is a macrophyte cover, sediments carried by lake currents also may settle. In our study, there was a distinct difference between average sedimentation rates at Sites A and B (250 and 75 g m^{-2} d^{-1}, respectively). At Site A, 30–40 cm thick soft sediments contained 0.5–0.6% TSN and 7–8% OM, while at Site B, there was almost no sediment and bottom gravels were exposed.

These preliminary investigations indicate that particulates from open water were the main nitrogen source to the reed-covered littoral zone, perhaps over 90% on a rough estimate. Reed-covered littoral zones may be a sink for internal TPN sources in the lake if these enriched areas do not release nitrogen back into open water.

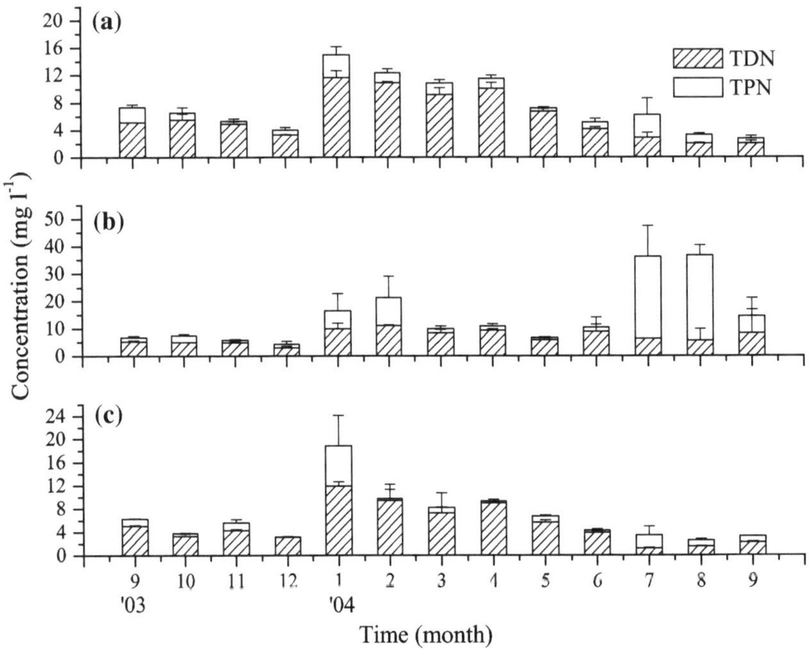

Fig. 5 Temporal variation of total dissolved nitrogen (TDN) and total particle nitrogen (TPN) in open water (**a**), reed belt (**b**), and bare littoral zone (**c**)

Spatial distribution of several forms of nitrogen

In the aquatic nitrogen cycle, energy utilization is more efficient and turnover is more rapid than in terrestrial systems (Heathwaite, 1993). Riparian zones may provide ideal conditions for nitrogen transformations, including ammonification, nitrification and denitrification (Fennessy & Cronk, 1997). Continuous deposition of plant and algae detritus provides rich carbon sources. Many studies have shown that ammonification in vegetated sediments is higher than in bare sediments (Herbert, 1999: 570, Table 5). Nitrification and denitrification in vegetated sediments also may be high (Christensen & Sørensen, 1986; Ottosen et al., 1999; Saunders & Kalff, 2001; Matheson et al., 2002). There is enough oxygen for nitrification in surface sediment of shallow vegetated zones, and denitrification may occur at greater depths. However, wetland vegetation transports atmospheric oxygen through the stem, to the root, and into adjacent soil, which may oxidize the reduced microzone around the root surface (Armstrong, 1964; Brix & Schierup, 1990). Two distinct layers are formed around a root surface favoring nitrification–denitrification (Reddy & Patrick, 1984). Trapped sediments provide ammonification sites; nitrification can utilize ammonification product in the aerobic water, surface sediments and root surface; and denitrification occurs in the underlying sediment, releasing dinitrogen or nitrous oxide. These processes occur simultaneously and are controlled, in part, by NH_4^+-N and NO_3^--N in the reed-covered littoral zone.

In the present study, there were opposite trends in NH_4^+-N and NO_3^--N dynamics in the reed covered littoral zone. NH_4^+-N concentrations decreased and NO_3^--N increased. In addition, other biotic factors influenced inorganic nitrogen concentrations, such as algal uptake and zooplankton and zoobenthos excretion. Moreover, NH_4^+-N is taken up by cyanobacteria more rapidly than NO_3^--N or NO_2^--N. Algae biomass in reed belt was not always higher than in open water, while the NH_4^+-N concentration gradient existed at all times, especially in the growing season. These results implied that nitrification was more intensive than ammonification, and nitrification

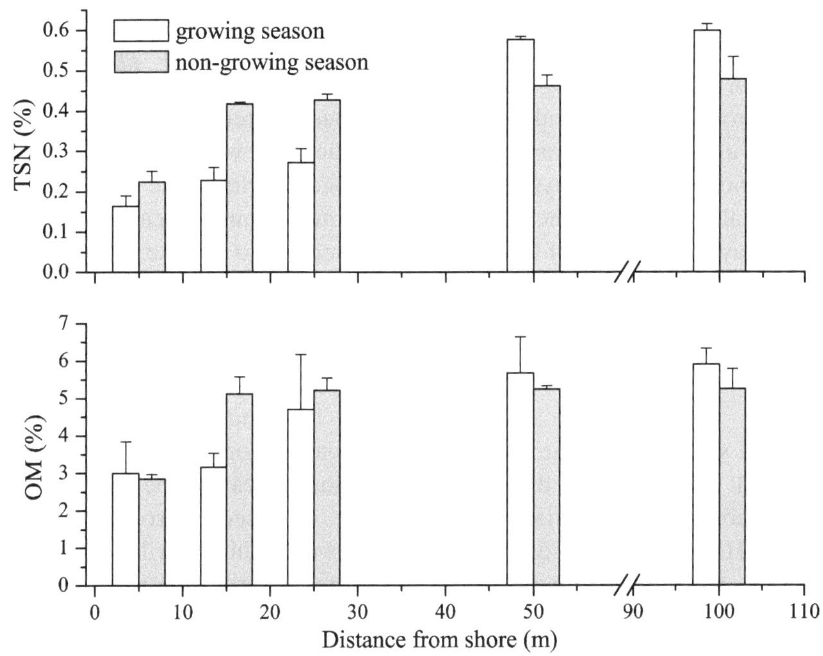

Fig. 6 Spatial variation of total soil nitrogen (TSN) and organic matter (OM) contents (mean ± SD) in surface soft sediment in the reed-covered littoral zone (A)

exceeded denitrification. Our data are inadequate to explain the transformations, and further biotic research is needed.

Inorganic nitrogen forms variations showed that the plant-covered littoral zone is a strong transformer of nutrients. TN and TDN in water column, combined with TSN and OM in surface sediment, provide clues about their fates. Spatial variation of TN concentrations in reed belt varied but generally was higher than, and sometimes equal to, that in open water, while TDN concentrations showed little spatial variation. The difference between TN and TDN suggested that particulate nitrogen dominated the spatial variation of TN. In summer, chl-*a* concentrations (100,000–300,000 μg l^{-1}) in reed belt were 500–1,000 times higher than those in open water. Second, there was no or inconspicuous secondary nitrogen release into open water from the littoral zone. Furthermore, periodic high concentrations of particulate nitrogen indicate that these particles mainly come from open water by intermittent wind forcing.

Denitrification is a key process of nitrogen removal from aquatic systems (Fennessy & Cronk, 1997; Martina et al., 1999). Spatial variation of inorganic nitrogen suggested that it was not intensive in the present study. To verify this, we investigated surface sediment (0–1 cm) and found an interesting phenomenon. TSN and OM contents in surface sediment showed synchronous spatial variation with both decreasing from open water to reed belt. The decrease indicates that nitrogen in settling sediment is lost in reed belt via release or denitrification. Because the dissolved nitrogen bulk in littoral zone was similar to that in open water, we deduced that nitrogen decrease in surface sediment was removed by denitrification. The sediment results further demonstrated that the reed-covered littoral zone could act as nitrogen sink of the lake, which played a vital role in removing internal nutrients.

Compared with the reed-covered littoral zone, spatial variation of nitrogen in bare zone was similar to that in open water. Lacking macrophytes, littoral zone erodes and particulates do not settle, so nutrients are not stored, and many biotic activities and biogeochemical cycles are limited. Altered physical conditions, when riparian zone is damaged, may result in loss of function (Fennessy & Cronk, 1997).

Temporal variation of inorganic nitrogen and TN

Littoral zone is often a small part of the aquatic system, and temporal variation of dissolved nitrogen fractions are controlled by variations in open water. TN and NH_4^+-N concentrations show winter maxima. These winter maxima possibly arise from (1) around lake, surface and subsurface catchment inputs are high in the non-growing season; (2) nutrients are concentrated in lower water levels; and (3) microbial activity is low due to low temperatures. NO_3^--N produced by nitrification, reaches peak value in spring often after a lag, since nitrification is sensitive to temperature. Below 15°C, nitrification rates decrease (Focht & Verstraete, 1977). In our study, TPN mainly came from suspended sediment and algae, and there were two maxima in reed belt (algae in summer, resuspended sediment in winter). While, in open water and bare zones, only one maxima in winter from suspended sediment was observed (Zhang et al., 2004).

Conclusions

Spatio-temporal distributions of different nitrogen forms were interpreted along gradients from open water to lakeshore in undulating littoral zones. In our two study zones, direct terrestrial nutrients accounted for little of the input, whereas more than 90% came from open water as resuspended sediments and algae. Between bare zone and reed-covered littoral zone, there were significant differences in hydrochemistry and physical conditions. The bare zone without soft sediment was almost homogeneous with open water, while the reed-covered littoral zone had abundant soft sediments and displayed spatial variations of inorganic nitrogen concentrations in water column. By comparison of spatial distribution of nitrogen in reed belt, the results showed that (1) nitrification was a key process controlling the nitrogen cycle in the reed-covered littoral zone, and most ammonification production was transformed in to nitrate; (2) little of the nitrogen enrichment to reed belt was released back into open water; (3) nitrification provided adequate nitrogen for denitrification; and (4) the reed-covered littoral zone was an important nitrogen sink. TSN and OM contents in surface sediments decreased from open water to reed belt. Denitrification appears to be significant in reed belt and most released nitrogen from sediments are removed, which further suggests that the reed covered littoral zone is a nitrogen sink and a self-balancing system.

Acknowledgements This research was funded by Key Project of Knowledge Innovation Program of CAS (KZCX1-SW-12), National Hi-Tech Research Program (2002AA601011-05), and National Basic Research Program of China (2006CB403306).

References

APHA, 1998. Standard Methods for the Examination of Water and Wastewater, 20th ed. American Public Health Association, NW, Washington DC.

Armstrong, W., 1964. Oxygen diffusion from the roots of some British bog plants. Nature (London) 204: 801–802.

Brix, H. & H. H. Schierup, 1990. Soil oxygenation in constructed reed beds: the role of macrophyte and soil atmosphere interface oxygen transport. In Cooper, P. & F. B. C. Findlater (eds), Constructed Wetlands in Water Pollution Control. Pergamon Press, London: 53–66.

Bratli , J. L., A. Skiple & M. Mjelde, 1999. Restoration of Lake Borrevannet – self-purification of nutrients and suspended matter through natural reed belts. Water Science and Technology 40: 325–332.

Christensen, P. B. & J. Sørensen, 1986. Temporal variation of denitrification activity in plant-covered littoral sediment from Lake Hampen, Denmark. Applied and Environmental Microbiology 51: 1174–1179.

Cirmo, C. P. & J. J. McDonnell, 1997. Linking the hydrologic and biogeochemical controls of nitrogen transport in near-stream zones of temperate-forested catchments: a review. Journal of Hydrology 199: 88–120.

Cooper, J. R., J. W. Gilliam, R. B. Daniels & W. P. Robarge, 1987. Riparian areas as filters for agricul-

tural sediment. Soil Science Society of America Journal 51: 416–420.
Detenbeck, N. E., C. A. Johnston & G. J. Niemi, 1993. Wetland effects on lake water quality in the Minneapolis/St. Paul metropolitan area. Landscape Ecology 8: 39–61.
Dou , H. S., P. M. Pu, S. Z. Zh, W. P. Hu & Y. Pang, 1995. An experimental study on culture of *Eichhornia crassipes* (Mart.) Solms on open area of Taihu Lake. Journal of Plant Resources and Environment 4: 54–60 (in Chinese).
Ebina, J., T. Tsutsui & T. Shirai, 1983. Simultaneous determination of total nitrogen and total phosphorus in water using peroxodisulfate oxidation. Water Research 17: 1721–1726.
Fennessy, M. S. & J. K. Cronk, 1997. The effectiveness and restoration potential of riparian ecotones for the management of nonpoint source pollution, particularly nitrate. Critical Review in Environmental Science and Technology 27: 285–317.
Focht, D. D. & W. Verstraete, 1977. Biochemical ecology of nitrification and denitrification. Advance in Microbial Ecology 1: 135–214.
Heathwaite, A. L., 1993. Nitrogen cycling in surface waters and lakes. In Burt, T. P., A. L. Heathwaite & S. T. Trudgill (eds), Nitrate: Processes, Patterns and Management. John Wiley & Sons Ltd., Chichester: 99–100.
Herbert, R. A., 1999. Nitrogen cycling in coastal marine ecosystems. FEMS Microbiology Reviews 23: 563–590.
Hu, W. P., J. Salomonsen, F. L. Xu & P. M. Pu, 1998. A model for the effects of water hyacinths on water quality in an experiment of physico-biological engineering in Tai Lake, China. Ecological Modelling 107: 171–188.
Johnston, C. A., G. D. Bubenzer, G. B. Lee, F. W. Madison & J. R. McHenry, 1984. Nutrient trapping by sediment deposition in a seasonally flooded lakeside wetland. Journal of Environmental Quality 13: 283–290.
Kleebery, A. & H. P. Kozerski, 1997. Phosphorus release in Lake Grosser Muggelsee and its implications for lake restoration. Hydrobiologia 342: 9–26.
Lao, J. C., 1996. Soil agrochemical analysis methods [M], 2nd ed. Agricultural Press, Beijing.
Martina, T. L., N. K. Kaushik, J. T. Trevors & H. R. Whiteley, 1999. Review: denitrification in temperate climate riparian zones. Water, Air, and Soil Pollution 111: 171–186.
Matheson, F. E., M. L. Nguyen, A. B. Cooper, T. P. Burt & D. C. Bull, 2002. Fate of ^{15}N-nitrate in unplanted, planted and harvested riparian wetland soil microcosms. Ecological Engineering 19: 249–264.
Mitsch, W. J., 1995. Restoration of our lakes and rivers with wetlands – an important application of ecological engineering. Water Science and Technology 31: 167–177.
Naiman, R. J. & H. Décamps, 1997. The ecology of interface: Riparian zones. Annual Review of Ecology and Systematics 28: 621–658.
Naiman, R. J., H. Décamps & F. Fournier (eds), 1989. The role of land/inland water ecotones in landscape management and restoration: a proposal for collaborative research. MAB Digest 4, UNESCO.
Ottosen, L. D. M., N. Risgaard-Petersen & L. P. Neilsen, 1999. Direct and indirect measurements of nitrification and denitrification in the rhizosphere of aquatic macrophytes. Aquatic Microbiol Ecology 19: 81–91.
Prentki, R. T., T. D. Gustafson & M. S. Adams, 1978. Nutrient movements in lakeshore marshes. In Good, R. E., D. F. Whigham & R. L. Simpson (eds), Freshwater Wetlands: Ecological Processes and Management Potential. Academic Press, New York: 169–194.
Présing, M., S. Herodek, T. Preston & L. Voros, 2001. Nitrogen uptake and the importance of internal nitrogen loading in Lake Balaton. Freshwater Biology 46: 125–139.
Pu, P. M. & W. P. Hu, 1996. Key strategies for controlling eutrophication of large shallow lakes in China. In 28th International Geographical Congress, Utrecht.
Qin, B. Q., W. P. Hu & W. M. Chen, 2004. Process and mechanism of environmental changes of the Taihu Lake. Science Press, Beijing.
Reddy, K. R. & W. H. Patrick, 1984. Nitrogen transformations and loss in flooded soils and sediments. CRC Critical Reviews in Environmental Control 13: 273–309.
Sager, P. E., S. Richman & H. J. Harris, 1985. Preliminary observations on the flux of carbon, nitrogen, and phosphorus in a Great Lakes coastal marsh. In Prince, H. H. & F. M. D'Itri (eds), Coastal Wetlands. Lewis Publishers, Inc., Chelsea, MI: 59–65.
Saunders, D. L. & J. Kalff, 2001. Denitrification rates in the sediments of Lake Memphremagog, Canada-USA. Water Research 35: 1897–1904.
Threlkeld, S. T., 1994. Benthic-pelagic interactions in shallow water columns: an experimentalist's perspective. Hydrobiologia 275/276: 293–300.
van der Valk, A. G., C. B. Davis, J. L. Baker & C. E. Beer, 1978. Natural fresh water wetlands as nitrogen and phosphorus traps for land runoff. In Greeson, P. E., J. R. Clark & J. E. Clark (eds), Wetland Functions and Values: The State of our Understanding. Proceeding of National Symposium on Wetlands, Lake Buena Vista, FL. American Water Resources Association, Minneapolis, MN: 457–467.
Wahbah, M. I. & M. B. Zughul, 2001. Temporal distribution of chlorophyll *a*, suspended matter, and the vertical flux of particles in Aqaba (Jordan). Hydrobiologia 459: 147–156.
Wang, W. D., D. L. Wang & C. Q. Yin, 2002. A field study on the hydrochemistry of land/inland water ecotones with reed domination. Acta Hydrochimica et Hydrobiologica 30: 117–127.
Watts, L. J., T. P. Rippeth & A. Edwards, 1998. The roles of hydrographic and biogeochemical processes in the distribution of dissolved inorganic nutrients in a cottish Sea-loch: consequences for the spring phytoplankton bloom. Estuarine, Coastal and Shelf Science 46: 39–50.

Xu, F. L., S. Tao, R. W. Dawson & Z. R. Xu, 2003. The distributions and effects of nutrients in the sediments of a shallow eutrophic Chinese lake. Hydrobiologia 429: 85–93.

Yin, C. Q. & Z. W. Lan, 1995. The nutrient retention by ecotone wetlands and their modification for Baiyangdian Lake restoration. Water Science and Technology 32: 159–167.

Zhang, Y. L., B. Q. Qin, W. M. Chen & L. C. Luo, 2004. A study on total suspended matter in Tai Lake. Resources and Environment in the Yangtze Basin (in Chinese) 13: 266–271.

Zhu, W. X. & M. M. Carreiro, 2004. Temporal and spatial variations in nitrogen transformations in deciduous forest ecosystems along an urban–rural gradient. Soil Biology & Biochemistry 36: 267–278.

Hydrobiologia (2007) 581:109–116
DOI 10.1007/s10750-006-0503-7

EUTROPHICATION IN LAKES

Variations in kinetics of alkaline phosphatase in sediments of eutrophic, shallow, Chinese lakes

Yiyong Zhou · Xiuyun Cao · Chunlei Song · Jianqiu Li · Guoyuan Chen · Liang Peng

Abstract Kinetics of alkaline phosphatase in sediments of a shallow Chinese freshwater lake (Lake Donghu) were investigated. Spatially, among 20 sites sampled, V_{max} and K_m values of alkaline phosphatase in surface sediments were higher in the zone adjacent to sites with the highest chlorophyll *a* concentrations. Vertically, there was a peak in V_{max} at intermediate sediment depths in addition to the expected maximum at the surface. Some inhibitors, such as $CuSO_4$, $ZnSO_4$ and Na_2WO_4, showed significantly different effects on kinetics of alkaline phosphatase in interstitial water and sediments. Moreover, alkaline phosphatase in interstitial water and sediments responded to Na_2WO_4 in different ways in Lake Taihu. These observations imply that the enzyme is immobilized in sediments, which became more stable with accelerated eutrophication, as suggested by highest alkaline phosphatase activity (APA) in sediments corresponding with highest water column chlorophyll *a* concentrations in Lake Donghu.

Keywords Alkaline phosphatase · Kinetics · Sediment · Eutrophication · Inhibitions · Immobilization

Introduction

Phosphorus often limits phytoplankton growth in freshwater systems, while organic phosphorus may account for larger part of phosphorus in lake sediments. For example, in a eutrophic lake (Lake Alserio, northern Italy), the results of the sedimentary phosphate fractionation showed that the most important P fraction was an organic fraction, and a digestion of the supernatant of the P-fraction bound to $CaCO_3$ allowed the detection of a large pool of org-P (Vicente et al., 2006). Therefore, regeneration of organic P in sediments is emphasized in aquatic ecology research. The phosphatase has been studied in relation to organic phosphorus decomposition in aquatic ecosystems (reviewed by Jansson et al., 1988). The relationship between alkaline phosphatase and substrate is of ecological interest because of implications for phosphorus cycling. However, most attempts to determine phosphatase in sediments have focused on activity. Kinetics of

Guest editors: B. Qin, Z. Liu & K. Havens
Eutrophication of shallow lakes with special reference to Lake Taihu, China

Y. Zhou (✉) · X. Cao · C. Song · J. Li · G. Chen · L. Peng
Institute of Hydrobiology, The Chinese Academy of Sciences, Wuhan 430072, China
e-mail: zhouyy@ihb.ac.cn

X. Cao · C. Song · G. Chen · L. Peng
Graduate School of the Chinese Academy of Sciences, Beijing 100039, China

phosphatase in sediments were influenced by discharge rate, water quality, concentration of dissolved orthophosphate and substratum in stream sediments (Marxsen & Schmidt, 1993), submerged macrophytes (Zhou et al., 2000) and P status in lake sediments (Zhou et al., 2001, 2002). Data on spatial and vertical variations in kinetics and properties of alkaline phosphatase in sediments are limited.

In this paper, alkaline phosphatase kinetics and its response to various inhibitors in a shallow freshwater lake (Lake Donghu) in China were described. The aims of the work were to establish whether enzyme regimes in sediments (1) were unique for the whole lake and depth, (2) shared a similarity in kinetic behavior with interstitial water, and (3) were related to trophic status in the lake.

Materials and methods

The experimental Lakes (Lake Donghu and Lake Taihu) are described elsewhere (Zhou et al., 2002; Qin et al., 2006). To evaluate the trophic state in Lake Donghu on an annual basis, surface water was collected monthly from sites 1, 12 and 21 (Fig. 1), from February 1997 through January 1999. To determine spatial variation of alkaline phosphatase kinetics, 20 sites were sampled in the largest basin (Fig. 1) in May 1995. Three sites in Lake Taihu were sampled in December 2004 to assay alkaline phosphatase activity (APA) in surface sediments. In May 1997, vertical samples were taken at five sites in Lake Donghu with Site 1 at the center and others (1A, 1B, 1C and 1D) located approximately 100 m away (Fig. 1).

Fig. 1 The map of Lake Donghu, showing the location of the sampling sites

For spatial profiles, sediments were sampled by Peterson dredge. Sediment columns were obtained using a hand-driven stainless steel corer 50 cm long with an internal diameter of 3.5 cm. For depth profiles, columns were sliced at 4 cm intervals. Samples were transferred to the laboratory for analysis. Interstitial water was extracted by centrifugation at 3,000 rpm for 30 min.

Total chlorophyll *a* was measured by acetone extraction. Lake surface water samples (0.4–0.7 l) were filtered through a Whatman GF/C filter and absorbance measured at 663 and 750 nm in 1-cm path-length glass cuvettes after overnight extraction in acetone (90% v/v) (Golterman & Clymo, 1969).

APA assays used *p*-nitro-phenylphosphate (*p*NPP), which is hydrolyzed at 37°C by alkaline phosphatase to yield *p*-nitrophenol; with this system, enzyme activity is indicated by an increase in light absorbance (Sayler et al., 1979). Sediment samples were slurried in Tris buffer (pH 8.9). *p*NPP was added to slurries at eight final concentrations ranging from 0.05 mmol l^{-1} to 10 mmol l^{-1}. Samples were incubated at 37°C. After 1 h, 1.6 ml of slurry were centrifuged at 3,000 rpm. One ml of the supernatant was mixed with 4 ml 0.1 M NaOH to stop the reaction. Absorption of the final solution was measured at 400 nm. *p*NPP was added to reagent blanks after NaOH addition. APA was converted to absolute units using a standard curve based on enzymatically hydrolyzed *p*-nitrophenol. V_{max} and K_m values were estimated by fitting linearized Michaelis–Menten equations per the Lineweave–Burk plot. APA in surface sediments was determined with a final substrate concentration of 6.0 mmol l^{-1}.

Kinetics of total APA in interstitial water was determined using a method adapted from Berman (1970). Substrate was added at eight final concentrations ranging from 0.01 mmol l^{-1} to 1.8 mmol l^{-1}. APA in interstitial water was determined with a final substrate concentration of 0.3 mmol l^{-1}. All samples were run in triplicate.

Several inhibitors were added to sediments and interstitial water. Final concentrations of inhibitors are described in results tables. Fisher's least significant difference (LSD) test and independent-samples T test were used to make comparisons among treatment levels for the inhibitor addition, using the SPSS statistical package.

Results

Seasonally, chlorophyll *a* concentration in surface water was highest at Site 1 (Fig. 2). APA kinetics in sediments exhibited spatial heterogeneity (Table 1). Higher V_{max} values were found at Sites 2, 6, 9, 10 and 12, which are adjacent to Site 1. However, in the northeastern zone represented by Sites 14, 15 and 19, both V_{max} and K_m values were markedly lower.

There were variations in vertical distribution of APA kinetics at different sites (Fig. 3). At Sites 1B and 1D, V_{max} decreased with depth, while, at Sites 1A, 1, and 1C, there was a peak in the middle layers. At the same time, highest values for K_m were observed in the surface sediment at Sites 1B, 1C, and 1D, and in the middle layers at Sites 1A and 1.

Responses of APA kinetic parameters in sediments and interstitial water to various inhibitors were examined (Table 2). In sediments, V_{max} decreased significantly with Cu^{2+} concentration ($p < 0.01$), while in interstitial water, it significantly increased ($p < 0.01$). It also increased markedly ($p < 0.01$) with tungstate concentration in interstitial water. With Zn^{2+} added, V_{max} increased significantly in sediments ($p < 0.01$) but decreased at higher concentration in interstitial water ($p < 0.01$).

Responses of K_m values were also variable in both sediments and interstitial water. In sediments, it decreased significantly with lower Cu^{2+} concentration but increased with higher concen-

Table 1 The kinetic parameters of alkaline phosphatase in the sediments of Lake Donghu

Sampling sites	V_{max} (SD) (μmol G^{-1} H^{-1})	K_m (SD) (mmol l^{-1})
1	147.0 (2.0)	0.32 (0.02)
2	557.3 (50.6)	3.18 (0.09)
3	208.1 (1.0)	0.56 (0.04)
4	139.7 (8.1)	0.66 (0.02)
5	62.5 (3.1)	0.81 (0.04)
6	424.3 (24.3)	3.74 (0.67)
7	258.9 (49.5)	3.38 (0.27)
8	143.4 (21.4)	0.80 (0.12)
9	458.3 (11.64)	1.39 (0.04)
10	305.6 (30.0)	1.03 (0.23)
11	74.4 (1.8)	0.47 (0.03)
12	293.4 (27.7)	0.61 (0.07)
13	133.2 (12.8)	0.77 (0.16)
14	64.5 (17.6)	0.33 (0.08)
15	68.6 (6.6)	0.28 (0.06)
16	109.6 (3.8)	2.19 (0.16)
17	141.8 (31.8)	0.69 (0.09)
18	144.3 (21.8)	2.67 (0.56)
19	94.6 (8.4)	0.74 (0.05)
20	169.4 (17.0)	1.00 (0.12)

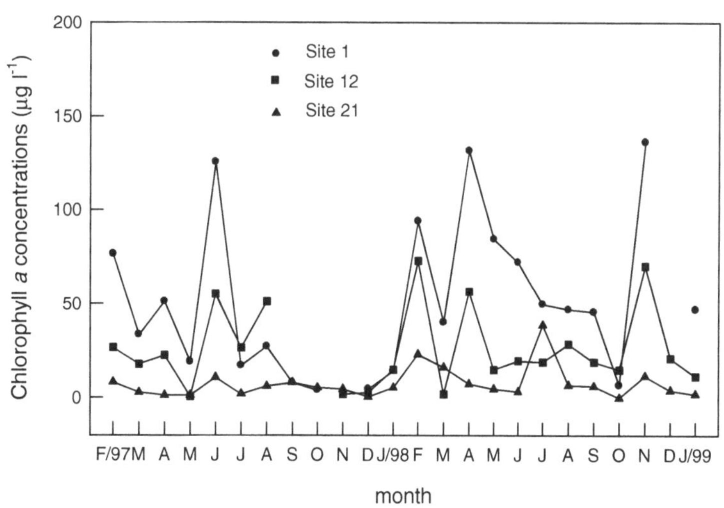

Fig. 2 Seasonal variation in chlorophyll *a* concentrations in surface water at three experimental stations in Lake Donghu

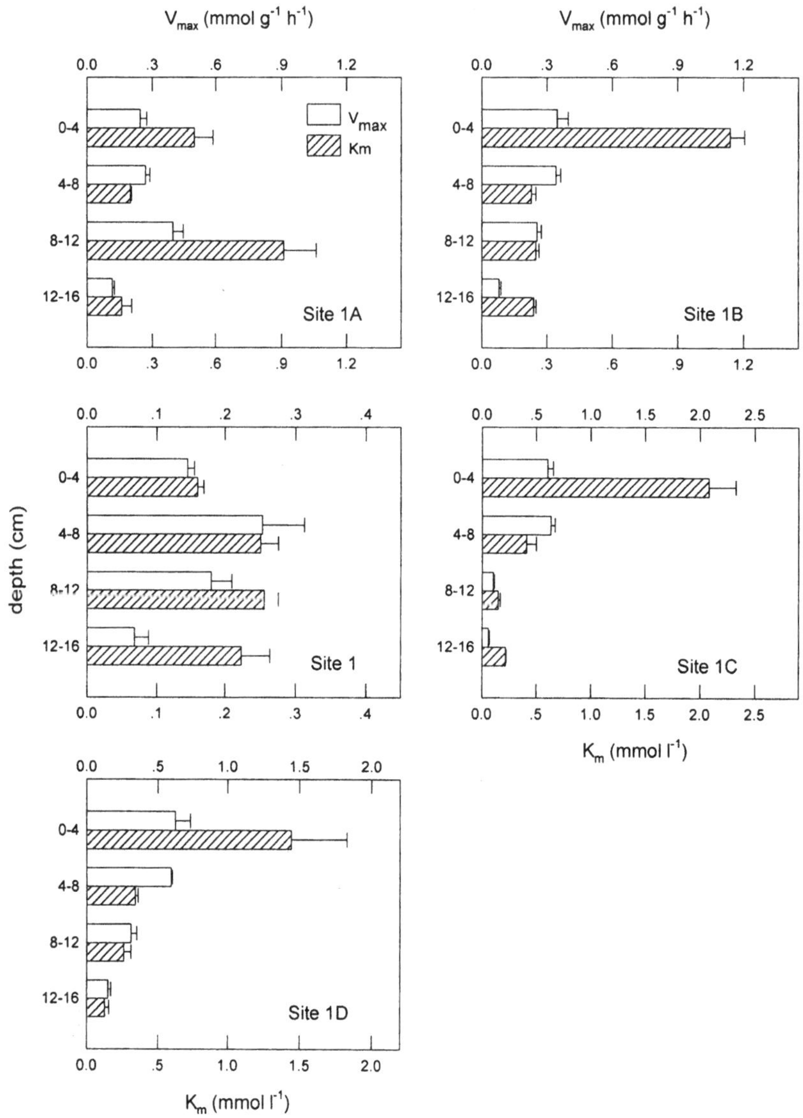

Fig. 3 Vertical distribution of V_{max} and K_m values of alkaline phosphatase activity (APA) in core sediments of Lake Donghu. Samples were taken on 12 May 1997

tration ($p < 0.01$). It also decreased significantly at lower tungstate concentration ($p < 0.05$) but increased significantly at higher concentration ($p < 0.01$) in sediments. While in interstitial water, it significantly decreased with tungstate concentrations ($p < 0.01$). With Zn^{2+} added, K_m significantly increased at lower concentration in sediments ($p < 0.05$), whereas in interstitial water, it greatly increased at higher concentration ($p < 0.01$).

In Lake Taihu, tungstate inhibited APA in sediment at Site T1 ($p < 0.01$) and interstitial water at Sites T1 and T3 ($p < 0.01$) at higher concentration. At Site T2, it stimulated APA in sediments at higher concentration, but inhibited APA in interstitial water at both concentrations ($p < 0.01$, Table 3).

Discussion

At Site 1, chlorophyll *a* in surface water was highest among experimental stations from 1997 to 1999 (Fig. 2), a situation commonly recorded in

Table 2 Effects of some inhibitors of alkaline phosphatase on kinetics of alkaline phosphatase activity (APA) in the surface sediments and the interstitial water in Lake Donghu

Sampling time	Inhibitors	Concentration (mmol l^{-1})	Surface sediments		Interstitial water	
			V_{max} (SD) (μmol G^{-1} H^{-1})	K_m (SD) (mmol l^{-1})	V_{max} (SD) (nmol l^{-1} min^{-1})	K_m (SD) (μmol l^{-1})
15 March, 1996	$CuSO_4$	0	157.5 (21.5)	3.25 (0.15)	51.9 (0.7)	4.11 (0.21)
		0.2	65.4 (6.6)	0.30 (0.06)	54.9 (0.5)	4.22 (0.29)
		2.0	52.5 (0.3)	4.78 (0.25)	105.5 (1.0)	1.65 (0.12)
		0	82.5 (0.8)	0.51 (0.02)	53.2 (0.9)	3.43 (0.23)
29 March, 1996	Na_2WO_4	0.05	90.8 (3.4)	0.40 (0.01)	67.9 (0.5)	1.81 (0.21)
		0.1	127.0 (37.7)	0.68 (0.02)	69 (0.5)	1.63 (0.11)
		0	143.7 (18.2)	0.54 (0.20)	46.7 (0.4)	1.59 (0.16)
8 February, 1996	$ZnSO_4$	0.2	291.9 (8.7)	1.07 (0.06)	44.6(0.6)	1.70 (0.10)
		2.0	287.4 (2.6)	0.48 (0.14)	38.6(0.2)	2.13 (0.08)

Table 3 Effects of tungsten on alkaline phosphatase activity (APA) in the surface sediments and interstitial water in Lake Taihu (Samples were taken on 14 December, 2004)

Sampling sites	Tungstate concentrations (mmol l^{-1})	APA in surface sediments (SD) (μmol G^{-1} H^{-1})	APA in interstitial water (SD) (nmol l^{-1} min^{-1})
T1	0	428.5 (6.0)	42.6 (1.5)
(31°32′4.6″ N, 120°13′22.6″ E)	0.16	157.0 (14.9)	35.2 (0.3)
T2	0	137.8 (16.3)	48.4 (1.0)
(31°32′1.7″ N,	0.08	144.3 (7.1)	39.8 (1.1)
120°13′18.7″ E)	0.16	309.2 (20.5)	35.9 (1.5)
T3	0	483.1 (53.5)	49.31(1.2)
(31°31′45.5″ N, 120°13′42.8″ E)	0.16	163.9 (13.4)	39.9 (0.6)

the lake (Zhou et al., 2002, 2004). Higher V_{max} and K_m values of APA in surface sediments were found at sites adjacent to Site 1, while, in the northeast, both were lower (Table 1). Significant inter-station variation in APA was observed in sediments of the Mandovi Estuary (Silva & Bhosle, 1990). Distinct areas of phosphatase activity also were seen in the Venice lagoon (Sabil et al., 1994). Variability of phosphatase activity in freshwater sediments may be related to heterogeneity within a site (Sayler et al., 1979). At sites along a eutrophication gradient in Nordruegensche Bodden (Baltic Sea, Germany), water column measurements revealed that turbidity, seston content, and chlorophyll *a* and inorganic nutrient concentrations increased from outer to inner parts of the Bodden. Sediment investigations confirmed this eutrophication gradient. Generally, hydrolytic enzyme activities increased with eutrophication (Koester et al., 1997). In Lake Donghu, the V_{max} values in sediment increased during the summer, in conjunction with lower K_m values in interstitial water that suggests a higher affinity for the substrate. The accumulation of organic matter in the sediment could be traced back to the breakdown of the algal bloom, which may stimulate APA with higher kinetic efficiency, by a combination of the higher V_{max} in sediments plus lower K_m values in interstitial water, in summer (Zhou et al., 2002). Our results highlight the connection of the kinetics of APA in sediments with phytoplankton abundance in surface water, indicating the induction of enzyme by organic matter.

The V_{max} of alkaline phosphatase was highest at the sediment surface (Sites 1B and 1D, Fig. 3). APA also was highest at the surface of marine (Kobori & Taga, 1979) and lake sediments (Sinke et al., 1991). In three lakes in Finland, the eleven hydrolytic enzyme activities were high into deep sediment layers indicating potential for turnover of organic matter in the permanently anoxic zones (Hakulinen et al., 2005). In the present study, there was a peak in the middle layers (Sites 1A, 1, and 1C, Fig. 3), showing an additional dimension for the degradation of organic matter mediated by alkaline phosphatase in lake sediments.

Spatial and vertical variations in V_{max} and K_m values of APA in sediments were similar (Table 1; Fig. 2). This may be interpreted with reference to organic matter and orthophosphate enrichment. At first, organic matter may stimulate V_{max} of APA in sediments. Surface sediment profiles were taken from the deepest part of six central Finnish lakes representing six different trophic states, and each profile was divided into three layers. Phosphatase activity and levels of organic substances were highest in the top layer of every sediment profile. Microbiological phosphate mineralization from organic substances may determine the internal phosphorus load in these lakes (Matinvesi & Heinonen-Tanski, 1992). In fish pond, inorganic phosphate added to the sediment scarcely restricted phosphatase activity. At the same time the enrichment with organic phosphorus compounds highly raised enzyme activity (Olah & Toth, 1978). In marsh sediments, the V_{max} of acid phosphatase followed the same trend as in situ activity. Sediment salinity and pH were negatively correlated with the enzyme activity, while soil organic matter content, clay content and sediment organic P were positively correlated (Huang & Morris, 2005). In tidal freshwater habitats adjacent to the Cooper River, acid phosphatase activity was highly correlated with the organic matter content of the sediment. The V_{max} of all phosphatases increased along the successional gradient. Trends in phosphatase activity and V_{max} correlated positively with plant biomass and negatively with concentrations of soluble reactive phosphorus in porewater, sediment extractable phosphorus, and total phosphorus (Huang & Morris, 2003); Secondly, organic matter may alter K_m of APA in sediments. Alkaline phosphatase kinetics in sediment associated with cage culture of *Oreochromis niloticus* was studied near Site 1 in Lake Donghu. Both V_{max} and K_m increased with the addition of fish feces (Zhou et al., 2001). A possible explanation is that humic substances would adsorb the substrates of enzymes. In the consecutive sapropel layers collected from bathyal sediments of the eastern Mediterranean Sea, the determination of exoenzyme activity with fluorescently labeled substrate analogues was impaired by the strong adsorption of up to 97% of the enzymatically liberated fluorophores to the sediment particles. High activities of aminopeptidase and alkaline phosphatase were detected even in a 124,000-year-old sapropel layer, whereas the activity of β-glucosidase was low in all layers. It had been assumed that the organic matter which constitutes the sapropels is highly refractory. Since a high adsorption capacity was determined not only for the low-molecular-weight compounds but also for DNA, the extraordinarily strong adsorption of structurally different substrates to the sapropel matrix appears to be the major reason for the long-term preservation of biodegradable carbon in this environment (Coolen & Overmann, 2000). This strong adsorption would weaken the affinity of substrates for enzymes, leading to increase in apparent K_m values. In addition, orthophosphate may be a competitive inhibitor for alkaline phosphatase, causing a significant increase in K_m values. Spatially, orthophosphate concentrations in interstitial water were highest at Site 1 among experimental stations in Lake Donghu in 1995 to 1996 (Zhou et al., 2002), it could act as a competitive inhibitor of extracellular phosphatase in lake (Chrost & Overbeck, 1987).

Alkaline phosphatase in sediments showed different responses to various inhibitors from those observed in interstitial water. For example, in interstitial water of lake Donghu, V_{max} increased with copper sulphate addition at both lower ($p < 0.05$) and higher ($p < 0.01$) concentrations (Table 2). Increased phosphatase activities were noted in green algae isolated from Lake Kinneret and pretreated with intermediary Cu^{2+} concentrations (Wynne & Pieterse, 2000). However, the enzyme was inhibited by zinc at higher concentration ($p < 0.01$, Table 2). Zhang et al. (2001) studied the kinetics of inactivation of alkaline phosphatase from green crab by zinc ions. The enzyme reversibly and quickly bound

Zn^{2+} and then underwent a slow, reversible inactivation and slow conformational change. In sediments, V_{max} and K_m decreased with Cu^{2+} at lower concentration ($p < 0.01$, Table 2). Accordingly, APA was affected by Cu^{2+} in soil polluted by heavy metals (Kuperman & Carriero, 1997; Kunito et al., 2001). The influence of Cu^{2+} and Zn on the activity and kinetics of acid phosphatase immobilized by two soil clays, kaolin or goethite, indicated that Cu decreased V_{max} of the enzymes, but increased affinity of the enzymes for the substrate (Huang & Shindo, 2001). In this context, alkaline phosphatase in sediment and interstitial water showed different kinetics with different inhibitors, implying enzyme immobilization in sediment. Concurrently, various inhibition patterns by Zn were observed for free and immobilized acid phosphatase at different pH values, likely related to the degree of deactivation by Zn (Huang & Shindo, 2000). In soils, kinetics of the immobilized enzyme conformed to Michaelis–Menten, and V_{max} was lower and K_m higher than those of the free enzyme (Rao & Gianfreda, 2000). Immobilized enzymes were more resistant to environmental changes compared to their soluble counterparts (Sabil et al., 1993). In Venice lagoon sediments within a shallow water area, phosphatase activity was prolonged by its insolubility (Sabil et al., 1994).

Shortly, there is a correlation between higher sediment APA in term of kinetics and chlorophyll *a* concentration in water of Lake Donghu. The kinetic properties of phosphatase and their distributions in sediments were linked to the process of lake eutrophication, which might become more stable upon immobilization in sediments with accelerated eutrophication.

Acknowledgements This work was supported by the grant (KZCX1-SW-12-II-02-02) from the Chinese Academy of Sciences (CAS), National Basic Research Program of China (2002CB412300) and funding from the National Science Foundation of China (39170165, 39670149 and 20177033). We also obliged to the funds (2002AA601013). The authors also gratefully acknowledge the support of the Donghu Lake Ecosystem Experimental Station of CAS for the help in sampling collection.

References

Berman, T., 1970. Alkaline phosphatase and phosphorus availability in Lake Kinneret. Limnology and Oceanography 15: 663–674.

Chrost, R. J. & J. Overbeck, 1987. Kinetics of alkaline phosphatase activity and phosphorus availability for phytoplankton and bacterioplankton in Lake Plusssee (North German eutrophic lake). Microbial Ecology, New York, NY 13(3): 229–248.

Coolen, M. J. & J. Overmann, 2000. Functional exoenzymes as indicators of metabolically active bacteria in 124,000-year-old sapropel layers of the eastern Mediterranean Sea. Applied and Environmental Microbiology 66(6): 2589–2598.

de Vicente, I., K. Cattaneo, L. Cruz-Pizarro, A. Brauer & P. Guilizzoni, 2006. Sedimentary phosphate fractions related to calcite precipitation in an eutrophic hardwater lake (Lake Alserio, northern Italy). Journal of Paleolimnology 35: 55–64.

Golterman, H. L. & R. S. Clymo, 1969. Method for chemical analysis of fresh water. In IBP Handbook No. 8. Blackwell Sci. Publ., Oxford, 1–188.

Hakulinen, R., M. A. Kahkonen & M. Salkinoja-Salonen, 2005. Vertical distribution of sediment enzyme activities involved in the cycling of carbon, nitrogen, phosphorus and sulphur in three boreal rural lakes. Water Research 39: 2319–2326.

Huang, Q. & H. Shindo, 2001. Comparison of the influence of Cu, Zn, and Cd on the activity and kinetics of free and immobilized acid phosphatase. Soil Science and Plant Nutrition 47: 767–772.

Huang, Q. & H. Shindo, 2000. Inhibition of free and immobilized acid phosphatase by zinc. Soil Science 165: 793–802.

Huang, X. & J. T. Morris, 2003. Trends in *phosphatase* activity along a successional gradient of tidal freshwater marshes on the cooper river, South Carolina. Estuaries 26(5): 1281–1290.

Huang, X. & J. T. Morris, 2005. Distribution of *phosphatase* activity in marsh *sediments* along an estuarine salinity gradient. Marine Ecology Progress Series 292: 75–83.

Jansson, M., H. Olsson & K. Pettersson, 1988. Phosphatases; origin, characteristics and function in lakes. Hydrobiologia 170: 157–175.

Kobori, H. & N. Taga, 1979. Occurrence and distribution of phosphatase in neritic and oceanic sediments. Deep-Sea Research 26A: 799–808.

Koester, M., S. Dahlke & L.-A. Meyer-Reil, 1997. Microbiological studies along a gradient of eutrophication in a shallow coastal inlet in the southern Baltic Sea (Nordruegensche Bodden). Marine Ecology Progress Series 152: 27–39.

Kunito, T., K. Saeki, S. Goto, H. Hayashi, H. Oyaizu & S. Matsumoto, 2001. Copper and zinc fractions affecting microorganisms in long-term sludge-amended soils. Bioresource Technology 79: 135–146.

Kuperman, R. G. & M. Carriero, 1997. Soil heavy metal concentrations, microbial biomass and enzyme activities in a contaminated grassland ecosystem. Soil Biology and Biochemistry 29: 179–190.

Marxsen, J. & H.-H. Schmidt, 1993. Extracellular phosphatase activity in sediments of the Breitenbach, a Central European mountain stream. Hydrobiologia 253: 207–216.

Matinvesi, J. & H. Heinonen-Tanski, 1992. Biodegradable substances in lake sediments and their relation to sediment microbiological activity and phosphorus recycling. Aqua Fennica 22: 193–200.

Olah, J. & E. O. Toth, 1978. The function of alkaline *phosphatase* enzyme in the phosphorus cycle of fertilized fishponds. Aquacult Hung 1: 15–23.

Qin, B., P. Xu, Q. Wu, L. Luo & Y. Zhang, 2006. Environmental issues of Lake Taihu, China. Hydrobiologia.

Rao, M. A. & L. Gianfreda, 2000. Properties of acid phosphatase-tannic acid complexes formed in the presence of Fe and Mn. Soil Biology and Biochemistry 32: 1921–1926.

Sabil, N., A. Cherri, D. Tagliapietra & M.-A. Coletti-previero, 1994. Immobilized enzymic activity in the Venice Lagoon sediment. Water Research 28: 77–84.

Sabil, N., D. Tagliapietra & M.-A. Coletti-Previero, 1993. Insoluble biodegradative potential of the Venice Lagoon. Environmental Technology 14: 1089–1095.

Sayler, G., S. M. Puziss & M. Silver, 1979. Alkaline phosphatase assay for freshwater sediments: Application to perturbed sediment system. Applied and Environmental Microbiology 38: 922–927.

Silva, C. D. & N. B. Bhosle, 1990. Phosphorus availability and phosphatase activity in the sediments of Mandovi estuary, Goa. Indian Journal of Marine Sciences 19: 143–144.

Sinke, A. J. C., A. A. Cornelese & T. E. Cappenberg, 1991. Phosphatase activity in sediments of the Loosdrecht lakes. Internationale Vereinigung für Theoretische und Angewandte Limnologie 24: 719–721.

Wynne, D. & A. J. H. Pieterse, 2000. The effect of copper on photosynthesis, nitrate reductase and phosphatase activities in Lake Kinneret phytoplankton. Limnology and Lake Management 55: 581–593.

Zhang, R. Q., Q. X. Chen, R. Xiao, L. P. Xie, X. G. Zeng & H. M. Zhou, 2001. Inhibition kinetics of green crab (*Scylla serrata*) alkaline phosphatase by zinc ions: A new type of complexing inhibition. Biochimica et Biophysica Acta 1545: 6–12.

Zhou, Y., J. Li & Y. Fu, 2000. Effects of submerged macrophytes on kinetics of alkaline phosphatase in Lake Donghu. I. Unfiltered water and sediments. Water Research 34: 3737–3742.

Zhou, Y., J. Li, Y. Fu & M. Zhang, 2001. Kinetics of alkaline phosphatase in lake sediment associated with cage-culture of *Oceochrolmis miloticus*. Aquaculture 203: 23–32.

Zhou, Y., J. Li, Y. Fu & M. Zhang, 2002. Temporal and spatial variations in kinetics of alkaline phosphatase in sediments of a shallow Chinese eutrophic lake (Lake Donghu). Water Research 36: 2084–2090.

Zhou, Y., J. Li, C. Song & X. Cao, 2004. Variations and possible source of potentially available phosphorus in a Chinese shallow eutrophic lake. Journal of Freshwater Ecology 19: 87–96.

Hydrobiologia (2007) 581:117–123
DOI 10.1007/s10750-006-0513-5

EUTROPHICATION IN LAKES

Environmental changes in Lake Taihu during the past century as recorded in sediment cores

Bin Xue · Shuchun Yao · Weilan Xia

Abstract The Lake Taihu drainage basin is an economically developed area with some of the highest population densities in China. The lake has deteriorated due to ecological destruction and eutrophication. Three short sediment cores from eastern, northeastern and southwestern Lake Taihu were collected. Total organic carbon (TOC), total nitrogen (TN), pigments, elements and particle size were analyzed for the purpose of understanding past trophic status and pollution levels. Sedimentation rates were based on ^{137}Cs or ^{210}Pb methods. Results indicated that sediment particle size became coarser since the 1920s, and the lake was contaminated by heavy metals, such as Cu and Zn, since the 1970s. A remarkable increase in eutrophication since the 1980s due to increased loading of untreated effluents from industry, agriculture and urbanization is reflected by total organic carbon, total nitrogen and pigments in the studied cores. However the onset times of eutrophication in different parts of Lake Taihu were not synchronous.

Guest editors: B. Qin, Z. Liu & K. Havens
Eutrophication of shallow lakes with special reference to Lake Taihu, China

B. Xue (✉) · S. Yao · W. Xia
Nanjing Institute of Geography and Limnology, Chinese Academy of Sciences (CAS), Nanjing 210008, China
e-mail: bxue@niglas.ac.cn

S. Yao
Graduate School of CAS, Beijing 100039, China

Keywords Lake Taihu · Sediment core · Eutrophication · Heavy metal

Introduction

Lake Taihu, situated in the southern Yangtze River Delta, is one of the five largest freshwater lakes in China (Shun & Huang, 1993). Anthropogenic activities in the basin can be traced to 7,000 B.P. (Chen & Wang, 1999; Chen et al., 1997). Human impacts have increased dramatically in the last two decades.

Lake sediments can serve as information archives of environmental changes through time (Mônica & Carlos, 2002), since each layer of buried sediments represents a record of past environmental conditions (Von Guten et al., 1997). Changing heavy metals and organic matter concentrations in dated sediment cores indicate the evolution of adjacent emission sources and terrestrial ecosystems. This evolution often is influenced by social changes, such as industrial development, deforestation, mining, population increase and energy consumption (Palanques et al., 1998). Pigments in lake sediments can also provide an effective means for reconstructing the paleoenvironmental history of lake ecosystems. As a result, sediment cores may provide a valuable record of anthropogenic and environmental changes.

The purpose of this study was to track trophic status and human impact in the past century as recorded in three short sediment cores taken from eastern, northeastern and southwestern Lake Taihu. By extending the record to the beginning of the last century, changes in water quality and the ecosystem may be identified.

Materials and methods

Cores were taken from two sites in Lake Taihu in 1995 using a gravity corer (Fig. 1). One core (W) was taken from western Lake Taihu; the other one (E) was taken from East Taihu Bay. In 2003, another core (THS) was taken from southern Lake Taihu. W station was near Wuxi, an industrial city located 2 km northeast of Wuli Bay. Wuli Bay receives industrial waste water and domestic sewage from Wuxi, and water flows from Wuli Bay to Lake Taihu in winter/spring (Zou et al., 1996). Site E receives a small input of waste water and has dense submerged macrophyte coverage, while THS receives only waste materials from small towns.

Cores W and E were sliced at 2 cm intervals for total organic carbon (TOC), total nitrogen (TN) and sediment pigments analysis. Core THS was sliced at 1 cm intervals for chemical analysis of sediments, physics and radioactivity dating.

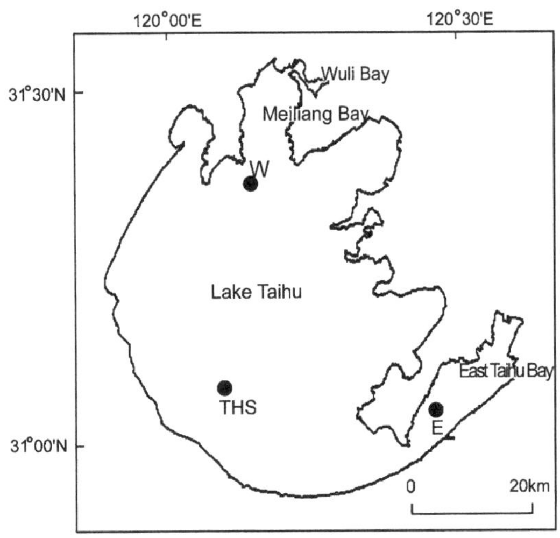

Fig. 1 Map of study area showing sampling sites

TOC was measured using concentrated sulfuric acid digestion with potassium dichromate, and TN was measured by the Kjeldahl technique (Institute of Soil Science, 1978; Professional Committee of agricultural chemistry of Chinese Society for Soil Sciences, 1983).

Samples were transferred quickly to air-tight, darkened containers for pigment analysis. Samples were stored at 4 °C before analysis. Pigments were determined using the method from Swain (1985). Samples were extracted using 90% acetone, and pigment concentrations were determined by measuring absorption at appropriate wavelengths with a spectrophotometer.

Samples for determination of heavy metals (Fe, Mn, Cu, Zn, Ni) were determined using EPA Method 3052 (HCl-HNO3-HF digestion) and ICP-AES (Leeman Labs Profile) (USEPA 1995). Accuracy was assessed with reference material from the National Institute of Standard and Technology (SRM 1646a).

Grain sizes of sediments were determined using a Malvern Mastersizer 2000. In addition, sediment age was estimated from available dates determined by ^{210}Pb or ^{137}Cs (Wan et al., 1987). ^{210}Pb and ^{137}Cs were determined using EG and G Ortec Gamma Spectrometry at Nanjing Institute of Limnology and Geography. ^{137}Cs was measured at 662 keV, while ^{210}Pb was determined via gamma emission at 46.5 keV and ^{226}Ra at 295 and 352 keV γ-rays emitted by its daughter isotope ^{214}Pb. A stable depositional rate was assumed.

Results

Sedimentation rate

Average sedimentaion rate for northern East Taihu Bay was 1.8 mm yr^{-1}, and average depositional flux was 0.13 g cm^{-2} yr^{-1}. Average depositional flux for southern East Taihu Bay was 0.24 g cm^{-2} yr^{-1}. Depositional flux was 0.08 g cm^{-2} yr^{-1} for western Lake Taihu (Shun & Huang, 1993). Average sedimentation rate for East Taihu Bay based on ^{210}Pb was 0.39 cm yr^{-1} according to Chang (1996), and our core E location was the same as Changs'. In 2002, a core from the same location as W had an average

Table 1 Sedimentation rates of W, E and THS

Core name	W		E		THS	
	^{210}Pb	^{137}Cs	^{210}Pb	^{137}Cs	^{210}Pb	^{137}Cs
Sedimentation rate (cm yr^{-1})	–	0.17 (Liu et al,2004)	0.39 (Chang, 1996)	–	0.34	0.41

sedimentation rate based on ^{137}Cs of 0.17 cm yr^{-1} (Liu et al., 2004). For core THS, the average sedimentation rate was 0.34 cm yr^{-1} based on ^{137}Cs, and 0.41 cm yr^{-1} using ^{210}Pb (Table 1).

Nutrients in the sediment

TOC, TN and C/N ratios for the three cores are shown in Fig. 2. Highest TOC and TN concentrations in core W occur in the top 4 cm, which represents about 20 years of deposits estimated from ^{137}Cs. Highest TOC and TN concentrations also occur in the top 4 cm in core E, representing about 10 years of deposits based on ^{210}Pb. In core THS, the TOC concentration maximum occur in the top 8 cm, representing about 20–24 years based on ^{210}Pb or ^{137}Cs. An abrupt increase in TN concentration in core THS occur in the top

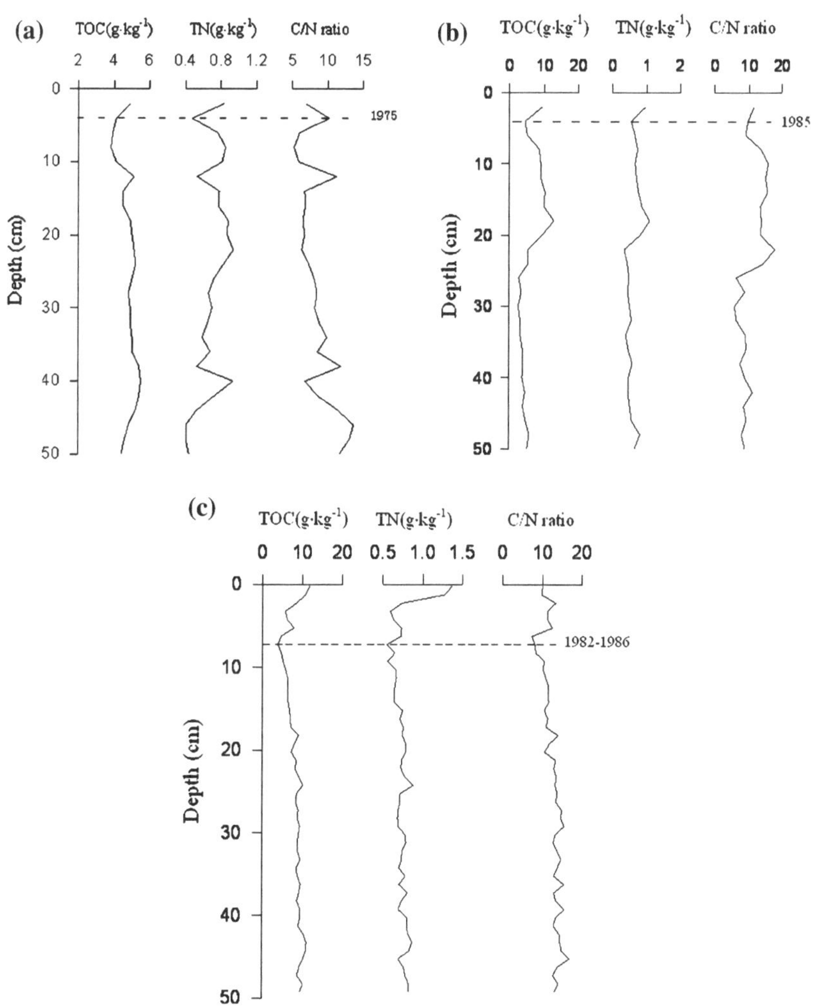

Fig. 2 Variation of TOC, TN and C/N in (**a**) W, (**b**) E and (**c**) THS core sediments

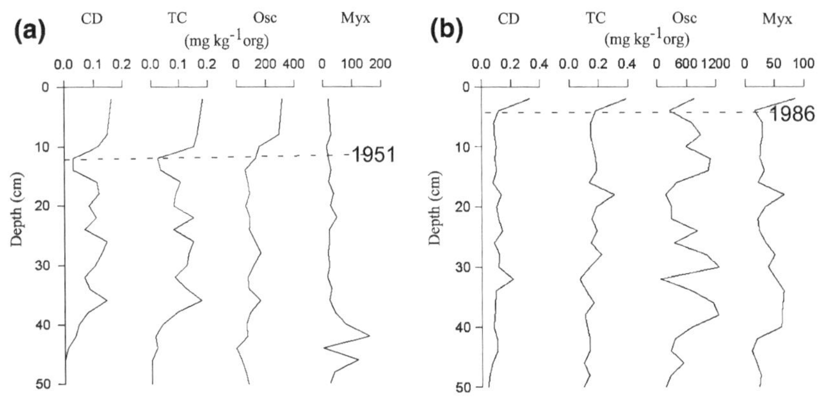

Fig. 3 Variation of pigments in (**a**) W and (**b**) E core sediments

3 cm, which represents about 7–9 years based on ^{210}Pb or ^{137}Cs.

In East Taihu Bay, C/N ratios range from 10 to 15 since the 1930s, while in core W C/N ratios are from 5 to 10 since 1975. C/N ratios vary little throughout the THS core. In general, lower C/N ratios are found in the top 2 cm of cores W and THS.

Pigments in the sediment

Pigments preserved in lake sediments contain detailed information about eutrophication history. An interpretation of major pigment types (e.g. chlorophyll and carotenoids) is often difficult and ambiguous. In contrast, pigments unique to blue-green algae are useful in tracing eutrophication development. Oscillaxanthin is a pigment unique to Oscillatoriaceae and has been found in only one genus other than Oscillatoria. However, myxoxanthophyll is found in most blue-green genera and all six families examined (Swain, 1985).

Results of pigment analysis in cores W and E are shown in Fig. 3. An increase in pigments is noted in the top 8 cm of core W. Based on ^{137}Cs, these sediments represent about 44 years. Distinct increases in pigments from core E are found in the top 4 cm. Based on ^{210}Pb, the top 4 cm represents about 9 years. The largest change in pigment concentration is in chlorophyll derivatives (CD) and total carotenoids (TC) in cores W and E. Differences exist between cores W and E. A large increase in myxoxanthophyll occurred in E core, and a dramatic rise in oscillaxanthin was found in W core.

Heavy metal pollution

Heavy metal concentrations and particle sizes in core THS are shown in Fig. 4. Mn, Cu, Zn, Ni and Pb concentrations decrease from bottom to depth

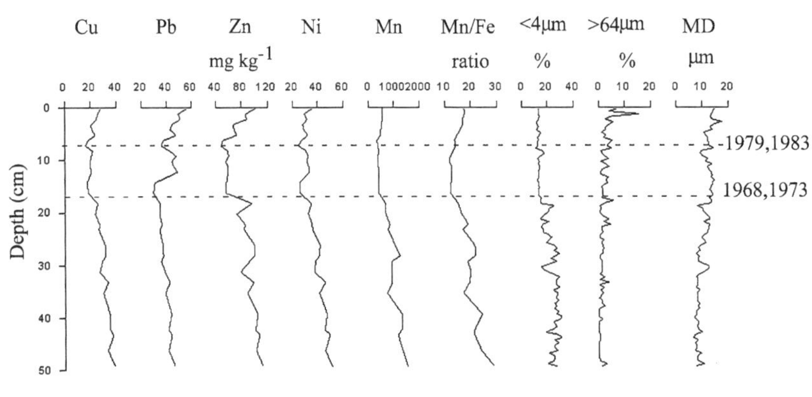

Fig. 4 Variation of metals in THS core sediments

17 cm, then become stable until depth 7 cm. Cu, Zn, Ni and Pb values shows the trend of increasing from depth 7 cm to 0 cm. Elevated organic carbon and Mn/Fe ratio also were observed, so heavy metals may be bound with organic matter or the Fe/Mn oxide fraction under oxydic conditions. However, coarse materials were found in these layers.

Others have reported that Hg originated partly from human pollution from the mid 1920s to 1970s at a site near W (Liu et al., 2004). Heavy metals, such as Cu, Zn, Pb, Hg, and As are indicative of human pollution. Source characteristics of heavy metals correspond to human activities and economic development in the Lake Taihu drainage basin (Liu et al., 2004).

Discussion

C/N ratios are useful for distinguishing between algal and vascular origins in organic sediments (Meyers et al., 1995). Dense submerged macrophytes may explain the high C/N ratios in East Taihu Bay. Lower C/N ratios in the top 2 cm of cores W and THS compared to the deep depth sediment indicate increasing algal origin.

Increasing values of total organic matter (TOM), TN and total phosphate (TP) concentrations were found in Lake Taihu surfical sediments dated between 1980 and the 1990s (Table 2; Hu, 2003). Increases in TOC and TN are also observed since the early 1980s in all the three cores. High TOC and TN in the three cores since the early 1980s agree with others concluding that accelerated eutrophication began in 1983/1984 in Lake Taihu (Chang, 1996). Rose et al. (2004) also got the same results from northern parts of Lake Taihu that nitrogen levels increase dramatically from the mid-1980s which is in agreement with N concentrations in lake water. However, estimates of nutrient enrichment onset may differ among cores if the stable depositional rate from ^{210}Pb or ^{137}Cs is accepted (i.e., about 1975 in core W, early 1980s in core THS and mid 1980s in core E).

The reason why Oscillaxanthin increase during 1940s and 1950s at site west while another variables do not increase until much later is unknown. Deep work should be done for this. Pigments are labile compounds, so the increases in the surfical sediment may indicate either better preservation via sediment anoxia or increased production. But changes of CD/TC profiles can help identify the cause for increased blue-green pigment concentrations, lower CD/TC ratios represent good preservation condition (Wu et al., 2004). Relatively lower CD/TC ratios in top sediments of cores W and E keep stable or show up-core increase trend. According Wu's method (Wu et al., 2004), increased production could be acceptable reason for the increase in blue-green algae.

Pigment enrichment in core W was caused by anthropogenic inputs due to increasing population and human activities in the watershed since the 1950s. However, the change was slow in the 1950s and 1970s, and inputs were primarily low amounts of agricultural and household wastes (Chang, 1996). Untreated effluents from the watershed have increased substantially in the last two decades. Total wastewater discharge in 1979 was 662 million tons, 1153 million tons in 1989, and 1387 million tons in 1999. Wastewater treatment facility capacity is only 328 million tons per year (Hu, 2003). This untreated effluent rapidly increased eutrophication in Lake Taihu. This

Table 2 Comparison of amount of total nitrogen, total phosphorus and total organic matter in the sediment of Lake Taihu over years

Year	TN(%)		TP(%)		TOM(%)	
	Range	Average	Range	Average	Range	Average
1960	–	0.067	–	0.044	0.54–6.23	0.68
1980	0.022–0.147	0.065	0.037–0.067	0.052	0.24–2.78	1.04
1990–1991	0.049–0.558	0.080	0.040–0.107	0.056	0.57–15.10	1.90
1995–1996	0.022–0.450	0.094	0.039–0.237	0.058	0.31–9.04	1.70
1997–1999	0.022–0.618	0.092	0.028–0.280	0.060	0.31–15.73	1.83

corresponds to increased pigment content in core E since 1986. Differences in onset time of sediment pigment enrichment may be due to the distance from river input locations. And Core W is located in northern Lake Taihu, where large amounts of untreated effluent are discharged, while E receives low waste water inputs.

Higher concentrations of Mn, Cu, Zn, Ni and Pb in the deepest layers may be related to differences in Mn/Fe ratio and fine-grained sediment. Largest concentration of heavy metals in finer sediments agrees with other observations (Salomons & Fostner, 1984). Rising heavy metal concentrations in core THS may indicate recent contamination. However, high Pb concentrations also were observed from 7–12 cm. Differences may indicate earlier discharges of Pb. And Pb, Cu, Ni and Zn were the main pollutants in some rivers investigated around Lake Taihu (Wang & Cao, 2002).

Conclusions

Combined with ^{210}Pb and ^{137}Cs dating and geochemical analyses, the sediment record of Lake Taihu demonstrates a complex history of environmental changes in the past century. Heavy metals concentration reveals contamination caused by human activities such as industry and urbanization. TOC and C/N ratios indicate changes in organic matter enrichment in sediment, such as enhanced algae contribution due to waste water and nutrient matter input to the lake. Changes in TN, TP and pigments from the cores confirm eutrophication of Lake Taihu. However the onset times of eutrophication in different parts of Lake Taihu were not synchronous. This means multi-core analyses would be better to understand eutrophication history of such a large shallow lake.

Acknowledgements The authors would like to thank colleagues in Lake Sedimentation and Environment Laboratory of NIGLAS for field and laboratory support. We are indebted to Dr Qu Wenchuan for reviewing of this manuscript. Thanks are due to Dr. Li Shiji for providing facilities. Financial support was from the CAS project (KZCX2-YW-319-2), NSFC (40472085) and a 973 project (2002CB412301).

References

Chang, W. Y. B., 1996. Major in environmental changes since 1950 and the onset of accelerated eutrophication in Lake Taihu, China. Acta Palaeontologica Sinica 35: 155–174 (in Chinese).

Chen, Z. Y., X. Q. Hong, S. Li, L. Wang & X. M. Shi, 1997. Archaeology-related environmental change: Yangtze Delta, China. Acta Geographic Sinica 52: 131–137 (in Chinese).

Chen, Z. Y. & Z. H. Wang, 1999. Yangtze Delta, China: Lake Taihu level variation since 1950's, response to sea level and human impact. Environmental Geology 37: 333–339.

Hu, B. B, 2003. Analysis on effect of water pollution in Lake Taihu basin on water quality of Lake Taihu, Shanghai Environmental Sciences 22: 1017–1021 (in Chinese).

Institute of Soil Science, Chinese academy of sciences, 1978. Physical and chemical analysis of soil. Shanghai scientific and Technical Publishers, Shanghai (in Chinese).

Liu, E. F., J. Shen, Y. X. Zhu, W. L. Xia & G. W. Zhu, 2004. Source analysis of heavy metals in surface sediments of Lake Taihu. Journal of Lake Sciences 16: 113–119 (in Chinese).

Meyers, P. A., M. J. Leenheer & R. A. Bourbonniere, 1995. Diagenesis of vascular plant organic matter components during burial in lake sediments. Aquatic Geochemistry 1: 35–52.

Mônica, A. L. S. & E. R. Carlos, 2002. Behavior of selected micro and trace elements and organic matter in sediments of a freshwater system in south-east Brazil. The Science of the Total Environment 292: 121–128.

Palanques, A., J. A. Sanchez-Cabeza & P. Masqué, 1998. Historical record of heavy metals in a highly contaminated Mediterranean deposit: the Besòs prodelta. Marine Chemistry 61: 209–217.

Professional Committee of agricultural chemistry of chinese society for soil sciences, 1983. Routine analysis method for soil agricultural chemistry, Beijing, 103 (in Chinese).

Rose, N. L., J. F. Boyle, Y. Du, C. Yi, X. Dai, P. G. Appleby, H. Bennionl, S. Cai & L. Yu, 2004. Sedimentary evidence for changes in the pollution status of Taihu in the Jiangsu region of eastern China. Journal of Paleolimnology 32: 41–51.

Salomons, W. & U. Fostner, 1984. Metals in the Hydrocycle. Springer-Verlag, Berlin: 349.

Shun, S. C. & Y. P. Huang, 1993. Taihu. Ocean publishing, Beijing, 271 (in Chinese).

Swain, E. B, 1985. Measurement and interpretation of sedimentary pigments. Journal of Freshwater Biology 15: 53–75.

USEPA 1995. Test methods for evaluating solid waste. Vol. IA: Laboratory manual physical/chemical methods (3rd edn.). U.S. Gov. Print. Office, Washington, DC, SW 846.

Von Guten, H. R., M. Sturm & R. N. Moser, 1997. 200-year record of metals in lake sediments and natural background concentrations. Environmental Science Technology 31: 2193–2197.

Wan, G. J., P. H. Santschi, M. Sturm, K. Farrenkothen, A. Lueck, E. Werth & Ch. Schuler, 1987. Natural (^{210}Pb, ^{7}Be) and fallout (^{137}Cs, ^{239,240}Pu, ^{90}Sr) radionuclides as geochemical tracers of sedimentation in Greifensee, Switzerland. Chemical Geology 63: 181–196.

Wu, J. L., K. G. Michael, X. Z. Jiang, W. L. Xia & S. M. Wang, 2004. Sedimentary geochemical evidence for recent eutrophication of Lake Chenghai, Yunnan, China. Journal of Paleolimnology 32: 85–94.

Zou, H. X., G. Y. Sheng, C. Sun & O. Y. Xu, 1996. Distribution of organic contaminants in Lake Taihu. Water Research 30: 2003–2008.

Hydrobiologia (2007) 581:125–134
DOI 10.1007/s10750-006-0504-6

EUTROPHICATION IN LAKES

Role of climate and agricultural practice in determining matter discharge into large, shallow Lake Võrtsjärv, Estonia

Peeter Nõges · Marju Kägu · Tiina Nõges

Abstract This article addresses how seasonal and annual matter discharge into large, shallow Lake Võrtsjärv (Estonia) is determined by climate and changes in agricultural practices. Climate variability involved increasing winter air temperatures and large inter-annual temperature and precipitation variation. Agriculture practices have transformed from high fertilizer usage (i.e. swine slurry) in the 1970s and 1980s, leading to high phosphate, ammonium, and BOD_5 loadings, to low loadings after the collapse of soviet-type agriculture in the early 1990s. The 28-year monthly record on river flow and concentrations of nutrients (N, P) and dissolved organic matter (BOD_5, COD_{Mn}) from four main tributaries was analysed with seasonal air temperature and precipitation data. Long-term trends in nutrient and organic matter loadings to Lake Võrtsjärv resulted from agriculture and climate changes. The change could be traced as a linear trend in loadings and a highly inter-correlated cluster of the slurry-related pollutants. Coincidental trends in air temperature and fertiliser use caused strong correlations between air temperature and pollutant loadings to the rivers, which turned non-significant after removing trends showing that the relationships were not based on year-to-year differences. Residual analysis revealed significant positive correlations between precipitation and annual loadings of ammonium, phosphates, and COD_{Mn}. Both components forming the load (water discharge and concentrations of substances) increased in wet years. The effect of high winter North Atlantic Oscillation Index (NAO) was expressed as more intensive river flow during winter months and decreased flow during the flood peak.

Keywords Large and shallow lake · Climate impact · Agricultural practice · Discharge · Nitrogen · Phosphorus · Dissolved organic matter

Guest editors: B. Qin, Z. Liu & K. Havens
Eutrophication of shallow lakes with special reference to Lake Taihu, China.

P. Nõges (✉) · M. Kägu · T. Nõges
Limnological Centre, Institute of Agricultural and Environmental Sciences, Estonian Agricultural University, 61101 Rannu, Tartumaa, Estonia
e-mail: peeter.noges@jrc.it

P. Nõges
Institute for Environment and Sustainability, European Commission – DG Joint Research Centre, via E. Fermi 1, 21020 Ispra (VA), Italy

M. Kägu
Institute of Zoology and Hydrobiology, University of Tartu, Vanemuise 46, 51014 Tartu, Estonia

Introduction

Studies on global climate have revealed that conditions in the Northern Hemisphere are warming, with wetter winters and drier summers (Magnuson et al., 1997). These changes are attributed to increasing content of green-house gases in the atmosphere, which may be reflected in the increasing tendency of the North Atlantic Oscillation Index (NAO) (Hurrell et al., 2001, 2003). The increase of NAO is most pronounced in winter during the last 30 years (Cook, 2003). The signal of the winter NAO can be traced in thermal (Gerten & Adrian, 2001), hydrological (Yoo & D'Odorico, 2002) and biological parameters (Straile et al., 2003) of lakes, where its persistence is modified mostly by lake morphology (Gerten & Adrian, 2001). Climate change affects the balance of heat, water and substances in lake catchments resulting in changes to the hydrological regime and loadings. Hydrological changes have the strongest impact on shallow lakes, where they cause large fluctuations in water volume and lake depth. In Lake Võrtsjärv (mean depth 2.8 m), mean annual amplitude of water level fluctuations is 1.4 m, and the maximum range reaches 3.2 m. The latter corresponds to a 1.4-times difference in lake area, 2.4-times difference in mean depth and 3-times difference in lake volume (Nõges & Nõges, 1999). Changes in lake depth affect average illumination of the mixed water column, resuspension intensity, nutrient release from sediments, and denitrification rate in non-stratified shallow lakes. These factors control growth and composition of phytoplankton, which is the first link of the pelagic food web. Thus, changing water level has been considered the leading factor controlling ecosystem dynamics of Lake Võrtsjärv, primarily through phytoplankton (Nõges et al., 2003).

The impact of climate change on discharge of substances into the lake also must be considered. To now, this aspect of climate change has been less studied compared to better-known impacts on water temperature, stratification and ice cover. Phosphorus, nitrogen and silica are the most important substances coming from the watershed and shaping lake ecosystems. In warm, dry years, nutrient loading usually decreases, while in-lake concentrations may increase due to increased evaporation and more intensive release from sediments at a low water stage (Magnuson et al., 1997). The direction of the changes caused by climatic alterations can differ between deep and shallow lakes as well as regions of different climatic conditions. For example, cold low-NAO winters in Sweden bring about lower nutrient input to lakes, while, in Great Britain, such winters are associated with increased nutrient loading (Monteich et al., 2000; Straile et al., 2003).

Recently, limnologists have studied connections between climate change and coloured dissolved organic matter (CDOM) loading. CDOM affects underwater light conditions (Arst, 2003) and is important for phytoplankton and food webs. This may be crucial in large, very shallow lakes, like Lake Võrtsjärv, where light conditions play a leading role among factors limiting phytoplankton development (Nõges et al., 1998). In addition, CDOM is a substrate for aquatic bacteria. Water in European lakes is getting darker, and this phenomenon has been connected with intensified CDOM leakage from peaty catchments in warmer conditions (Freeman et al., 2001). There is, however, a current debate regarding effects of global warming and climate change on terrestrial organic matter export to aquatic environments (Tranvik & Jansson, 2002) since, e.g., northern boreal lakes generally have high CDOM concentrations (Curtis, 1998). It is clear that knowledge about the nature of climate change and its impacts is far from complete. The response of lakes to climate changes via temperature regime and stratification is complex and unpredictable, and different lakes may react in individual ways. Eutrophication may either diminish or magnify effects of climate change.

Answering questions regarding climate change impacts on lakes often is hindered by the scarcity of reliable, long-term limnological data. Changes in agricultural practices also may affect loadings and either mask or magnify climate change effects. The nearly 30-year record of monthly meteorological, hydrological, and chemical data available for Lake Võrtsjärv and its main tributaries offers a unique opportunity to study seasonal and annual dynamics of loading of different substances into the lake. This paper addresses how seasonal and annual matter discharge is determined by climate and changes in agricultural practices in large, shallow Lake Võrtsjärv (Estonia).

Site description

Lake Võrtsjärv is a large, shallow lake (270 km^2, mean depth 2.8 m) located in central Estonia (58°05′–58°25′ N, 25°55′–26°10′ E). Its catchment area (3374 km^2; Fig. 1) belongs to the southern Boreal forest zone. Forests cover 30–40% of the southern catchment, reaching 50–60% in the Väike Emajõgi Valley. The largest wetland areas cover the eastern coastal region and the mouth the Väike Emajõgi and Õhne rivers. The proportion of arable land is the lowest (10–20%) in the northern part and highest (30–40%) in the western and eastern parts of the catchment (Järvet et al., 2004). Agricultural activities are the main anthropogenic impact on the lake. In the 1980s, large areas, especially in the R. Tänassilma sub-catchment, were fertilized with swine slurry causing increased nutrient loading (Järvet & Laanemets, 1989, 1990). Improvements in wastewater treatment since the 1990s has decreased point source loadings from Valga (pop. 14,000) and Viljandi (pop. 21,000), located on the watershed boundary, and other smaller settlements. Abandonment of several large cattle farms has contributed to this decrease. The lake is

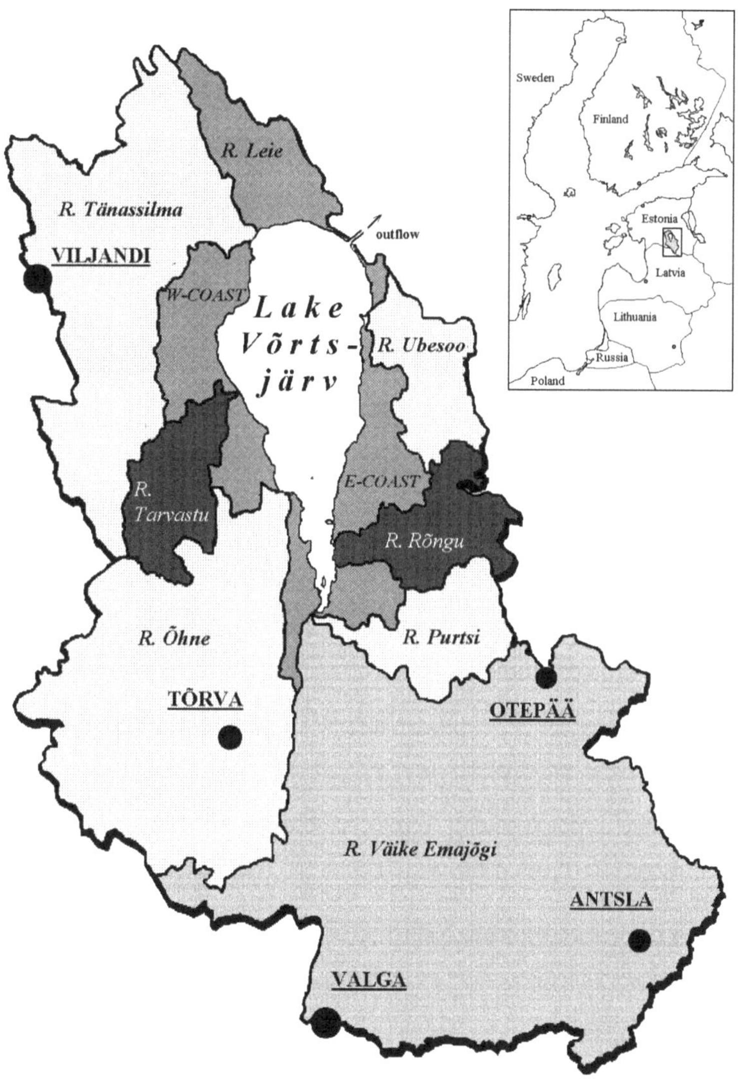

Fig. 1 Location map and division of Lake Võrtsjärv catchment area

ice-covered from mid-November to mid-April (average 135 days). The flow regime of inflowing rivers is natural, and inflow usually peaks in April with about one-quarter of its annual total inflow. Water retention time is about 1 year, and Lake Võrtsjärv is highly eutrophic, characterised by a mean total nitrogen concentration of 1.6 mg l^{-1}, total phosphorus of 54 μg l^{-1}, and chlorophyll *a* of 24 μg l^{-1} (Haberman et al., 1998). During the ice-free period, Secchi depth commonly is less than 1 m, mainly due to resuspended sediments.

Material and methods

Nutrient concentration data (different forms of N and P), biochemical oxygen demand (BOD_5), chemical oxygen demand by permanganate (COD_{Mn}), and water colour (Pt-Co°) in rivers were measured from 1975 to 2002 in the State Monitoring Programme, and data were obtained from the Information Centre of the Estonian Ministry of Environment. Yearly riverine loadings (*L*) were calculated based on daily water discharges and monthly concentrations of substances:

$$L = \sum_{i=1}^{12} W_{ki} \cdot C_{ki}$$

W_{ki}—volume of monthly runoff
C_{ki}—mean monthly concentration

NAO data from 1886 to 2002 were obtained from http://www.cru.uea.ac.uk/cru/data/nao.htm. Winter NAO (NAO_w), calculated as the December–February mean, was used in this analysis. Air temperature (1894–2002), precipitation (1866–2002), lake water level (1923–2002), and riverine discharges were measured routinely during meteorological and hydrological surveys by the Estonian Institute of Hydrology and Meteorology.

Statistica for Windows (v 6.0) was used for statistical analyses. Data were detrended linearly using the Time Series Analysis module. Spearman rank correlation was used throughout the analysis.

Results

Twenty-eight-years of data on annual nutrient loading were analyzed from four Lake Võrtsjärv sub-catchments: Tänassilma, Tarvastu, Õhne, and V. Emajõgi (Fig. 1) together with precipitation and air temperature data. Several peak loadings of pollutants visible in the long-term dynamics were related directly to extreme wet years (Fig. 2). For example, rainy years 1981 and 1998 were distinguished by increased runoff of COD_{Mn} and phosphates, while 1990 was characterized by that of COD_{Mn} and nitrates (Fig. 2). Besides rainy years, peaks of ammonium, phosphate and BOD_5 (latter not shown) in 1986 and 1987 coincided with cold winters (Fig. 2). Time-series of winter air temperature data (Jan.–Mar.) can be

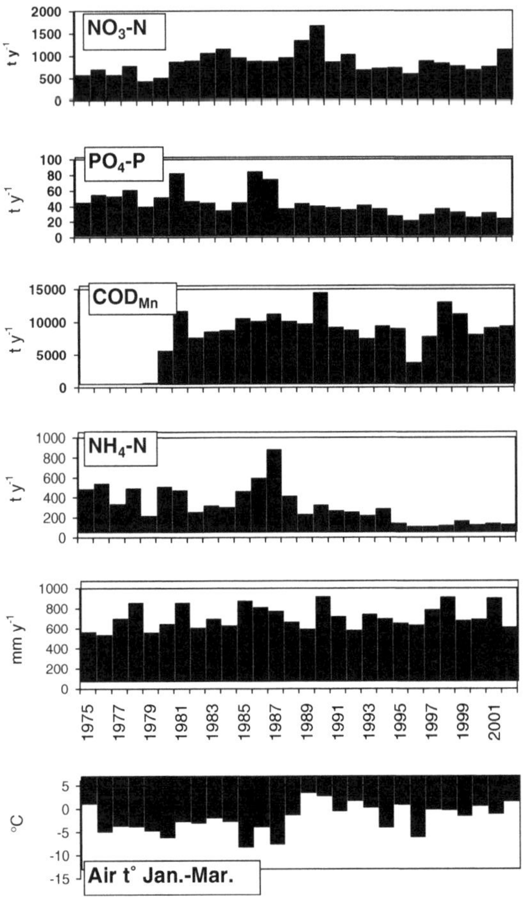

Fig. 2 Long-term annual loading of nitrate and ammonium nitrogen (NO_3^--N and NH_4^+-N, respectively), phosphate phosphorus (PO_4^{3-}-P) and permanganate oxygen demand (COD_{Mn}) from four sub-catchments of Lake Võrtsjärv (Tänassilma, Tarvastu, Õhne, and V. Emajõgi), annual amount of precipitation (Prec.) and winter air temperature measured at Tartu hydrometeorological station

roughly subdivided into two periods: 1975–1988 and 1989–2002 (Fig. 6). In the second period, temperature was notably higher. At the same time, loading of some substances, phosphates and ammonium in particular, showed a similar relationship: average loadings were different before and after the late 1980s. Presumably, this trend, which coincidentally was synchronized with the climate shift and caused, in part, by agricultural practice changes, may result in a negative relationship between temperature and substance loadings and bias other relationships. Correlation differences before and after trend removal were analyzed to distinguish between effects caused by year-to-year variations of (mostly climatic) indices and those caused by coinciding trends. Annual loading data was analyzed with annual and seasonal data on climatic variables: air temperature and precipitation. Trend removal caused several changes in the correlations (Table 1):

(1) all correlations between air temperature and loadings of substances became insignificant,
(2) correlations between phosphates, ammonium and biochemical oxygen demand weakened,
(3) statistical relationships between loadings and precipitation strengthened,
(4) correlations between loadings of chemical oxygen demand and other substances strengthened.

Second, emphasis was placed on two sides of the loading formation: river discharge and concentration of substances and their relationships with NAO. Relationships were analyzed between detrended NAO_w, discharge, and concentrations of CDOM and nutrients (N and P) in two major inflows to Lake Võrtsjärv (R. Väike Emajõgi and R. Õhne). These two rivers account for 60% or more of total discharge and nutrient loading to the lake. Figure 3a, b show the monthly inflow regime, and both rivers show peak flow in April. From January to March, discharge (Q) of both rivers was related positively to NAO_w, while the relationship was negative in April–May (Fig. 3c).

Water colour and COD_{Mn} were used as proxies for CDOM. These two parameters were related strongly to each other ($r = 0.57$ in the R. Õhne and $r = 0.66$ in the R. Väike Emajõgi; $p < 0.00001$). NAO_w showed strong correlations with both CDOM proxies. From January to March, correlations were mostly positive ($p < 0.05$ in February and March) and turned mostly negative for the rest of the year (Fig. 4a,

Table 1 Spearman correlations between climatic variables and loadings of substances to Lake Võrtsjärv (1975–2002) before and after removing linear trends from all data series

Direction of change	Between variables	Spearman r before detrending	Spearman r after detrending
Weakening	T_{1-3} - PO_4^{3-}	–0.438*	–0.177
	T_{1-3} - NH_4^+	–0.448*	–0.303
	T_{4-6} - BOD_5	–0.472*	0.027
	T_{4-6} - PO_4^{3-}	–0.507**	–0.301
	T_{4-6} - NH_4^+	–0.418*	–0.174
	T_{7-9} - BOD_5	–0.391*	–0.174
	T_{7-9} - PO_4^{3-}	–0.512**	–0.269
	T_{7-9} - NH_4^+	–0.377*	–0.033
	BOD_5 - PO_4^{3-}	0.875**	0.552**
	BOD_5 - NH_4^+	0.862**	0.432*
	NH_4 - PO_4^{3-}	0.857**	0.589**
Strengthening	P_{7-9} - NH_4^+	0.322	0.391*
	P_{7-9} - COD_{Mn}	0.374	0.420*
	P_{7-9} - PO_4^{3-}	0.383*	0.522**
	P_{1-12} - PO_4^{3-}	0.152	0.605**
	COD_{Mn} - NH_4^+	0.368	0.517**
	COD_{Mn} - PO_4^{3-}	0.263	0.517**
	COD_{Mn} - NO_3^-	0.437	0.629**
Unchanged	P_{1-12} - COD_{Mn}	0.570**	0.554**

T—temperature, P—amount of precipitation (numbers in the foot index show the period in months to which the data pertain); loadings: PO_4^{3-}—phosphates, NO_3—nitrates, NH_4^+—ammonium, BOD_5–biochemical oxygen demand, COD_{Mn}—permanganate oxygen demand

* $p < 0.05$, ** $p < 0.01$

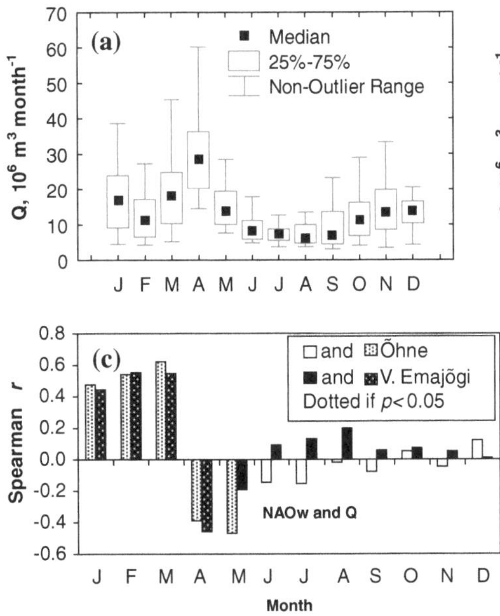

Fig. 3 Monthly discharges (Q) of rivers Õhne (**a**) and Väike Emajõgi (**b**) and their Spearman correlations with the NAO index for winter (**c**). All series detrended

b). Strongest negative correlations occurred in both rivers and for both parameters in July. In both rivers, water colour and COD_{Mn} were related positively to discharge throughout the year, and strong correlations occurred in March, June and August–November (Fig. 4c, d).

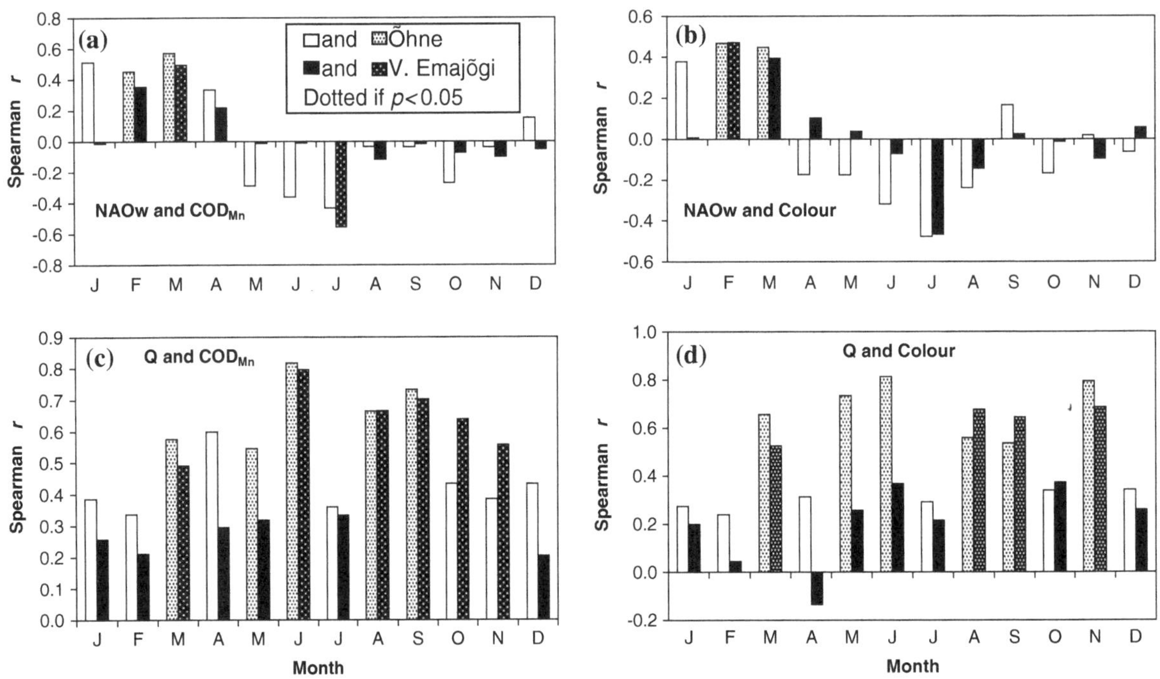

Fig. 4 Monthly Spearman correlations of permanganate oxygen demand (COD_{Mn}) and water colour with the winter NAO (NAO_w: **a** & **b**, respectively), and discharge (Q: **c** & **d**, respectively) in two major rivers in Lake Võrtsjärv catchment. All series detrended

Increased nutrient concentrations in January were characteristic of high-NAO_w years. For total nitrogen (TN) and, especially, nitrates, the positive relationship with NAO_w continued in February and March (Fig. 5a, b). For the rest of the year, the relationship with NAO_w turned negative for TN, while nitrates showed a positive relationship from July to September. For phosphates and total phosphorus (TP), the correlation with NAO_w remained mostly negative from February to July-August (strongest in April; Fig. 5c, d). In autumn, the correlation with NAO_w was mostly positive for TP and variable for phosphates. Increased discharge generally brought about higher nitrogen concentrations in rivers (Fig. 6a, b) with exceptions for the flood peak in April and the end of the year (November or December). Strong flushing of nitrates from the watershed by increased runoff in summer and autumn especially was pronounced in the R. Väike Emajõgi. The dependence of phosphorus concentration on river flow was more variable (Fig. 5c, d): in both rivers the correlation was negative in February and March, and strongest positive correlations occurred in August for phosphates and September–October for TP.

Discussion

Analysis showed that correlations between the original series of air temperature and substance loadings were caused by common trends in variables and not an inherent causal relationship. The trend in winter air temperatures, expressed as a step-wise change since the late 1980s, perhaps could be attributed to climate change or climate warming observed over large areas of the northern hemisphere (IPCC, 2001). Climate investigations in Estonia show that air temperature increased by 0.3–0.5°C from 1966 to 1998 compared to the earlier 90-year period, and the largest increase (>1°C) has taken place in March (Jaagus, 1999). Changes in winter temperatures may cause changes in catchment processes and affect loadings of substances to the lakes. Year-to-year variation in winter air temperature influences nitrate concentration in lakes in the English

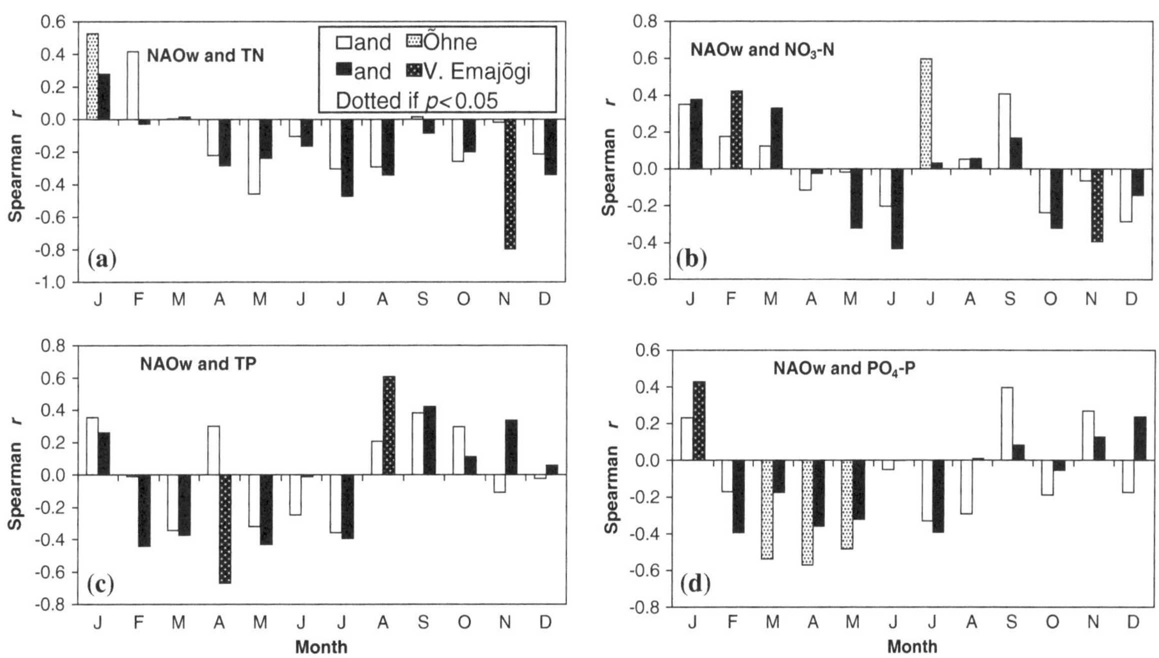

Fig. 5 Monthly Spearmen correlations of the winter NAO (NAO_w) with (**a**) total nitrogen (TN), (**b**) nitrate (NO_3^--N), (**c**) total phosphorus (TP), and (**d**) phosphate (PO_4^{3-}-P) concentrations in two major rivers in Lake Võrtsjärv catchment. All series detrended

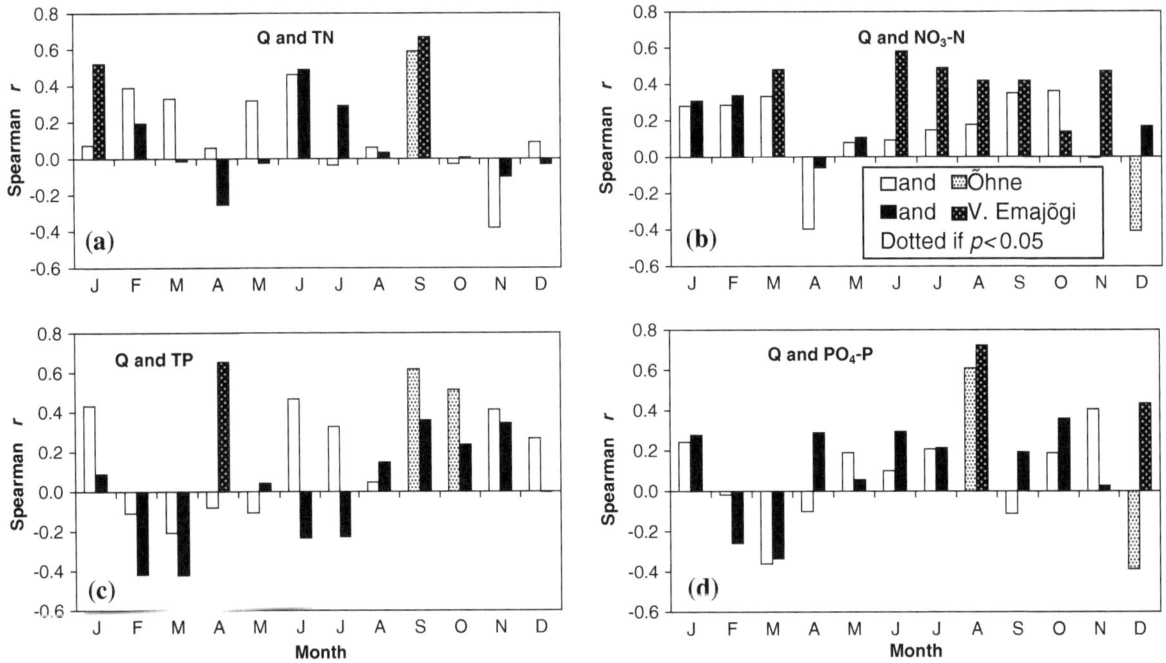

Fig. 6 Monthly Spearmen correlations of the discharge (Q) with (**a**) total nitrogen (TN), (**b**) nitrate (NO_3^--N), (**c**) total phosphorus (TP), and (**d**) phosphate (PO_4^{3-}-P) concentrations in two major rivers in Lake Võrtsjärv catchment. All series detrended

Lake District, since warm winters lead to a reduced nitrate concentration (George et al., 2004). As an extreme example, a 2° temperature increase melted the permafrost in the Lake Schwarzensee catchment, a remote high-alpine lake in Austria, and totally changed the chemical regime of the lake. Conductivity and silica concentration doubled, and pH increased dramatically (Psenner & Schmidt, 1992).

However, climate variability had a clear influence on loadings via precipitation changes. Despite several cases (1981, 1990, 1998) where effects of rainy years on loadings was visible in the time-series (Fig. 2), relationships remained insignificant without detrending the data. Residual analysis revealed significant positive correlations between precipitation and annual loadings of ammonium, phosphates, and COD_{Mn}. Ammonium and COD_{Mn} loadings showed strongest relationships with rainfall from July to September, and phosphate loading with annual precipitation.

After detrending, relationships between BOD_5, phosphates, and ammonium became weaker. Increased runoff of these pollutants is typical for agricultural catchments fertilised with slurry and, hence, changes in slurry application were expressed as a common trend in their loading rates. Experiments carried out on a sloping site using purpose-built facilities (Smith et al., 2001a, b) showed that application of cattle slurries to the silty clay loam soil increased loss of solids, ammonium nitrogen, and particulate and soluble phosphorus in surface water flow compared to control plots receiving inorganic fertiliser only, or no treatment, but had little effect on nitrate losses by this route. Rainfall events immediately after manure applications may cause large run-off of nutrients (Jarvis et al., 1987; Smith et al., 2001a), which may explain the synchronous increase of ammonium and phosphate loadings observed in the present study in some rainy years (e.g. in 1981).

In the original data, COD_{Mn} loading showed a different behaviour compared to other substances: it did not correlate with any others and had a significant positive relationship only with annual precipitation. After trend removal, the latter did not change, but correlations between loadings of all substances became highly significant. COD_{Mn} is represented in runoff waters from the Lake Võrtsjärv catchment mainly by humic substances, which give a brownish colour to the water. Dynamics of COD_{Mn} loadings reflected natural changes in catchment hydrology and was not affected by changes in agriculture, which caused different behaviour of other substances. After trend removal, dynamics of the slurry-related components became more natural, while the role of precipitation as the most important climatic parameter became a clearer manifestation.

NAO is one of the often-used proxy indicators allowing complex atmospheric circulation patterns to be expressed as a simple index. It combines effects of changes in temperature, affluence, and wind patterns. The relationship between NAO_w and river flow is established through precipitation in winter and spring but, first, by a shift in the timing and height of the flood peak. Milder winters (increased air temperature in February from –7.9 to –5.5°C during 1950–1997) and changed precipitation patterns have influenced mean annual water discharge, which has resulted in more intensive flow during colder seasons and decreased runoff in summer (Mander et al., 2000). During mild winters (high NAO_w), recurrent snowmelts keep water flowing throughout the winter, thus reducing peak discharges in spring (Fig. 3). Since temperature affects timing of the flood peak and, to a lesser extent, annual discharge (if differences in evaporation are considered), its effect on annual loading remained marginal and could not be revealed in our data. Concentrations of substances in rivers were related positively to river flow (Figs. 4c, d, 6). Thus, both components forming the load (water discharge and concentrations of substances) increased in wet years. As mentioned, summer and autumn rains (July–September) correlated most strongly with annual loadings of substances. Runoff of substances in this period, characterised usually by low flow in rivers (Fig. 3), was sensitive to precipitation, evidenced by strong correlations between water discharge and concentrations. Periods of negative correlations between water flow and concentrations occurred only for nutrients and were observed mostly in winter and spring. For phosphorus, the negative relationship in February and March may be attributed to the winter low flow period when most rivers switch to groundwater feeding (Järvet, 2001). Negative correlations between water flow and nitrogen concentration coincided with the April flood peak and may be caused by dilution.

Conclusions

(1) Long-term changes in loadings of nutrients and organic matter to Lake Võrtsjärv resulted from changes in agricultural practices as well as climate change.
(2) Agriculture changes involved the transition from high loadings of phosphates, ammonium, and BOD_5 in the 1970s and 1980s from fertilisers, in particular swine slurry, to low loadings after the collapse of soviet-type agriculture in the early 1990s. Changes are traced as a linear trend in loadings and an inter-correlated cluster of slurry-related pollutants.
(3) Climate variability consisted of an increasing trend in winter air temperatures and large inter-annual fluctuations of temperature and precipitation.
(4) Coinciding trends in air temperature and fertiliser use caused strong correlations between air temperature and pollutant loadings to rivers. All these relationships were insignificant after removing the trends, showing that the relationships were based on a coincidental common trend and not year-to-year differences.
(5) Residual analysis revealed significant positive correlations between precipitation and annual loadings of ammonium, phosphates, and COD_{Mn}. Both components forming the load (water discharge and concentrations of substances) increased in wet years.
(6) The effect of high winter NAO was expressed in more intensive river flow during winter months and decreased flow during the flood peak.

Acknowledgements Funding was provided by the Estonian Ministry of Education (No. 0362480s03), Estonian Science Foundation grants 5738 and 5425, and the EC project CLIME (No. EVK1-CT-2002-00121). Data collection in state monitoring program was supported by the Estonian Ministry of Environment.

References

Arst, H., 2003. Optical Properties and Remote Sensing of Multicomponental Water Bodies. Springer-Verlag, 238.

Cook, E., 2003. Multi-proxy reconstructions of the North Atlantic Oscillation (NAO) index: a critical review and a new well verified NAO index reconstruction back to AD 1400. In Hurrell, J. W., Y. Kushnir, G. Ottersen & M. Visbeck (eds), The North Atlantic Oscillation: Climatic Significance and Environmental Impact. American Geophysical Union, Washington DC, 63–79.

Curtis, P. J., 1998. Climatic and hydrologic control of DOM concentration and quality in Lakes. In Hessen, D. O. & L. J. Tranvik (eds), Aquatic Humic Substances, Vol. 133, Springer-Verlag, 93–104.

Freeman, C., C. D. Evans, D. T. Monteih, B. Reynolds & N. Fenner, 2001. Export of organic carbon from peat soils. Nature 412: 785.

George, D. G., S. C. Maberly & D. P. Hewitt, 2004. The influence of the North Atlantic Oscillation on the physical, chemical and biological characteristics of four lakes in the English Lake District. Freshwater Biology 49(6): 760–774.

Gerten, D. & R. Adrian, 2001. Differences in the persistency of the North Atlantic Oscillation signal among lakes. Limnology & Oceanography 46: 448–455.

Haberman, J., P. Nõges, E. Pihu, T. Nõges, K. Kangur & V. Kisand, 1998. Characterization of Lake Võrtsjärv. Limnologica 28(1): 3–11.

Hurrell, J. W., Y. Kushnir & M. Visbeck, 2001. The North Atlantic Oscillation. Science 291: 603–608.

Hurrell, J. W., Y. Kushnir, G. Ottersen, M. Visbeck, 2003. An overview of North Atlantic Oscillation. In Hurrell, J. W., Y. Kushnir, G. Ottersen & M. Visbeck (eds), The North Atlantic Oscillation: Climatic Significance and Environmental Impact. American Geophysical Union, Washington DC, 7–36.

IPCC, 2001. Summary for Policymakers, the third assessment report: 1–18.

Jaagus, J. 1999. Uusi andmeid Eesti kliimast. (New data about the climate in Estonia). In Jaagus, J. (ed), Uurimusi Eesti kliimast. Publicationes Instituti Geographici Universitatis Tartuensis 85: 28–40.

Järvet, A., 2001. Veekogude klimaatilised aastaajad. Publicationes Instituti Geographici Universitatis Tartuensis 90: 48–81.

Järvet, A., R. Karukäpp & I. Arold, 2004. Location and physico-geographical conditions of the catchment area. In Haberman, J., E. Pihu & A. Raukas (eds), Lake Võrtsjärv. Estonian Encyclopedia Publishers, Tallinn: 11–28.

Järvet, A. & A. Laanemets, 1989. Võrtsjärv on ohus. Eesti Loodus 5: 308–312.

Järvet, A. & A. Laanemets, 1990. Võrtsjärv kui looduslik puhastusseade. Eesti Loodus 11: 708–712.

Jarvis, S. C., M. Sherwood, J. H. A. M. Steenvoorden, 1987. Nitrogen losses from animal manures: from grazed pastures and from applied slurry. In: van der Meer, H. G. et al. (eds), Animal Manure On Grassland And Fodder Crops. Martinus Nijhoff Publishers, The Netherlands, 195–212.

Magnuson, J. J., K. E. Webster, R. A. Assel, C. J. Bowser, P. J. Dillon, J. D. Eaton, H. E Evans, E. J. Fee, R. I. Hall, L. R. Mortsch, D. W. Schindler & F. H. Quinn, 1997. Potential effects of climate changes on aquatic ecosystems: Laurentian Great Lakes and Precambrian Shield Region. In Cushing, C. E. (ed.), Freshwater Ecosystems and Climate Change in North America, A Regional Assessment. John Wiley & Sons, New York, 7–53.

Mander, Ü., A. Kull, V. Kuusemets & T. Tamm, 2000. Nutrient runoff dynamics in a rural catchment: Influence of land-use changes, climatic fluctuations and ecotechnological measures. Ecological Engineering 14: 405–417.

Monteich, D. C., C. D. Evans & B. Reynolds, 2000. Are temporal variations in the nitrate content of UK upland freshwaters linked to the North Atlantic Oscillation? Hydrological Processes 14: 1745–1749.

Nõges, T. & P. Nõges, 1999. The effect of extreme water level decrease on hydrochemistry and phytoplankton in a shallow eutrophic lake. Hydrobiologia 408/409: 277–283.

Nõges, T., V. Kisand, P. Nõges, A. Põllumäe, L. Tuvikene & P. Zingel, 1998. Plankton seasonal dynamics and its controlling factors in shallow polymictic eutrophic Lake Võrtsjärv, Estonia. International Review of Hydrobiology 83: 279–296.

Nõges, T., P. Nõges & R. Laugaste, 2003. Water level as the mediator between climate change and phytoplankton composition in a large shallow temperate lake. Hydrobiologia 506(1): 257–263.

Psenner, R. & R. Schmidt, 1992. Climate-driven pH control of remote alpine lakes and effects of acid deposition. Nature 356: 1–3.

Smith, K. A., D. R. Jackson & T. J. Pepper, 2001a. Nutrient losses by surface run-off following the application of organic manures to arable land. 1. Nitrogen. Environmental Pollution 112: 41–51.

Smith, K. A., D. R. Jackson & P. J. A. Withers, 2001b. Nutrient losses by surface run-off following the application of organic manures to arable land. 2. Phosphorus. Environmental Pollution 112: 53–60.

Straile, D., D. M. Livingstone, G. A. Weyhenmeyer & D. G. George, 2003. The response of freshwater ecosystems to climate variability associated with the North Atlantic Oscillation. In Hurrell J. W, Y. Kushnir, G. Ottersen & M. Visbeck (eds), The North Atlantic Oscillation: Climatic Significance and Environmental Impact. American Geophysical Union, Washington DC, 263–279.

Tranvik, L. J. & M. Jansson, 2002. Climate change and terrestrial export of organic carbon. Nature 415: 861–862.

Yoo, J. C. & P. D'Odorico, 2002. Trends and fluctuations in the dates of ice break-up of lakes and rivers in Northern Europe: the effect of the North Atlantic Oscillation. Journal of Hydrology 268: 100–112.

Hydrobiologia (2007) 581:135–140
DOI 10.1007/s10750-006-0505-5

EUTROPHICATION IN LAKES

Stable isotopic composition of nitrate in Lake Taihu, China, and major inflow rivers

Amy Townsend-Small · Mark J. McCarthy · Jay A. Brandes · Longyuan Yang · Lu Zhang · Wayne S. Gardner

Abstract Identification of sources of nutrient pollution is a first step towards remediation of eutrophication in aquatic ecosystems. The stable isotope nitrogen-15 (^{15}N) is a natural indicator of nitrogen (N) source and biogeochemistry. We sampled Lake Taihu, a hyper-eutrophic lake in eastern China, and major inflow rivers during winter and spring of 2004 to determine concentration and $\delta^{15}N$ of nitrate (NO_3^-). Nitrate concentrations in rivers and the lake were higher, in most cases, in spring than in winter. $\delta^{15}N$ of NO_3^- was not correlated with NO_3^- concentration, indicating that concentrations alone are insufficient to describe N sources. Results show that riverine N inputs in winter are influenced by discharge of human sewage into rivers and the lake. In spring, however, wastewater inputs to the lake appear to be balanced by fertilizers, atmospheric, and/or N_2 fixation sources. Rain NO_3^- concentrations were seasonally high and isotopically enriched compared to potential sources, indicating that rain may be a significant or even dominant source of N to the lake during the rainy season. $\delta^{15}N$ values show that urbanized areas of the lake have more sewage-derived N than those areas dominated by agriculture, aquaculture, or industry. This observation has important implications for human health, since Lake Taihu is a source of drinking and irrigation water as well as fish for human consumption.

Keywords Nitrate · Nitrogen-15 · Lake Taihu · China · Stable isotopes · Eutrophication

Guest editors: B. Qin, Z. Liu & K. Havens
Eutrophication of shallow lakes with special reference to Lake Taihu, China

A. Townsend-Small (✉) · M. J. McCarthy · J. A. Brandes · W. S. Gardner
Marine Science Institute, The University of Texas at Austin, 750 Channel View Drive, Port Aransas, TX 78373, USA
e-mail: amy@utmsi.utexas.edu

L. Yang · L. Zhang
Nanjing Institute of Limnology and Geography, Chinese Academy of Sciences, 73 East Beijing Road, Nanjing 21008, China

Introduction

Eutrophication is a serious environmental problem in many aquatic ecosystems around the world, including Lake Taihu, located in eastern China in the Changjiang River delta. Excess nutrient inputs into Lake Taihu have resulted in seasonally high population densities of phytoplankton and macrophytes with detrimental effects on water quality and fish populations (Pu & Yan, 1998; Qin et al., 2007). High nutrient inputs can be concurrent with loading of other toxins, such as heavy metals and harmful bacteria that can be associated with wastewater (Wu et al., 1999).

Nitrogen (N) has two naturally occurring stable isotopes, ^{14}N and ^{15}N, which are present as 99.64 and 0.36% of all N atoms, respectively (Faure, 1986). Nitrogen isotope ratios usually are expressed in delta notation with respect to the N isotope ratio of air; therefore, samples containing an N isotope ratio close to that of air will have $\delta^{15}N$ values near zero (Mariotti, 1984). For this reason, inorganic fertilizers derived from atmospheric N_2 and biomass of N_2–fixing plants and microorganisms have a $\delta^{15}N$ value around zero (Peterson & Fry, 1987). Most inorganic fertilizers have a $\delta^{15}N$ of nitrate (NO_3^-) between –1 and 2‰ (Macko & Ostrom, 1994), whereas organic fertilizers such as manure may produce NO_3^- with higher $\delta^{15}N$ values (Jin et al., 2004). N cycling processes such as denitrification and ammonia volatilization discriminate against heavier ^{15}N, leaving remaining substrate enriched in ^{15}N relative to initial composition and the end product. Dissolved N compounds derived from human waste can be enriched in ^{15}N as a result of the sewage treatment process and the high trophic status of humans. Nitrate derived from wastewater has a $\delta^{15}N$ from 10 to 20‰, whereas natural soil nitrate $\delta^{15}N$ values usually range from 2 to 8‰ (McClelland & Valiela, 1998). It is important to note that $\delta^{15}N$ of NO_3^- is not an unambiguous N source indicator in lakes (Bedard-Haughn et al., 2003), especially since fertilizers and natural soil NO_3^- may become isotopically enriched in ground waters and rivers before they reach the lake (DeSimone & Howes, 1996; Kellman & Hillaire-Marcel, 1998). However, isotopic differences of N sources and changes in ^{15}N-content of those sources can be used to identify seasonal N contributions to aquatic ecosystems, especially when considered in the context of other nutrient monitoring and flux studies.

Site description

Lake Taihu (Fig. 1) is a large (2338 km^2), shallow (average depth 1.9 m) lake located in the Changjiang River delta in southeastern China (Qin et al., 2007). The lake provides drinking water, irrigation water, and fish for the approximately 36 million people residing in the catchment (Qin et al., 2007). Lake Taihu suffers from eutrophication due to point and non-point nutrient sources (Guo et al., 2004). Meiliang Bay and

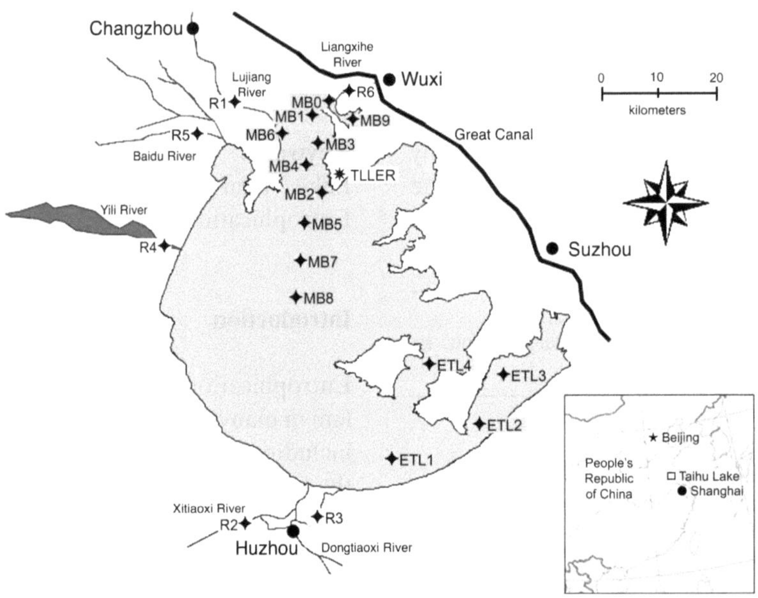

Fig. 1 Map of Lake Taihu and major inflow rivers (outflow rivers not shown) with inset showing location within China. Sampling locations within the lake and in major inflow rivers are shown. Also depicted are major cities in the area and TLLER, where rainwater was collected

Wuli Bay, located in the urbanized northern part of the lake, are hyper-eutrophic and characterized by annual cyanobacterial blooms (Chen et al., 2003). The northwest and southern portions of the lake are high-density urban areas. The landscape to the east and west of the lake is largely industrial and agricultural. Annual rainfall is about 1250 mm, 75% of which falls during the summer monsoon season from May to October (Pu & Yan, 1998). In 2004, rainfall rates were highest in the summer months, peaking at 181 mm for July, but only about 55% of total rain occurred during the monsoon season (L. Yang, personal communication). Annual river inflow is about 80×10^8 m^3 and the residence time of lake water is 5 months (Qin et al., 2007).

Materials and methods

Samples were taken from Lake Taihu and major inflow rivers in January and May of 2004 (Fig. 1). Most sampling sites were concentrated in Meiliang Bay (MB, Fig. 1). Some samples were taken from East Taihu (ETL), an area dominated by rooted macrophytes and aquaculture enclosures. Rainwater was sampled once during each field season at the Taihu Laboratory for Lake Ecosystem Research (TLLER; Fig. 1). Water samples were filtered in the field through 0.2 μm nylon syringe filters into glass vials and frozen. Samples not frozen immediately were filtered and acidified with 1 μl ml^{-1} of H_2SO_4, then neutralized prior to isotopic analysis.

Nitrate concentrations were determined by cadmium reduction using a flow-injection autoanalyzer (Lachat QuikChem 8000) at The University of Texas Marine Science Institute (UTMSI). Water samples were analyzed in triplicate for $\delta^{15}N$ of NO_3^- using the "denitrifier" method (Sigman et al., 2001). Nitrous oxide produced via this method was analyzed on a Finnegan MAT DELTA^{plus} with a ThermoFinnegan Trace GC and a PreCon interface, also at UTMSI. The precision of this method for $\delta^{15}N$ analysis is 0.5‰ or greater.

Results

Dissolved NO_3^- plus nitrite (NO_2^-) concentrations and $\delta^{15}N$ of NO_3^- values determined at each site during each sampling season are shown in Table 1. $NO_2^- + NO_3^-$ concentrations were higher in spring than winter at all but four sites (Table 1). Concentrations were not correlated to $\delta^{15}N$ of NO_3^- ($r^2 = 0.0884$, $n = 40$).

$\delta^{15}N$ of NO_3^- also showed seasonal differences. In winter, $\delta^{15}N$ of NO_3^- values in rivers draining urban areas were generally higher than those from rivers draining agricultural or industrial areas. In January, the Liangxihe (R6), Lujiang (R1), and Dongtiaoxi Rivers (R3), which drain the urban watersheds of Wuxi, Changzhou, and Huzhou, respectively, all exhibited $\delta^{15}N$ values above 10‰ (Table 1). The other three rivers sampled in January had $\delta^{15}N$ values below 7.5‰. During spring, however, only the Dongtiaoxi River (R3) had a $\delta^{15}N$ of NO_3^- above 10‰.

In winter, lake sites located near river mouths reflected the $\delta^{15}N$ of river water. In general, sampling stations near river mouths had higher $\delta^{15}N$ of NO_3^- values, and isotopic signatures decreased with distance from rivers. Stations in central Lake Taihu had low $\delta^{15}N$ values, reflecting either an increased contribution of rain (4.4‰) in the central lake or dilution of enriched riverine N. In spring, NO_3^- in Meiliang Bay maintained high $\delta^{15}N$ levels. Stations near the middle of the lake had higher $\delta^{15}N$ in spring than winter.

Stations in East Taihu had consistent $\delta^{15}N$ values, between 6.3 and 7.6‰. Maximum $\delta^{15}N$ in East Taihu during January occurred near the Dongtiaoxi River mouth. In spring, two areas of East Taihu had NO_3^- levels too low to measure $\delta^{15}N$, but overall $\delta^{15}N$ of NO_3^- was comparable from season to season at this location.

Rainwater $\delta^{15}N$ of NO_3^- varied little with sampling season (Table 1), although rainwater NO_3^- concentrations were very high in the summer (145.8 μM, Table 1).

Discussion

In January, rivers on the northwest and southern parts of the lake that drain large urban areas had elevated $\delta^{15}N$ of NO_3^- values, indicating the predominance of sewage-derived N. In comparison, rivers draining agricultural or industrial

Table 1 Nitrate and nitrite concentrations and $\delta^{15}N$ of NO_3^- values for water samples collected at Lake Taihu during January and May 2004

	JANUARY 2004		MAY 2004	
	$NO_2^- + NO_3^-$ (μM)	$\delta^{15}N$ (‰)	$NO_2^- + NO_3^-$ (μM)	$\delta^{15}N$ (‰)
R1	55.1	19.4	93.0	5.7
R2	59.1	7.3	130.4	7.1
R3	74.7	11.1	212.9	13.8
R4	184.1	6.3	131.8	8.6
R5	28.1	5.9	56.7	6.7
R6	84.9	10.1	22.9	7.4
ETL1	49.8	7.6	59.0	3.9
ETL2	7.45	6.8	0.8	undetectable
ETL3	6.95	6.5	0.7	undetectable
ETL4	30.0	6.3	52.5	7.1
Rain water	17.6	4.4	145.8	6.0
MB0	127.0	9.7	96.1	7.1
MB1	84.1	4.0	169.7	8.3
MB2	115.7	6.1	218.0	27.3
MB3	109.7	3.0	228.5	10.0
MB4	111.8	16.2	193.0	11.7
MB5	110.9	6.5	201.0	6.7
MB6	118.2	9.7	89.0	1.6
MB7	85.4	6.0	186.8	8.4
MB8	84.7	3.9	193.5	9.9
MB9	68.9	5.4	329.0	5.9

areas, such as those on the southwest side of the lake, had lower isotopic signatures, presumably due the to dominance of fertilizer or plant-derived N. In spring, river $\delta^{15}N$ of NO_3^- was lower at all locations, with only one value above 10‰. This result may indicate the additional contribution of fertilizers or N_2 fixation to river NO_3^- in northwest or southern rivers during the growing season, especially when considered in the context of higher springtime riverine NO_3^- concentrations. Alternatively, the reduction in $\delta^{15}N$ and increase in NO_3^- concentration in rivers during spring may be due to an increased contribution of high concentration, low $\delta^{15}N$ NO_3^- rainwater to rivers, since there is an increase in rainfall at this time.

As expected in a monsoonal climate, river discharge in the Lake Taihu basin is higher in spring than winter (Qin et al., 2007). For January, concentration and discharge weighted average riverine $\delta^{15}N$ values were 93% successful at predicting the average $\delta^{15}N$ of NO_3^- in Meiliang Bay and the central lake (weighted riverine average $\delta^{15}N = 6.6‰$; lake average $\delta^{15}N = 7.1‰$), indicating that river discharge was the primary control on lake $\delta^{15}N$ in the winter months. Such integration of river values in lake water was not unexpected, since Taihu is a large lake with a long residence time. However, in May the same calculation was only 77% successful at predicting lake $\delta^{15}N$ values (weighted riverine average $\delta^{15}N = 7.5‰$; lake average $\delta^{15}N = 9.7‰$), indicating that biological processes that fractionate N isotopes are important controls on lake isotopic composition in spring and summer. Also, since there is a greater turnover of lake water in spring due to higher inflow, it is expected that springtime river and rainfall values become less effective predictors of lake conditions.

During both sampling seasons, some sites in Meiliang Bay had higher $\delta^{15}N$ values than river or rain sources, indicating fractionation during N cycling processes. Denitrification rates in Lake Taihu decrease with distance from inflow rivers (McCarthy et al., 2007), so high denitrification rates may explain locally high $\delta^{15}N$ values in Meiliang Bay. Samples from May were taken during initiation of the annual phytoplankton bloom, and isotopic discrimination during uptake may explain some of the high $\delta^{15}N$ values observed in spring in Meiliang Bay. However, primary production in Meiliang Bay appears to be limited by phosphorus, and inorganic N is plentiful (McCarthy et al., 2007), indicating minimal isotopic discrimination of the DIN pool during N uptake. On the other hand, $\delta^{15}N$ values from East Taihu were within the range of natural soil NO_3^- and rainwater. This area lacks major inflow rivers and is used for aquaculture (Qin et al., 2007). More sampling of end-members, especially aquaculture sources, is needed to determine the major N sources to this part of the lake, but $\delta^{15}N$ values from this study indicate that sewage is minimally present.

Rain could be a major source of N to Lake Taihu and the surrounding watershed, considering the high NO_3^- concentrations observed in spring. Atmospheric NO_3^- derived from volatilization or nitrification of fertilizers usually is depleted in ^{15}N, with reported values as low as –15‰ (Wada & Hattori, 1991, Macko & Ostrom, 1994). Lightning is another source of isotopically light fixed atmospheric N, with $\delta^{15}N$ of NO_3^- values reported from 0 to –15‰ (Macko & Ostrom, 1994). Seasonal variation in $\delta^{15}N$ of NO_3^- in rain has been attributed to changes in atmospheric NO_3^- sources, with the dominant sources in winter and summer being fossil fuel combustion ($\delta^{15}N$ around 0‰) and soil ($\delta^{15}N$ around –5‰) NO_3^-, respectively (Freyer, 1991; Hastings et al., 2003). The $\delta^{15}N$ of NO_3 in rain observed at Lake Taihu (4.4 to 6‰) is more enriched than these expected sources, implying either a greater influence of fossil fuel combustion than previously observed, or a progressive isotopic enrichment of atmospheric fixed N as rain clouds move east over the European and Asian land masses. Enriched rainwater N also complicates the use of stable N isotopes to identify N sources, since the observed rainwater $\delta^{15}N$ values fall within the range of natural soil N. Factors influencing the N concentration and isotopic composition of rain in the Lake Taihu area warrant further study, especially since nitrogen oxides contribute to the formation of acid rain and since it represents a non-point source of pollution that is difficult to attenuate.

This study indicates that seasonal fertilizer application in the Lake Taihu watershed and rain are the major sources of NO_3^- in the summer months. In the winter months, sewage is a major source of NO_3^- to urbanized areas of the lake. This finding raises important questions about water quality, since most drinking water in China is derived from rivers and lakes. Wastewater discharge into these water bodies has been linked to illnesses such as typhoid and other bacterial or parasitic infections in China (Wu et al., 1999). Future studies should include N loading estimates, quantification of sewage and fertilizer discharge, and greater spatial and temporal coverage of $\delta^{15}N$ of NO_3^- measurements, especially rainfall, in order to completely characterize the NO_3^- budget for this system. An annual record of the isotopic composition of NO_3^- in lake water combined with a complete water and N budget would be a useful next step in the characterization of N cycling and sources in this ecosystem.

Acknowledgements This project was supported by National Natural Science Foundation of China (No. 40371104) and National Science Foundation of the United States. We thank the many faculty members, staff, and students at TLLER who assisted with project coordination and sampling. Patricia Garlough of the University of Texas Marine Science Institute assisted with stable isotopic analyses. The manuscript benefited from the comments of three anonymous reviewers.

References

Bedard-Haughn, A., J. W. van Groenigen & C. van Kessel, 2003. Tracing ^{15}N through landscapes: potential uses and precautions. Journal of Hydrology 272: 175–190.

Chen, Y., B. Qin, K. Teubner & M. T. Dokulil, 2003. Long-term dynamics of phytoplankton assemblages: *Microcystis*-domination in Lake Taihu, a large shallow lake in China. Journal of Plankton Research 25: 445–453.

DeSimone, L. A. & B. L. Howes, 1996. Denitrification and nitrogen transport in a coastal aquifer receiving wastewater discharge. Environmental Science and Technology 30: 1152–1162.

Faure, G., 1986. Principles of Isotope Geology. Wiley & Sons Publishers, New York.

Freyer, H. D, 1991. Seasonal variation of $^{15}N/^{14}N$ ratios in atmospheric nitrate species. Tellus 43B: 30–44.

Guo, H. Y., X. R. Wang & J. G. Zhu, 2004. Quantification and index of non-point source pollution in Lake Taihu region with GIS. Environmental Geochemistry and Health 26: 147–156.

Hastings, M. G., D. M. Sigman & F. Lipschultz, 2003. Isotopic evidence for source changes of nitrate in rain at Bermuda. Journal of Geophysical Research-Atmospheres 108: doi: 10.1029/2003JD003789.

Jin, Z., C. Yingxu, F. Wang & N. Ogura, 2004. Detection of nitrate sources in urban groundwater by isotopic and chemical indicators, Hangzhou City, China. Environmental Geology 45: 1017–1024.

Kellman, L. & C. Hillaire-Marcel, 1998. Nitrate cycling in streams: using natural abundance of $NO_3^-\delta^{15}N$ to

measure in-situ denitrification. Biogeochemistry 43: 273–292.
Macko, S. A. & N. E. Ostrom, 1994. Pollution studies using stable isotopes. In Lajtha, K. & R. H. Michener (eds), Stable isotopes in ecology and environmental science. Blackwell Scientific Publications, Oxford, 45–62.
Mariotti, A., 1984. Natural ^{15}N abundance measurements and atmospheric nitrogen standard calibration. Nature 311: 251–252.
McCarthy, M. J., P. J. Lavrentyev, L. Yang, Y. Chen, L. Zhang, B. Qin & W. S. Gardner, 2007. Nitrogen dynamics and microbial food web structure during a summer cyanobacterial bloom in a subtropical, shallow, well-mixed, eutrophic lake (Lake Taihu, China). Hydrobiologia 581: 195–207.
McClelland, J. W. & I. Valiela, 1998. Linking nitrogen in estuarine producers to land-derived sources. Limnology and Oceanography 43: 577–585.
Peterson, B. J. & B. Fry, 1987. Stable isotopes in ecosystem studies. Annual Review of Ecology and Systematics 18: 293–320.
Pu, P. & J. Yan, 1998. Lake Taihu - a large shallow lake in the East China Plain. Journal of Lake Sciences 10 (supplement): 1–12 (in Chinese).
Qin, B., P. Xu, Q. Wu & L. Luo, 2007. Environmental issues of Lake Taihu, China. Hydrobiologia 581: 3–14.
Sigman, D. M., K. L. Casciotti, M. Andreani, C. Barford, M. Galanter & J. K. Böhlke, 2001. A bacterial method for the nitrogen isotopic analysis of nitrate in seawater and freshwater. Analytical Chemistry 73: 4145–4153.
Wada, E. & A. Hattori, 1991. Nitrogen in the Sea: Forms, Abundances, and Rate Processes. CRC Press, Boca Raton.
Wu, C., C. Maurer, Y. Wang, S. Xue & D. L. Davis, 1999. Water pollution and human health in China. Environmental Health Perspectives 107: 251–256.

Hydrobiologia (2007) 581:141–150
DOI 10.1007/s10750-006-0523-3

EUTROPHICATION IN LAKES

Distribution and chemical fractionation of heavy metals in recent sediments from Lake Taihu, China

Ji Shen · Enfeng Liu · Yuxin Zhu · Shouyun Hu · Wenchuan Qu

Abstract Lake Taihu is one of the most contaminated lakes in China. Surface sediment data show that the northern area of the Lake has the worst heavy metals pollution, and high heavy metal concentrations were attributed to discharge of untreated and partially treated industrial waste water from cities to the north of the lake. To study geochemical features and pollution history of heavy metals, total content and chemical fractionations of Cu, Fe, Mn, Ni, Pb, and Zn were analyzed for core sediments from western Lake Taihu using the speciation extraction procedure, proposed by the Commission of the European Communities Bureau of Reference (BCR), together with grain size and organic carbon measurements. Results show that sediments are composed of organic-poor clayey-fine silts for Cores MS and DLS, and have similar geochemical features shown by heavy metals. Cu, Fe, Ni, and Zn mainly are associated with the residue fraction, Mn is concentrated in the exchangeable-carbonate and residue fractions, and Pb is concentrated in the Fe–Mn oxide fraction and organic-sulfide fraction. The fractions of Ni, Pb, and Zn bound to Fe–Mn oxide show significant correlations with Mn from the Fe–Mn oxide fraction, and the organic-sulfide fractions of Cu, Mn, Ni, Pb, and Zn are correlated with TOC. The increase of Cu, Mn, Ni, Pb and Zn content and percentage of extractable fractions in the upper layers of the sediments are correlated with anthropogenic input of heavy metals due to rapid industrial development. This coincides with rapid economic development in the Taihu basin since late 1970s. Heavy metals in the surface sediments have certain potential biological toxicity as shown by the higher SEM/AVS ratio.

Keywords Heavy metals · Sediments · Distribution · Fractionation · Lake Taihu

Introduction

Lake Taihu, the third largest fresh water lake in China, is situated in the center of the Changjiang River Delta (Qin et al., 2007) (Fig. 1). With intensive development of agriculture and industry in this densely populated area since the 1950s, especially since 1978, untreated or partially treated wastewater from industry, agriculture and sewage increasingly has been discharged into Lake Taihu, which induced degeneration of lake ecology and water quality. The quality of Lake Taihu

Guest editors: B. Qin, Z. Liu & K. Havens
Eutrophication of shallow lakes with special reference to Lake Taihu, China

J. Shen (✉) · E. Liu · Y. Zhu · S. Hu · W. Qu
Nanjing Institute of Geography and Limnology, Chinese Academy of Sciences, Nanjing 210008, China
e-mail: Jishen@niglas.ac.cn

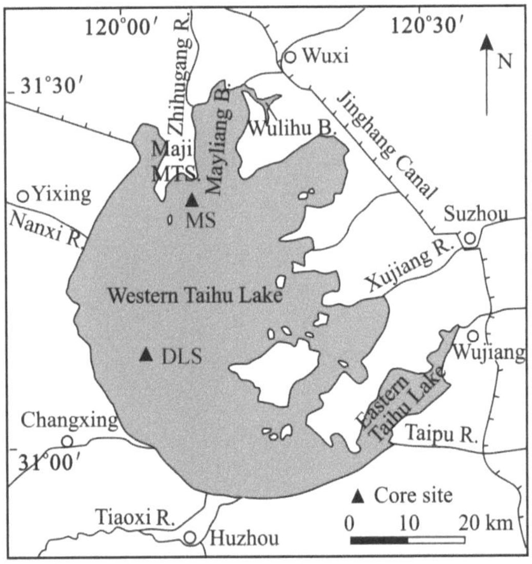

Fig. 1 Location of Lake Taihu and core site

water degraded 2–3 levels from the 1970s to 1990s (Lu, 1998), which caused about 1 billion RMB economic loss per year during the late 1980s, accounting for 3.12% of the local GNP (Ge, 1992).

As one of many environmental problems in Lake Taihu, heavy metal pollution has drawn great attention in recent years. A few studies have been conducted on heavy metal pollution of surface sediments (Dai & Sun, 2001; Qu et al., 2001; Yuan et al., 2002a), which mainly address heavy metal total concentration. However, work on chemical fractionation of heavy metals and pollution history of the sediments are rare (Wang et al., 2002; Rose et al., 2004). There is a lack of systemic investigation on distribution and chemical forms of heavy metals in sediment cores, which are an important record of historical pollution, biogeochemical recycling and toxicity for heavy metals (Luoma, 1983).

In the present study, two sediment cores in western Lake Taihu were obtained. Total concentration of heavy metals and their chemical forms deduced from the BCR method (proposed by the Community of the European Bureau of Reference) (Quevauviller et al., 1997) in the sediment cores are presented. The aim of this study is to understand geochemical characteristics of the heavy metals and pollution features in the sediments and their correlations with economic development and human activities in the Lake Taihu catchment.

Material and methods

Sample collection

According to sedimentary features and the degree of pollution in the sediments (Sun & Huang, 1993; Qu et al., 2001), two sediment cores (MS and DLS) were obtained in northern and southern Lake Taihu, respectively, in October 2002 using a gravity corer (Fig. 1). The sediment exhibited no bioturbation and can be used to reconstruct environmental evolution. Sediments were sliced into sections at 0.5 cm intervals. Samples were freeze dried using a LABCONCO freeze drier. Physical and chemical data from the upper 13 and 7 cm sediments for Cores MS and DLS, respectively, were used.

Analysis methods

TOC concentration in samples was determined using a CNHO-S elemental analyzer (Leeman Company). Particle size was measured with a Mastersizer–2000 laser diffraction particle size analyzer (Malvern Instruments Ltd.). Analysis errors for experiments were within ±5%.

For heavy metal analysis, the HCl–HNO_3–HF digestion method was applied. About 125 mg of sediment, ground in an agate mortar to pass 100-μm sieve, was digested with 6.0 ml HNO_3, 0.5 ml HCl, and 3.0 ml HF at 180 ± 5°C for 15 min in a microwave oven (Berghof MWS–3 Digester). Residue was dissolved with 0.5 ml $HClO_4$ by braising in a heating block at <200°C and diluted to 25 ml with double-distilled deionized water. The solution was analyzed for heavy metals and major elements using inductively coupled plasma-atomic emission spectrometry (ICP-AES).

Chemical fractionations of Cu, Fe, Mn, Ni, Pb and Zn were determined using the BCR sequential extraction procedure (Quevauviller et al., 1997; Tokalioğlu et al., 2000) (Fig. 2). Only the three extractable fractions (exchangeable-carbonate, Fe–Mn oxide, and organic-sulfide fraction) were analyzed. Heavy metals concentrations in the extracted solution were analyzed by ICP-AES.

Sediments were dated using the ^{137}Cs methods described by Rosales-Hoza et al. (2003).

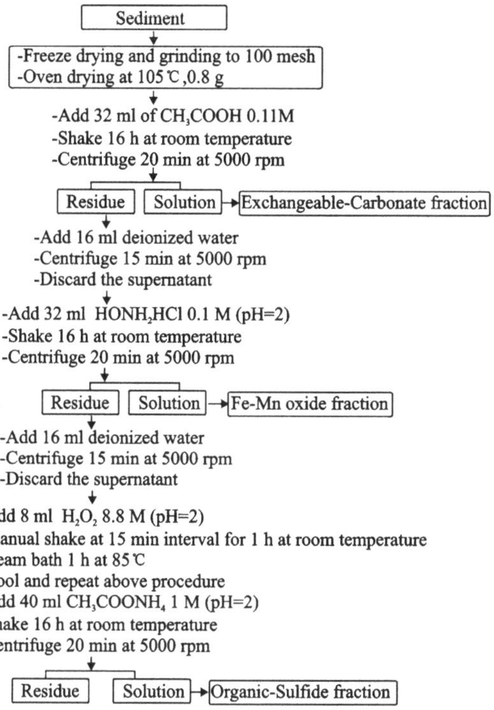

Fig. 2 Heavy metals sequence extraction procedure of BCR method

Accuracy of heavy metals analysis method

Four parallel analyses of standard sediment reference material (GSD–9 from Institute of Geophysical and Geochemical Exploration, Chinese Academy of Geological Sciences) were conducted to verify the accuracy of the heavy metal digestion method. The maximum error of all results was <3% of the standard value, showing that the HCl–HNO_3–HF digestion method and ICP-AES analysis have reasonable precision and reproducibility.

Three lacustrine samples from Lake Taihu were measured in parallel analyses to verify the accuracy of the BCR procedure due to the lack of suitable standard reference material. Blanks were also used for background correction. Results are shown in Table 1. Maximum errors of the results are within 10%, except 23.0 and 18.6% for the exchangeable-carbonate and organic-sulfide fractions of Pb in samples MS–5 and TH-2, respectively, and 24.4% for the Fe–Mn oxide fraction of Zn in sample MS-5. Errors for Pb and Zn may be related to contamination of certain procedure during the sequential extraction process, such as losing of the liquid during shaking etc (Tokalioğlu et al., 2000).

Table 1 Error analysis of BCR sequential extraction method for sedimentary samples of Lake Taihu

Elements		MS-5				TH-1				TH-2			
		AC	S.D.	CV	ME	AC	S.D.	CV	ME	AC	S.D.	CV	ME
Cu	F1	3.05	0.07	2.19	2.52	2.01	0.05	2.35	2.71	3.54	0.03	0.75	0.85
	F2	3.62	0.01	0.16	0.18	2.97	0.05	1.52	1.57	3.21	0.14	4.21	4.36
	F3	5.28	0.16	3.08	3.35	5.05	0.12	2.28	2.31	10.0	0.30	3.00	3.35
Fe	F1	250	10.4	4.17	4.79	595	14.2	2.38	2.45	1590	17.4	1.10	1.17
	F2	5070	41.2	0.81	0.90	5570	66.7	1.20	1.23	8390	246	2.93	3.38
	F3	989	53.4	5.40	5.81	1310	46.7	3.55	3.77	2190	175	7.99	8.52
Mn	F1	302	4.48	1.48	1.71	192	2.78	1.45	1.52	823	4.34	0.53	0.59
	F2	92.1	3.10	3.37	3.83	49.3	0.50	1.01	1.11	208	4.01	1.93	1.97
	F3	36.2	1.35	3.73	4.09	33.1	1.00	3.03	3.37	64.8	4.74	7.31	7.32
Ni	F1	3.39	0.06	1.62	1.67	2.10	0.07	3.45	4.00	2.96	0.01	0.39	0.45
	F2	4.00	0.06	1.38	1.58	2.86	0.06	1.92	2.21	3.49	0.19	5.48	6.11
	F3	4.59	0.11	2.30	2.40	3.46	0.04	1.17	1.35	4.12	0.06	1.38	1.54
Pb	F1	1.59	0.34	21.3	23.0	0.82	0.07	8.15	9.89	0.42	0.00	0.81	0.55
	F2	8.97	0.51	5.72	6.39	3.74	0.08	2.09	2.41	2.96	0.09	3.04	3.04
	F3	14.19	0.51	3.62	4.13	9.08	0.24	2.59	2.61	11.9	1.95	16.41	18.6
Zn	F1	12.3	0.56	4.56	5.19	3.27	0.07	2.04	2.21	4.12	0.15	3.52	3.56
	F2	16.8	3.59	21.4	24.4	4.48	0.12	2.60	2.83	5.49	0.20	3.70	4.25
	F3	9.10	0.48	5.30	5.86	3.68	0.33	8.89	9.96	6.40	0.21	3.21	3.28

AC, average content (mg kg^{-1}); S.D, standard deviation; CV, coefficient of variation; ME, maximum error

F1, exchangeable-carbonate fraction; F2, Fe–Mn oxide fraction; F3, organic-sulfide fraction

Due to good precision and reproducibility for the digestion and BCR methods, the residue fraction concentration of heavy metals can be represented by the difference of total concentration and extractable fractions of each heavy metal.

Results

Dating results

The chronological framework of Cores MS and DLS were determined by ^{137}Cs dating. Average sedimentation rates were 1.7 mm $year^{-1}$ and 0.9 mm $year^{-1}$, respectively, for the two cores (Liu et al., 2005). These results are similar to those from other researchers (Sun & Huang, 1993). The ages for Core MS and DLS are shown in Fig. 3.

Heavy metals, grain size and TOC in Cores MS and DLS

Grain size and TOC in the cores are shown in Fig. 3. Sediments were dominated by silt (4–64 μm) and clay (<4 μm). The grain size becomes finer gradually from 14 to 4 cm depth and coarser from 4 cm to the surface in Core MS; and it becomes finer gradually from 7 to 2.5 cm and coarser from 2.5 cm to the surface in Core DLS.

TOC content was <2% for most of the surface sediments (Sun & Huang, 1993). TOC was around 0.62–1.35% in Cores MS and DLS and increases near the surface (Fig. 3).

Concentrations of the major elements (Al, Mg, K and etc) increased from 14 to 4 cm and decreased from 4 cm to the surface in Core MS, having consistent variation with the finer fraction (clay and fine silt). For Core DLS, major elements increased from 7 to 2.5 cm and decreased from 2.5 cm to the surface, also having consistent variation with the finer fraction.

Heavy metals, such as Mn, Ni, Pb and Zn, also had similar variation patterns as major elements in the 14–4 cm interval for Core MS, and 7–2.5 cm for Core DLS, except for higher Mn and Fe in the middle part of the core. In contrast to major elements and the finer fraction, heavy metals generally increased in the top 4–0 cm and 2.5–0 cm for the two sediment cores.

Chemical forms of heavy metals in Cores MS and DLS

Chemical fractionation compositions of Cu, Fe, Mn, Ni, Pb and Zn determined by BCR are illustrated in Fig. 4. Cu, Fe, Ni and Zn were

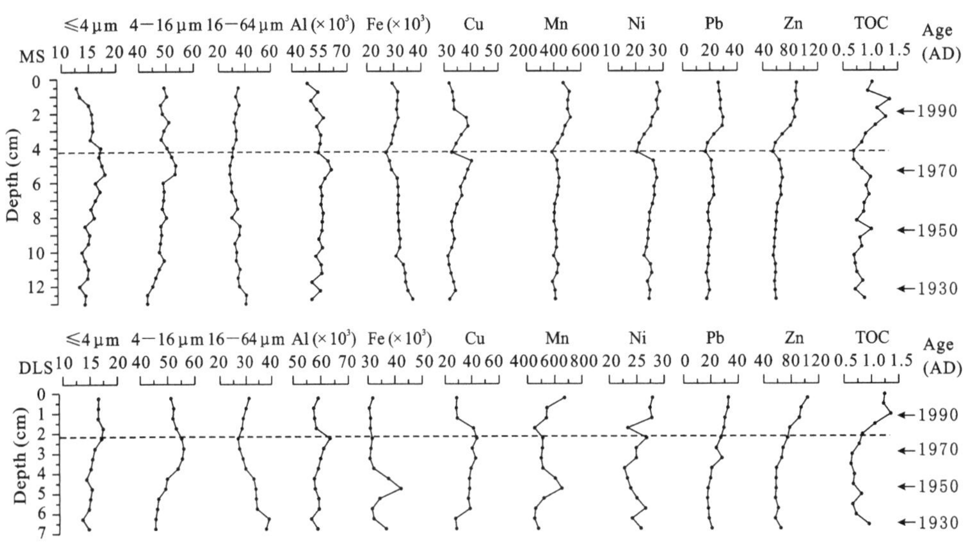

Fig. 3 The variation of grain size (%), heavy metals (mg kg^{-1}) and TOC (%) in Cores MS and DLS

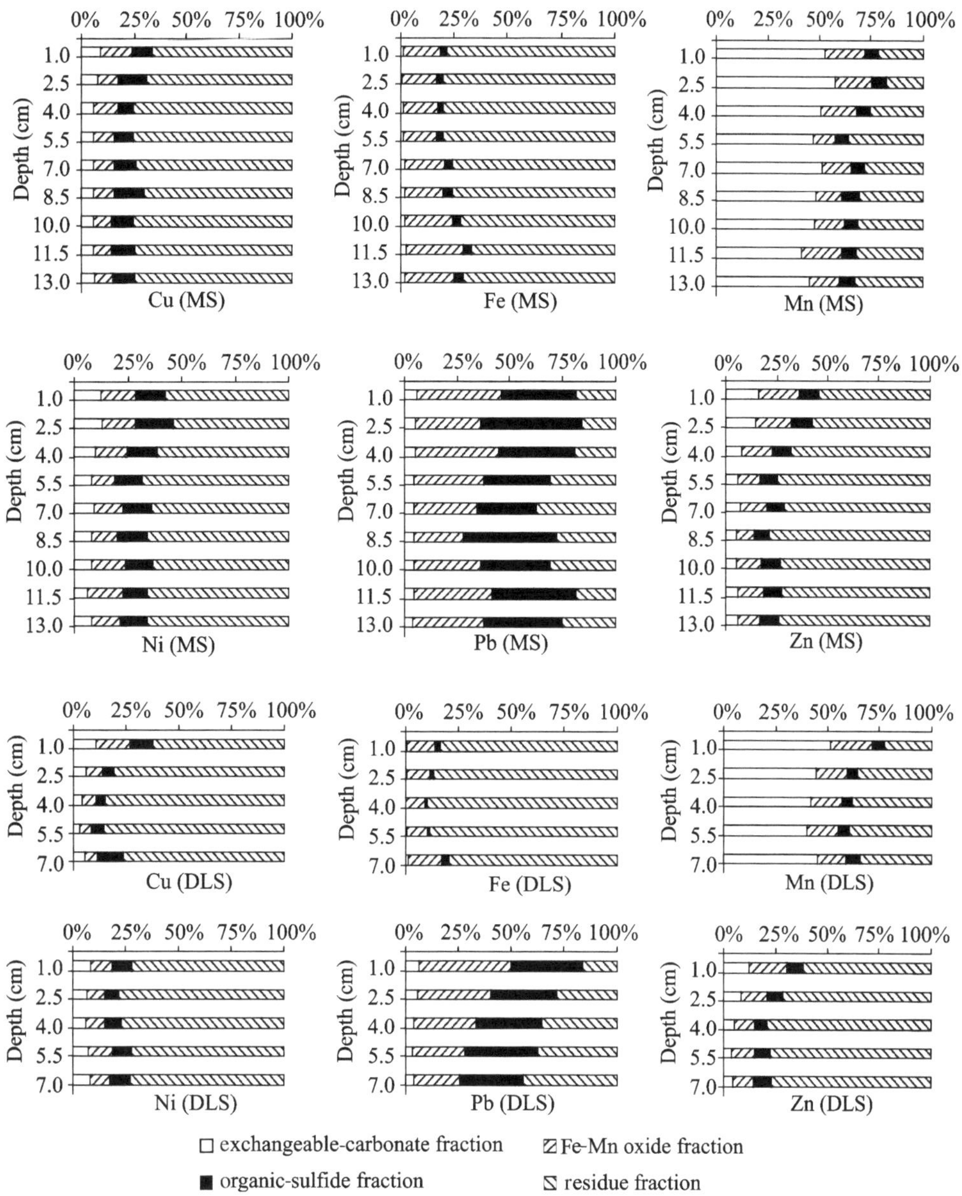

Fig. 4 The chemical fractionation percentages of Cu, Fe, Mn, Ni, Pb and Zn in the sediments of Cores MS and DLS (with 1.5 cm interval)

concentrated in the residue fraction, and average percentages were 72.9%, 75.6%, 62.9% and 69.7% in Core MS, and 77.8%, 85.4%, 72.9% and 73.3% in Core DLS, respectively. Mn was present in the exchangeable-carbonate fraction, and its percentages were 48.9 and 44.3% in Cores MS and DLS, respectively, whereas Pb was present in both the Fe–Mn oxide fraction and organic-sulfide fraction.

Percentages of extractable fractions of Cu, Mn, Ni, Pb and Zn, including exchangeable-carbonate, Fe–Mn oxide and organic-sulfide fractions, had a consistent variation in the vertical sequence for both cores, which were stable at 14–4 and 7–2.5 cm, but increased from 4 to 0 cm and from 2.5 to 0 cm for Cores MS and DLS, respectively (Fig. 4). The percentages of extractable fractions of Cu, Mn, Ni, Pb and Zn in the top 4 and 2.5 cm

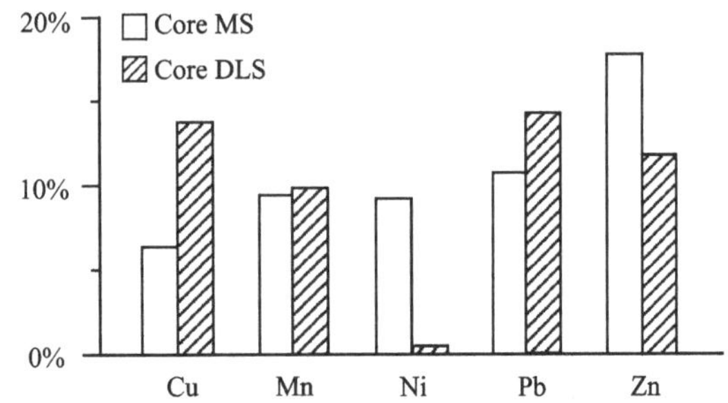

Fig. 5 The increased percentages of Cu, Mn, Ni, Pb and Zn in extractable fractions in the top 4 and 2.5 cm of Cores MS and DLS compared with that in the deeper layers

of Cores MS and DLS increased 0.5–17.8% comparing with that in the deeper layers (Fig. 5). The average percentage of Pb, Ni, Cu, Zn and Mn in the organic-sulfide fraction were 35.8%, 13.5%, 10.3%, 9.3% and 6.8% in Core MS, and 32.3%, 8.7%, 7.9%, 7.6% and 5.9% in Core DLS.

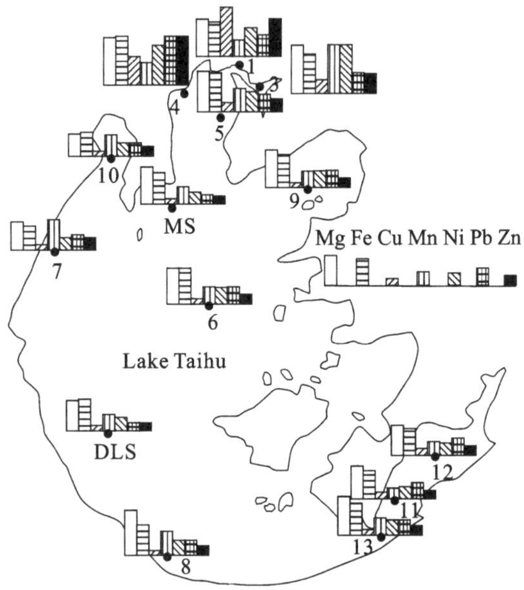

Fig. 6 The relative concentration of heavy metals in each site compared with the average of the 14 sites of Lake Taihu surface sediments (the data of site 1–13 from Qu et al., 2001)

Discussion

Pollution features of the surface sediment

There are two main sources for heavy metals in sediments: natural sources and anthropogenic sources. Heavy metals content commonly has good correlation with major elements and finer fractions in unpolluted sediments, but human pollution can alter this relationship (Abraham, 1998; Covelli & Fontolan, 1997). Therefore, variation patterns of major elements, heavy metals and their correlation with grain size composition can be used to evaluate heavy metals pollution (Rubio et al., 2000). Al, Fe, Mg and K mainly come from natural versus anthropogenic sources and can be used as reference elements for heavy metal pollution evaluation (Covelli & Fontolan, 1997; Soto-Jiménez & Páez-Osuna, 2001).

Heavy metals in surface sediments of Core MS and DLS and the other 12 sites from Qu et al. (2001) were analyzed for spatial distribution of heavy metals in Lake Taihu. Contents of Fe, Mg and other major elements in the sediments of the 14 sites are similar, but Cu, Mn, Ni, Pb and Zn in the sediments of sites 1–5 have higher concentrations versus Fe and Mg (Al and K not determined at sites 1–13) than other sites (Fig. 6). Concentrations of the three extractable fractions of Cu, Ni and Zn in surface sediments also were higher in the northern lake compared with other areas (Wang et al., 2002). These indicate that sediments in northern Lake Taihu, such as Meiliang Bay and Wuli Bay, are polluted with heavy metals. Pollution in the southern lake is much lighter (Qu et al., 2001; Wang et al., 2002). Larger amounts of waste water with higher concentrations of heavy metals from industries in Changzhou and Wuxi discharged into Wuli and Meiliang Bays via the Zhihugang and Liangxihe is the source of heavy metal pollution in the northern Lake Taihu (Huang et al., 2001; Qu et al., 2001).

Geochemical features of the core sediment

For total heavy metal concentrations, Cu, Mn, Ni, Pb and Zn have similar variation with major elements (Al, Mg and K) and finer fractions before the late 1970s. However, there is a sharp contrast between Cu, Mn, Ni, Pb and Zn with major elements and finer fractions after the late 1970s for Cores MS and DLS. This indicates that heavy metals in the upper sediments may come primarily from anthropogenic sources (Covelli & Fontolan, 1997; Abraham, 1998). High content of Fe and Mn in the lower part of Core DLS may be related to redox conditions since little industrial waste water was discharged into southern Lake Taihu during the 1950s.

Chemical fractionation compositions of Cu, Fe, Mn, Ni, Pb and Zn were similar for both cores, and similar to those in surface sediments (Wang et al., 2002), but different from other sediments (Tokalioğlu et al., 2000; Tüzen, 2003). Percentages of exchangeable-carbonate fraction of Cu, Fe, Ni, Pb and Zn were <10% and are influenced by physico-chemical properties of the sediments. The exchangeable fraction of heavy metals in sediments generally is low and has little contribution to the percentage of the exchangeable-carbonate fraction (Alvarez et al., 2001). The percentage of the exchangeable-carbonate fraction mainly is influenced by adsorption quantity of the carbonate fraction since the carbonates in sediments contain significant heavy metals (Tokalioğlu et al., 2000). However, the carbonate content of Lake Taihu is low (Yuan et al., 2002b), which may explain the relatively low percentage of the exchangeable-carbonate fraction of Cu, Fe, Ni, Pb and Zn in the sediments. But, the percentage of Mn in the exchangeable-carbonate form was up to 40.9–57.9% (Wang et al., 2002). This may be related to the chemical features of Mn (Liu et al., 1984; Tokalioğlu et al., 2000). Mn can easily be adsorbed on the surface of fine granules than other ions; the oxidation of Mn^{2+} is more relaxed than Fe^{2+} and can be precipitated until higher pH (8.5–10), which is lower than 8.1 in Lake Taihu sediments (Sun & Huang, 1993). So Mn can be easily associated with carbonates (Liu et al., 1984). Another feature is that the radius of Mn^{2+} (0.91) is similar to that of Ca^{2+} (1.08) and Mg^{2+} (0.8), so the replacement of Ca^{2+} and Mg^{2+} by Mn^{2+} is easier. All these may be the reasons for higher Mn content in the exchangeable-carbonate forms in the sediments.

The Fe–Mn oxide fraction of heavy metals, particularly those bound by Fe–Mn oxide or precipitated as hydroxide, is an electrovalent bond form and can be re-released into the water under reductive conditions. The average percentage of Cu, Fe, Mn, Ni and Zn in this fraction is 10–19.4%, but Pb is as high as 34.2%. The carbonate form of Pb is loosely bound and may change with environmental conditions (Li et al., 2001), but it can form compounds with Fe–Mn oxide easily. The higher content of Pb in the Fe–Mn oxide fraction in Lake Taihu sediment versus others (Tokalioğlu et al., 2000; Tüzen, 2003) may be related to the better oxidative environment in Lake Taihu sediments. The relationship between Cu, Ni, Pb, Zn of the Fe–Mn oxide fraction and content of Fe and Mn in the Fe–Mn oxide fraction are presented in Table 2. Ni, Pb and Zn display significant correlation with Mn for Core MS, and Cu, Pb and Zn also show good correlation with Mn for Core DLS. No significant relationship between Ni, Pb, Zn and Fe were found. This is consistent with (Oakley et al., 1981), where hydrous Mn oxides

Table 2 Correlation coefficients of the heavy metal chemical forms and with TOC

Correlation coefficients	Core MS					Core DLS				
	Cu	Mn	Ni	Pb	Zn	Cu	Mn	Ni	Pb	Zn
r_1	−0.77*	0.04	0.24	−0.52	−0.41	−0.25	−0.12	0.33	−0.07	0.08
r_2	0.42	1.00	0.89**	0.77*	0.85**	0.87	1.00	−0.03	0.89	0.95*
r_3	0.74	0.81**	0.89**	0.90**	0.81**	0.80	0.68	0.76	0.77	0.94*

** Correlation is significant at 0.01 level; *Correlation is significant at 0.05 level

r_1, the correlation coefficients of Fe–Mn oxide fraction concentrations of Cu, Mn, Ni, Pb, Zn and that of Fe

r_2, the correlation coefficients of Fe–Mn oxide fraction concentrations of Cu, Mn, Ni, Pb, Zn and that of Mn

r_3, the correlation coefficients of organic-sulfide fraction concentrations of Cu, Mn, Ni, Pb, Zn and TOC exhibited more extensive isomorphic substitution than amorphous Fe oxides and showed greater conditional equilibrium constants for heavy metals than Fe oxides.

Heavy metals bound with sulfide can be re-released into the water under oxidizing conditions. Sulfide in Lake Taihu sediments is low due to seasonal variation of redox conditions and the generally weak oxidation or reduction conditions in surface sediment (Sun & Huang, 1993). The correlation between TOC and the organic-sulfide fractions of Cu, Mn, Ni, Pb and Zn was high for both cores (Table 2). This indicates that contents of the organic-sulfide fraction of Cu, Mn, Ni, Pb and Zn are influenced mainly by organic content in Lake Taihu. In aquatic systems, organic matter exhibits high selectivity for divalent ions. The average percentages of Pb, Ni, Cu, Zn and Mn in the organic-sulfide fraction are consistent with the order of binding strength for metal ions onto organic matter (Cu > Pb > Ni > Zn > Mn) (Föstner & Wittmann, 1988). Copper mainly exists in sulfide form in nature (Li et al., 2001; Hlavay & Polyák, 2002), and the low percentage in the organic-sulfide fraction in the cores may be due to low sulfide content.

Seasonal variation of redox conditions between weak oxidation or reduction and mineralization of organic matter in Lake Taihu surface sediment can influence the chemical forms of heavy metals in the exchangeable-carbonate, Fe–Mn oxide and organic-sulfide fractions (Saeki et al., 1993; Tack et al., 1996). Higher content of the extractable fractions of Cu, Mn, Ni, Pb and Zn in the upper part of the sediment may have potential toxicity. The ratio of SEM (amount of simultaneously extracted metals when AVS is extracted) to AVS (acid volatile sulfides, the concentration of sulfide extracted by 1 M cold hydrochloric acid) can be used to evaluate heavy metals toxicity in sediments (Edward, 1998; Li et al., 2004). SEM/AVS in northern Lake Taihu sediment ranges from 1.3 to 5.9, so heavy metals have potential toxicity (Edward, 1998; Li et al., 2004). This also indicates that, although the organic-sulfide fraction of the heavy metals mainly is influenced by organic content, sulfide also has some influence on heavy metal activity due to seasonal variation of redox conditions at the water-sediment interface.

Pollution history of the sediment

Chemical fractionation of heavy metals in the cores also can be used to evaluate pollution in the lake. Extractable forms of heavy metals from anthropogenic sources have a higher proportion of heavy metal content than natural sources (Föstner & Wittmann, 1988; Li et al., 2001). Lower concentrations and percentages of extractable forms of Cu, Mn, Ni, Pb and Zn in sediments before the late 1970s suggests that heavy metals mainly came from natural sources, which is consistent with underdeveloped industry in the catchment; higher ones in recent years suggest that heavy metals were anthropogenic partially.

Wuxi is a typical city representative of economic development in the catchment. The gross domestic product of Wuxi increased annually by 30% since the 1980s, accompanied by a large amount of waste water discharge. The main industries in the region are chemical, paper, textile, leather and electroplating, which account for 93.8% of the wastewater discharge (Huang et al., 2001). Wastewater from the industries may be the main source of Cu, Mn, Ni, Pb and Zn in the sediments. Pollution history in the two cores is coincident with distinct heavy metals pollution in Meiliang Bay (Rose et al., 2004). Pollution history of heavy metals recorded in the sediments of Lake Taihu also is consistent with rapid economic development characteristics of the Taihu lake catchment.

Conclusion

Cu, Mn, Ni, Pb and Zn have higher content in surface sediments of northern Lake Taihu and were anthropogenic more than other areas of the lake. The main chemical form for Cu, Fe, Ni and Zn in the sediments was the residual fraction. Mn is mainly present in the exchangeable-carbonate and residue fractions. The organic-sulfide and Fe–Mn oxide fractions were important for Pb. Organic matter and Mn in the Fe–Mn oxide fraction play an important role in binding heavy metals in the sediments. High concentrations of Cu, Mn, Ni, Pb and Zn, and their extractable fractions in recent sediments indicate that heavy metals were anthropogenic and have potential toxicity. Heavy metals pollution history was recent for both cores, which coincides with rapid economic development in the Taihu catchment.

Acknowledgements The work was financially supported by Chinese Academy of Sciences (No. KZCX1-SW-12, KZCX3-SW-348). We would like to thank all the people who participated in the sample collection and experiments analyses. Special thanks to Weilan Xia for dating of the two cores. We also thank Hongxi Pan and Guangwei Zhu for their help with the experiments and Sumin Wang for his valuable suggestions and comments.

References

Abraham, J., 1998. Spatial distribution of major and trace elements in shallow reservoir sediments: an example from Lake Waco, Texas. Environmental Geology 36: 349–363.

Alvarez, M., M. E. Malla & D. A. Batistoni, 2001. Comparative assessment of two sequential chemical extraction schemes for the fractionation of cadmium, chromium, lead and zinc in surface coastal sediments. Fresenius Journal of Analytical Chemistry 369: 81–90.

Covelli, S. & G. Fontolan, 1997. Application of a normalization procedure in determining regional geochemical baselines. Environmental Geology 30: 34–45.

Dai, X. & C. Sun, 2001. The characteristics of heavy metals distribution and pollution in sediment from Lake Taihu. Shanghai Environmental Sciences 20: 71–74 (in Chinese).

Edward, R., 1998. Predicting the toxicity of sediment–associated trace metals with SEM:AVS and dry weight-normalized concentrations: A critical comparison. Environmental Toxicology and Chemistry 17: 972–974.

Föstner, U. & G. T. W. Wittmann, 1988, Metal Pollution in the Aquatic Environment. China Ocean Press, Beijing (Chinese translation edition translated by Wang, Z. & Z. Yao).

Ge, J., 1992. Lost assessment of Lake Taihu caused by water pollution. Environmental Science 2: 68–72 (in Chinese).

Hlavay, J. & K. Polyák, 2002. Investigation on the pollution sources of bottom sediments in the Lake Balaton. Microchemical Journal 73: 65–78.

Huang, Y., C. Fan, P. Pu, J. Jiang & Q. Dai, 2001. The Water Environment and Pollution Control of Lake Taihu. The Science Press, Beijing (Chinese).

Li, J., J. Song & X. Wang, 2004. Acid volatile sulfides and simultaneous extracted metals in the surficial sediments of Wuli Lake, Lake Taihu, Journal of Lake Sciences 16: 77–80 (in Chinese).

Li, X., Z. Shen, W. H. Wai Onyx & Y. Li., 2001. Chemical forms of Pb, Zn and Cu in the sediment profiles of the Pearl River Estuary. Marine Pollution Bulletin 42: 215–223.

Liu, E., J. Shen & Y. Zhu, 2005. Geochemical Records and Comparative Study of the Sediments in the stern Lake Taihu. Scientia Geographica Sinica 25: 102–107 (in Chinese).

Liu, Y., L. Cao, Z. Li, H. Wang, T. Chu & J. Zhang, 1984. Element Geochemistry. The Science Press, Beijng (in Chinese).

Luoma, S. N., 1983. Bioavailability of trace metals to aquatic organisms – a review:Science of the Total Environment 28: 1–22.

Lu, Y., 1998. The agricultural non-point pollution in Taihu Basin and sustainable development. Environmental Science Trends 2: 1–4 (in Chinese).

Oakley, S. M., P. O. Nelson & K. L. Williamson, 1981. Model of trace-metal portioning in marine sediments: Environmental Science and Technology 15: 474–480.

Qin, B. Q., P. Z. Xu, Q. L. Wu, L.C. Luo & Y. L. Zhang, 2007. Environmental issues of Lake Taihu, China. Hydrobiologia 581: 3–14.

Qu, W., M. Dickman & S. Wang, 2001. Multivariate analysis of heavy metal and nutrient concentrations in sediments of Lake Taihu, China. Hydrobiologia 450: 83–89.

Quevauviller, Ph., G. Rauret, J. F. López-Sánchez, R. Rubio, A. Ure & H. Muntau, 1997. Certification of trace metal extractable contents in a sediment reference material (CRM 601) following a three-step sequential extraction procedure. Science of the Total Environment 205: 223–234.

Rosales-Hoza, L., A. B. Cundyb & J. L. Bahena-Manjarrez, 2003. Heavy metals in sediment cores from a tropical estuary affected by anthropogenic discharges: Coatzacoalcos estuary, Mexico. Estuarine, Coastal and Shelf Science 58: 117–126.

Rose, N. L., J. F. Boyle, Y. Du, C. Yi, X. Dai, P. G. Appleby, H. Bennion, S. Cai & L. Yu, 2004. Sedimentary evidence for changes in the pollution status of Taihu in the Jiangsu region of eastern China. Journal of Paleolimnology 32: 41–51.

Rubio, B., M. A. Nombela & F. Vilas, 2000. Geochemistry of major and trace elements in sediments of the Ria de Vigo (NW Spain): An assessment of metal pollution. Marine Pollution Bulletin 40: 968–980.

Saeki, K., M. Okazaki & S. Matsumoto, 1993. The chemical phase changes in heavy metals with drying and oxidation of the lake sediments. Water Research 27: 1243–1251.

Soto-Jiménez, M. F. & F. Páez-Osuna, 2001. Distribution and Normalization of Heavy Metal Concentrations in Mangrove and Lagoonal Sediments from Mazatlán Harbor (SE Gulf of California). Estuarine, Coastal and Shelf Science 53: 259–274.

Sun, S. & Y. Huang, 1993. Lake Taihu. China Ocean Press, Beijing (in Chinese).

Tack, F. M., O. W. J. J. Callewaert & M. G. Verloo, 1996. Metal solubility as a function of pH in a contaminated dredged sediment affected by oxidation. Environmental Pollution 91: 199–208.
Tokalioğlu, Ş., Ş. Kartal & E. Latif, 2000. Determination of heavy metals and their speciation in lake sediments by flame atomic absorption spectrometry after a four-stage sequential extraction procedure. Analytica Chimica Acta 413: 33–40.
Tüzen, M., 2003. Determination of trace metals in the River Yeşilırmak sediments in Toka, Turkey using sequential extraction procedure. Microchemical Journal 74: 105–110.
Wang, H., C. Wang & Z. Wang, 2002. Speciations of heavy metals in surface sediment of Lake Taihu. Environmental Chemistry 21: 430–435 (in Chinese).
Yuan, X., J. Chen, J. Ji, Y. Tao & W. Qu, 2002b. Compositions and the formative conditions of sediments in Lake Taihu,Eastern China. Journal of Nanjing University (Nature Sciences) 38: 756–765 (in Chinese).
Yuan, X., J. Chen, J. Ji, Y. Tao & R. Wang, 2002a. Characteristics and environmental changes of pollution elements in Taihu sediments and soils near the lake. Acta Sedimentologica Sinica 20: 427–434 (in Chinese).

Hydrobiologia (2007) 581:151–159
DOI 10.1007/s10750-006-0495-3

EUTROPHICATION IN LAKES

Vegetation and soil properties in restored wetlands near Lake Taihu, China

Jinwei Lu · Hongjun Wang · Weidong Wang · Chengqing Yin

Abstract Riparian wetlands are important components of the lake ecosystem, and they play essential roles in maintaining system health. Remediation of degraded lakeshore wetlands is an essential component of lake restoration. A study was conducted to investigate the restoration of lakeshore wetlands, which were converted to rice fields and then abandoned for 2, 5, 10 and 15 years, near Lake Taihu. Soil samples (0–20 cm and 20–40 cm) were taken and plant species were investigated. The carbon content in the soil had increased significantly, rising from 0.71% to 1.85% between 2 and 15 years. Organic matter accumulation improved soil texture, and water stable aggregate content (>0.25 mm) and soil porosity increased. Total nitrogen in the soil increased from 0.06% to 0.13%, and total Kjeldahl nitrogen increased from 124.4 mg kg^{-1} to 351.5 mg kg^{-1}. Total phosphorus in the soil increased from 0.045% to 0.071%, and the Olsen-*P* value increased from 5.13 mg kg^{-1} to 16.0 mg kg^{-1}. Results showed that phosphorous did not increase as much as nitrogen. In the vegetation restoration process, plant species composition moved towards a natural wetland community, and spatial heterogeneity and landscape diversity increased. The richness of plant biodiversity increased rapidly in the first 2 years, then more slowly in later restoration stages. The wetlands recovery process may be complicated by interactions of biota and soil and hydrological conditions.

Keywords Lakeshore · Wetlands · Restoration · Dynamics · Process · Vegetation · Soil nutrients · Lake Taihu

Guest editors: B. Qin, Z. Liu & K. Havens
Eutrophication of shallow lakes with special reference to Lake Taihu, China

J. Lu · H. Wang · W. Wang · C. Yin (✉)
SKLEAC, the Research Center for Eco-Environmental Sciences, Chinese Academy of Sciences, Beijing 100085, China
e-mail: cqyin@263.net

J. Lu · H. Wang
Graduate School of the Chinese Academy Sciences, Beijing 100039, China

C. Yin
Northeast Institute of Geography and Agricultural Ecology, CAS, Jinlin Province, Changchun 130000, China

Introduction

Eutrophication in lakes and reservoirs is a worldwide environmental problem (Vollenweider, 1992), particularly since the 1980s in China due to rapid economic development (Jin et al., 1990). Riparian wetlands are important components of the lake ecosystem, and they play essential roles for maintaining system health (Naiman & Décamps, 1997). These wetlands are effective buffers, retaining water and nutrients, improving water quality, and providing diverse and dynamic habitats for wildlife (Naiman et al., 1994; Yin & Lan, 1995; Tabacchi et al., 1998; Coveney et al., 2002). Therefore, remediation and restoration of lakeshore wetlands is an essential component of lake restoration.

Due to historical lack of recognition of the ecological value of riparian wetlands, many wetlands were destroyed in the last 50 years. Most of the destruction is a result of conversion to agriculture and other economic uses (National Bureau of Statistics of China, 2001; Chen & Chen, 2002). Approximately 160 km^2 of riparian areas are cultivated along Lake Taihu (Hu & Jiang, 2004; Xiong, 2004). Losing the buffering capacity of riparian zones enhanced eutrophication, and, in recent years, wetland restoration has been implemented extensively in China (Chen, 2001). Hydrology, soil, and vegetation are interconnected components of wetlands and undergo dynamic changes during restoration. This paper describes soil and vegetation changes in the riparian wetlands and discusses the effectiveness and feasibility of natural restoration. This information will be useful for future wetland restoration efforts.

Study area, materials and methods

Description of study area

The study area is located at the lakeshore of northeast Lake Taihu. The land in this region is flat with adjacent undulating steep hills. Natural riparian wetlands were changed into paddy fields and other agricultural uses from the 1960's to 1980's. After 1988, with a local policy for protecting the lake and an increase in labor costs, some paddy fields were abandoned. Vegetation recolonized these paddy fields, and the weir was destroyed, allowing water to move freely into the wetlands. In investigation years 2003 and 2004, the lake water table was low, so the wetlands were not inundated.

Field investigation and sampling

Experiment design

Five sections of paddy fields, with an area ~0.15 ha and an age series of 2-, 5-, 10-, and 15-years since abandonment, and a reference wetland were selected in the riparian areas of Lake Taihu in the Dafu Town of Wuxi (Fig. 1). Since it was difficult to find an undisturbed natural wetland as a reference site, an uncultivated and less disturbed 30-year old riparian wetland was chosen. All of the abandoned lands in the study areas were flat and subject to similar hydrological conditions. The age of abandoned sites were

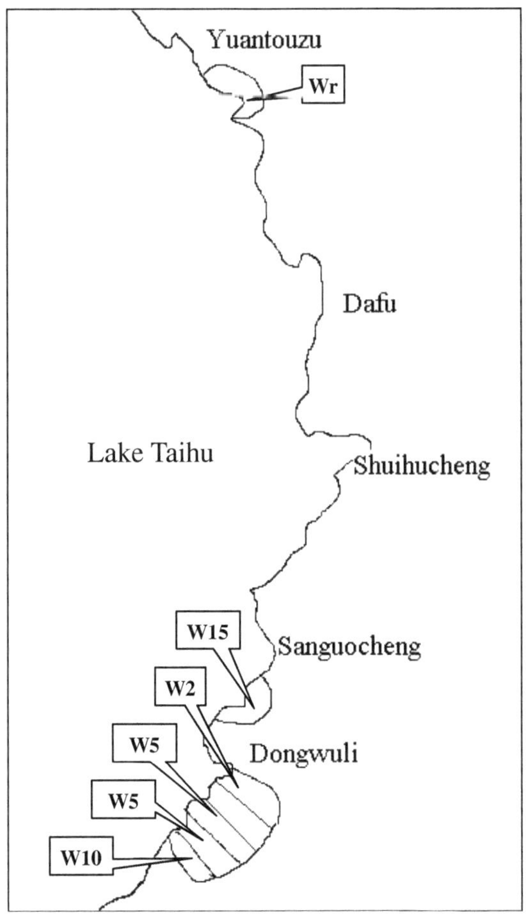

Fig. 1 Map of the distribution of restored wetlands in the studied area in Lake Taihu. Numbers refer to the numbers of years after they were abandoned as rice fields. W indicates wetland, and Wr indicates reference wetland

determined from field surveys and interviews with local landowners. Each wetland was numbered according to the time since abandonment as paddy fields (e.g. W5 indicates a wetland in its fifth year or growing season since abandonment as a paddy field). Wr means reference wetland.

Investigation and measurement of restored vegetation

Vegetation plays a critical role in wetland restoration and may indicate wetland value and function. In 2003 and 2004, four vegetation investigations were conducted in five different wetlands. Belts were identified according to vegetation characteristics. Six 1 × 1 m quadrats were chosen randomly in each wetland belt. Species, coverage, diversity and biomass were recorded in each quadrat every three months from March to October. Six 0.5 × 0.5 m quadrats also were chosen for ratio analysis of biomass between upland and low land parts in the same season. Height and density of the dominant vegetation were measured in three of these quadrats selected at random (Song, 2001). Data on hydrology, vegetation, soil and other conditions in the study area were collected as reference.

Soil analysis

Soil was collected from five wetlands in different stages of restoration. Three to five sample sites were selected in each vegetation belt. The soil profile was divided into two horizontal layers: 0–20 cm and 20–40 cm, and five samples were collected at each sampling site. Soil samples were air-dried and hand-sieved through a 20-mesh sieve screen to remove roots and debris. Sieved soil samples were sealed in a plastic bag for pH, biologically available N, and Olsen-P analysis. Remaining soil was crushed and sieved through a 100-mesh sieve screen, then stored for soil organic matter (SOM), total N (TN), and total P (TP) analysis. Soil bulk density and porosity were evaluated with a cutting ring. All soil samples were analyzed with standard methods for soil analysis (Soil Science Society of America and American Society of Agronomy, 1996).

Statistical analysis

Locating sites abandoned for different periods of time but similar characteristics is difficult. Therefore, data interpretation techniques for the four durations of abandonment were used. While the experiments were pseudo-replicates, the distance between sampling sites was sufficient to ensure adequate variation within each site.

Similarities among communities at each site and the reference natural wetland were determined using the Sorensen similarity coefficient (SC) (Song, 2001) calculated as $SC = 2\omega/(a + b)$, where ω is the number of species found in both the natural and a given abandoned site; a is the total number of species found in the reference natural wetland, and b is the total number of species found in a given abandoned site. SAS software was used to analyze experimental data, and all significant differences are at a probability of 0.05 or less, unless otherwise noted.

Results

C, N, and P changes in restored wetlands soil

An increasing trend was observed for TOC, TN, KN, TP, and Olsen-P content along the succession gradient defined by 2-, 5-, 10-, and 15-year-old paddy field abandonment sites and the reference wetland. TOC ranged from 0.73% to 2.76% in the 0–20 cm layer and 0.31–0.46% in the 20–40 cm layer (Fig. 2). There was lower

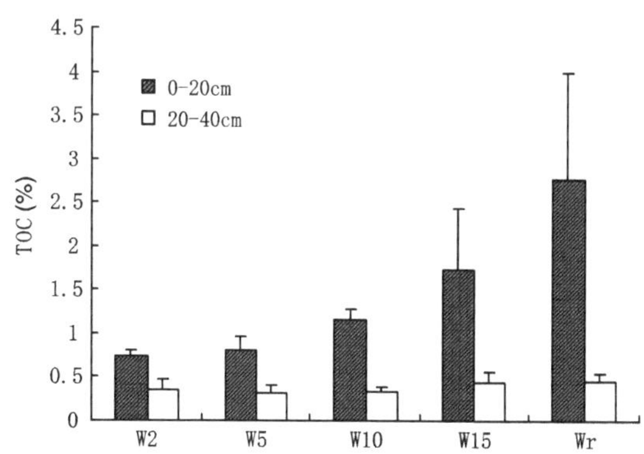

Fig. 2 Changes of soil Organic C in different restored wetlands

TOC content in the 2- to 5-year periods and higher content in the 10- to 15-year period and lay obvious accumulation of TOC at 0–20 cm layer. One-way ANOVA showed a significant difference in the 0–20 cm layer, but not in the 20–40 cm layer.

TN and KN ranged from 0.039% to 0.098% and 124.35 mg kg^{-1} to 351.48 mg kg^{-1}, respectively, in the 0–20 cm layer and 0.020–0.044% and 52.5–83.1 mg kg^{-1}, respectively, in the 20–40 cm layer (Fig. 3). There lay while an obvious decrease trend of TN content in the earlier restoration periods and an increase trend in the later restoration period. One-way ANOVA showed significant difference in KN in the 0–20 cm layer, and no significance difference in the 20–40 cm layer or TN in either layer.

TP and Olsen-P ranged from 0.046% to 0.070% and 5.13 mg kg^{-1} to 15.99 mg kg^{-1}, respectively, in the 0–20 cm layer and 0.031–0.037% and 1.87–5.57 mg kg^{-1}, respectively, in the 20–40 cm layer (Fig. 4). One-way ANOVA analysis showed no significant difference in Olsen-P and TP in both 0–20 cm layer 20–40 cm layer. The results showed that although the content of TP and Olsen-P indicate an increasing trend, there were no significant change of TP and Olsen-P in restored wetlands in all.

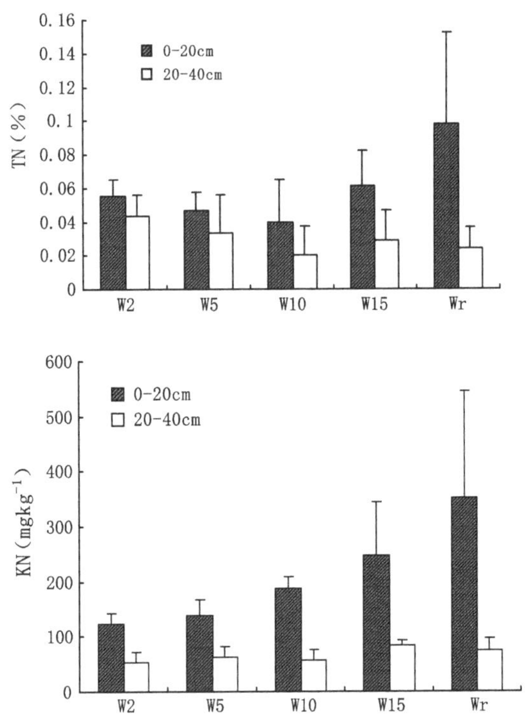

Fig. 3 Changes of soil total N and K N in different restored wetlands

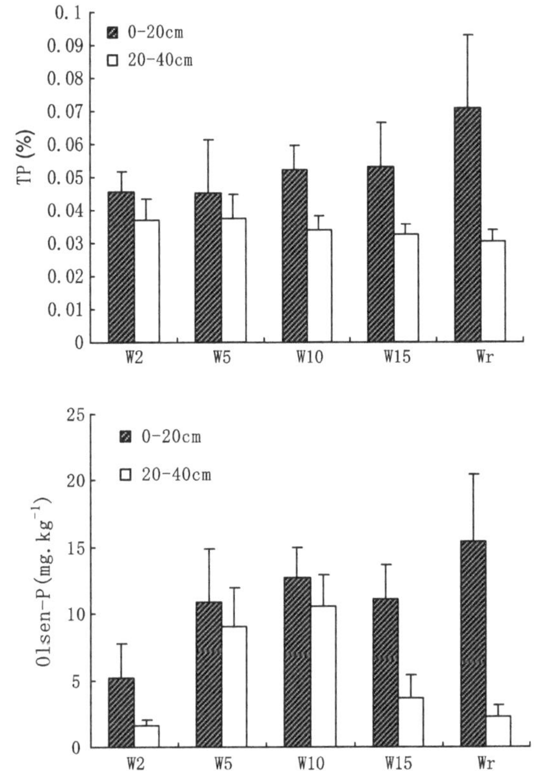

Fig. 4 Changes of total P and Olsen-P in different restored wetlands

Soil nutrient (TN, TP, TOC, KN, and Olsen-P) concentrations were lower in the 20–40 cm layer than the 0–20 cm layer. This indicates that nutrients are stored primarily in recent sediment deposits.

Table 1 Changes of soil bulk density, porosity and water stable aggregate content in different restored wetlands in Lake Taihu, East China

Time abandoned	Soil bulk density	Soil porosity (%)	WSA (>0.25 mm)
2 years	1.06 g cm^{-3}	59.9	20.2%
5 years	0.87 g cm^{-3}	71.2	30.6%
10 years	0.74 g cm^{-3}	72.0	44.3%
15 years	0.74 g cm^{-3}	72.0	54.8%
Reference wetland	0.58 g cm^{-3}	77.9	57.1%

Table 2 The species colonized after rice fields abandonment

Species	Length of land abandonment (years)				Natural reference wetland
	2	5	10	15	
Phragmites communis	0	1	1	1	1
Miscanthus sacchariflorous	0	1	1	1	1
Zizania latifolia stapf L.	0	0	1	1	0
Cyperus iria L.	0	0	0	1	0
Eleocharis yakoscensis	0	0	1	0	0
Polygonum amphibium	0	0	0	1	0
Polygonum perfoliatum L.	0	0	1	1	1
Juncus setchueusis Bush.	1	1	0	0	0
Murdannia triquetra (Wall) Bruckn	0	1	0	0	0
Lycopus lucidus Turcz var. *hirtus* Regel	0	1	1	1	0
Poa annua	1	1	0	0	0
Poa faberi Rendle	1	1	0	0	0
Glgcine sojasieb et Zucc	1	0	0	0	0
Solanum nigrum	1	0	0	0	0
Setaria viridis (L.) Beauv	1	1	0	0	0
Lactuca indica. L.	1	0	0	0	0
Artemisia lavandulaefolia DC	1	1	0	0	0
Artemisia annua	0	1	0	0	0
Gynura crepidioides	0	1	0	0	0
Clematis	0	0	1	0	0
Humulus scandens	1	1	1	1	1
Bidens bipinata	0	0	1	1	1
Cocculus trilobus (Thrund.) DC.	0	0	0	0	1
Rosa cymosa Tratt	1	1	1	0	0
Ampelopsis Michx	0	0	0	1	1
Sapium P. Br	0	0	0	0	1
Chinese Fevervine scandens	0	0	1	1	1
Total number of species	10	13	11	10	9

Soil texture changes in restored wetlands

Table 1 shows soil bulk density, soil water content, and soil porosity along the succession gradient defined by paddy fields abandoned for 2, 5, 10, and 15 years and the natural reference wetland. Soil bulk density ranged from 0.58 g cm^{-3} to 1.06 g cm^{-3} with a decreasing progression with time abandonment. During the wet season, water content increased from 50% to 300% along the succession gradient. In the dry season, land that had been abandoned longer had more water than more recently abandoned paddy fields. There also was an increase in soil water stable aggregates (>0.25 mm) in abandoned paddy fields along the succession gradient, and the proportion of large water stable aggregate at later stages was higher than earlier stages of succession. These results indicate that restoration of abandoned paddy fields improves soil quality and texture.

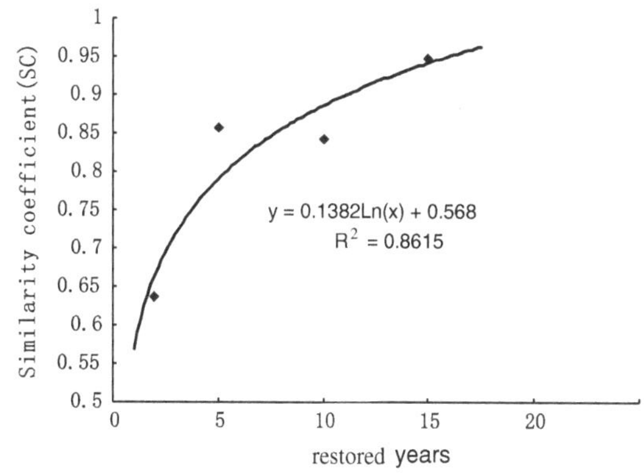

Fig. 5 Similarity coefficient (SC) of vegetation in different restored wetlands

Vegetation changes in restored wetlands

Dominant plant species changed in the first 10 years after the paddy fields were abandoned (Table 2). In the first 2 years, abandoned land became covered by a plant community dominated by *Poa annua, Humulus scandens, Gynura crepidioides*, and *Glgcine sojasieb*. This situation was

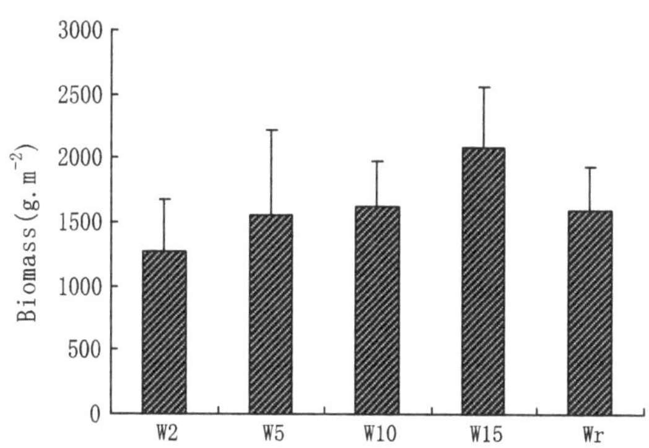

Fig. 6 Changes of vegetation biomass in different restored wetlands

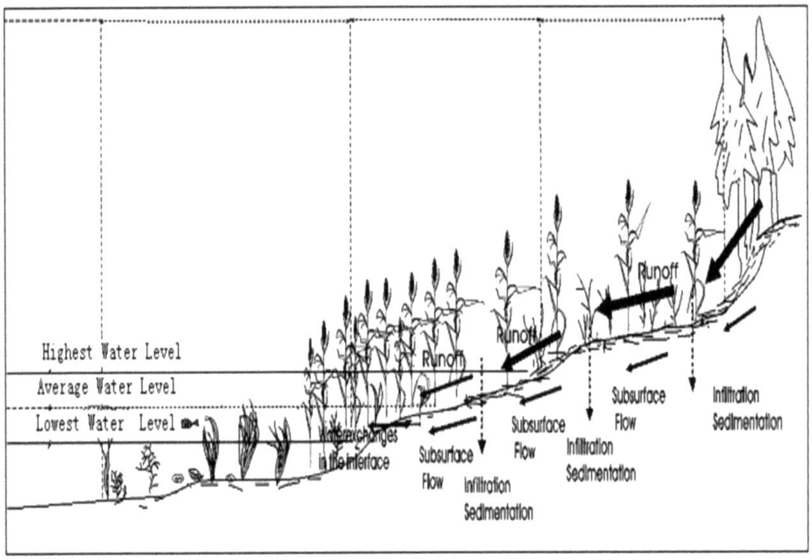

Fig. 7 The sketch map of the function of riparian wetlands

common in most abandoned paddy fields. During the early stage of abandonment (2–5 years), the number of plant species increased from 10 at the 2-year abandonment site to 13 at the 5-year site. Dominant plant species were *Humulus scandens, Miscanthus saccharifloous, Lycopus lucidus* Turcz, Var. *hirlus* Regel, *Poa annua*, and *Poa orinosae.* At the 5-year site, some wetland species, such as *Miscanthus saccharifloous, Zizania latifolia stapf,* and *Polygonum amphibium*, increased significantly, but some annual and biennial species and weeds still existed. After 5 years, some initial annual species, such as *Poa annua, Gynura crepidioides*, and *Glgcine sojasieb* declined. The number of species decreased from 13 at 5-years to 10 and 9 in the 10- and 15-year fields, respectively. Species composition after 15 years of abandonment was similar to the reference wetland, and *Phragmites communis, Miscanthus saccharifloous*, and *Zizania latifolia stapf* were dominant. SC was 0.64 at the 2-year site and 0.94 at the 15-year site compared with the reference wetland (Fig. 5). There was a significant relationship between SC and number of years since abandonment ($r^2 = 0.8615$). Dry biomass increased from 1270 g m^{-3} to 2100 g m^{-3} over the whole period (Fig. 6). One-way ANOVA showed a significant difference in dry biomass in wetland restoration, although there was no significance difference between W5 and W10. That showed there lay an obvious trend of dry biomass in the wetland restoration process. Plant community changes after land abandonment suggest that vegetation self-regenerates and species diversity recovers by returning paddy fields or lakeshore land to a more natural state.

Discussion

Riparian zones are critical in maintaining biological connections along environmental gradients (as indicated in Fig. 7). They reduce runoff from upland and increase infiltration; they are a physical buffer to trap sediments and sediment-bound pollutants; they are also a biological buffer to remove nutrients via plant uptake.

Vegetation is important in riparian ecosystems, and wetlands converted from abandoned paddy field showed secondary succession. Height and density of colonized plants increased along the succession series. With longer abandonment time, species composition increased in similarity and stability and reflected functional characteristics of a natural riparian wetland vegetation community. Wetland species, such as *Phragmites communis, Miscanthus saccharifloous, and Zizania latifolia stapf* dominated in abandoned paddy fields and colonized gradually over 15 years. Other plant species may be more competitive in other environments (Cooper et al., 1987). Changes in the similarity index and aboveground biomass indicated an increasing trend during restoration. Comparison of vegetation changes and related indices between restored wetlands showed a general trend towards a mature community or ecosystem, which has also been observed in terrestrial systems (Peng, 1995; Comin et al., 2001). In this study, we observed that groundwater table fluctuation might limit riparian wetland ecosystem succession. In some situations, only a high groundwater table allowed abandoned paddy fields or degraded wetlands to be restored to natural riparian wetlands (Andrew & Baldwin, 2002; Castelli et al., 2000).

Soil restoration is important to consider in evaluating wetland restoration. Results showed that soil quality improved with time due to accumulation of TOC, TN, TP, and improvement of soil texture. Similar results were reported in terrestrial ecosystems (Browman et al., 1990; Peng, 1995; Kosmas et al., 2000; Comin et al., 2001; Singh et al., 2001; Lee et al., 2002; Bruland & Richardson, 2004, 2005). Soil pH in riparian areas ranged from 4.90 to 5.28 in the 0–20 cm layer and 5.68 to 5.92 in the 20–40 cm layer in Fig. 5. In this study, TOC, TN, TP, KN, and Olsen-P increased with time and eventually reached similar concentrations to natural, undisturbed wetlands. Recovery of vegetation is key to successful soil restoration.

The results suggest a faster increase for soil TOC, KN, TP and Olsen-P in the early vegetation succession stage followed by a slower rate in later stages, while TN decreased in early stages. Nutrient accumulation probably resulted from decomposition of plant debris and retention of nutrients from upland areas. The recovery of vegetation and litter decomposition result in lower soil pH during restoration and create a suitable environment for soil C, N, P accumulation. Although nutrient accumulation in riparian restoration may result in leaching into lakes by subsurface flows, it is negligible compared with controlling non-point source pollution (Naiman & Décamps, 1997).

Low TOC accumulation in the early restoration period mainly was due to high soil TN concentration, which caused a higher decomposition rate of vegetation litter. The TN decrease early in the succession period may be due to growth of wetland vegetation, high nitrogen fertilizer consumption in the paddy soil before land abandonment, and ammonia volatilization and denitrification. Results also showed that soil TOC, TN, TP, KN, and Olsen-P increased towards levels found in natural wetlands, but restoration of degraded ecosystems likely would need a long time to complete the process. Vegetation restoration in Lake Taihu degraded riparian wetlands probably takes 15 years or more, and restoration of soil characteristics may take more time. In comparison to other ecosystems, restoration of degraded tidal wetlands needs about 6 years (Comín et al., 2001). Craft (2000) suggested that restoration of salt marsh vegetation requires 15 years, and restoration of soil C, N and bulk density needs 25 years. Restoration of degraded pasture may require 30 years (Aronson et al., 1993; Grantz et al., 1998; Li, 2003). Restoration of degraded forest ecosystems may take 50–100 years (Peng, 1995).

Restored vegetation not only protects surface soil from erosion and improves soil texture, but it also traps nutrient-rich material carried by runoff from uplands to riparian areas. Organic C, N and P accumulation in soil facilitated dominant wetland species in later succession stages. Vegetation restoration increased the roughness of land and coverage. Improvement in soil texture increases soil porosity and infiltration, facilitates accumulation of soil nutrients, and provides a suitable environment for microbe activity. Implications are that more non-point source pollutants may be treated in restored wetlands. It is expected that, when water level reaches inundation conditions, wetlands still remove pollutants, but is not a pollution source to the lake.

Conclusions

Changing land use in abandoned paddy fields may exert a profound influence on soil, vegetation, and the environment. With increasing time, vegetation in abandoned fields increased in the maturity of species composition and biomass. Results also showed that soil TOC, TN, TP, KN, and Olsen-P improved with time. Soil and vegetation facilitated one another in the restoration process. It may take 15 years or more for vegetation and soil characteristics to reach levels in natural wetlands.

Acknowledgments This research was supported by Chinese Academy of Sciences (No. KZCX1-SW-12) and the National Hi-Tech Research Program (No. 2002AA601011-05). We gratefully thank Yaohui Hu, Zexiang Lei, Nian'gen Du and Hongmei Hao for their assistance with sampling and analysis.

References

Andrew, E. C. & H. Baldwin, 2002. Responses of wetland plants to ammonia and water level. Ecological Engineering 18: 257–264.

Aronson, J., C. Floret, E. Le Floc'h, C. Ovalle & R. Pontanier, 1993. Restoration and rehabilitation of degraded ecosystems in arid and semiarid lands. I. A view from the south. Restoration Ecology 1: 8–17.

Browman, R. A., J. D. Reeder & R. W. Lober, 1990. Changes in soil properties in Central Plain rangeland soil after 3, 20, and 60 years of cultivation. Soil Science 150: 851–857.

Bruland, G. L. & C. J. Richardson, 2004. A spatially explicit investigation of phosphorus sorption and related soil properties in two riparian wetlands. Journal of Environmental Quality 33: 785–794.

Bruland, G. L. & C. J. Richardson, 2005. Spatial variability of soil properties in created, restored, and paired natural wetlands. Soil Science Society of America Journal 69: 273–284.

Castelli, R. M., J. C. Chambers & R. J. Tausch, 2000. Soil–water relations along a soil-water gradient in Great Basin riparian meadows. Wetlands 20: 251–266.

Chen, H. S., 2001. Restoration project of the ecosystem in Lake Taihu. Resources and Environment in the Changjiang River Basin 10(2): 173–178 (in Chinese).

Chen, J. Y. & S. L. Chen, 2002. Challenges to estuary and shoreline in China. Marine Geology 18(1): 1–5 (in Chinese).

Comin, F. A., J. A. Romero, H. Oliver & M. Menéndez, 2001. Restoration of wetlands from abandoned paddy fields for nutrient removal, and biological community and landscape diversity. Restoration Ecology 9(2): 201–208.

Coveney, M. F., D. L. Stites & E. F. Lowe, 2002. Nutrient removal from eutrophic lake water by wetland filtration. Ecological Engineering 19: 141–159.

Cooper, J. R., J. W. Gilliam & R. B. Daniels, 1987. Riparian areas as filters for agricultural sediment. Soil Science Society of American Journal 51(2): 416–420.

Craft, C., 2000. Co-development of wetland soil and benthic invertabrate communities following salt marsh creation. Wetland Ecology and Management 8(2): 197–207.

Grantz, D. A., D. L. Vaughn, R. J. Farber, B. Kim, L. Ashbaugh, T. VanCuren, R. Campbell, D. Bainbridge & T. Zink, 1998. Transplanting native plants to revegetate abandoned farmlands in the western Mojave Desert. Journal of Environmental Quality 27: 960–967.

Hu, Y. & L. G. Jiang, 2004. Returning farmland to lake-restoration of the kidney of Yangtse River. Chinese National Geography 11: 138–146.

Jin, X. S., H. Liu & Q. Y. Tu, 1990. Lake eutrophication in China. Environmental Sciences Press of China (in Chinese).

Kosmas, C., S. T. Gerontidis & M. Marathianou, 2000. The effect of land use change on soil and vegetation over various lithological formations on Lesvos (Greece). Catena 40: 51–68.

Lee, C. S., Y. H. You & G. R. Robinson, 2002. Secondary succession and natural habit restoration in abandoned fields of central Korea. Restoration Ecology 10: 306–314.

Li, Y. Q., 2003. Vegetation dynamics and soil characteristics during grassland succession process in abandoned land. MS degree dissertation thesis, Inner Mongolia Agricultural University, 36 pp.

Naiman, R. J. & H. Décamps, 1997. The ecology of interfaces: riparian zones. Annual Review Ecological Systematics 28: 621–658.

Naiman, R. J., H. Décamps & F. Fournier (eds), 1994. The role of land/inland water ecotones in landscape management and restoration: a proposal for collaborative research. MAB digest 4, UNESCO.

National Bureau of Statistics of China, 2001. Development Report of China in 2001: the ninth five-year plan. Chinese Statistics Press, Beijing, 352 pp (in Chinese).

Peng, S. L., 1995. The restoration and rehabilitation of the degraded ecosystems in the southern part of subtropical China. In Chen, L. Z. & W. L. Chen (eds), Studies on Degraded Ecosystems in China. China's Science and Technology Publishing House, Beijing, 94–113.

Singh, K. P., T. N. Mandal & S. K. Tripathi, 2001. Patterns of restoration of soil physicochemical properties and microbial biomass in different landslide sites in the hilly forest ecosystem of Nepal Himalaya. Ecological Engineering 17: 385–401.

Soil Science Society of America and American Society of Agronomy, 1996. Methods of Soil Analysis: Chemical Methods. SSSA Incorporation and ASA Incorporation, Madison, NJ, USA.

Song, Y. C., 2001. The Species Composition of Vegetation Communities. Vegetation Ecology. East China Normal University Press, 62–89.

Tabacchi, E., D. L. Correll, R. Hauer, G. Pinay, A.-M. Planty-Tabacchi & R. C. Wissmar, 1998.

Development, maintenance and role of riparian vegetation in the river landscape. Freshwater Biology 40: 497–516.

Vollenweider, R. A., 1992. Coastal marine eutrophication: principles and control. In Vollenweider, R. A., R. Marchetti & R. Viviani (eds), Marine Coastal Eutrophication, Elsevier Science Publications, 1–20.

Xiong, Z. P., 2004. The analysis for the causes of floods in China and the counter measures for reducing the damage. China Water Resources, 16 pp.

Yin, C. Q. & Z. W. Lan, 1995. The nutrient retention by ecotone wetlands and their modification for Baiyangdian Lake restoration. Water Science and Technology 32(3): 159–167.

Hydrobiologia (2007) 581:161–165
DOI 10.1007/s10750-006-0518-0

EUTROPHICATION IN LAKES

Quantitative studies on phosphorus transference occuring between *Microcystis aeruginosa* and its attached bacterium (*Pseudomonas* sp.)

Lijuan Jiang · Liuyan Yang · Lin Xiao · Xiaoli Shi · Guang Gao · Boqiang Qin

Abstract Phosphorus release from *Microcystis aeruginosa* and attached bacterium (*Pseudomonas* sp.) isolated from Lake Taihu was examined using a phosphorus isotope tracer in order to investigate the phosphorus transference between the two species. Our results reveal that the amount of phosphorus released form ^{32}P-saturated *M. aeruginosa* is determined by its growth phase and most of phosphorus is assimilated by *Pseudomonas* finally while the amount of phosphorus released from ^{32}P-saturated *Pseudomonas* is also determined by the growth phase of *M. aeruginosa* and most of them are assimilated by *M. aeruginosa*. The results suggest that phosphorus transference occurs between *M. aeruginosa* and its attached *Pseudomonas* . This process makes microenvironment of mucilage of *M. aeruginosa* attached bacteria maintain relative high amounts of phosphorus. Attached bacteria may be a temporary phosphorus bank to the growth of *M. aeruginosa*, and assimilation of phosphorus by *M. aeruginosa* becomes easy when *M. aeruginosa* is in lag growth phase. Thus, the phosphorus exchange between *M. aeruginosa* and attached *Pseudomonas* in microenvironment may be important to microfood web and cyanobacteria bloom.

Keywords *Microcystis aeruginosa* · Attached bacterium · *Pseudomonas* sp. · Phosphorus · Lake Taihu · ^{32}P release

Guest editors: B. Qin, Z. Liu & K. Havens
Eutrophication of shallow lakes with special reference to Lake Taihu, China

L. Jiang · L. Yang (✉) · L. Xiao · X. Shi
State Key Laboratory of Pollution Control and Resource Reuse, School of the Environment, Nanjing University, 22 Hankou Road, Nanjing 210093, P. R. China
e-mail: yangly@nju.edu.cn

L. Yang · X. Shi · G. Gao · B. Qin
Nanjing Institute of Geography and Limnology, CAS, 73 East Beijing Road, Nanjing 210008, P. R. China

Introduction

Microcystis aeruginosa is the dominant cyanobacterium in Lake Taihu, a shallow hypereutrophic lake in eastern China with extensive cyanobacterial algal blooms in summer and autumn (Dokulil et al., 2000). *M. aeruginosa* forms large mucilaginous colonies, usually colonized by the great deal of bacteria. There is a mutualistic relationship between *M. aeruginosa* and attached bacteria (Whitton, 1973). This symbiotic relation may be favorable to *Microcystis* cells (Steppe et al., 1996). *M. aeruginosa* constitute a microhabitat where attached bacteria are sheltered from grazing by embedding in the mucilage. The microenvironment allows attached bacteria to associate with the primary producer of organic carbon for its growth. In this microenvironment, heterotrophic bacteria are important: (1) as contributors of CO_2 and possibly sources of nitrogen, phosphorus and trace elements to *M. aeruginosa*; (2) as primary consumers of algal excretions; and (3) as decomposers. Some bacteria may promote blooms while others have algicidal and lysis effects and are involved in termination and decomposition of blooms (Caiola, 1991; Pellegrini et al., 1997; Lovejoy et al., 1998; Van Hannen et al., 1999; Manage et al., 2000; Manage et al., 2001). Thus, the metabolic relationship of *Microcystis* with attached bacteria is important for understanding bloom dynamics.

Phosphorus is the major limiting nutrient for primary producer, including *M. aeruginosa* in Lake Taihu (Dokulil et al., 2000). Pilot studies were conducted on the phosphorus metabolism of *M. aeruginosa* and its attached bacteria, respectively (Shi et al., 2003; Shi et al., 2004; Zou et al., 2005). The information of interaction of cyanobacterium with its attached bacterium for phosphorus transference was little. In this paper, a phosphorus isotope tracer was used to study phosphorus uptake and release from *M. aeruginosa* and its attached bacterium, *Pseudomonas*, to probe their interactions and phosphorus transformation.

Materials and methods

Organism and cultivation

Microcystis aeruginosa was obtained from Institute of Hydrobiology, Chinese Academy of Sciences. It was grown in the modified MA medium (pH = 8.6) in which inorganic phosphorus was replaced by β-Na_2-glycerophosphate (Oh & Lee, 2000). Cultures were incubated at 25°C under illumination intensity of 2200 lx with a 14:10 h of light:dark cycle.

Pseudomonas sp. was isolated from the mucilage of *Microcystis* in Lake Taihu, China and grown in complex medium: glucose 6.0; NH_4Cl 1.0; NaCl 2.0; $MgSO_4 \cdot 7H_2O$ 0.2; K_2HPO_4 1.0; KH_2PO_4 1.0; yeast extract 0.5 (all in grams per liter of deionized water). pH was adjusted to 7.2–7.5 before the complex medium was autoclaved.

Precultivation

This study was conducted with phosphorus-starved *M. aeruginosa*, ^{32}P-saturated *M. aeruginosa*, phosphorus-starved *Pseudomonas* and ^{32}P-saturated *Pseudomonas*. *M. aeruginosa* and *Pseudomonas* in exponential growth phase were harvested by centrifugation at 10,000*g* for 15 min respectively, washed with sterile deionized water and resuspended in MA medium without phosphorus. Phosphorus-starved cells were directly obtained after incubation for 3 days. Phosphorus-starved cells were collected by centrifugation and transferred to MA medium in which β-Na_2-glycerophosphate was replaced with 12 mg l^{-1} $Na_2H^{32}PO_4$, then were incubated for 3 days to obtain ^{32}P-saturated cells of *M. aeruginosa* and *Pseudomonas* respectively.

Transference of phosphorus from ^{32}P-saturated *M. aeruginosa* to *Pseudomonas*

Attachment of bacteria to *M. aeruginosa* mucilage made it difficult to measure ^{32}P content in bacteria cells or in cyanobacteria cells. Considering this difficulty, a dialysis membrane bag was used to separate *Pseudomonas* from *M. aeruginosa*. However, low molecular weight substances, such as phosphate and small organic compounds, could move across the dialysis membrane.

Phosphorus-starved *Pseudomonas* cells were harvested by centrifugation at 10,000*g* for 15 min, and washed with sterile deionized water and resuspended in a 500 ml beaker containing 200 ml MA medium without phosphorus to 1.0×10^7 cell ml^{-1}. ^{32}P-saturated *Microcystis* cells were harvested by the same way except they were resuspended in the dialysis bag to 5.2×10^6 cell ml^{-1}. This dialysis bag was put in the beaker so then *M. aeruginosa* and *Pseudomonas* cells were separated, while phosphate could migrate across the dialysis membrane. They were incubated at 25°C under a 14:10 h of light:dark cycle.

In order to probe the change of intracellular phosphorus, the cells of *M. aeruginosa* or *Pseudomonas* in certain volume were separated respectively via centrifugation at 10000 *g* for 15 min and rinsed several times with MA medium until rinse solution radioactivity was close to background. These cells were digested with 0.2 ml 60% $HClO_4$ and 0.4 ml 30% H_2O_2 at 80°C for 60 min. Then, the digested liquids and the upper aqueous solution separated by centrifugation were respectively put into scintillation vials containing 5 ml Triton X-100 toluene scintillator to analyze the radioactivity with a Beckman LS9800 liquid scintillation counter. Actual phosphorus concentration was corrected using the radioactive decay rate. All samples had three replicates.

In order to measure the biomass of *M. aeruginosa* and *Pseudomonas*, their optical densities were measured with at 460 nm and 380 nm, respectively. Then, the amounts of ^{32}P in *M. aeruginosa* or *Pseudomonas* were calculated.

Transference of phosphorus from ^{32}P-saturated *Pseudomonas* to *M. aeruginosa*

The experimental procedures were same as described above except that phosphorus-starved *M. aeruginosa* was incubated in a 500 ml beaker, while ^{32}P-saturated *Pseudomonas* was incubated in dialysis bag.

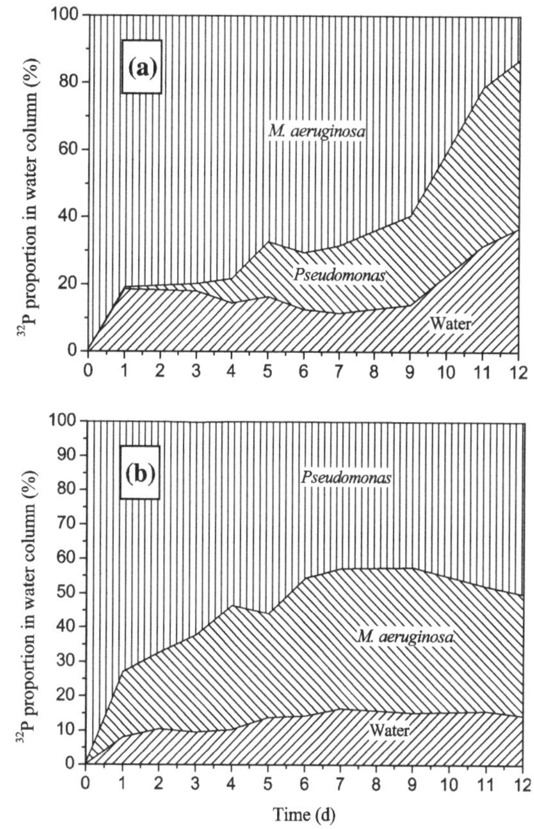

Fig. 1 The transference of ^{32}P released from *M. aeruginosa* (**a**) and from *Pseudomonas* sp. (**b**) among water column.

Results

Transference of phosphorus from ^{32}P-saturated *M. aeruginosa* to *Pseudomonas*

The transference of phosphorus from *M. aeruginosa* to *Pseudomonas* was investigated in a culture system of phosphorus-starved *Pseudomonas* and ^{32}P-saturated *M. aeruginosa* which were separated by a dialysis membrane (Fig. 1a). The growth phases of *M. aeruginosa* comprised four distinct phases of the lag (0–1 day), exponential (1–4 day), stationary (4–6 day) and decline phases (6–12 day) while the growth phases of *Pseudomonas* was in the exponential phase. The total ^{32}P in *Pseudomonas* always increased though the increase was slight for *Pseudomonas* in initial 4-days. Because of the growth of *Pseudomonas* and its incessant assimilation of ^{32}P, the ^{32}P content in *Pseudomonas* increased after *M. aeruginosa* was in the stationary phase. ^{32}P content in *Pseudomonas* increased to 50% of the total ^{32}P on the 12th day when *M. aeruginosa* was in the decline phase.

During the whole experiment period, the intracellular ^{32}P content of *M. aeruginosa* decreased at all times. The ^{32}P concentration in the aqueous solution remained relative steadily from the 1st day to the 9th day, then the sharp increase happened because *M. aeruginosa* in decline growth phase released a large amount of phosphorus to aqueous solution.

Transference of phosphorus from ^{32}P-saturated Pseudomonas to *M. aeruginosa*

Not only phosphorus could move from *M. aeruginosa* to *Pseudomonas*, but also from

Pseudomonas to *M. aeruginosa*. When phosphorus-starved *M. aeruginosa* was present, partitioning of ^{32}P released from ^{32}P-saturated *Pseudomonas* was observed (Fig. 1b). The growth of *M. aeruginosa* exhibited the lag phase (0–2 day), exponential phase (2–9 day) and decline phase (9–12 day) while *Pseudomonas* grew slightly at all times. Most of ^{32}P released from *Pseudomonas* was assimilated by *M. aeruginosa*.

Intracellular ^{32}P content of *M. aeruginosa* increased in the lag and exponential phases, but decreased in decline phase which decrease of ^{32}P content in *M. aeruginosa* was due to the growth of *Pseudomonas* though this decrease was not rapid. ^{32}P concentration in the aqueous solution increased in the first day and then remained relative steadily in the following days.

Discussion

In this experimental system, phosphorus was main factor for the growth of *M. aeruginosa* and *Pseudomonas* and the latter was also controlled by organic carbon. When the intracellular phosphorus content of *M. aeruginosa* was luxury, it grew by utilization of the intracellular phosphorus in its lag and exponential phases (Okada & Sudo, 1982), thus, only 20% of intracellular phosphorus was released form *M. aeruginosa* to aqueous solution and to *Pseudomonas*. However, *M. aeruginosa* released a lot of phosphorus in its stationary and decline phases while it could simultaneously provide *Pseudomonas* with organic carbon and trace elements (Sommaruga & Robarts, 1997; Worm, 1998; Brunberg, 1999). Thus, the growth of *Pseudomonas* was better and the assimilated phosphorus was more when *M. aeruginosa* was in decline phase than that in stationary phase.

After the amount of *M. aeruginosa* was low enough, its second growth became possible. *M. aeruginosa* could use phosphorus in *Pseudomonas* to grow because *Pseudomonas* had a geat deal of phosphorus which was assimilated from overripe *M. aeruginosa* formerly (Fig. 1b). Compared with the phosphorus transference form *M. aeruginosa* to *Pseudomonas*, phosphorus movement form *Pseudomonas* to *M. aeruginosa* was easy and a great deal of ^{32}P was assimilated by *M. aeruginosa* in lag and exponential phases. When *M. aeruginosa* was in the decline phase, its intracellular ^{32}P then turned back to *Pseudomonas* again. This process made soluble phosphorus in aqueous solution stable. Thus, phosphorus released form *M. aeruginosa* could store in attached bacteria and *Pseudomonas* might be a temporary phosphorus bank to *M. aeruginosa* in the microenvironment for its growth.

Conclusion

Phosphorus transference from *Pseudomonas* to *M. aeruginosa* occurs when *M. aeruginosa* is in grow phase while phosphorus transference from *M. aeruginosa* to *Pseudomonas* happens when *M. aeruginosa* is in decline phase. *Pseudomonas* may serve as a temporary phosphorus bank to *M. aeruginosa* in the microenvironment. Thus, phosphorus transference between cyanobacteria and their attached bacteria is decided by their growth phases and phosphorus amount and this exchange could keep an amount of phosphorus in the microenvironment of cyanobacteria and their attached bacteria which is important to microfood web and cyanobacteria bloom.

Acknowledgements We thank F. Wang (Modern Analysis Center, Nanjing University) for their technical assistances. This study was funded by Chinese Academy of Sciences (No. KZCX2-311), "973" Project (No. 2002CB412307) and National Natural Science Foundation of China (No. 40371102, 40501078).

References

Brunberg, A. K., 1999. Contribution of bacteria in the mucilage of *Microcystis* spp. to benthic and pelagic bacterial production in a hypereutrophic lake. FEMS Microbiology Ecology 29: 13–22.

Caiola, M. G., 1991. *Bdellovibrio*-like bacteria in *Microcystis aeruginosa*. Algological Studies 64: 369–376.

Dokulil, M., W. Chen & Q. Cai, 2000. Anthropogenic impacts to large lakes in China: the Tai Hu example. Aquatic Ecosystem Health and Management 3: 81–94.

Lovejoy, C., J. P. Bowman & G. M. Hallegraeff, 1998. Algicidal effects of a novel marine *Pseudoalteromonas* isolate (Class Proteobacteria, Gamma Subdivision) on harmful algal bloom species of the genera

Chattonella, Gymnodinium, and *Heterosigma*. Applied Environmental Microbiology 64: 2806–2813.

Manage, P. A., Z. Kawabata & S. Nakano, 2000. Algicidal effect of the bacterium *Alcaligenes denitrificans* on *Microcystis* spp.. Aquatic Microbial Ecology 22: 111–117.

Manage, P. A., Z. Kawabata & S. Nakano, 2001. Dynamics of cyanophage-like particles and algicidal bacteria causing *Microcystis aeruginosa* mortality. Limnology 2: 73–78.

Oh, H. M. & S. J. Lee, 2000. Microcystin production by *Microcystis aeruginosa* in a phosphorus-limited chemostat. Applied and Environmental Microbiology 66: 176–179.

Okada, M. & R. Sudo, 1982. Phosphorus uptake and growth of bule-green alga, *Microcystis aeruginosa*. Biotechnology and Bioengineering 24: 143–152.

Pellegrini, S., L. Allievi, B. Lolli & M. G. Caiola, 1997. *Bdellovibrio* isolation from the Lake Varese (Italy). Annali Di Microbiologia ED Enzimologa 47: 121–129.

Shi, X., L. Yang, X. Niu & L. Xiao, 2003. Intraintracellular phosphorus metabolism of *Microcystis aeruginosa* under various redox potential in darkness. Micobiological Research 158: 345–352.

Shi, X., L. Yang, F. Wang, L. Xiao, L. Jiang, Z. Kong, G. Gao & B. Qin, 2004. Growth and phosphate uptake kinetics of *Microcystis aeruginosa* under varying environmental conditions. Journal of Environmental Sciences 16: 88–92 (in Chinese).

Sommaruga, R. & R. D. Robarts, 1997. The significance of autotrophic and heterotrophic picoplankton in hypertrophic ecosystems. FEMS Microbiology Ecology 24: 187–200.

Steppe, T. F., J. B. Olson, H. W. Paerl, R. W. Litaker & J. Belnap, 1996. Consortial N2 fixation: a strategy for meeting nitrogen requirements of marine and terrestrial cyanobacterial mats. FEMS Microbiology Ecology 21: 149–156.

Van Hannen, E. J., G. Zwart & H. J. Laanbroek, 1999. Changes in bacterial and eukaryotic community structure after mass lysis of filamentous cyanobacteria associated with viruses. Applied and Environmental Microbiology 65: 795–801.

Whitton, B. A., 1973. Interactions with other organisms. In Carr, N. G. & B. A. Whitton (eds), The Biology of Blue-green Algae. Blackwell, Oxford: 415–433.

Worm, J., 1998. Dynamics of heterotrophic bacteria attached to Microcystis spp. (Cyanobacteria). Aquatic Microbial Ecology 14(1): 19–28.

Zou, D., L. Xiao, L. Yang & Y Wan, 2005. Effects of Phosphorus Sources of different forms on phosphorus metabolism of *Microcystis aeruginosa* and adhesive *Pseudomonas* sp.. Environmental Science 26(3): 118–121 (in Chinese).

Hydrobiologia (2007) 581:167–176
DOI 10.1007/s10750-006-0517-1

EUTROPHICATION IN LAKES

Intracellular phosphorus metabolism and growth of *Microcystis aeruginosa* in dark/light cycles under various redox potential difference conditions

Xiaoli Shi · Liuyan Yang · Lijuan Jiang · Fanxiang Kong · Boqiang Qin · Guang Gao

Abstract Phosphorus metabolism and growth of *M. aeruginosa* were studied under three different conditions of diel fluctuation in redox potential. Redox potential in the culture increased in light and decreased in dark in all treatments except one, when cysteine was added in darkness. Total phosphorus content in *M. aeruginosa* decreased in darkness and increased in light during exponential growth but increased continuously in the stationary phase. Conversely, polyphosphate (PolyP) accumulated in darkness but was lost in the light. Low redox potential in darkness promoted PolyP accumulation. Polyglucose and soluble orthophosphate may provide energy and phosphorus, respectively, for PolyP synthesis. PolyP was important to *M. aeruginosa* survival during poor growth conditions. If the redox potential difference in the dark/light cycle was large, *M. aeruginosa* initially grew faster, but soon lost viability.

Keywords *Microcystis aeruginosa* · Growth · Phosphorus metabolism · Dark/light cycle · Redox potential difference · Lake Taihu

Introduction

In Lake Taihu, *Microcystis aeruginosa* has formed dense blooms in summer since the 1980s (Chen et al., 2003). Due to oxygen-producing photosynthesis, *Microcystis* usually are associated with aerobic environments, and, consequently, physiological characterization has focused on aerobic metabolism. However, this characterization discounts the possibility that they are found in environments that are permanently anoxic or become anoxic in the dark (Stal & Moezelaar, 1997).

Microcystis is colonial and possesses gas vacuoles, hollow proteinaceous bodies that provide cell buoyancy. In Lake Taihu, when the water column is stable, colonies accumulate at the surface and form surface blooms. Wind may concentrate colonies into dense "scums" on the leeward shore, forming microbial mats, which may experience periodic anoxia. During daytime, oxygenic photosynthesis leads to oxygen supersaturation. In the

Guest editors: B. Qin, Z. Liu & K. Havens
Eutrophication of shallow lakes with special reference to Lake Taihu, China

X. Shi · L. Yang (✉) · F. Kong · B. Qin · G. Gao
Nanjing Institute of Geography and Limnology, Chinese Academy of Sciences, Nanjing 210008, P.R. China
e-mail: yly@niglas.ac.cn

X. Shi · L. Yang · L. Jiang
State Key Laboratory of Pollution Control and Resource Reuse, School of the Environment, Nanjing University, Nanjing 210093, P.R. China

dark, cyanobacteria continue to respire, and oxygen diffusion into the mat usually is insufficient to prevent anoxia (Revsbech et al., 1983).

Microbial mats exhibit diel oxygen fluctuations due to cyanobacterial metabolism. In light, these organisms grow photoautotrophically, while *Microcystis* switches to chemotrophic metabolism in dark, at the expense of endogenous glycogen accumulated in light (Heyer & Krumbein, 1991). However, a cohesive, overall pattern of dark cyanobacteria metabolism has not emerged (Richardson & Castenholz, 1987).

Phosphorus is a major growth-limiting nutrient in aquatic systems, and, unlike nitrogen, there is no large atmospheric source that is biologically available. All organisms require phosphorus for growth and metabolism (Bostrom et al., 1988; Currie & Kalff, 1984; Howarth, 1988). Phosphorus in microalgae is contained in internal phosphorus pools of polyphosphate (PolyP), soluble orthophosphorus (Pi), structural and other organic phosphorus (John & Flynn, 2000). Cyanobacteria store inorganic phosphate as PolyP granules (Bental et al., 1991). However, little is known about phosphorus regulation in cyanobacteria (Lawrence et al., 1998). Previous studies have found that phosphorus metabolism of *M. aeruginosa* may inhibit continuous anaerobic dark conditions, which are unfavorable for *M. aeruginosa* growth (Shi et al., 2003). *M. aeruginosa* retaining viability under dark anaerobic/light aerobic cycles also may relate to phosphorus metabolism. However, to our knowledge, no studies on phosphorus metabolism of *M. aeruginosa* under dark/light cycle with high redox potential differences have been conducted. Here, we studied phosphorus metabolism and *M. aeruginosa* growth under diel fluctuations in redox potential.

Materials and methods

Organism and cultivation

Axenic *M. aeruginosa* cultures were obtained from the Institute of Hydrobiology of the Chinese Academy of Sciences and grown in batch culture in modified MA medium where inorganic phosphorus was replaced with β-Na_2glycerophosphate (Oh et al., 2000). Media were adjusted to pH 8.6 with 2 mol l^{-1} NaOH and incubated at 25°C under 40 μE m^{-2} s^{-1} light intensity provided by fluorescent lamps with a 14:10 h light/dark cycle. *M. aeruginosa* density was counted on a microscope with a 0.1 mm^3 counter, and about 300 cells were counted. Cell numbers were converted to dry weight using a 1.32×10^{-7} mg dry weight per cell (Johnston et al., 1994).

Experimental methods

Studies were conducted with *M. aeruginosa* cells harvested during exponential growth. Cells were washed twice and resuspended in the medium described above (MA). Aliquots were transferred to serum bottles with screw caps and a butyl rubber inlayer. Bottles were wrapped in black cloth and 0, 0.01, and 0.04% (m/v) cysteine (reducing agent) were introduced through serum caps with a syringe. After one day, all bottles were unwrapped and serum caps were opened. Then, *M. aeruginosa* was incubated under illumination intensity of 40 μE m^{-2} s^{-1} for another day. The cycle then was repeated. Therefore, days 1, 3, 5 and 7 had air-tight dark cultivation periods with different concentrations of cysteine, while days 2, 4, 6 and 8 were open light cultivation. All experiments were carried out with triplicate bottles. Cell density, polyglucose, PolyP and soluble orthophosphate content in *M. aeruginosa*, as well as phosphate concentration in the water, were assayed daily. A microelectrode was used to monitor initial and final redox potential.

Analysis

M. aeruginosa cells were harvested by centrifugation for physiological assay. Samples were digested by autoclaving in 10% persulphate for 30 min at 121°C for total phosphorus measurement (Grillo & Gibson, 1979). Orthophosphate in *M. aeruginosa* was extracted with 10% cold trichloroacetic acid. For PolyP, 40 ml of culture was centrifuged and digested with 1 ml alkaline hypochlorite reagent (5.4%) for 45 min at 25°C. Residue after centrifugation was washed twice, then extracted twice with distilled water. PolyP was precipitated with ethanol and measured after hydrolysis to orthophosphate at

100°C (Harold, 1963; Rao et al., 1985). Later, all phosphorus forms were converted to orthophosphate, and all samples were measured spectrophotometrically with an automated ascorbic acid reduction method (Greenberg, 1985). All phosphorus forms were expressed in milligrams of phosphorus per unit dry weight (DW) for *M. aeruginosa*. Polyglucose was extracted from cells, harvested from 20 ml culture with 1 ml 10% trichloroacetic acid, precipitated by adding three volumes of alcohol, and determined using anthrone reagent (Roe & Dailey, 1966).

Results

Redox potential decreased from 252 to 52 mV with 0.01% cysteine and to –50 mV with 0.04% cysteine at the beginning of the experiment (Fig. 1). Redox potential increased in light and declined in darkness in all treatments except the first dark period. Average redox potential differences were 67 mV, 83.5 mV and 276 mV, respectively, when 0, 0.01 and 0.04% cysteine were added in darkness.

M. aeruginosa cell density increased in all treatments, and high redox potential differences in the diel cycle enhanced *M. aeruginosa* growth, with maximum cell density reaching 7.5×10^6 cell ml^{-1} on the 7th day and 7.8×10^6 cell ml^{-1} on the 5th day when adding 0.01% and 0.04% cysteine in darkness, respectively (Fig. 2). Compared with the control, *M. aeruginosa* had a short growth period and lost viability after reaching maximum cell density.

Phosphate concentration in the aqueous phase increased in darkness and declined in light in all treatments initially, but this pattern ceased on the 5th day, when cysteine concentrations were 0.01% and 0.04% in the dark (Fig. 3). However, total phosphorus in *M. aeruginosa* increased (Fig. 4). There were no significant differences observed among the three treatments ($P > 0.05$), so enlarging the redox difference in the dark/light cycle did not have an obvious effect on phosphorous metabolism.

Soluble orthophosphate in *M. aeruginosa* declined initially, with increasing steepness at high cysteine concentrations, then increased in light and decreased in darkness (Fig. 5). Polyglucose

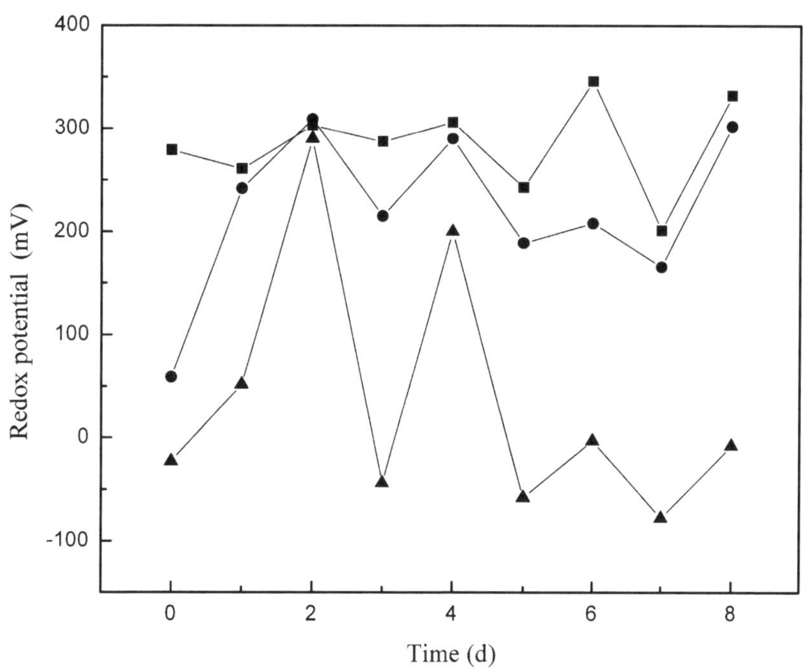

Fig. 1 Redox potential under different treatments. ■, Control; ●, 0.01% of cysteine in darkness; ▲, 0.04% of cysteine in darkness

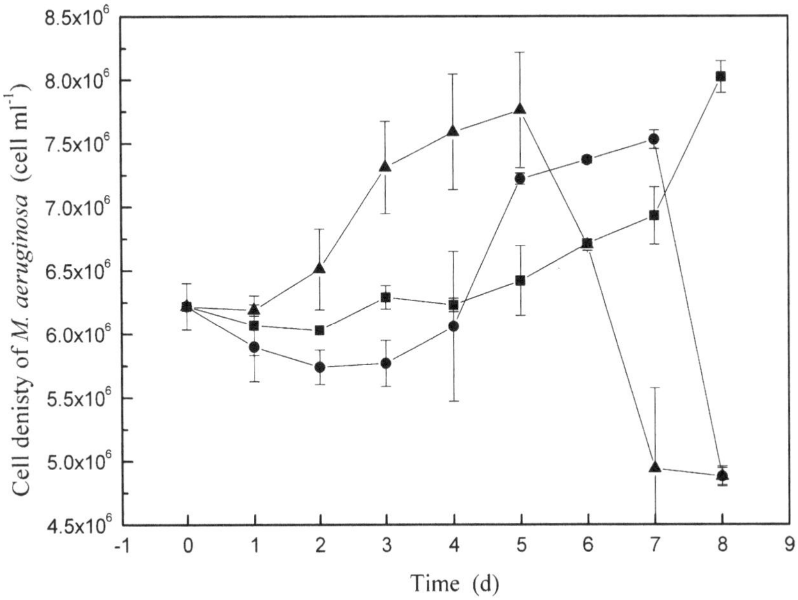

Fig. 2 Growth of *M. aeruginosa* under all the treatments. ■, Control; ●, 0.01% of cysteine in darkness; ▲, 0.04% of cysteine in darkness. Error bars represent standard derivation

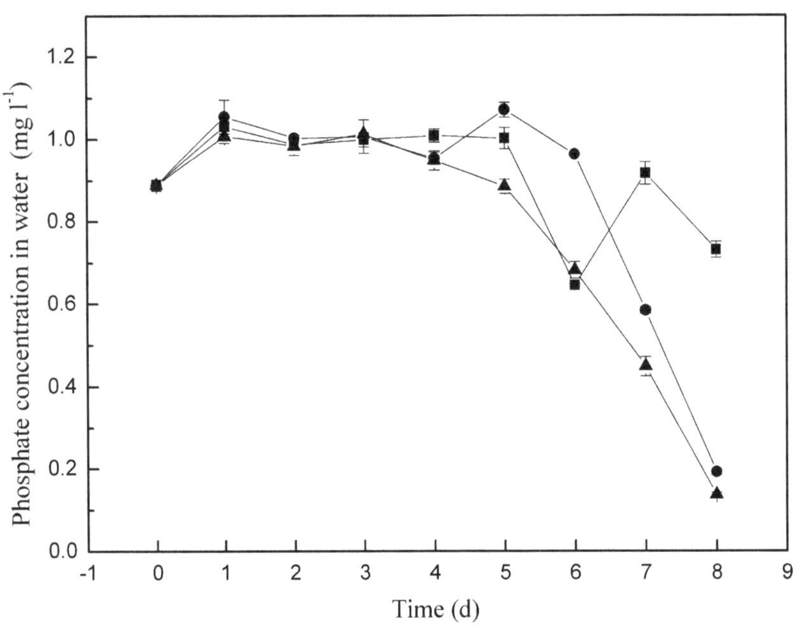

Fig. 3 Variation of phosphate concentration in water. ■, Control; ●, 0.01% of cysteine in darkness; ▲, 0.04% of cysteine in darkness. Error bars represent standard derivation

showed a similar pattern as soluble orthophosphate content (Fig. 6). Adding reducing agent promoted PolyP accumulation in *M. aeruginosa*, with PolyP reaching 0.038 and 0.043 mg g^{-1} DW on the 3rd day when cysteine concentrations were 0.01% and 0.04%, respectively. PolyP decreased in the light and increased in the dark (Fig. 7).

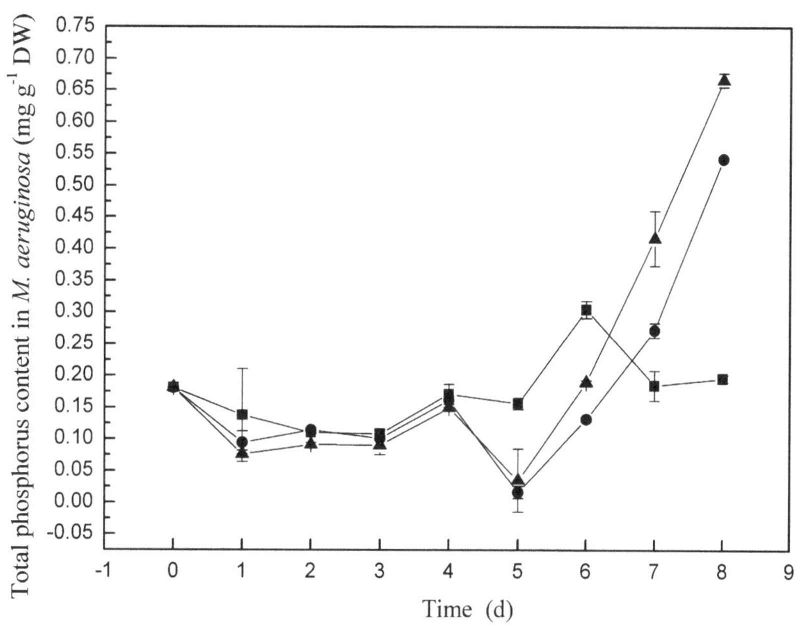

Fig. 4 Total phosphorus content in *M. aeruginosa*. ■, Control; ●, 0.01% of cysteine in darkness; ▲, 0.04% of cysteine in darkness. Error bars represent standard derivation

Discussion

Phosphorus metabolism of *M. aeruginosa* under dark/light cycles with different redox potential difference

M. aeruginosa does not form microbial mats or dense scums, which could cause anoxia, in the laboratory, so phosphorus metabolism of *M. aeruginosa* generally is studied under aerobic cultivation conditions. Phosphate uptake rate is considered a function of ambient phosphate and cell quota, and Droop's equation describes the relationship between cell quota and population growth rate (Fuhs et al., 1972). Phosphorus in *M. aeruginosa* decreased in darkness, increased in light, and increased during the stationary phase when reducing agent was added in darkness (Fig. 4). Thus, *M. aeruginosa* may release phosphorus in darkness and absorb phosphorus in light.

M. aeruginosa may synthesize PolyP when cells are added to fresh medium. PolyP in *M. aeruginosa* increased if redox potential was low. After a short period of growth, PolyP in *M. aeruginosa* increased in darkness and declined in the light (Fig. 7), opposite the trend of TP and polyglucose (Fig. 4). Synthesis of PolyP may help maintain phosphorous balance in the cell. PolyP accumulation coinciding with decreased polyglucose and soluble orthophosphate shows that these compounds may provide energy and phosphorus for PolyP synthesis. *M. aeruginosa* may accumulate polyglucose during daylight and consume it at night, and previous research showed that polyglucose fermentation can provide enough energy for PolyP synthesis (Shi et al., 2003). An experiment with yeast also indicated that endogenous substrate is adequate to supply energy for PolyP synthesis (Schuddemat et al., 1989).

Inorganic polyphosphates are polymers of orthophosphate (Pi) with a phosphoanhydride bond, which has free energy close to that of ATP. The exact physiological function of PolyP remains uncertain, although various biological functions have been demonstrated, including as a reservoir of energy and phosphate, a chelator of divalent cations, a capsule material, and a "channelling" agent in the phenomenon of bacterial transformation (Magrath & Quinn, 2000; Keasling & Hupf, 1996; Kornberg, 1995). In general, PolyP concentration is low during exponential growth

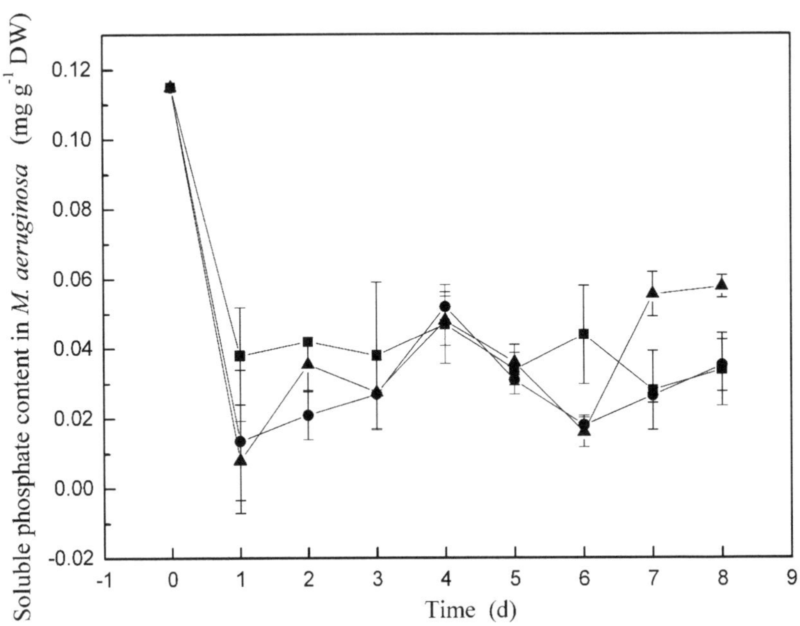

Fig. 5 Soluble phosphate content in *M. aeruginosa* under all the treatments. ■, Control; ●, 0.01% of cysteine in darkness; ▲, 0.04% of cysteine in darkness. Error bars represent standard derivation

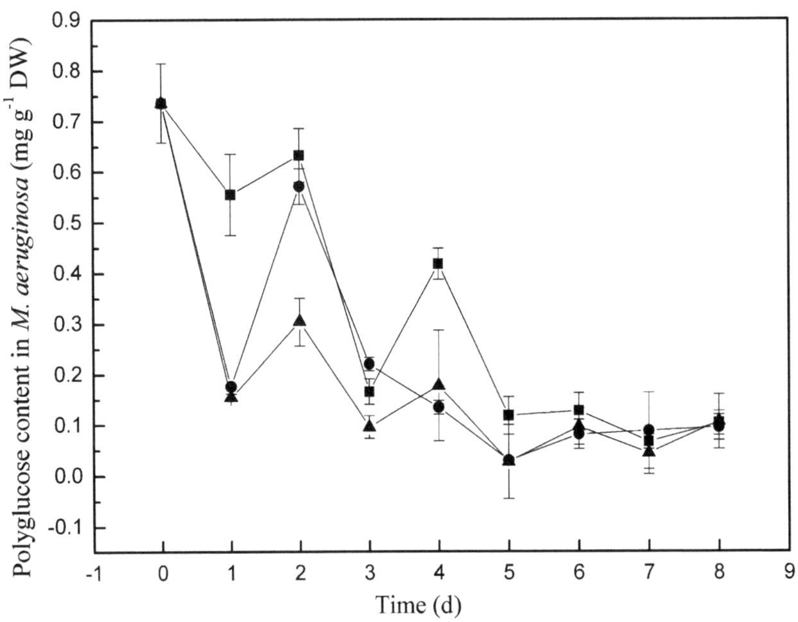

Fig. 6 Polyglucose content in *M. aeruginosa* under all the treatments. ■, Control; ●, 0.01% of cysteine in darkness; ▲, 0.04% of cysteine in darkness. Error bars represent standard derivation

but may increase when the stationary phase begins or when growth is arrested due to nutrient imbalance (Noegel & Gotschlich, 1983). Pi released and accumulated by the cells would be sequestered as PolyP to maintain Pi and H^+ concentration (Harold, 1963). Thus, rapid

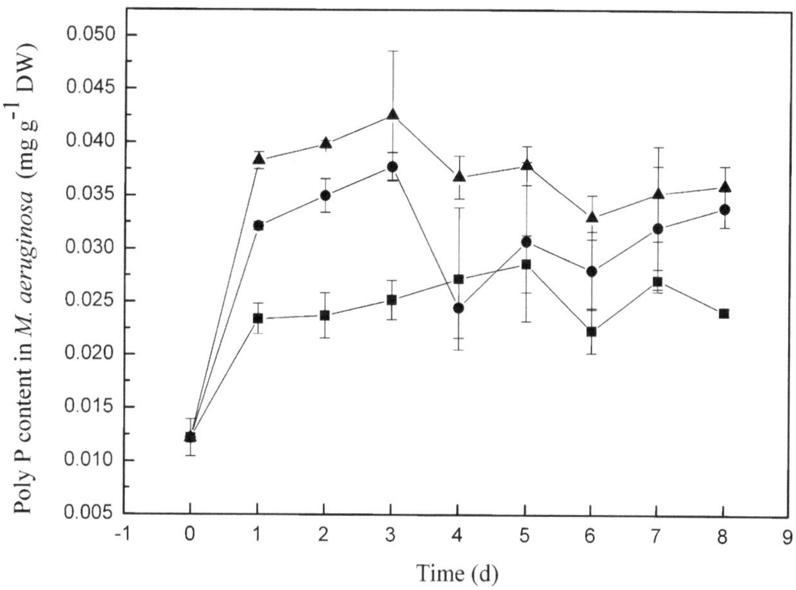

Fig. 7 PolyP content in *M. aeruginosa* under all the treatments. ■,Control; ●, 0.01% of cysteine in darkness; ▲, 0.04% of cysteine in darkness. Error bars represent standard derivation

orthophosphate release may stimulate PolyP synthesis. PolyP is involved in controlling Pi and serves, in effect, as a metabolic buffer.

Prokaryotes are composed of cytoplasm, cell surface, periplasm and plasma membrane. PolyP granules can be found in all of these, and different locations of PolyP may play different roles. Electron microscopy indicated that PolyP granules in *M. aerugnosa* were located in the cytoplasm (Shi et al., 2003). PolyP granules may be used for phosphorus and energy storage in our experiments (Kulaev et al., 1999).

M. aeruginosa growth in the dark/light cycle under different redox potential difference

Some cyanobacterial strains can survive in dark, reducing conditions. *Oscillatoria terebriformis*, a cyanobacterium, moved each night (Richardson & Castenholz, 1987). *O. chalybea* blooms are correlated with lowered partial pressure of O_2 accompanied by increased sulfide at increasing depth in lakes (Leventer & Eren, 1970). The presence of *O. redekei* in the lower strata of some lakes was attributed to reducing conditions of sulfide (Tash, 1967). Thus, *Oscillatoria* can survive in dark, reducing conditions, and some *Oscillatoria* species are dependent on lowered redox potential.

M. aeruginosa can endure low redox potential in darkness, which stimulated slight growth and reduced mortality in darkness (Shi, 2003). Large redox potential difference may promote *M. aeruginosa* growth initially, then cause cell death (Fig. 2). When adding 0.01% and 0.04% cysteine, average redox potential differences between light and dark were 83 mV and 276 mV, respectively. In this case, *M. aeruginosa* lost viability at days 5 and 7, respectively. In previous research, we found that large cell density in culture may result in high diel redox potential difference, and, when initial cell density reached 1.5×10^7, 6×10^7 and 1×10^8 cell ml^{-1}, average diel redox potential differences were 65, 330, and 450 mV, respectively (Fig. 8). During the 18-day experiment, *M. aeruginosa* grew and had no evident mortality (data not shown). Since cysteine had no other chemical effect besides lowing redox potential (Weller et al., 1975), we assumed that, only when *M. aeruginosa* reached a certain cell density, could it survive low redox potential. *M. aeruginsoa* seemed to thrive to

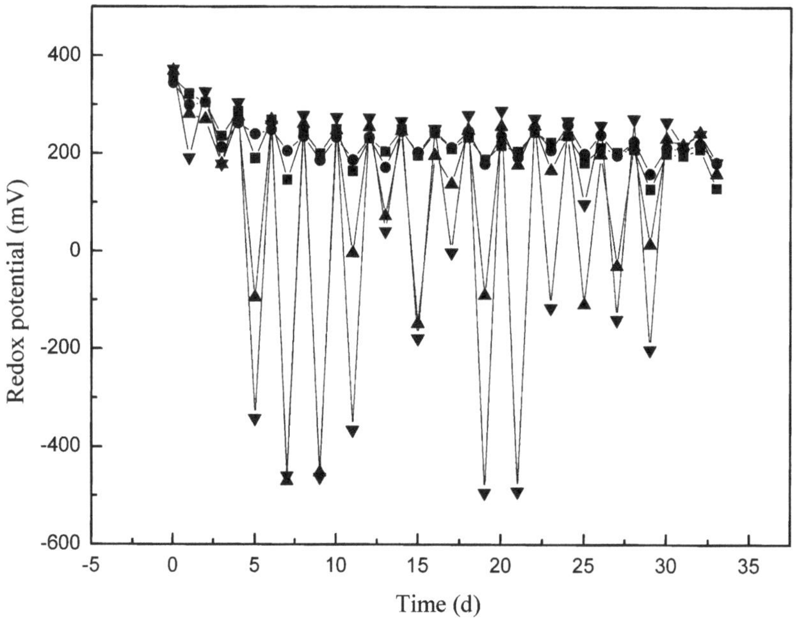

Fig. 8 Redox potential of the medium with different cell density of *M. aeruginosa*. ■, 3×10^6 cell ml^{-1}; ●, 1.5×10^7 cell ml^{-1}; ▲, 6×10^7 cell ml^{-1}; ▼, 1×10^8 cell ml^{-1}

resist the large diel redox potential difference since redox potential increased in the fist darkness period when the reducing agent was added, but could not hold back the designed trend. In our experiment, cell density of 6.2×10^6 cell ml^{-1} was too low to endure 83 and 276 mV diel redox potential difference. In Lake Taihu, large diel redox potential differences are observed in microbial mats or scums, and, in that case, cell density is high enough to adapt to such circumstances. Thus, large diel rodox potential difference may promote algae growth, but shows no negative effect.

In autumn, microbial mats of *Microcystis* sink to the sediment. *M. aeruginosa* colonies remain on the sediment as vegetative cells. These winter colonies are viable and constitute an inoculum for the following year, some reentering the water column in spring (Preston et al., 1980). Redox potential of sediment in Lake Taihu changes with depth from about -100 to + 100 mV (Fig. 9). Sediment suspension often occurs under wind and waves in large shallow lakes. Redox potential at the sediment surface increases at this time and decreases during calm periods. Therefore, *M. aeruginosa* in the sediment regularly experiences variation in redox potential.

Microcystis was a small proportion of biomass in non-stratified enclosures but was dominant in diurnally and continuously stratified enclosures (Nakano et al., 2001). Although thermal stratification was required for *Microcystis* growth, it is unclear why it was important. We hypothesize that stratification may lead to high diel redox potential difference, which is more favorable for *Microcystis* than other species, and, therefore, it is dominant in eutrophic environments.

Anoxia was vital for survival of *O. terebriformi* during dark periods, because fast degradation of glycogen under aerobic conditions would result in exhaustion of energy reserves within a few hours (Moezelaar et al., 1996). However, PolyP accumulation under low redox potential in dark may be important to *M. aeruginosa* for survival in unfavorable conditions. Therefore, the ability to accumulate PolyP in the dark and negative redox potential may be an advantage in low light, organically rich, and low-redox habitats. Finally, the phenomenon reported here may be important in understanding mechanisms of phosphorous

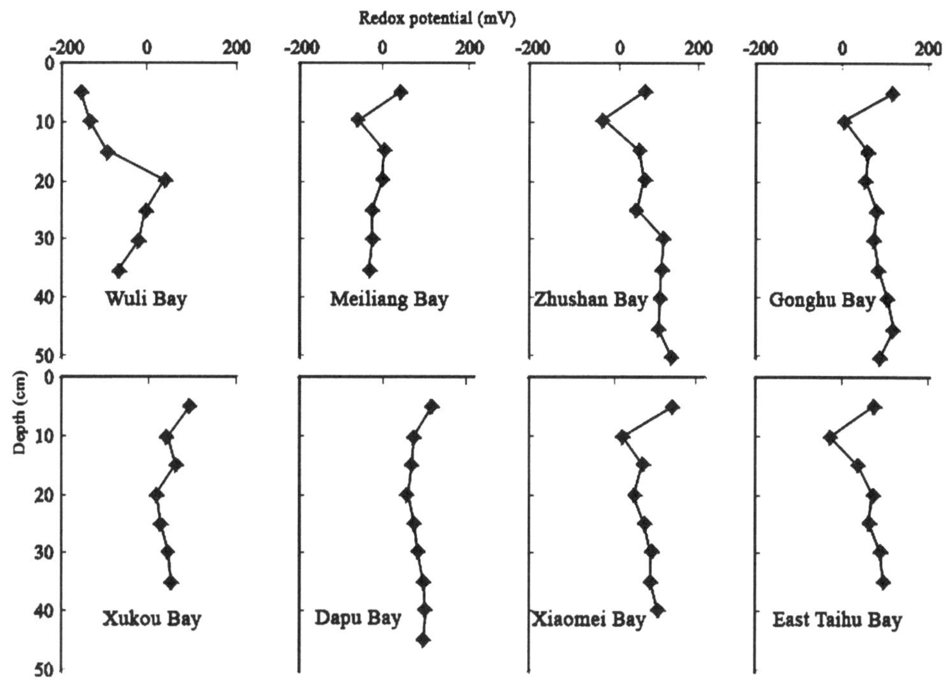

Fig. 9 Redox potential in sediment of main region of Lake Taihu. Data from (Zhu et al., 2004)

metabolism in *Microcystis* during high diel redox potential difference and explain why mats and *Microcystis* in sediment can remain viable for several months in Lake Taihu.

Acknowledgements This research was funded by Chinese Academy of Sciences (KZCX2-403), "973" Project (2002CB412307) and National Natural Science Foundation of China (No. 40371102, 40501078).

References

Bental, M., U. Pick, M. Avron & H. Degani, 1991. Polyphosphate metabolism in the alga *Dunaliella salina* studied by ^{31}P-NMR. Biochimca et Biophysica Acta 1092: 21–28.

Bostrom, B., G. Pertsson & B. Broberg, 1988. Bioavailability of different phosphorus forms in freshwater systems. Hydrobiologia 170: 133–155.

Chen, Y. W., B. Q. Qin, K. Teubner & M. T. Dokulil, 2003. Long-term dynamics of phytoplankton assemblages: Microcystis-domination in Lake Taihu, a large shallow lake in China. Journal of Plankton Research 25: 445–453.

Currie, D. J. & J. Kalff, 1984. The relative importance of bacterioplankton and phytoplankton in phosphorus uptake in freshwater. Limnology and Oceanography 29: 311–321.

Fuhs, G. W., S. D. Demmerle, E. Canelli & M. Chem, 1972. Characterization of phosphorous-limited planktonic algae. In Nutrients and Eutrophication. American Society of Limnology and Oceanography, Special Symposium 1: 113–132.

Greenberg, A. E., 1985. Standard Methods for the Examination of Water and Wastewater. American Public Health Association, Washington DC.

Grillo, J. F. & J. Gibson, 1979. Regulation of phosphate accumulation in the unicellular Cyanobacteria *Synechococcus.* Journal of Bacteriology 140: 508–517.

Harold, F. M., 1963. Inorganic polyphosphate of high molecular weight from aerobacter aerogenes. Journal of Bacteriology 86: 885–887.

Heyer, H. & W. E. Krumbein, 1991. Excretion of fermentation products in dark and anaerobically incubated cyanobacteria. Archives of Microbiology 155: 284–287.

Howarth, R. W., 1988. Nutrient limitation of net primary production in marine ecosystems. Annual Review of Ecology and Systematics 19: 89–110.

John, E. H. & K. J. Flynn, 2000. Modelling phosphate transport and assimilation in microalgae; how much complexity is warranted? Ecological Modelling 125: 145–147.

Johnston, N. A. L., V. S. Campagna, P. R. Hawkins & R. J. Banens, 1994. Response of the eastern rainbowfish (Melanotaenio duboulay) to toxic *Microcystis aeruginosa.* In Jones, G. J. (ed.), Cyanobacterial Research. CSIRO Cataloguing-in-Publication Entry, Australia: 187–193.

Keasling, J. D. & G. A. Hupf, 1996. Genetic manipulation of polyphosphate metabolism affects cadmium tolerance in *Escherichia coli.* Applied and Environmental Microbiology 62: 743–746.
Kornberg, A., 1995. Inorganic polyphosphate: toward making a forgotten polymer unforgettable. Journal of Bacteriology 177: 491–496.
Kulaev, I., V. Vagabov & T. Kulakovskaya, 1999. New aspects of inorganic polyphosphate metabolism and function. Journal of Bioscience and Bioengineering 88: 111–129.
Lawrence, B. A., C. Suarez, A. DePina, E. Click, N. H. Kolodny & M. Allen, 1998. Two internal pools of soluble polyphosphate in the Cyanobacterium Synechocystis sp. Strain PCC 6308: an in vivo ^{31}P NMR spectroscopic study. Archives of Microbiology 169: 195–200.
Leventer, H. & J. Eren, 1970. Taste and odor in the reservoirs of the Israel National Waster System. In Shuval, H. I. (ed.), Developments in Water Quality Research. Ann Arbor Humphrey Science Publishers, Mich: 19–37.
Magrath, J. W. & J. P. Quinn, 2000. Intracellular accumulation of polyphosphate by the yeast *Candida humi cola* G-1 in response to acid pH. Applied and Environmental Microbiology 9: 4068–4047.
Mozelaar, R., S. M. Bijvank & L. J. Stal, 1996. Fermentation and sulfur reduction in the mat-building Cyanobacterium *Microcoleus chthonoplastes*. Applied and Environmental Microbiology 62: 1752–1758.
Nakano, S., K. Hayakawa, J. Frenette, T. Nakajima, C. M. Jiao, S. Tsujimura & M. Kumagai, 2001. Cyanobacterial blooms in a shallow lake: a large-scale enclosure assay to test the importance of diurnal stratification. Archiv fur Hydrobiologie 150: 491–509.
Noegel, A. & E. C. Gotschlich, 1983. Isolation of a high molecular weight polyphosphate from *Neisseria gonorrhoeae*. Journal of Experimental Medicine 157: 2049–2060.
Oh, H. M., S. J. Lee, M. H. Jang & B. D. Yoon, 2000. Microcystin production by *Microcystis aeruginosa* in a phosphorus-limited chemostat. Appied and Environmental Microbiology 66: 176–179.
Preston, T., W. D. P. Stewart & C. S. Reynolds, 1980. Bloom-forming cyanobacterium *Microcystis aeruginosa* overwinters on the sediment. Nature 288: 365–367.
Rao, N. N., M. F. Roberts & A. Torriani, 1985. Amount and chain length of polyphosphates in *Escherichia coli* depend on cell growth conditions. Journal of Bacteriology 162: 242–247.
Revsbech, N. P., B. B. Jorgensen, T. H. Blackburn & Y. Cohen, 1983. Microelectrode studies of the photosynthesis and O_2, H_2S and pH profiles of a microbial mat. Limnology and Oceanography 28: 1062–1074.
Richardson, L. L. & R. W. Castenholz, 1987. Enhanced survival of the Cyanobacterium *Oscillatoria terebriformis* in darkness under anaerobic conditions. Applied and Environmental Microbiology 53: 2151–2158.
Roe, J. H. & R. E. Dailey, 1966. Determination of glycogen with the anthrone reagent. Analytical Biochemistry 15: 245–250.
Schuddemat, J., R. De Boo, C. C. M. Van leeuwen, P. J. A. Brock & J. Van Steveninck, 1989. Polyphosphate synthesis in yeast. Biochimica et Biophysica Acta 1010: 191–198.
Shi, X. L., L. Y. Yang, X. J. Niu, L. Xiao, Z. M. Kong, B. Q. Qin & G. Gao, 2003. Intracellular phosphorus metabolism of *Microcystis aeruginosa* under various redox potential in darkness. Microbiological Research 158: 345–352.
Stal, L. J. & R. Moezelaar, 1997. Fermentation in cyanobacteria. FEMS Microbiology Review 21: 179–211.
Tash, C. J., 1967. Summary. In Bartsch, A. F. (ed.), Proceedings of the Symposium on the Environmental Requirements of Blue-green Algae. Federal Water Pollution Control Administration, Oreg: 103–106.
Weller, D., W. Doemel & T. Brock, 1975. Requirement of low oxidation-reduction potential for photosynthesis in a blue-green alga (*Phormidium* sp.). Archives of Microbiology 104: 7–13.
Zhu, G. W. & C. X. Fan, 2004. Physicochemical characters of sediment in Lake Taihu. In Qin B. Q., W. P. Hu & W. M. Chen (eds), Evolution Process and Mechanism of Water Environment of Lake Taihu. Science Publication Company, Beijing, 170–187 (in Chinese).

Hydrobiologia (2007) 581:177–188
DOI 10.1007/s10750-006-0511-7

EUTROPHICATION IN LAKES

The bacterioplankton of Lake Taihu, China: abundance, biomass, and production

Guang Gao · Boqiang Qin · Ruben Sommaruga · Roland Psenner

Abstract Abundance, biomass and production of pelagic bacteria were examined over one year at monthly sampling intervals across a trophic profile in Meiliang Bay, Lake Taihu. With the lowest density in the open lake, the bacterial abundance showed a clear trend in relation to trophic status. The carbon content per cell was higher in autumn and winter, and the opposite was true for bacterial biomass. Bacterial ^{3}[H]-*TdR* and ^{14}[C]-Leu incorporation rates, cell production, turnover times and carbon production varied during the annual cycle at different sites. The ratio of bacterial production to primary production was high, independently of the method used, indicates that the microbial food web in Lake Taihu is an important component of the total food web of the lake and dominated by external inputs.

Keywords Eutrophication · Trophic status · Bacterial biomass · Bacterial production · Lake Taihu

Guest editors: B. Qin, Z. Liu & K. Havens
Eutrophication of shallow lakes with special reference to Lake Taihu, China

Electronic Supplementary Material Supplementary material is available for this article at http://dx.doi.org/10.007/s10750-006-0511-7 and is accessible for authorized users

G. Gao (✉) · B. Qin
Nanjing Institute of Geography and Limnology, Chinese Academy of Sciences, 73 East Beijing Road, Nanjing 210008, P.R. China
e-mail: guanggao@niglas.ac.cn

R. Sommaruga · R. Psenner
Institute of Ecology, University of Innsbruck, Innsbruck, Austria

Introduction

Bacteria play a pivotal role in aquatic ecosystems. They use dissolved organic matter (DOM), which is mineralized and transformed into biomass, forming the base of the microbial food web (Azam et al., 1983). Measurement of bacterial cell numbers and biomass is, thus, a central aspect in the study of plankton.

By using fluorescent dyes and epifluorescence microscopy for direct counting (Hobbie et al., 1977; Porter & Feig, 1980; reviewed by Raymond, 1994), combined with automated image analysis (Psenner, 1991a, b) or electron microscopy (Krambeck et al., 1981) for measuring bacterial size distribution and cell volumes, reliable estimates of bacterial biomass can be made. Using the incorporation of radio-labeled precursors into DNA (Thymidine) and proteins (Leucine) (Fuhrman & Azam, 1980, 1982; Kirchman et al., 1985, 1986), as well as other approaches, e.g., the frequency of dividing cells (Davis & Sieburth,

1984), realistic estimates of in situ bacterioplankton production can be achieved.

Eutrophication and contaminant enrichment can directly and indirectly affect structure and functioning of microbial food webs, because any changes in trophic state of a water body would affect the relative significance of microorganisms and alter the entire pelagic food web (Weisse, 1991; Sommaruga, 1995). Bacterial abundance (Bird & Kalff, 1984) and production (Cole et al., 1988) correlate with phytoplankton biomass and primary production. Therefore, bacterial biomass and production will be enhanced with increased nutrient loading and primary production (Weisse, 1991). Also, the abundance and biomass of other microbial food web components (such as autotrophic picoplankton, small eukaryotic algae, heterotrophic nanoflagellates, ciliates and metazoan microplankton) will increase, which will decrease the relative importance of the microbial food web to carbon flux (Weisse, 1991).

In this paper, we reported a study on selected microbial parameters along a trophic gradient in Lake Taihu, a large shallow lake in eastern China. The major aim of these studies was to evaluate the relationships between the bacterioplankton abundance, cell volume and biomass, production and the trophic state of lake. We tested the hypothesis that bacteria and lake eutrophication are tightly coupled and bacterial growth in lakes largely occurs at the expense of allochthonous carbon sources other than phytoplankton production.

Materials and methods

Sampling sites

Lake Taihu is the third largest fresh water lake in China and is influenced by intensive blooms of blue-green alga during summer. In Meiliang Bay, at the northern end of Lake Taihu at the inflow of the main tributary, a north-south profile across the bay and a transverse profile from west to east were chosen (Fig. 1). These five stations represented different trophic levels, including a highly contaminated area at the river mouth (#0) and several sites in Meiliang Bay with nutrient levels decreasing with distance from the river mouth (3#, #5, and site #2 in front of the Taihu Station). The site with the lowest influence from nutrient inputs by the river is the open area (#8) in the center of the lake.

Sampling

Samples were taken with a 5 l Patalas–Schindler sampler. A 50 ml sub-sample was transferred into a glass bottle (75 ml) containing 2.5 ml of pre-filtered (pore size, 0.2 μm) formaldehyde (final concentration 2%). The glass bottles were washed with acid, rinsed with particle-free distilled water and autoclaved one day before use. After return to the laboratory, samples were stored in a refrigerator at 4°C until slides for enumeration and image analysis were prepared.

Bacterial abundance

For bacteria abundance, a well-mixed sub-sample was diluted with particle-free distilled water (1:40) and stained with 4′, 6′-diamidino-2-phenylindole (DAPI, Sigma) at a final concentration of 2 μg ml^{-1} for 7 min (modified from Porter & Feig, 1980; Raymond et al., 1994; Posch et al., 2001). After staining, samples were filtered onto black polycarbonate filters (0.2 μm pore size, 25 mm diameter, Poretics™) and embedded in nonfluorescent immersion oil (Cargille type A, Cargille Laboratories, Inc., USA). On each slide, at least 400 bacteria were counted at 1600× with a Zeiss Axiovert 135 M epifluorescence microscope equipped with a HBO 50 W mercury lamp and a filter set for UV excitation (BP 365, FT 395, LP397 nm, Zeiss filter set 01) (Zeiss, Germany).

Bacterial biomass

Bacterial biomass is the product of cell abundance and carbon content per cell. Measurement of cell sizes and carbon contents is described elsewhere. Briefly, cell length and width were measured microscopically using a DAPI concentration of 1 μg ml^{-1} and a sensitive camera coupled to an image analysis system. Carbon

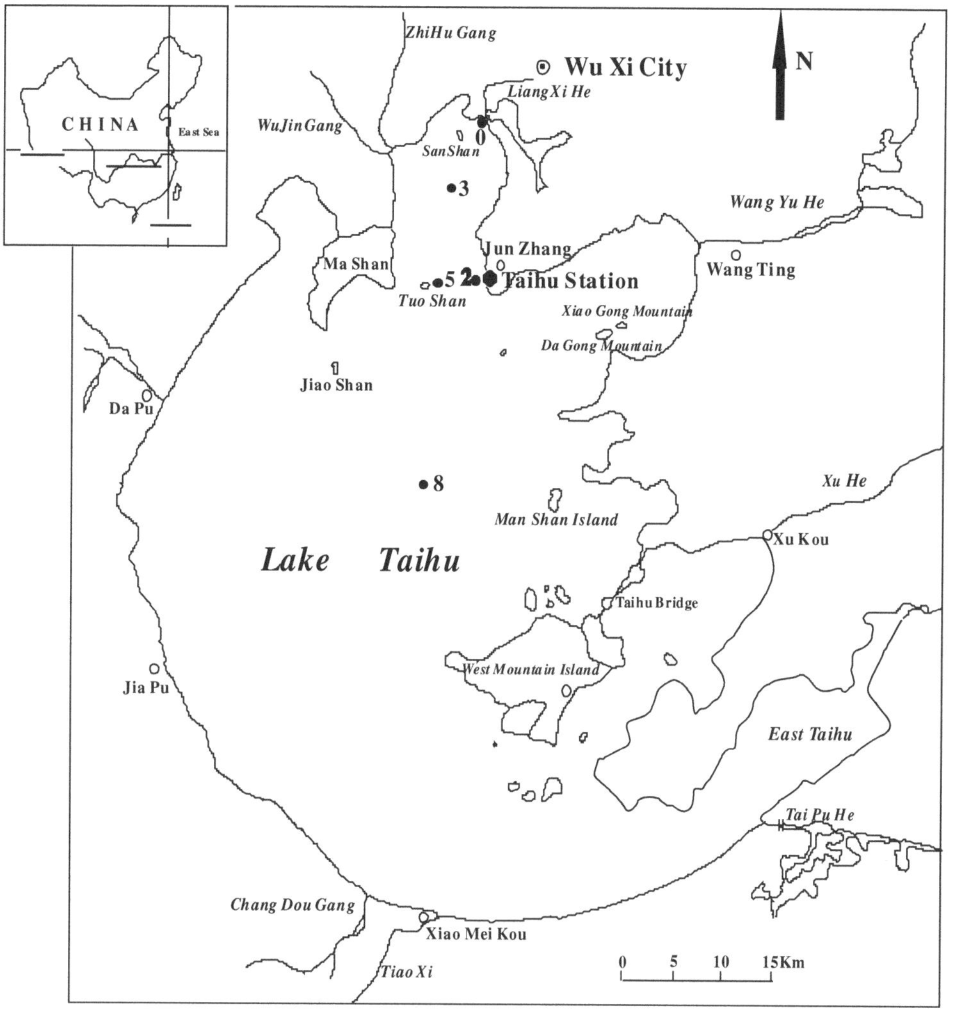

Fig. 1 The location of the sampling sites in Lake Taihu

content was inferred by an empirical formula established under comparable staining procedures (Loferer-Krößbacher et al., 1998; Posch et al., 2001).

Metabolically active bacteria

A 500 ml sub-sample was transferred into a glass bottle prepared as above. After return to the laboratory, 1 ml of 2-(*p*-iodophenyl)-3-(*p*-nitrophenyl)-5-phenyl tetrazolium chloride (INT) stock solution (2 mg ml^{-1}) was added to 10 ml of sample in 20 ml polyethylene vials (INT final conc. ca. 0.02%). All samples were incubated at in situ temperature in a water bath for 1 h without any substrate addition. After incubation, samples were fixed with pre-filtered (pore size, 0.2 μm) formaldehyde (final concentration 2%). Samples were stored in a refrigerator at 4°C at least 1 h and no longer than 24 h before preparation (modified from Zimmermann et al., 1978; Dufour & Colon, 1992; Thom et al., 1993). Other procedures were the same as for bacterial abundance estimations. A blue filter and halogen lamp were used for INT measurements.

Bacterial production (BP)

A dual-label incubation method was used to measure incorporation of thymidine (*TdR*) and leucine (Leu) by adding [^{3}H]-*TdR* and [^{14}C]-Leu to a single sample and collecting cold-TCA-insoluble material (modified from Chin-Leo & Kirchman, 1988; Kirchman & Hoch, 1988). For each sample, triplicates of 10 ml each and one control (already containing 0.5 ml of 0.2 μm pre-filtered formalin at a final concentration of 2%) were used. In each sub-sample, 100 nM (final concentration) of [^{14}C]-Leu (specific activity, 308 mCi mmol^{-1}, Amersham Life Science, England.) was added, and samples were incubated for 40 min in a water bath at in situ temperature. After 40 min, 10 nM (final concentration) of [^{3}H]-*TdR* (specific activity, 81 Ci mmol^{-1}, Amersham Life Science, England.) was added, and samples were incubated for an additional 20 min. The incubation was stopped by adding 0.5 ml formalin at a final concentration of 2%. After incubation, samples were filtered through cellulose nitrate filters (Sartorius^{TM}, 0.2 μm pore size, 25 mm diameter). Then, 5 ml of ice-cold 5% TCA were added for 5 min to extract macromolecular fractions. The filter was rinsed 3 times with 5 ml of ice-cold 5% TCA. Radioactivity incorporated into cellular material was assayed by a Beckman liquid scintillation counter (LS-8000IC).

According to the literature (Simon & Azam 1989, Wetzel & Likens 2000), the following formulas and conversion factors were used to calculate bacterial [^{3}H]-*TdR* and [^{14}C]-Leu incorporation, cell production and carbon production.

$$[^{3}\text{H}] - TdR.\text{incorporation}\ [\text{n mol}\ TdR\ \text{l}^{-1}\ \text{h}^{-1}] = \frac{U \times 4.5 \times 10^{-13}}{S \times T \times V}$$

where 4.5×10^{-13} = the number of curies per dpm; U = dpm of the filter; S = Specific activity of the *TdR* [Ci mmol^{-1}]; T = Incubation time [h]; V = Volume of filtered sample [l].

Cell production [cells l^{-1} h^{-1}] = 2×10^{18} cells mol^{-1} *TdR*

Carbon production [μg C l^{-1} h^{-1}] = TI × CP × CC

where TI = *TdR* incorporation [moles*TdR* l^{-1} h^{-1}]; CP = Cell production [cells mol^{-1}]; CC = Cellular carbon content [μg C cell^{-1}].

$$[^{14}\text{C}]\ \text{- Leu. incorporation}\ [\text{n mol Leu l}^{-1}\ \text{h}^{-1}] = \frac{U \times 4.5 \times 10^{-13}}{S \times T \times V}$$

where 4.5×10^{-13} = the number of curies per dpm; U = dpm of the filter; S = Specific activity of the Leu [Ci mmol^{-1}]; T = Incubation time [h]; V = Volume of filtered sample [l].

Cell production [cells l^{-1} h^{-1}] = 0.07×10^{18} cells mol^{-1} Leu.

BCP [μg C l^{-1} h^{-1}] = mol. Leu (inc.) × (7.3/100)$^{-1}$ × 131.2 × ID × (C/protein)

where: 131.2 = molecular weight of leucine

(7.3/100)$^{-1}$ = mole% of leucine in protein, hence 0.073

C/protein = ratio of cellular carbon to protein, best estimate = 0.86

ID = Isotope dilution; here ID = 2

Results

Some physical and chemical parameters in different parts of Lake Taihu during May 1998 to May 1999 are presented in Electronic Supplementary material. Due to the shallowness of the lake and strong winds, the water column remained mixed. However, nutrient concentrations, Chl *a* concentration, and other physical and chemical parameters varied among the sampling sites and showed a clear trophic gradient from the river mouth (hypertrophy), Meliang Bay (eutrophy) to the open lake (meso/eutrophy).

Bacterial abundance

The distribution patterns of bacterial abundance varied significantly along the trophic gradient of this lake (One-way ANOVA, $P = 0.005$). At the hypertrophic river mouth, the highest bacterial abundance was determined. With the mean value of 15.4×10^{9} l^{-1}, the bacterial abundance was ca. 55% higher than the mean in eutrophic Meiliang Bay and almost twice as high as the meso/eutrophic open lake (Fig. 2).

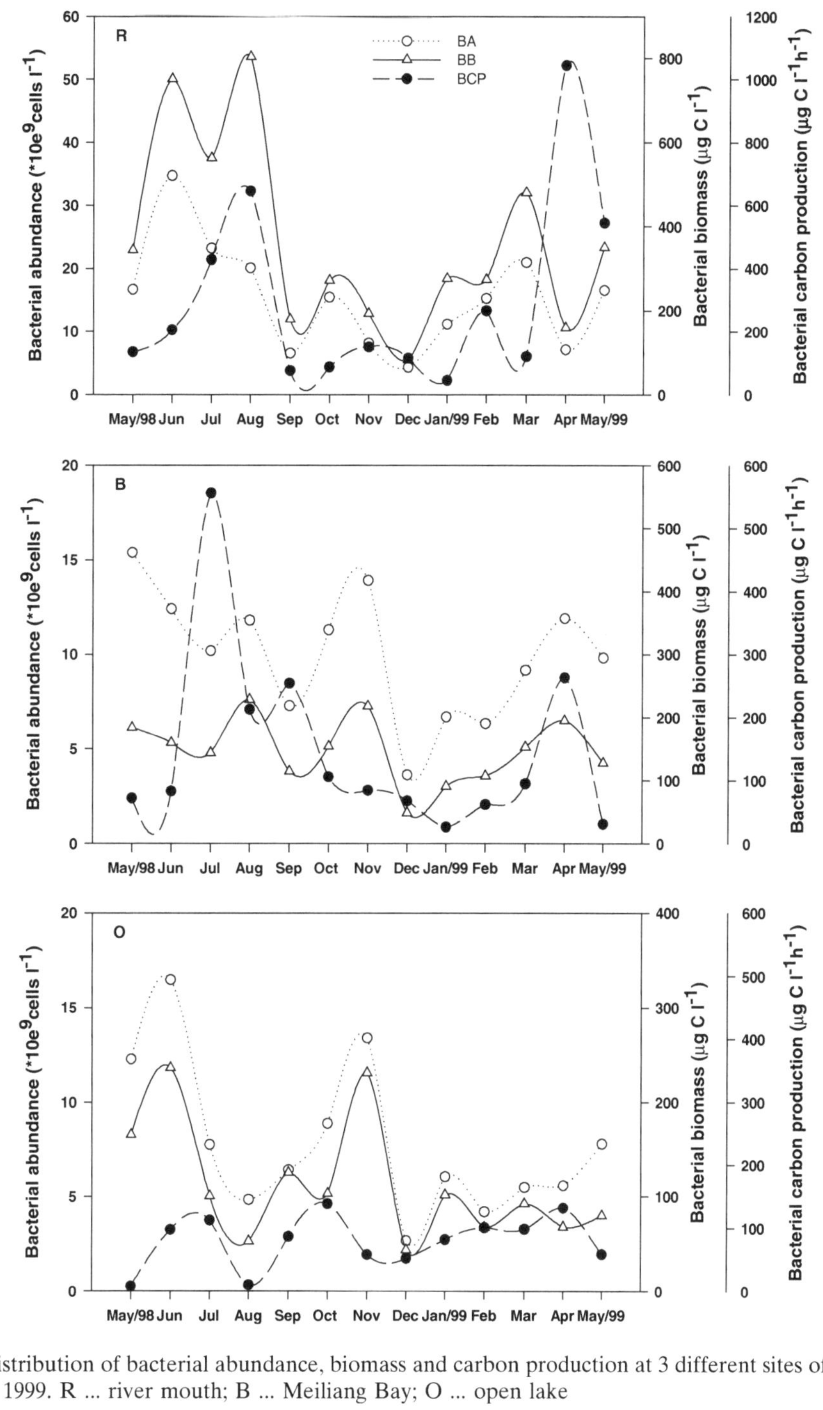

Fig. 2 Seasonal distribution of bacterial abundance, biomass and carbon production at 3 different sites of Lake Taihu from May 1998 to May 1999. R ... river mouth; B ... Meiliang Bay; O ... open lake

Bacterial biomass

The bacterial biomass also markedly increased along the trophic gradient of Lake Taihu (Fig. 2). At the hypertrophic river mouth, the mean bacterial biomass was 364 ± 225 µg C l^{-1}, i.e. ca. 1.5 times and 2.2 times higher than that at eutrophic Meiliang Bay and the open lake, respectively.

The difference between the highest and lowest value was ca. tenfold at the river mouth, while only ca. fivefold were recorded both in the Meiliang Bay and the open lake. No significant differences for seasonality were found in different trophic area of this lake (One-way ANOVA, $P > 0.05$).

Active bacteria

A highly significant positive correlation was found between the number of active bacteria and the total bacterial abundance in all parts of this lake, although this relationship varied among the habitats (Fig. 3.). In the hypertrophic river mouth, the mean ratio of active bacteria versus total bacterial abundance (34%) was slightly higher than that in the eutrophic Meiliang Bay (29%) and the meso/eutrophic open lake (28%). Variation of the percentage of active cells between the different trophic sites was smaller than the variation in abundance and biomass.

Bacterial production

The bacterial [^{3}H]-*TdR* incorporation rate also increased along the trophic gradient, and showed a significantly different over time (One-way ANOVA, $P < 0.0001$, $n = 65$) and space (One-way ANOVA, $P = 0.04$, $n = 65$), although the distribution patterns were similar in all of the sites. At the river mouth, with the maximum value of 14603 pmol l^{-1} h^{-1}, the annual mean value of [^{3}H]-*TdR* incorporation rate was ca. 1.2 times than Meiliang Bay and 2.2 times higher than the open lake, respectively (Table 1).

While the turnover times estimated from cell production were extremely fast and highly variable among the sampling sites (Fig. 3). At the hypertrophic river mouth, the mean turnover times was only ca. 87% less than the open lake, although the cell production was ca. 3 times higher (Fig. 3). Differences of turnover times within the temporal (One-way ANOVA, $P > 0.05$, $n = 65$) and spatial (One-way ANOVA, $P > 0.05$, $n = 65$) distribution were not significant.

Bacterial carbon production at different sites showed a similar distribution pattern as bacterial [^{3}H]-*TdR* incorporation rate, particularly at the river mouth and in Meiliang Bay (Fig. 2). With a mean of 141.8 ± 52.8 μg C l^{-1} h^{-1} at the river mouth, bacterial carbon production was ca. 2.4 and 4.5 times higher than means at Meiliang Bay and the open lake, respectively.

Temporal distribution of bacterial incorporation rates at different sites were similar for both [^{14}C]-Leu and [^{3}H]-*TdR*, even though the value for [^{14}C]-Leu was much higher (Figs. 3 and 4). No significant differences were found between bacterial cell production based on either method (Paired *t*-test, $P > 0.05$). However, turnover times estimated from [^{3}H]-*TdR* and [^{14}C]-Leu at the river mouth (One-way ANOVA, $P = 0.002$, $n = 39$) and Meiliang Bay (One-way ANOVA, $P = 0.001$, $n = 39$) were significantly different, although cell production based on both methods was not significantly different. In the open lake, no significant difference was found (One-way ANOVA, $P > 0.05$, $n = 39$).

While a clear positive correlation between [^{3}H]-*TdR* and [^{14}C]-Leu measurement was found, bacterial carbon production measured by [^{14}C]-Leu was ca. 2 times higher than that of [^{3}H]-*TdR* (Fig. 4). Since both methods are based on measuring some aspect of macromolecular synthesis (DNA for *TdR*, Protein for Leu), it is possible to convert these incorporation rates to rates of macromolecular synthesis and, in turn, to biomass production. But this conversion depends on detailed information about several cellular components, thus bacterial cell size and biomass were measured at the same time (Figs. 3 and 5).

Discussion

Factors affecting bacterial abundance and activity

Bacterioplankton assemblages often are dominated by inactive or dormant cells, and only a small fraction of bacterial cells appear to be metabolically active in aquatic systems (Stevenson, 1978; Mason et al., 1986). However, the fraction of active cells varies in time and space (Zimmerman et al., 1978; Quinn, 1984). Over a broad range of aquatic systems, the total number

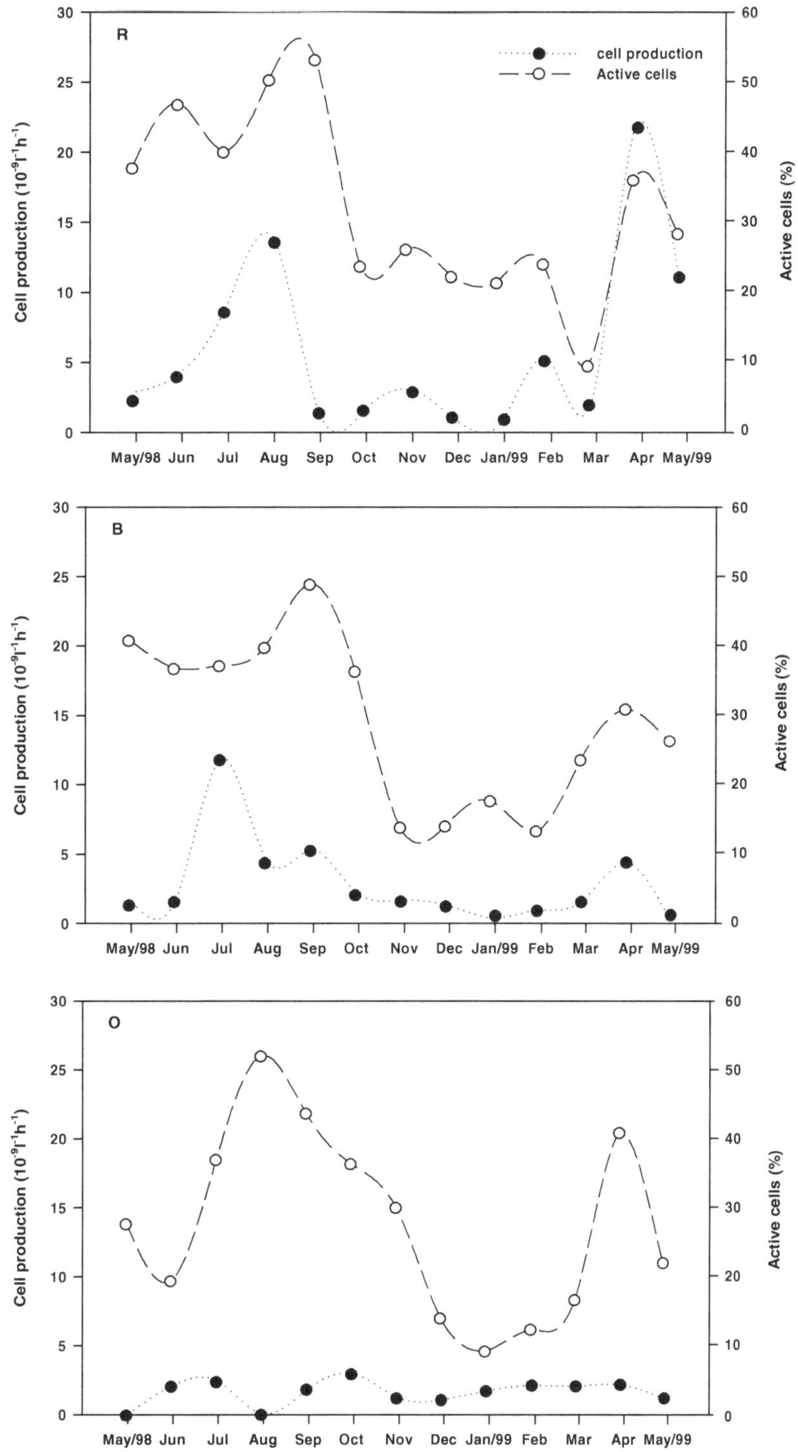

Fig. 3 Seasonal distribution of cell production rates and percentage of active cells at 3 different sites of Lake Taihu from May 1998 to May 1999. R ... river mouth; B ...Meiliang Bay; O ... open lake

Table 1 Leucine and Thymidine incorporation rates (pmol l^{-1} h^{-1}) at different sites of Lake Taihu

		TdR	Leu	*TdR*/Lue
R	Mean ± S.E	2878.8 ± 1069.0	92.2 ± 25.3	31.2
	Range	343.3–14603.2	14.0–323.9	24.5–45.1
B	Mean ± S.E	1338.8 ± 409.3	45.6 ± 12.6	29.4
	Range	195.2–4755.8	8.1–172.4	24.1–27.6
O	Mean ± S.E	886.5 ± 150.0	24.7 ± 3.5	35.9
	Range	68.1–2124.1	2.5–43.0	27.2–49.4

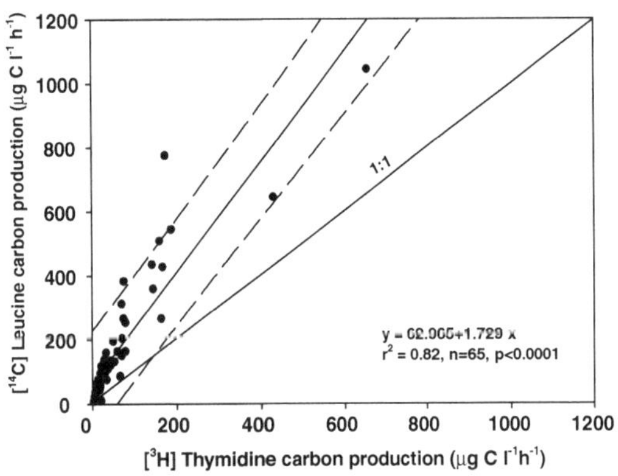

Fig. 4 Correlation of bacterial carbon production measured by [^{3}H] *TdR* and [^{14}C] leu in Lake Taihu from 1998 to 1999. $r^2 = 0.82$, $n = 65$, $p < 0.0001$. The dashed line shows the 95% confidence limits

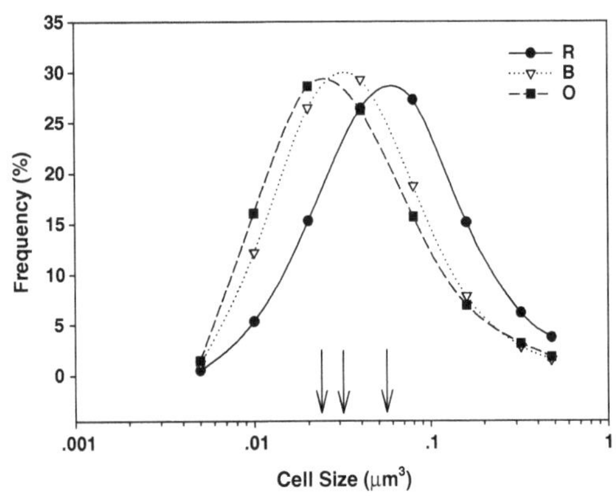

Fig. 5 The distribution of bacterial cell size at 3 different sites of Lake Taihu. R ... river mouth; B ... Meiliang Bay; O ... open lake

of bacteria varied by three orders of magnitude, active bacteria varied by four orders of magnitude, and the proportion of active cells increased systematically along gradients of enrichment from ultra-oligotrophic open ocean areas (<5%) to highly productive estuaries (>50%) (Del Giorgio & Scarborough, 1995).

With lowest bacterial abundance in the open lake, bacterial density showed a clear trend in relation to trophic status in Lake Taihu (Fig. 2), corresponding to the overall trend of abundance increase with system enrichment (Weisse, 1991). Bacterial abundance in the highly contaminated river mouth, however, was only slightly higher and not significantly different from those observed in eutrophic Meiliang Bay. Similar results were obtained by Azam et al. (1983) and Bird & Kalff (1984), who also found that fluctuations of bacterial biomass are small across trophic gradients. The low rate of increase in total bacterial density may imply that total bacterial biomass tends to be more uniform among systems, and their relative contribution to community biomass declines with increasing system productivity (Fuhrman et al., 1989). This indicates that structural and functional characteristics of the 'microbial loop' may be operating differently in stressed versus unstressed ecosystems (Munawar & Weisse, 1989).

With a mean ratio of ca. 34%, 29% and 28% in the river mouth, Meiliang Bay and open lake, respectively, the ratio of metabolically active bacteria to total bacterial abundance does not always correspond to these environmental gradients, although a significant positive correlation between total bacterial abundance and active bacteria existed in all parts of Lake Taihu (Fig. 3). However, data on total and active bacteria from the literature for aquatic systems show that our data are in the range of pelagic systems in general (Del Giorgio & Scarborough, 1995). The estimated average proportion of active cells in this work was similar to the mean for published data on lake bacteria (23%) (Del Giorgio & Scarborough, 1995), although some papers show that this ratio can reach 100% in a hypertrophic lake (Sommaruga, 1995). In most aquatic systems, there is a pool of rapidly growing cells embedded in a larger matrix of inactive bacteria, and the relative size of the active and inactive pools varies systematically along gradients of enrichment (Del Giorgio & Scarborough, 1995). This large variance among systems may also reveal that there are differential control mechanisms for active and inactive bacteria in different ecosystems.

Although Billen (1990) suggested that the lack of simultaneous changes in phytoplankton and bacterioplankton abundance on a seasonal scale is caused, in many cases, by the delayed response of bacteria to phytoplankton development, it is reasonable that the number of active cells is more variable among systems than the total number of bacteria, and that only metabolically active bacterioplankton are responsible for growth, nutrient uptake and transformation of organic carbon substrates (Del Giorgio & Scarborough, 1995). Moreover, this variation may be masked by a rather large and constant pool of inactive bacteria. Thus, the function of bacterioplankton, based on the total number of bacteria, may have to be revised to accommodate large variations in the proportion of metabolically active cells (Del Giorgio & Scarborough, 1995). However, like other planktonic components, bacterioplankton density and production are correlated positively to system productivity, such as nutrient concentration, dissolved organic carbon, chlorophyll *a*, and primary production (Cole et al., 1988).

Leucine and Thymidine incorporation

Although there are still technical problems, incorporation of [^{3}H] thymidine (*TdR*) into DNA and [^{14}C] leucine (Leu) into protein are considered the most reliable methods (Fuhrman & Azam, 1980; Chin-Leo & Kirchman, 1988). However, accurate estimates of bacterial production depend on use of an appropriate conversion factor (Chin-Leo & Kirchman, 1988), which may vary (even by more than a factor of 100) over short time and space scales (Kirchman et al., 1986; Chin-Leo & Kirchman, 1988; Smits & Riemann, 1988). In this case, observed variation in thymidine incorporation could be due to changes in conversion factors rather than bacterial production (Kirchman & Hoch, 1988). So, it is necessary to ensure that observed changes in *TdR* incorporation are due to real changes in bacterial production, not in the conversion factor or some artifact (Chin-Leo & Kirchman, 1988).

When both *TdR* and Leu incorporation are measured in a single incubation by the dual-label approach, agreement between independent measurements made simultaneously should separate real changes from artifacts more effectively than a single-label approach (Fig. 4). Thus, we are confident that observed variations reflect real changes in the metabolism of bacterial assemblages (Kirchman et al., 1986). In addition, incorporation rates determined from dual-label incubations differed only slightly from those obtained from single-label incubations (Chin-Leo & Kirchman, 1988). The dual-label procedure, however, can reduce the time and material needed to perform simultaneous measurements of *TdR* and Leu incorporation, while minimizing errors.

Physico-chemical parameters directly or indirectly affecting bacterial production vary over time and relatively short distances (Kirchman & Hoch, 1988), which may cause methodological problems, especially in attempting to measure bacterial production. For this reason, the dual-label approach was used to measure *TdR* and Leu incorporation. In spite of problems inherent to each method, *TdR* and Leu incorporation rates covaried and gave the same qualitative picture of bacterial production in different parts of Lake Taihu (Table 1, Fig. 4). Although bacterial carbon production data lay within the broad range of the literature, bacterial carbon production measured by Leu was ca. 2 to 3 times higher than by *TdR*. This may indicate that conversion factors in Lake Taihu vary by a factor of two or so and differences may be due to methodology. Servais (1992) suggested that bacterial production measured by *TdR* seems to be systematically underestimated compared with those measured by Leu.

Bacterial production and its coupling with primary production

Bacterial production can be used to estimate average growth rate of bacterial assemblages and as an indicator of bacterial response to fluctuations in environmental conditions (Chin-Leo & Kirchman, 1988). Bacterial abundance and production are correlated positively with phytoplankton biomass and primary production in a broad range of freshwater and marine systems, and bacterial production varied from 2% to >80% of primary production (Bird & Kalff, 1984; Cole et al., 1988). However, bacterial abundance and production do not increase proportionally with chlorophyll concentration as well as primary production (Robarts et al., 1994), although a positive empirical relationship was found between bacterial abundance and chlorophyll concentration in fresh and marine waters (Bird & Kalff, 1984).

Lake Taihu data show that bacterioplankton production was higher at the river mouth than in the open lake (Table 1, Fig. 3). Bacterial production (both measured by *TdR* and Leu) and phytoplankton biomass at the river mouth and in Meiliang Bay were correlated positively, but not in the open lake. The bacterial production to primary production ratio varied among the sampling sites (One-way ANOVA, $P = 0.023$, $n = 65$) and increased along the trophic gradient with a mean value ca. 2 and 4 times higher than in Meiliang Bay and the open lake, respectively. At the same time, no significant correlation between primary production and bacterial secondary production based on [^{3}H]-*TdR* and [^{14}C]-Leu measurement was found in any part of Lake Taihu.

The mean ratio of bacterioplankton production to primary production (Table 2) presented a similar picture to that in humic-rich reservoirs, where bacterioplankton production was usually ca. 2 to 4 times higher than primary production (Balogh & Vörös, 1997). These authors found that lakes receiving large amounts of humic compounds contain larger standing stocks of pelagic bacteria than less humic lakes, suggesting a large non-phytoplankton carbon source. Our results support the hypothesis that bacterial growth in lakes largely occurs at the expense of allochthonous carbon sources rather than phytoplankton production, especially at the river mouth and in Meiliang Bay, where organic matter is entering from the river. Moreover, the ratio of bacterioplankton to primary production decreases with trophic status (Table 2), which is in agreement with other results (Cole et al., 1988), who associated increasing eutrophication and bacterioplankton production, and a decreasing ratio of bacterioplankton biomass and production to Chl *a*. They suggested that the relative importance of bacterioplankton in lake ecosystems seem to decrease markedly with trophic status of the studied ecosystems (Cole et al. 1988).

Conclusions

With the highest value in the hypertrophic river mouth, the abundance, biomass and production of pelagic bacteria markedly increased along the trophic gradient of Lake Taihu (the open lake, Meiliang Bay, and the river mouth). A positive correlation between bacterial production (measured by *TdR* and Leu) and phytoplankton biomass was found in the river mouth and Meiliang Bay, but not in the open lake. The bacterial to primary production ratio increased with trophic status. No significant correlation between primary production and bacterial secondary production was found in any part of Lake Taihu. Our results support the hypothesis that bacteria and lake eutrophication are tightly

Table 2 Comparison of primary production and bacterial secondary production (μg C l^{-1} h^{-1}) at different sites of Lake Taihu from May 1998 to May 1999

		PP	*TdR*	Leu	*TdR*/PP	Leu/PP
R	Mean ± S.E	79.1 ± 8.1	141.8 ± 52.8	297.1 ± 81.5	1.8 ± 0.5	4.0 ± 1.0
	Range	37.0–138.4	17.0–654.5	45.1–1044.1	0.2–5.9	0.5–11.6
B	Mean ± S.E	76.0 ± 9.4	41.2 ± 7.9	147.1 ± 25.6	0.8 ± 0.2	3.1 ± 0.7
	Range	19.8–309.1	4.2–173.8	16.2–775.3	0.1–4.7	0.2–19.7
O	Mean ± S.E	76.1 ± 9.3	25.8 ± 4.7	79.8 ± 11.4	0.5 ± 0.1	1.5 ± 0.4
	Range	18.9–135.6	1.8–65.3	8.2–138.7	0.01–1.6	0.1–5.3

* The primary production (PP) was calculated from Chl *a* concentration by using the empirical formula: PP = –0.63 + 0.197 × Chl *a*, ($r = 0.90$) (Huang et al., 2001), where PP (mg O_2 m^{-2} d^{-1}), Chl *a* (mg m^{-3}), by using the factor of 2.67 to convert the O_2 unit to C unit (Sorokin, 1999)

coupled and bacterial growth in lakes largely occurs at the expense of allochthonous carbon sources other than phytoplankton production (Balogh & Vörös, 1997), especially at the river mouth and in Meiliang Bay. Moreover, the ratio of bacterioplankton to primary production decreases with trophic status (Table 2)

Acknowledgements This study was supported by the key Projects of CAS (No. KZCX1-SW-12, KZCX-SW-01-15). G. Gao was supported by a PhD scholarship of OeAD, the Austrian Exchange Service.

References

Azam, F., T. Fenchel, J. S. Field, J. S. Gray, L. A. Meyer-Reil & F. Thingstad, 1983. The ecological role of water-column microbes in the sea. Marine Ecology Progress Series, 10: 257–263.

Billen, G., 1990. Delayed development of bacterioplankton with respect to phytoplankton: a clue for understanding their trophic relationships. Archiv für Hydrobiologie Beihefte Ergebnisse der Limnologie 34: 191–201.

Bird, D. F. & J. Kalff, 1984. Empirical relationship between bacterial abundance and chlorophyll concentrations in fresh and marine waters. Canadian Journal of Fisheries and Aquatic Sciences 41: 1015–1023.

Chin-Leo, G. & D. L. Kirchman, 1988. Estimating bacterial production in marine water from the simultaneous incorporation of thymidine and leucine. Applied and Environmental Microbiology 54: 1934–1939.

Cole, J. J., S. Findlay & M. L. Pace, 1988. Bacterial production in fresh and saltwater ecosystems: a cross–system overview. Marine Ecology Progress Series 43: 1–10.

Davis, P. G. & J. M. Sieburth, 1984. Estuarine and oceanic microflagellate predation of actively growing bacteria: estimation by frequency of dividing – divided bacteria. Marine Ecology Progress Series 19: 237–246.

Del Giorgio, P. A. & G. Scarborough, 1995. Increase in the proportion of metabolically active bacteria along gradients of enrichment in freshwater and marine plankton: implications for estimates of bacterial growth and production rates. Journal of Plankton Research 17: 1905–1924.

Dufour, P. & M. Colon, 1992. The tetrazolium reduction method for assessing the viability of individual bacterial cells in aquatic environments: improvements, performance and applications. Hydrobiologia 232: 211–218.

Fuhrman, J. A. & F. Azam, 1980. Bacterioplankton secondary production estimates for coastal waters of British Columbia, Antarctica, and California. Applied and Environmental Microbiology 39: 1085–1095.

Fuhrman, J. A. & F. Azam, 1982. Thymidine incorporation as a measure of heterotrophic bacterial production in marine surface waters: evaluation and field results. Marine Biology 66: 109–120.

Fuhrman, J. A., T. D. Sleeter, C. A. Carlson & L. M. Proctor, 1989. Dominance of bacterial biomass in the Sargasso Sea and its ecological implications. Marine Ecology Progress Series 57: 207–217.

Hobbie, J. E., R. J. Daley & S. Jasper, 1977. Use of nucleopore filters for counting bacteria by fluorescence microscopy. Applied and Environmental Microbiology 33: 1225–1228.

Huang, Y. P., C. X. Fan, P. M. Pu, J. F. Jiang & Q. Y. Dai, 2001. The Water Environment and Pollution Control in Lake Taihu. Sciences Press, Beijing (in Chinese).

Kirchman, D. L., E. Knees & R. Hodson, 1985. Leucine incorporation and its potential as a measure of protein synthesis by bacteria in natural aquatic systems. Applied and Environmental Microbiology 49: 599–607.

Kirchman, D. L. & M. P. Hoch, 1988. Bacterial production in the Delaware Bay estuary estimated from thymidine and leucine incorporation rates. Marine Ecology Progress Series 45: 169–178.

Kirchman, D. L., S. Y. Newell & R. E. Hodson, 1986. Incorporation versus biosynthesis of leucine: implications for measuring rates of protein synthesis and biomass production by bacteria in marine systems. Marine Ecology Progress Series 32: 47–59.

Krambeck, C., H. J. Krambeck & J. Overbeck, 1981. Microcomputer assisted biomass determination of plankton bacteria on scanning electron micrographs. Applied and Environmental Microbiology 42: 142–149.

Loferer-Krößbacher, M., K. P. Witzel & R. Psenner, 1998. Determination of bacterial cell dry mass by transmission electron microscopy and densitometric image analysis. Applied and Environmental Microbiology 64: 688–694.

Mason, C. A., G. Hamer & J. D. Bryers, 1986. The death and lysis of microorganisms in environmental processes. FEMS Microbiology Review 39: 373–401.

Munawar, M. & T. Weisse, 1989. Is the 'microbial loop' an early warning indicator of anthropogenic stress? Hydrobiologia 188/189: 163–174.

Porter, K. G. & Y. S. Feig, 1980. The use of DAPI for identifying and counting aquatic microflora. Limnology and Oceanography 25: 943–948.

Posch, T., M. Loferer-Krößbacher, G. Gao, A. Alfreider, J. Pernthaler & R. Psenner, 2001. Precision of bacterioplankton biomass determination: a comparison of two fluorescent dyes, and of allometric and linear volume-to-carbon conversion factors. Aquatic Microbial Ecology 25: 55–63.

Psenner, R., 1991a. Detection and sizing of aquatic bacteria by means of epifluorescence microscopy and image analysis. Microscope Analysis 26: 13–15.

Psenner, R., 1991b. Determination of bacterial cell volumes by image analysis. Verhandlungen Internationale Vereinigung Für Theoretische und Angewandte Limnologie 24: 2605–2608.

Quinn, J. P., 1984. The modification and evaluation of some cytochemical techniques for the enumeration of

metabolically active heterotrophic bacteria in the aquatic environment. Journal of Applied Bacteriology 57: 51–57.
Raymond, L., J. R. Kepner & J. R. Pratt, 1994. Use of fluorochromes for direct enumeration of total bacteria in environmental samples: past and present. Microbiology Review 58: 603–615.
Robarts, R. D., M. T. Arts, M. S. Evans & M. J. Waiser, 1994. The coupling of heterotrophic bacterial and phytoplankton production in a hypertrophic, shallow prairie lake. Canadian Journal of Fisheries and Aquatic Sciences 51: 2219–2226.
Servais, P., 1992. Bacterial production measured by ^{3}H-thymidine and ^{3}H-leucine incorporation in various aquatic ecosystems. Archiv für Hydrobiologie Beihefte Ergebnisse der Limnologie 37: 73–81.
Simon, M. & F. Azam, 1989. Protein content and protein synthesis rates of planktonic marine bacteria. Marine Ecology Progress Series 51: 201–213.
Smits, J. D. & B. Riemann, 1988. Calculation of cell production from [^{3}H]thymidine incorporation with freshwater bacteria. Applied and Environmental Microbiology 54: 2213–2219.
Sommaruga, R., 1995. Microbial and classical food web: A visit to a hypertrophic lake. FEMS Microbiology Ecology 17: 257–270.
Sorokin, Y. I., 1999. Aquatic microbial ecology. Backhuys Pbulishers, Leiden, Netherlands.
Stevenson, L. H., 1978. A case for bacterial dormancy in aquatic systems. Microbial Ecology 4: 127–133.
Thom, S. M., R. W. Horobin, E. Seidler & M. R. Barer, 1993. Factors affecting the selection and use of tetrazolium salts as cytochemical indicators of microbial viability and activity. Journal of Applied Bacteriology 74: 433–443.
V.-Balogh, K. & L. Vörös, 1997. High bacterial production in hypertrophic shallow reservois rich in humic substances. Hydrobiogia 342/343: 63–70.
Weisse, T., 1991. The microbial food web and its sensitivity to eutrophication and contaminant enrichment: a cross–system overview. Internationale Revue der gesamten Hydrobiologie 76: 327–337.
Wetzel, R. G. & G. E. Likens, 2000. Limnological Analyses, 3 edn. Springer-Verlag, New York.
Zimmerman, R., R. Iturriaga & J. Becker-Birck, 1978. Simultaneous determination of the total number of bacteria and number thereof involved in respiration. Applied and Environmental Microbiology 36: 926–935.

Hydrobiologia (2007) 581:189–193
DOI 10.1007/s10750-006-0497-1

EUTROPHICATION IN LAKES

Floating-leaved macrophyte (*Trapa quadrispinosa* Roxb) beds have significant effects on sediment resuspension in Lake Taihu, China

Peisheng Huang · Boping Han · Zhengwen Liu

Abstract *Trapa quadrispinosa* Roxb is a common floating-leaved macrophyte in China. In this study, the effects of *Trapa* on sediment resuspension in Lake Taihu, a large, shallow, eutrophic lake in eastern China, were investigated using sediment traps. The study was conducted at stations with and without *Trapa* beds from 7 September to 18 October, 2003. Results showed that sediment resuspension rates at the station without *Trapa* were significantly higher than those at the station with *Trapa*. During the study period (41 days), 10,970 g dw m^{-2} of surface sediment was resuspended within the *Trapa* beds and 29,903 g dw m^{-2} in the pelagic zone. These equate to phosphorus fluxes of 7.4 g m^{-2} to the water column at the station with *Trapa* and 16.1 g m^{-2} at the station without *Trapa*. Floating-leaved macrophytes, such as *Trapa*, may be important in sediment resuspension and internal phosphorus loading in shallow lakes.

Guest editors: B. Qin, Z. Liu & K. Havens
Eutrophication of shallow lakes with special reference to Lake Taihu, China

P. Huang · B. Han · Z. Liu (✉)
Institute of Hydrobiology, Jinan University,
Guangzhou 510632, China
e-mail: zliu@niglas.ac.cn

P. Huang · Z. Liu
Nanjing Institute of Geography and Limnology,
Chinese Academy of Sciences, Nanjing 210008, China

Keywords Lake Taihu · Sediment resuspension · Floating-leaved macrophytes

Introduction

Many shallow lakes have become eutrophic as the result of anthropogenic influences. Eutrophication often results in decreased water transparency and subsequent declines in macrophyte abundance. Macrophytes are important to nutrient cycling in aquatic systems, particularly in shallow lakes, in many ways: nutrient incorporation in plant biomass, inhibiting phosphorus release from sediments, enhancement of sedimentation and reducing sediment resuspension. Sediment resuspension may redistribute settled materials, increase water turbidity and, thus, affect light conditions, nutrient fluxes, photosynthesis, phytoplankton and zooplankton community structure and abundance, fish feeding, and growth of submerged macrophytes (Kristensen et al., 1992; Søndergaard et al., 1992; Carrick et al., 1993; Ogilvie & Mitchell, 1998; Scheffer, 1999; Saulnier & Mucci, 2000; Madsen et al., 2001). Hence, factors controlling sediment resuspension rates have substantial consequences for the lake ecosystem.

Sediment resuspension has direct and indirect effects on aquatic macrophytes. Conversely, aquatic macrophytes can reduce sediment resuspension by attenuating wave and current energy and, thus, increasing net sedimentation rate (Madsen et al., 2001). Effects of submerged macrophytes on sediment resuspension have been the focal point of intensive research (Madsen et al., 2001; Crossley et al., 2002). Effects of emergent and submerged macrophytes on sediment resuspension also have been studied in recent years (Dieter, 1990; Horppila & Nurminen, 2001). However, similar studies have rarely applied to floating-leaved macrophytes, common in shallow lakes. In this study, sediment traps were used to evaluate effects of floating-leaved macrophytes on sediment resuspension in large, shallow, eutrophic Lake Taihu (China).

Methods and materials

The study site was located along the northern coast of Lake Taihu, one of the largest freshwater lakes in China (Qin et al., 2007, Fig. 1). The study was carried out from 7 September to 18 October, 2003, at different stations within a stand of a floating-leaved macrophyte, *Trapa quadrispinosa* Roxb. The station with *Trapa* was located inside the *Trapa* stand ~20 m from the outer edge, and the station without *Trapa* was outside the *Trapa* stand ~30 m from the stand edge. Plant coverage was about 80% of the lake surface within the *Trapa* stand during the study.

Triplicate sediment traps were placed at each station to determine the gross sedimentation rate. The traps were constructed from a cylindrical stainless steel pipe and had an inside diameter of 60 mm and a length: width ratio of 5:1. The tops of the traps were 40 cm above the sediments. Each trap was attached to a fine rope, which was fastened to an anchor and a subsurface buoy.

The traps were recovered after 10–14 days of deployment. In the laboratory, trap contents were removed and dried at 60°C to constant weight. Loss on ignition (LOI) of the entrapped material was determined by ignition at 550°C. On each sampling date, triplicate surface (2 cm) sediment samples were taken from each station with a corer and analyzed for LOI and total phosphorus (TP). For TP analysis air-dry sediment samples were digested with sulphuric acid, then P was determined using the acidic molybdate-ascorbic acid (AMAA) method (Murphy & Riley, 1962). Seston samples were taken with a tube sampler, then filtered onto GF/C filters and analyzed for suspended solids and LOI.

Sediment resuspension rate was calculated using the method of Gasith (1975), which is referred to as the "label approach" by Floderus

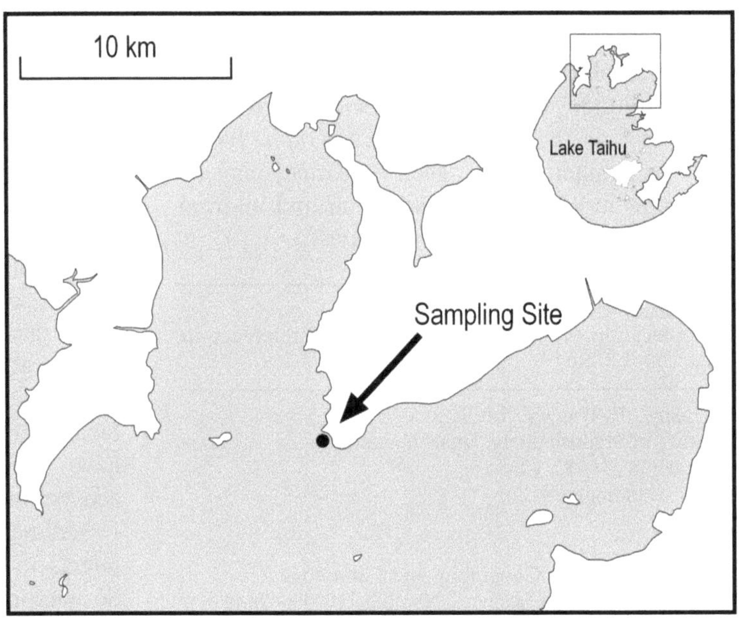

Fig. 1 Lake Taihu and the sampling site

(1989) and has been modified by several authors (Bloesch, 1994). The method uses the equation:

$$R = S\frac{(f_S - f_T)}{(f_R - f_T)}$$

where R is resuspended sediment (g dry weight); S is entrapped settling flux (g dry weight); f_S is the organic fraction (%) of S; f_R is the organic fraction (%) of R and f_T is the organic fraction (%) of seston (T) suspended in the water column. In general, organic matter is lower in resuspended sediments than fresh material (Bloesch, 1994). According to Blomqvist & Hakanson (1981), this method is reliable if the organic content of seston (f_T) is significantly different from surface sediment (f_R). The rate of P resuspension at each station was estimated using the calculated resuspension rate and P concentration in surface sediments.

ANOVA (SPSS 10.0) was used to test for significant differences between the data of the stations with and without *Trapa*.

Results

The organic fraction of suspended seston ranged from 21.3% to 50.3% (Table 1) and was significantly higher than the surface sediments (f_R) ($P < 0.001$), suggesting that the method can be used reliably in this study. The organic fraction of entrapped material was above 12% at the station with *Trapa* and significantly higher than the station without *Trapa* ($P < 0.001$). For surface sediments, the organic fraction at the *Trapa* station was also significantly higher than the no *Trapa* station (Table 1).

Gross sedimentation rates ranged from 128 to 726 and 423–1503 g dw m^{-2} day^{-1} for the station with and without *Trapa*, respectively (Fig. 2). Highest values for both stations were recorded 7–18 October, 2003. The sedimentation rate at the station with *Trapa* was significantly lower than the station without *Trapa* ($P < 0.001$). The resuspension rate at the *Trapa* station ranged from 105 to 566 g dw m^{-2} day^{-1} and was significantly lower than the no *Trapa* station, which ranged from 328 to 1505 g dw m^{-2} day^{-1} (Fig. 2). At both stations the sedimentation rate was related to the sediment resuspension rate ($R^2_{Trapa} = 0.999$, $R^2_{No\ Trapa} = 0.996$).

Total P concentration in surface sediments at the *Trapa* station had a narrow range and was always above 0.64 mg g dw^{-1}, while the station without *Trapa* was always below 0.64 mg g dw^{-1} (Table 2). The resuspension rate for TP ranged from 68 to 362 and 85 to 963 mg day^{-1} m^{-2} at the *Trapa* and no *Trapa* stations, respectively. During the study period (41 days), 10,970 g dw m^{-2} was resuspended within *Trapa* beds and 29,903 g dw m^{-2} in the pelagic zone. With the sediments, 7.4 g P m^{-2} entered the water column at the station with *Trapa* and 16.1 g P m^{-2} for the station without *Trapa*.

Discussion

In shallow waters, the concentration of suspended particles in the water column depends on the continuous processes of sedimentation and

Table 1 The organic fraction (%, ±95% confidence limits) of surface sediment, entrapped material and suspended seston in both sampling station during the study period

	17 September	27 September	7 October	18 October
Sediments				
Station with *Trapa*	8.6 ± 0.2	8.2 ± 1.0	8.0 ± 1.0	7.9 ± 0.0
Station without *Trapa*	6.7 ± 0.5	5.9 ± 0.4	7.0 ± 0.3	7.9 ± 1.0
Entrapped material				
Station with *Trapa*	12.9 ± 0.2	14.4 ± 0.6	12.2 ± 1.0	12.5 ± 2.5
Station without *Trapa*	10.8 ± 0.1	10.3 ± 0.4	8.5 ± 1.5	8.0 ± 0.1
Seston				
Station with *Trapa*	34.6 ± 1.7	32.6 ± 6.4	26.8 ± 1.1	31.3 ± 5.2
Station without *Trapa*	50.3 ± 2.3	22.3 ± 8.9	21.3 ± 3.1	38.1 ± 0.6

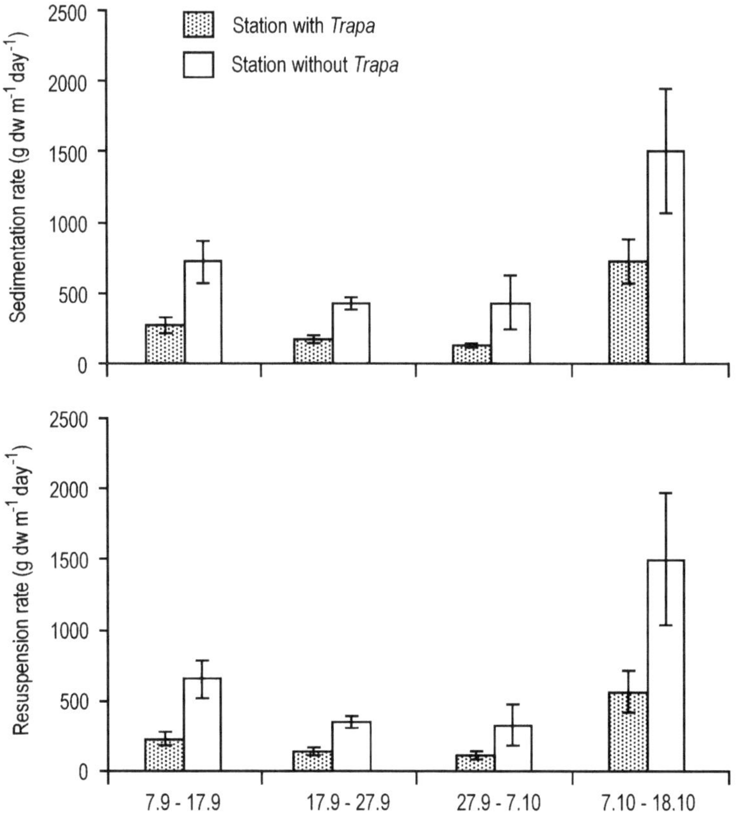

Fig. 2 Sedimentation and resuspension rates at stations with and without a floating-leaved macrophyte *Trapa quadrispinosa* (mean ± SD)

Table 2 Concentration of total phosphorus (mg g dw^{-1}) in the surface sediments (mean ± 95% confidence limits)

	17 September	27 September	7 October	18 October
Station with *Trapa*	0.75 ± 0.07	0.71 ± 0.08	0.65 ± 0.04	0.64 ± 0.01
Station without *Trapa*	0.40 ± 0.18	0.24 ± 0.13	0.63 ± 0.14	0.64 ± 0.09

resuspension. Macrophyte coverage is an important factor influencing sedimentation and resuspension in lakes (Dieter, 1990; Horppila & Nurminen, 2001; Madsen et al., 2001). In Lake Hiidenvesi, sediment resuspension in a stand of emergent *Typha angustifolia* was significantly lower than outside the stand (Horppila & Nurminen, 2001). In the present study, the mean resuspension rate within the *Trapa* stand was 36% of that in the open water, suggesting that this floating-leaved macrophyte attenuates sediment resuspension.

The study area was located along the northern coast of Lake Taihu, which is exposed to frequent southwest wind. *Microcystis* blooms often accumulate in this region and subsequently settle within the *Trapa* beds. This may be responsible in part for the high organic fraction in surface sediments and entrapped material at the station with *Trapa* (Table 1). Rooted macrophytes assimilate P mainly from sediments and, therefore, may reduce sediment P concentration (Chen & Barko, 1988). However, P concentration in sediments at the station with *Trapa* was higher than the station without *Trapa*. Reduced resuspension within the *Trapa* beds and organic matter burial, including settled *Microcystis* cells and macrophyte detritus, may explain this observation.

Large shallow lakes are susceptible to resuspension (Carper & Bachmann, 1984). The high regression coefficient between resuspension and sedimentation rates in Lake Taihu suggests that resuspension supplies particles to the water column. This study demonstrated that *Trapa* reduces resuspension and internal P loading. Re-establishment of macrophytes is a strategy for restoration of eutrophic shallow lakes. However, floating-leaved macrophytes usually are ignored in theoretical study and management. Growth of floating-leaved macrophytes is not dependent on light in the water column and may present an alternative for vegetation restoration in eutrophic, turbid lakes, such as Lake Taihu, at least in some areas.

Acknowledgements The study was supported jointly by the Key Project of CAS (KZCX1-SW-12), "863" Project (2002AA601011-04), a NSFC Project (40371103) and Jinan Science Fundation. The authors are grateful to H. Pan and H. Wang for their assistance in sampling and chemical analyses, to B. Gu and M. McCarthy for his valuable comments and correcting English. The comments of anonymous reviewers were helpful and acknowledged.

References

Bloesch, J., 1994. A review of methods used to measure sediment resuspension. Hydrobiologia 284: 13–18.

Blomqvist, S. & L. Hakanson, 1981. A review on sediment traps in aquatic environments. Archiv für Hydrobiologie 91: 101–132.

Carper, G. L. & R. W. Bachmann, 1984. Wind resuspension of sediments in a prairie lake. Canadian Journal of Fisheries and Aquatic Sciences 41: 763–7.

Carrick, H. J., F. J. Aldridge & C. L. Schelske, 1993. Wind influences phytoplankton biomass and composition in a shallow, productive lake. Limnology and Oceanography 38(6): 1179–1192.

Chen, R. J. & J. W. Barko, 1988. Effects of freshwater macrophytes on sediment chemistry. Journal of Freshwater Ecology 4: 279–289.

Crossley, M. N., W. C. Dennison, R. R. Williams & A. H. Wearing, 2002. The interaction of water flow and nutrients on aquatic plant growth. Hydrobiologia 489: 63–70.

Dieter, C. D., 1990. The importance of emergent vegetation in reducing sediment resuspension in wetlands. Journal of Freshwater Ecology 5: 467–473.

Floderus, S., 1989. The effect of sediment resuspension on nitrogen cycling in the Kattegat-variability in organic matter transport.–PhD Thesis, Dept of Physical Geography, Uppsala University. UNGI Rapport 71. 21 pp.

Gasith, A., 1975. Tripton sedimentation in eutrophic lakes—simple correction for the resuspended matter. Verhandlungen der Internationalen Vereinigung für Limnologie 19: 116–122.

Horppila, J. & L. Nurminen, 2001. The effect of an emergent macrophyte (*Typha angustifolia*) on sediment resuspension in a shallow north temperate lake. Freshwater Biology 46: 1447–1455.

Kristensen, P., M. Søndergaard & E. Jeppesen, 1992. Resuspension in a shallow eutrophic lake. Hydrobiologia 228: 101–109.

Madsen, J. D., P. A. Chambers, W. F. James, E. W. Koch & D. F. Westlake, 2001. The interaction between water movement, sediment dynamics and submersed macrophytes. Hydrobiologia 444: 71–84.

Murphy, J. & J. P. Riley, 1962. A modified single solution method for the determination of phosphate in natural water. Analytica Chimica Acta 27: 31–36.

Ogilvie, B. G. & S. F. Mitchell, 1998. Does sediment resuspension have persistent effects on phytoplankton? Experimental studies in three shallow lakes. Freshwater Biology 40: 51–63.

Qin, B. Q, P. Z. Xu, Q. L. Wu, L. C. Luo & Y. L. Zhang, 2007. Environmental issues of Lake Taihu, China. Hydrobiologia 581: 3–14.

Saulnier, I. & A. Mucci, 2000. Trace metal remobilization following the resuspension of estuarine sediments: Saguenay Fjord, Canada. Applied Geochemistry 15: 203–222.

Scheffer, M., 1999. The effect of aquatic vegetation on turbidity: how important are the filter feeders? Hydrobiologia 408/409: 307–316.

Søndergaard, M., P. Kristensen & E. Jeppesen, 1992. Phosphorus release from resuspended sediment in the shallow and wind-exposed Lake Arresø, Denmark. Hydrobiologia 228: 91–99.

Hydrobiologia (2007) 581:195–207
DOI 10.1007/s10750-006-0496-2

EUTROPHICATION IN LAKES

Nitrogen dynamics and microbial food web structure during a summer cyanobacterial bloom in a subtropical, shallow, well-mixed, eutrophic lake (Lake Taihu, China)

Mark J. McCarthy · Peter J. Lavrentyev · Longyuan Yang · Lu Zhang · Yuwei Chen · Boqiang Qin · Wayne S. Gardner

Abstract Nitrogen dynamics and microbial food web structure were characterized in subtropical, eutrophic, large (2,338 km^2), shallow (1.9 m mean depth), and polymictic Lake Taihu (China) in Sept–Oct 2002 during a cyanobacterial bloom. Population growth and industrialization are factors in trophic status deterioration in Lake Taihu. Sites for investigation were selected along a transect from the Liangxihe River discharge into Meiliang Bay to the main lake. Water column nitrogen and microbial food web measurements were combined with sediment–water interface incubations to characterize and identify important processes related to system nitrogen dynamics. Results indicate a gradient from strong phosphorus limitation at the river discharge to nitrogen limitation or co-limitation in the main lake. Denitrification in Meiliang Bay may drive main lake nitrogen limitation by removing excess nitrogen before physical transport to the main lake. Five times higher nutrient mineralization rates in the water column versus sediments indicate that sediment nutrient transformations were not as important as water column processes for fueling primary production. However, sediments provide a site for denitrification, which, along with nitrogen fixation and other processes, can determine available nutrient ratios. Dissimilatory nitrate reduction to ammonium (DNRA) was important, relative to denitrification, only at the river discharge site, and nitrogen fixation was observed only in the main lake. Reflecting nitrogen cycling patterns, microbial food web structure shifted from autotrophic (phytoplankton dominated) at the river discharge to heterotrophic (bacteria dominated) in and near the main lake.

Guest editors: B. Qin, Z. Liu & K. Havens
Eutrophication of shallow lakes with special reference to Lake Taihu, China

M. J. McCarthy (✉) · W. S. Gardner
Marine Science Institute, The University of Texas at Austin, 750 Channel View Drive, Port Aransas, TX 78373, USA
e-mail: markm@utmsi.utexas.edu

P. J. Lavrentyev
Department of Biology, The University of Akron, Akron, OH 44325, USA

L. Yang · L. Zhang · Y. Chen · B. Qin
Nanjing Institute of Geography and Limnology, Chinese Academy of Sciences, 73 East Beijing Road, Nanjing, Jiangsu 210008, P.R. China

Keywords Lake Taihu · Eutrophication · Nitrogen · Cyanobacterial blooms · Denitrification

Introduction

Nitrogen (N) and phosphorus (P) often limit primary productivity in aquatic systems. Unlike P, dissolved inorganic N (DIN) occurs in different forms and oxidation states. DIN is assimilated and converted to organic N by phytoplankton, other plants, or bacteria. Organic N can be regenerated as ammonium (NH_4^+) or dissolved organic N (DON) compounds, reassimilated by plants or heterotrophs, or oxidized by bacteria.

DIN composition and ratios can affect phytoplankton community structure. For example, NH_4^+ is conducive to cyanobacteria (Blomqvist et al., 1994; Hyenstrand et al., 1998; Dokulil & Teubner, 2000; Jacoby et al., 2000). Unlike diatoms, cyanobacteria are poor competitors for nitrate (NO_3^-; Hyenstrand et al., 1998). Some cyanobacteria can fix atmospheric dinitrogen (N_2), which can be regenerated as bioavailable DIN compounds (Shapiro, 1990; Downing et al., 2001). Cyanobacteria dominance often is regarded as an indicator of eutrophication (Dokulil & Teubner, 2000).

Nitrification coupled with denitrification converts biologically available N forms (NH_4^+ and NO_3^-, respectively) to N_2 gas and may reduce effects of excessive N inputs and eutrophication (Seitzinger, 1988). Nitrate for coupled denitrification derives from organic matter mineralization to NH_4^+ followed by nitrification (Seitzinger, 1988). Dissimilatory NO_3^- reduction to NH_4^+ (DNRA) is an alternative pathway for NO_3^- in sediments (Tobias et al., 2001; An & Gardner, 2002). The relative partitioning between NO_3^- reduction pathways (denitrification versus DNRA) is important since denitrification removes fixed N from the system while DNRA returns it as bioavailable NH_4^+ (Tobias et al., 2001).

The microbial food web (MFW) includes heterotrophic, autotrophic, and mixotrophic prokaryotes and eukaryotes. Multiple cascading trophic interactions within the MFW can affect the biogeochemical N cycle and have feedback effects on trophic conditions in coastal waters (Lavrentyev et al., 1998). MFW structure also is an important factor regulating algal community dynamics in eutrophic, cyanobacteria-dominated systems (Elser, 1999). MFW processes should be included in nutrient-food web models to understand complex processes in aquatic environments (Edwards et al., 2000).

Evaluating internal nutrient cycling and transformations, and conditions enhancing them, in lakes and other aquatic systems help managers address effects of excessive nutrient inputs and eutrophication, such as cyanobacteria dominance. Eutrophication studies in freshwater lakes often focus on P, since it is the most common limiting nutrient. This situation describes the current status of research on Lake Taihu, a large (2,338 km^2), subtropical, shallow (1.9 m mean depth), well-mixed, and eutrophic lake in China (Pu & Yan, 1998). Studies in this lake have been limited to monitoring nutrient sources and dominant compounds (i.e. nutrient concentrations).

Water column NH_4^+ regeneration and uptake, benthic nutrient fluxes and N sinks (i.e. denitrification), water column versus benthic N cycling, and MFW structure are important factors related to trophic status and water quality. These issues have not been addressed in Lake Taihu. The size, depth, trophic status, latitude, and basin land use for Lake Taihu are similar to Lake Okeechobee (Florida, USA; Havens et al., 2001). General information about Lake Taihu geography is presented elsewhere in this issue (Qin et al., 2007).

Lake Taihu was oligotrophic as recently as the 1950s, but increased nutrient inputs related to population and economic growth have led to eutrophication (Cai et al., 1997; Chen et al., 2003). Most pollutants come from rivers discharging into Meiliang Bay and other parts of the lake (Huang, 2000). Meiliang and Wuli Bays (Fig. 1) suffer from severe eutrophication (Huang, 2000). Nutrient concentrations decrease with distance

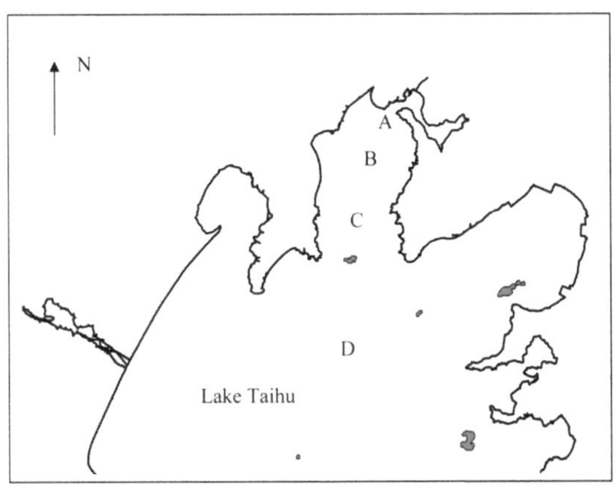

Fig. 1 Map of northern Lake Taihu showing locations of sampling stations in Meiliang Bay. A is the "river" site, B is "inner bay", C is "outer bay", and D is "main lake"

from river outflows (Cai et al., 1997), but whole-lake nutrient concentrations have increased by an order of magnitude (Huang, 2000).

Wind-driven mixing and shallow depth prevent stratification in the lake, and bottom waters remain oxic (Dickman et al., 1998). Phytoplankton diversity has decreased since 1981, but cyanobacteria populations (*Microcystis* and *Anabaena*) have increased (Pu & Yan, 1998) and can comprise 85% of summer phytoplankton biomass (Chen et al., 2003). Annual blooms have clogged intakes at municipal waterworks, interrupted domestic and industrial water supply, disrupted tourism and capture fisheries, and caused cultured fish stock losses (Pu & Yan, 1998). Wind-driven sediment resuspension inhibits photosynthesis below 0.5–1 m depth. However, *Microcystis* use gas vesicles to maintain buoyancy and have a competitive advantage over other phytoplankton by avoiding light limitation (Shapiro, 1990; Chen et al., 2003).

The primary goal of this study was to characterize, for the first time, N dynamics relative to MFW structure in this understudied, eutrophic system. Specific objectives were to: (1) measure water column NH_4^+ regeneration and potential uptake rates; (2) measure sediment N fluxes and transformation rates; (3) compare sediment N transformation rates with water column N cycling rates; and (4) relate microbial plankton composition and distribution to N cycling rates.

Materials and methods

Site description

Four stations were selected along the Liangxihe River discharge gradient from Meiliang Bay to the main lake (Fig. 1). Stations were located at the Liangxihe River discharge into Meiliang Bay (river; "A" on Fig. 1), inner Meiliang Bay (inner bay; "B" on Fig. 1), outer Meiliang Bay (outer bay; "C" on Fig. 1), and the main lake ("D" on Fig. 1). These stations correspond to monitoring stations (0, 1, 3, and 7, respectively) along a larger transect detailed in previous work (Cai et al., 1997; Chen et al., 2003). Sampling occurred on 26 Sept 2002 (outer bay and main lake) and 2 Oct 2002 (river and inner bay). Data from triplicate incubation chambers were averaged for each measured parameter or flux. Standard errors were calculated to determine replicate variation and are reported with means as plus/minus (±) one standard error (SE).

Water column characteristics

Water depth was estimated using a wooden pole and tape measure. Turbidity was estimated using a ~30 cm Secchi disc, and surface water temperature was measured with a mercury thermometer. Dissolved oxygen concentration (DO) was measured using Winkler titration, and chlorophyll *a* concentration (chl) was measured using hot ethanol extraction followed by spectrophotometry (Chen et al., 2003).

Water column nutrient {NO_3^- + nitrite (NO_2^-), NO_2^-, and ortho-phosphate (o-PO_4^{3-})} samples were filtered with 0.2 μm syringe filters (Osmonics) and frozen in 14 ml snap-cap tubes (Falcon). NH_4^+ samples were filtered (0.2 μm syringe filter) and frozen in 8 ml glass vials (Wheaton). Frozen samples were carried to the University of Texas Marine Science Institute (UTMSI) for flow-injection analysis of NO_3^- + NO_2^-, NO_2^-, and PO_4^{3-} (Lachat QuikChem 8000) and high performance liquid chromatographic (HPLC) NH_4^+ analysis (Gardner et al., 1995a). Total N (TN) and total P (TP) were determined by persulfate oxidation and spectrophotometry (Jin & Tu, 1990). Total P was oxidized to PO_4^{3-} at 120°C. TN and TP recovery efficiencies were 98.4% and 99.7%, respectively.

Water column NH_4^+ regeneration and uptake

Water from each site was collected in 2 l plastic bottles, returned to the Taihu Laboratory for Lake Ecosystem Research (TLLER), enriched with 99.8% $^{15}NH_4Cl$ (Isotec; 16 μmol l^{-1} final isotope concentration), and partitioned into triplicate light and dark 70 ml tissue culture bottles (Corning). Dark bottles were wrapped with aluminum foil. Initial samples were collected immediately after enrichment and mixing, filtered (0.2 μm syringe filter), and frozen in 8 ml glass vials (Wheaton). Bottles were incubated in a mesh bag suspended from the TLLER pier into surface water. Intermediate samples were collected after 4 (river and inner bay) or 11 h (outer bay and main lake), filtered, and frozen. Intermediate samples were collected to coincide with day/night transition. Final samples were collected after 20 h, filtered, and frozen. All frozen samples were hand-carried to UTMSI for analysis of total NH_4^+ concentration and atom % ^{15}N using HPLC (Gardner et al., 1995a). Ammonium regeneration and potential uptake rates were calculated from these data using the Blackburn/Caperon isotope dilution model (Blackburn, 1979; Caperon et al., 1979). NH_4^+ uptake rates in this study are qualified as "potential", and no instances of complete isotope substrate depletion were observed.

Sediment–water interface incubations

Triplicate sediment cores (depth 15–20 cm) were collected from each site using a coring device allowing acrylic tube (30.5 cm length, 7.6 cm inside diameter, 0.32 cm wall thickness) insertion into the sediment with minimal disturbance and overlying water retention. Sediment cores with overlying water were sealed immediately at both ends with butyl caps and electrical tape. Near-bottom water from each site was collected into two ~20 l carboys using a submersible pump.

Cyanobacteria formed a "scum" covering the water surface at all sampling stations. The water column also was turbid from wind-driven sediment resuspension. Thus, the sediment–water interface was expected to receive little or no light, and intact sediment cores with overlying water were wrapped with aluminum foil to prevent light effects. Wind-driven turbulence and a shallow water column also maintain bottom water normoxia. Therefore, carboys with bottom water were aerated using an aquarium air pump to maintain oxic conditions in overlying water.

Cores were installed into a flow-through incubation system (Lavrentyev et al., 2000; McCarthy & Gardner, 2003) consisting of aerated bottom water, a multi-channel proportioning pump (Technicon), transmission tubing (Teflon), and an acetol plunger with Viton o-ring. The plunger was positioned ~5 cm above the sediment surface to give ~230 ml of overlying water volume (An & Gardner, 2002). This system maintained a gas-tight setting with inflow (flow rate ≅ 0.072 l h^{-1}) from the bottom water carboy and outflow from positive displacement of overlying water. The cores were placed in a water bath maintained at in situ temperature and allowed to reestablish steady-state overnight.

After the pre-incubation period, discreet inflow (from bottom water carboy) and outflow samples were collected daily for 2 days for nutrient analyses as described above. Nutrient fluxes (in μmol m^{-2} h^{-1}) were calculated by: $(C_o - C_i) \times f/a$, where C_o is the outflow concentration in μmol l^{-1}, C_i is the inflow concentration, f is the flow rate (0.072 l h^{-1}), and a is the sediment surface area (0.0045 m^2).

Dissolved gas samples were collected from core outflow by overflowing 15 ml ground-glass stopper test tubes (Chemglass; 19.8 cm length and 1 cm ID), injecting 200 μl 50% $ZnCl_2$, and inserting the ground-glass stopper quickly while twisting to prevent air bubble entrapment. Parafilm was wrapped around the top of the tubes, which were stored under water in 4 l bottles (Nalgene) to prevent large temperature changes. Samples were hand-carried to UTMSI for dissolved gas (N_2, O_2, and argon) analysis via membrane inlet mass spectrometry (MIMS; Kana et al., 1994; An et al., 2001; An & Gardner, 2002; McCarthy & Gardner, 2003). Net N_2 flux and sediment O_2 demand (SOD; O_2 flux) were calculated from the flux formula above. In the absence of N fixation, net N_2 flux provides a reasonable estimate for "actual" denitrification rates.

Bottom water was enriched with 98% $Na^{15}NO_3^-$ (~100 μmol l^{-1} final isotope concentration) after the second sampling day. Samples were collected for two more days for $^{28}N_2$, $^{29}N_2$, and $^{30}N_2$ analysis via MIMS (An et al., 2001; An & Gardner, 2002). These data allowed isotope pairing (Nielsen, 1992) analysis of potential denitrification (sum of $^{28}N_2$, $^{29}N_2$, and $^{30}N_2$ fluxes and N fixation) driven by ^{14}N and ^{15}N pools and simultaneous evaluation of N fixation. Potential denitrification, ^{14}N and ^{15}N denitrification, and N fixation calculations are detailed in An et al. (2001). Ammonium isotope ratios obtained by HPLC (Gardner et al., 1995a) allowed calculation of $^{15}NH_4^+$ concentrations and, using the above flux formula, $^{15}NH_4^+$ production (potential DNRA; An & Gardner, 2002).

MIMS uses a quadrupole mass spectrometer (QMS) to detect dissolved gases in water. The QMS ion source ionizes gases and produces O^+ ions, which react with N_2 to form nitric oxide (NO; Eyre et al., 2002). This scavenging results in a lower N_2 signal at high O_2 concentrations, and vice versa, and may lead to over- or underestimation of denitrification rates. The error, however, is machine dependant (Eyre et al., 2002), and the extent of this effect was small (0.13%) on the QMS used in this study (McCarthy & Gardner, 2003). Also, sulfide can inhibit the final step in denitrification (nitrous oxide (N_2O) → N_2) and lead to N_2O release (Nedwell & Dong, 2002). Since MIMS uses N_2:Ar to determine N_2 and a liquid N_2 trap to remove interferences (Kana et al., 1994), denitrification leading to N_2O release is not detected and may cause rate underestimation.

Water column MFW structure

Heterotrophic bacteria and phototrophic picoplankton were quantified from 1% formalin-fixed samples using epifluorescence microscopy (EFM; Olympus BX-40). Heterotrophic bacteria were stained with DAPI (Porter & Feig, 1980) and sonicated to count attached bacteria (Velji & Albright, 1993). Attached bacterial abundance was the calculated difference between total cell number (treated samples) and total free-suspended cells (untreated samples). Bacterial cells were measured with a SPOT-2 digital camera and Image Pro 4.5 software. Bacterial and picocyanobacterial biovolumes were converted to carbon following Loferer-Krossbacher et al. (1998) and Menden-Deuer & Lessard (2000), respectively.

Heterotrophic nanoflagellates (HNF) were preserved with 1% formaldehyde, concentrated onto black 0.8 μm polycarbonate membrane filters, and counted using EFM following dual-staining with FITC/DAPI (Sherr et al., 1993). Lugol's iodine preserved microplankton were settled overnight in 20–50 ml chambers and counted with an Olympus IX-70 inverted differential interference contrast (DIC) microscope. Linear dimensions of 30–90 individuals (fewer for less abundant taxa) were measured at 400–600× and converted to volumes using appropriate geometric shapes. Tintinnid volume was determined using the same approach as for aloricate protists, since they were clearly visible under DIC. Phytoplankton and HNF volumes were converted to carbon following Menden-Deuer & Lessard (2000). Ciliate and rotifer volumes were converted following Putt & Stoecker (1989) and Fahnenstiel et al. (1998), respectively.

Results

Water column characteristics

Site data for sampling stations are provided in Table 1. All sites except 'river' were deeper than the mean lake depth. Mean DO (8.4 ± 0.2 mg O_2 l^{-1}) was near saturation (8.6; Colt, 1984) for the mean temperature (22.7 ± 0.4°C). Nutrient concentrations decreased along the transect except PO_4^{3-} concentrations were low at all sites. TN:TP was about eight-fold higher at 'river' versus the other sites.

Water column NH_4^+ regeneration and uptake

Water column NH_4^+ regeneration and potential uptake rates (hereafter "regeneration" and "uptake") were an order of magnitude higher at 'river' than the other sites (Fig. 2). Light/dark uptake differences were observed at all sites. Light/dark differences in regeneration were observed at 'river' and 'outer bay'. Light uptake was higher than light regeneration at all sites.

Sediment–water interface nutrient fluxes

Benthic nutrient fluxes were higher at 'river' than the other sites (Table 2). Mean sediment DIN flux for all sites (129 ± 34 μmol N m^{-2} h^{-1}) was almost two orders of magnitude higher than PO_4^{3-} flux (1.4 ± 0.6 μmol P m^{-2} h^{-1}). Mean NO_2^- (-7.2 ± 3.3 μmol N m^{-2} h^{-1}) and NO_3^- (-13.1 ± 9.6 μmol N m^{-2} h^{-1}) fluxes for all sites were negative due to high fluxes into the sediment at 'river'. Lake Taihu sediments were a source of NH_4^+ (149 ± 47 μmol N m^{-2} h^{-1}).

Table 1 Site characteristics for Lake Taihu sampling stations

Parameter	Units	River	Inner Bay	Outer Bay	Main Lake
Lat (N)	d, m, s	31, 32, 17	31, 29, 37	31, 26, 16	31, 21, 0
Long (E)	d, m, s	120, 13, 17	120, 11, 51	120, 11, 12	120, 10, 58
Depth	m	1.4	2.6	2.9	2.8
Temp	°C	23.8	22.4	22.4	22.0
pH	su	8.65	9.04	9.08	8.55
Secchi	m	0.35	0.35	0.50	0.40
NH_4^+	μmol l^{-1}	185	3.4	2.9	0.07
NO_3^-	μmol l^{-1}	73.3	6.4	0.14	0.35
NO_2^-	μmol l^{-1}	13.8	0.6	0.02	0.06
NH_4^+:NO_x		2.1	0.5	17.9	0.2
PO_4^{3-}	μmol l^{-1}	0.18	0.25	0.22	0.04
TN	μmol l^{-1}	135	20.7	20.6	27.4
TP	μmol l^{-1}	1.5	1.7	2.5	2.0
TN:TP		89	12	8	13
Chl *a*	μg l^{-1}	85.4	4.5	3.9	14.0
DO	Mg l^{-1}	8.3	9.0	8.3	8.0

Lat (N) = north latitude in degrees, minutes, seconds (d, m, s). Long (E) = east longitude. Temp = temperature. NH_4^+ = ammonium. NO_3^- = nitrate. NO_2^- = nitrite. NO_x = NO_3^- + NO_2^-. PO_4^{3-} = ortho-phosphate. TN = total N. TP = total P. Chl *a* = chlorophyll *a*. DO = dissolved oxygen

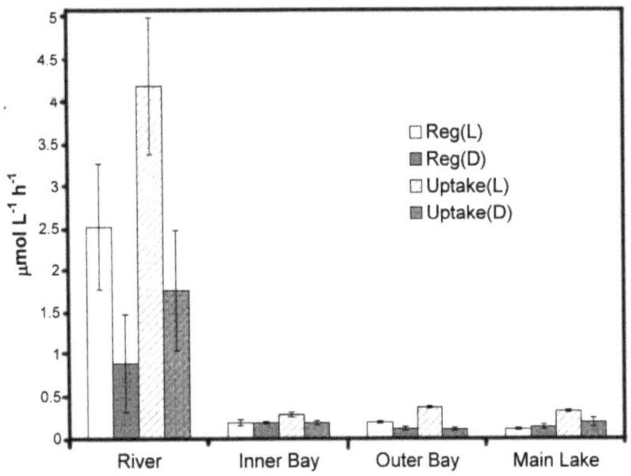

Fig. 2 Light (L) and dark (D) water column ammonium (NH_4^+) regeneration (Reg) and potential uptake rates plus/minus one standard error

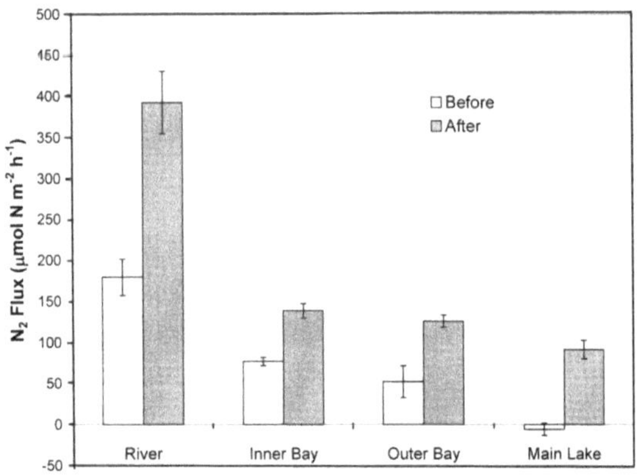

Fig. 3 Sediment–water interface net dinitrogen (N_2) flux before and after isotope ($^{15}NO_3^-$) addition plus/minus one standard error

SOD, denitrification, and N_2 fixation

Before isotope addition, SOD (in μmol O_2 m^{-2} h^{-1}) ranged from 750 ± 0.2 at 'outer bay' to 1350 ± 83 at 'river' and was similar at 'inner bay' (950 ± 32) and 'main lake' (970 ± 23). Net N_2 flux (in μmol N m^{-2} h^{-1}) decreased along the transect before and after $^{15}NO_3^-$ addition (Fig. 3). N fixation (after $^{15}NO_3^-$ addition) was observed only at 'main lake' (14 ± 10 μmol N m^{-2} h^{-1}). Thus, net N_2 fluxes before isotope addition likely reflect "actual" denitrification rates at the other sites.

Potential denitrification was equivalent to net N_2 flux after $^{15}NO_3^-$ addition (Fig. 4) except at 'main lake', where N fixation was observed. Denitrification fueled by $^{15}NO_3^-$ and the ^{14}N pool

Table 2 Lake Taihu sediment–water interface nutrient fluxes (μmol N or P m^{-2} h^{-1}) plus/minus one standard error

Nutrient	River	Inner Bay	Outer Bay	Main Lake
NH_4^+	490 ± 31	40 ± 19	20 ± 0.5	45 ± 1.4
NO_3^-	–83 ± 6.8	5.8 ± 11	12 ± 0.1	13 ± 1.2
NO_2^-	–31 ± 3.8	–2.0 ± 0.4	0.08 ± 0.05	4.2 ± 0.5
DIN	380 ± 27	44 ± 7.2	33 ± 0.6	63 ± 3.1
PO_4^{3-}	5.2 ± 0.2	1.2 ± 0.5	0.8 ± 1.0	–1.5 ± 0.9

See Table 1 legend for nutrient abbreviations

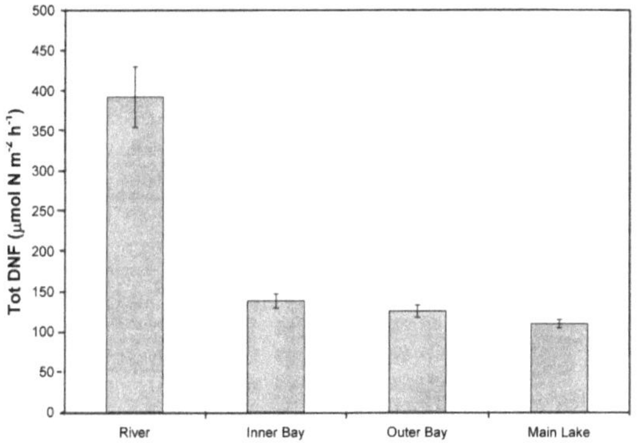

Fig. 4 Sediment–water interface potential (Tot) denitrification (DNF) after isotope ($^{15}NO_3^-$) addition plus/minus one standard error

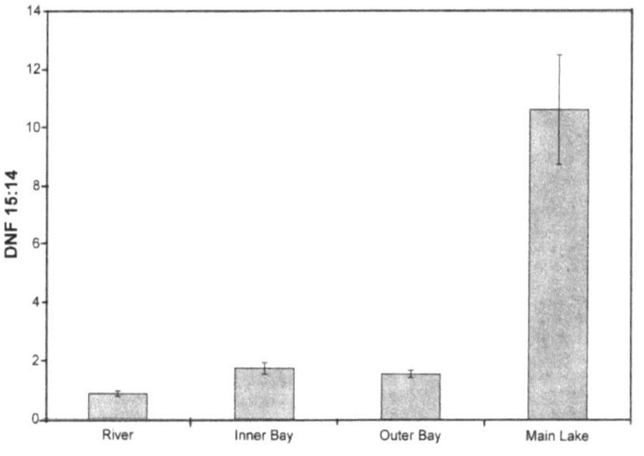

Fig. 5 Ratio of denitrification (DNF) fueled by ^{15}N versus ^{14}N nitrate (15:14) pools plus/minus one standard error

were similar at 'river', but the ^{15}N pool accounted for more potential denitrification than the ^{14}N pool at the other sites (Fig. 5).

Potential DNRA

Potential DNRA (in μmol N m^{-2} h^{-1}) was observed at all sites in Lake Taihu (data not shown). Potential DNRA rates ranged from 0.4 ± 0.1 at 'outer bay' to 250 ± 80 at 'river' and were 9.5 ± 2.4 at 'inner bay' and 3.0 ± 1.9 at 'main lake.

Water column MFW structure

Cyanobacteria were abundant at all sites during sampling (Table 3). Maximum phytoplankton biomass was observed at 'river' and dominated by large (>10,000 μm^3) euglenophytes (~80%; Fig. 6). The "edible" (to microbial grazers) fraction (i.e. < 200 μm^3) was mostly cryptophytes. Bacteria also reached maximum abundance (5×10^7 cells ml^{-1}) at 'river', but >67% were aggregated or attached to particles, and the median cell volume was minimal (0.033 μm^3). Rotifers (mostly *Polyarthra remata*) accounted for >50% of micrograzer biomass (HNF + ciliates + rotifers). The choreotrich ciliate, *Strobilidium* spp., and colorless nano-chysophytes dominated heterotrophic protists.

Meiliang Bay sites (inner bay and outer bay) had lower phytoplankton biomass consisting of green algae, cyanobacteria, and diatoms (Table 3; Fig. 6). Heterotrophic biomass in Meiliang Bay also was lower versus 'river', but to a lesser extent than for bacterioplankton alone, which were represented by large cells (median vol. = 0.085 μm^3). The colonial peritrich ciliate *Epystilis rotans*, chrysophytes, and rotifers *Asplanchna priodonta* and *Brachionus* spp. dominated the microbial grazers.

The main lake site had higher phytoplankton biomass than Meiliang Bay sites and was dominated by *Anabaena flos-aquae* and *Synechococcus*-like picocyanobacteria (>50%; Fig. 6). Bacterial biomass was lowest (cell volume 0.05 μm^3), but micrograzer biomass was high, primarily due to the tintinnid ciliate *Codonella cratera* and a diverse HNF assemblage, including chysophytes, kinetoplastids, and choanoflagellates.

Discussion

Water column nutrients and phytoplankton

Nutrients in Lake Taihu followed the expected pattern of decreasing concentrations with distance from river outflow. Possible explanations include dilution via physical transport (i.e. wind-driven currents), higher nutrient uptake versus regeneration rates along the transect, and nutrient sinks, such as sediment burial and denitrification.

High TN:TP ratios at 'river'(Table 1) support P-limitation observations (Cai et al., 1997; Huang, 2000; Chen et al., 2003). However, TN:TP ratios below Redfield (~16) at the other sites

Table 3 Phytoplankton (Phyto) and microbial food web (MFW) characteristics during Lake Taihu sampling

			River	Inner Bay	Outer Bay	Main Lake
MFW	$\mu\mu g\ C\ l^{-1}$	Phytoplankton	6930	149	55.4	518
		Bacteria	577	292	263	157
		HNF	20.8	8.94	6.97	15.1
		Ciliates	35.3	13.5	10.3	52.9
		Rotifers	63.8	3.57	5.75	2.62
Phyto.	%phyto	Chlorophytes	4.90	41.8	27.4	33.5
		Cyanobacteria	1.90	28.6	37.2	59.7
		Diatoms	2.00	6.60	16.7	5.87
		Euglenophytes	78.9	0	0	0
		Varia (i.e. cryptophytes)	12.2	23.0	18.7	0.96
Cyanos	%phyto	*Anabaena*	0.14	14.5	6.01	38.9
		Synechococcus	0.2	6.8	29.9	11.7
		Aphanizomenon	0	0	0	3.68
		Total N-fixers	0.33	21.3	35.9	54.3
		Microcystis	0.80	7.33	1.27	0.14
Ratios	1	Bacteria:NH_4^+	0.22	6.13	6.48	160.2
	2	Chl:DIN	0.02	0.03	0.09	2.08
	3	Chl:NH_4^+	0.03	0.09	0.09	14.28
	4	Chl:NO_3^-	0.08	0.05	1.96	2.86
	5	Chl:PO_4^{3-}	15.3	0.58	0.56	11.3
	6	Diatom:NO_3^-	3.18	8.5	216	278
	7	Diatom:PO_4^{3-}	41.8	7.02	4.43	78.5
	8	Heterotrophs:Chl	0.10	2.13	5.16	0.44
	9	MZP:total phyto	0.01	0.11	0.29	0.11
	10	MZP:edible phyto	0.56	0.63	0.58	0.54
	11	Protists:bacteria	0.10	0.08	0.07	0.43
	12	Light uptake:Chl	49.0	63.0	95.3	23.1
	13	Light reg:hetero C	3.61	0.59	0.66	0.48
	14	Dark reg:hetero C	1.28	0.59	0.39	0.60

% phyto = proportion of phytoplankton groups to total phytoplankton; Cyanos = cyanobacteria; C = Carbon; HNF = heterotrophic nanoflagellates; Chl = chlorophyll; MZP = microzooplankton; Reg = regeneration; Hetero = heterotrophic. See Table 1 legend for nutrient abbreviations

indicate surplus P and conditions conducive to cyanobacteria (Jacoby et al., 2000). Nutrient concentrations in Meiliang Bay are lowest in

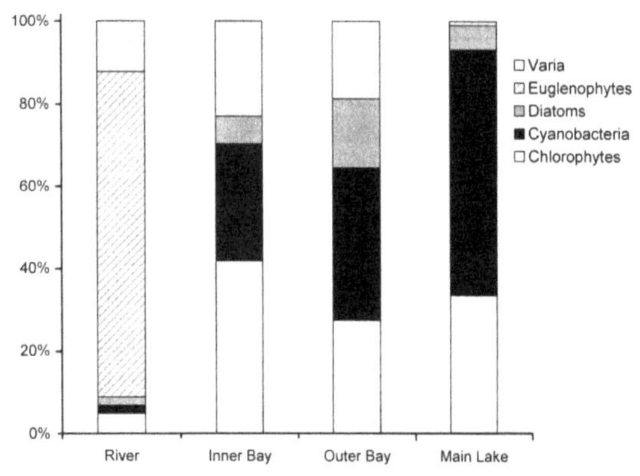

Fig. 6 Proportions of chlorophytes, cyanobacteria, diatoms, euglenophytes, and varia (mostly cryptophytes) in total phytoplankton biomass

summer (wet season) and highest in winter (dry season), when biological activity decreases and inverse water flow occurs (Cai et al., 1997). Thus, nutrient limitation status may change in different seasons.

Nutrient concentrations and algal standing stocks are important factors affecting cyanobacteria dominance (Downing et al., 2001). Transect NH_4^+:NO_x ratios were >1 at 'river' and 'outer bay' and <1 at 'inner bay' and 'main lake' (Table 1). Cyanobacteria percentage biomass, relative to total phytoplankton biomass, was highest at 'main lake' (Table 3), where the NO_3^- proportion peaked and NH_4^+ likely was depleted by cyanobacteria. Diatom percentage biomass peaked at 'outer bay', where highest NH_4^+:NO_x was observed, likely due to NO_3^- uptake by diatoms (Hyenstrand et al., 1998). Percentage cyanobacteria biomass was greater than for diatoms at all sites except 'river'. Phytoplankton biomass at 'river' was dominated (79%) by *Euglena* spp., which are common in hypereutrophic systems (Hewson et al., 2001).

Water column NH_4^+ regeneration and uptake

Water column NH_4^+ cycling rates in Lake Taihu were within ranges observed in other eutrophic systems. High regeneration and uptake (Fig. 2) at 'river' are not surprising given high DIN concentration and phytoplankton biomass. Similar rates were observed in the Mississippi River plume (Bode & Dortch, 1996) and hypereutrophic Lake Maracaibo (Gardner et al., 2000). Uptake in Lake Okeechobee (0.67 μmol l^{-1} h^{-1}; Gu et al., 1997) was lower than the rate at 'river' but in the same range as those from Meiliang Bay and the main lake. Regeneration and uptake at sites other than 'river' were higher than oligotrophic Great Lakes (Gardner et al., 1995b) but lower than south Texas coastal waters (McCarthy et al., unpub. data).

Light/dark uptake differences were observed at all sites (Fig. 2). This result supports the idea that N is not limiting, since phytoplankton uptake depends less on light during N stress (Cochlan et al., 1991). However, light/dark uptake differences are less pronounced in N limited versus N replete systems (Cochlan et al., 1991). Light-to-dark uptake ratio was ~1.7 (less pronounced) at 'inner bay' and 'main lake' and 2.4 and 3.5 (more pronounced) at 'river' and 'outer bay', respectively. This observation suggests possible N limitation at 'inner bay' and 'main lake'. Lack of light/dark regeneration differences at these sites also is consistent with N limitation (Fig. 2; Gardner et al., 2000), since N-starved heterotrophs may retain N for biomass rather than release it. Light/dark uptake differences also may reflect autotrophic versus heterotrophic dominance, especially when those differences are more pronounced ('river' and 'outer bay').

Sediment–water interface nutrient fluxes

Cyanobacteria may acquire sufficient P to sustain all pelagic growth from sediments, thus avoiding competition with other phytoplankton (Blomqvist et al., 1994). Low water column PO_4^{3-} concentrations (Table 1) and PO_4^{3-} flux out of the sediments (Table 2) at 'river' and 'inner bay' indicate rapid P uptake and support P limitation observations at these sites. No sediment P flux at 'outer bay' and negative flux at 'main lake' (Table 2) are consistent with water column TN:TP observations below Redfield and suggest N limitation or co-limitation in outer Meiliang Bay and the main lake.

Positive NH_4^+ flux at all sites (Table 2) indicates high organic matter inputs and rapid remineralization (including DNRA). Cyanobacteria in Lake Taihu may be fueled in part by sediment NH_4^+ flux into the water column, especially at 'main lake' where DIN concentration was below 0.5 μmol l^{-1} (Table 1). Negative NO_3^- flux at 'river' may reflect denitrification and/or DNRA fueled by water column NO_3^-. Positive NO_3^- flux at 'outer bay' and 'main lake' can be explained by sediment nitrification. Small positive NO_2^- fluxes at these sites may result from incomplete nitrification and/or denitrification, since NO_2^- is an intermediate in both processes (Zumft, 1997).

N_2 fixation

Nitrogen fixation was not observed in Meiliang Bay (including 'river'). Before isotope addition, net N_2 flux at 'main lake' was negative, but standard error bars overlapped with zero, indicating that denitrification and N fixation were in balance. This site had the highest water column N-fixer biomass (Table 3). *Anabaena* comprised nearly 40% of total phytoplankton biomass, and 'main lake' is the only site where *Aphanizomenon* was observed. Phytobenthos were not evaluated, so it cannot be assumed that these N-fixers were responsible for observed sediment N fixation. However, high water column N-fixer abundance and sediment N fixation further support possible N limitation at this site (Dokulil & Teubner, 2000).

Denitrification

Sediments provide a site for denitrification, which can drive aquatic systems toward N limitation and regulate nutrient supply ratios (Seitzinger, 1988). Denitrification in Meiliang Bay may be a factor in observed evidence for N limitation in the main lake. If excess N is denitrified before physical transport via wind-driven currents moves it to the main lake, then it follows that organisms in the main lake may have to rely on other N sources, such as fixation. Lack of N fixation at Meiliang Bay sites (including 'river') indicates that net N_2 fluxes before isotope addition (Fig. 3) may reflect "actual" denitrification rates, assuming no N_2O release from incomplete denitrification. Increases in net N_2 flux after $^{15}NO_3^-$ addition (Fig. 3) indicate that potential denitrification may be substrate limited, especially at 'outer bay' and 'main lake', where water column NO_3^- was low (Table 1). Another possible explanation for increased net N_2 flux after isotope addition is that the sediment core incubations eliminated wind-driven turbulence and resuspension, thus allowing settlement of suspended particles and development of a more defined redox gradient. After 2 days of incubation, redox conditions may have become inhibitory for nitrifiers and stimulative for denitrifiers, which could take advantage of increased substrate (as $^{15}NO_3^-$) to replace NO_3^- supply from decreased nitrification.

Decreases in potential denitrification along the transect may reflect diminishing organic matter availability (Fig. 4). Denitrification fueled by $^{15}NO_3^-$ versus the $^{14}NO_3^-$ pool (from water column or coupled to nitrification) shows substrate stimulation (Fig. 5), especially at sites further from river discharge. Decreasing water column NO_3^- concentrations along this gradient may result from lower nitrification and lead to reduced nitrification/denitrification coupling and increased dependence on water column NO_3^-.

DNRA

Since NH_4^+ is a primary factor in cyanobacterial dominance, it is important to characterize N transformations capable of maintaining NH_4^+ in the system. DNRA is associated with sulfidic estuarine or marine sediments (Tiedje, 1988; An & Gardner, 2002) but also can comprise up to 30% of NO_3^- reduction in lake sediments (Tiedje, 1988). Potential denitrification to DNRA partitioning ratios (DNF:DNRA) in Lake Taihu were 1.6 at 'river', 14.6 at 'inner bay', 315 at 'outer bay', and 35 at 'main lake' (data not shown). Since DNRA depends more on organic matter availability than NO_3^- concentration (Tiedje, 1988), higher DNF:DNRA may reflect organic matter limitation of DNRA away from river discharge. While potentially a meaningful process at 'river', DNRA is not as important as fixed N removal via denitrification at the other sites.

Water column versus sediment NH_4^+ regeneration

Depth-averaged water column NH_4^+ regeneration (water column regeneration converted from μmol N l^{-1} h^{-1} to μmol N m^{-2} h^{-1} given depth) and sediment N recycling rates (sum of denitrification and net DIN flux, which includes DNRA; Table 4) indicates that Lake Taihu nutrient mineralization was four to six times higher in the water column versus sediments. Thus, primary production appears to be driven primarily by

Table 4 Comparison of depth averaged water column (WC) and sediment (Sed) nitrogen regeneration (reg) in Lake Taihu

	River	Inner Bay	Outer Bay	Main Lake
DNF	180	77	52	8
DIN flux	380	44	33	63
Sed reg	560	121	85	71
WC reg	3520	480	560	310
WC:sed	6.3	4.0	6.6	4.4
WC:DIN	9.3	10.9	17.0	4.9

Fluxes in μmol N m^{-2} h^{-1}. Sediment regeneration determined by adding denitrification (DNF) and net dissolved inorganic N (DIN) flux before $^{15}NO_3^-$ addition. DNF at 'main lake' determined by adding net N_2 flux before $^{15}NO_3^-$ addition to calculated nitrogen fixation rate water column processes. This dependence is more pronounced if denitrification is excluded from the sediment calculation (WC Reg:DIN; Table 4), since denitrification results in N removal from the system.

MFW structure and N cycling

MFW structure shows a general shift from autotrophic to heterotrophic dominance along the river to lake transect. Nutrient concentration and cycling rate comparisons with MFW structure indicate increasing N limitation along the transect (ratio #'s 2, 3, 4, and 6; Table 3). Interestingly, the light uptake-to-chl ratio (#12) was lowest at 'main lake', but N fixers may have accumulated sufficient N and, thus, did not respond to ^{15}N addition. Diatom biomass-to-NO_3^- ratio (#6) indicates severe N limitation at 'outer bay' and 'main lake'. High regeneration-to-heterotrophic biomass ratios (#'s 13 and 14) at 'river' may indicate heterotrophic bacterial NH_4^+ recycling. Recycling at all other sites, particularly 'main lake', likely is driven by grazing, since bacteria may become N limited as indicated by a higher bacteria-to-NH_4^+ ratio (#1).

The microzooplankton (ciliates and rotifers)-to-"edible" phytoplankton biomass ratio (#10) remained remarkably constant along the transect, indicating tight trophic coupling between these groups, despite sharp variation in the microzooplankton-to-total phytoplankton biomass ratio (#9). Strong grazing pressure on nanoplankton may help cyanobacteria by eliminating competitors for limiting nutrients. Grazers often are ineffective at preying on colonial cyanobacteria due to colony size, potential toxicity, and/or gelatinous envelopes (Dokulil & Teubner, 2000). However, grazers prey on competing phytoplankton (i.e. diatoms) and transform organic N to NH_4^+ via excretion (Shapiro, 1990). Although large colonial cyanobacteria and euglenophytes may not be involved in direct herbivorous trophic transfers within the MFW, the latter group are mixotrophs (i.e. can combine autotrophic and heterotrophic modes of nutrition). Therefore, nutrient limitation is less likely to affect these taxa, since they can obtain nutrients by grazing on bacteria and picocyanobacteria.

Low chl:PO_4^{3-} (#5) and diatom:PO_4^{3-} ratios (#7) at 'inner bay' may be due to top-down pressure rather than N limitation. Greater than 1,000 cladocerans l^{-1} were present at this site. Higher heterotrophic C-to-chl ratio (#8), predominance of mesoplankton-sized ciliate colonies (less vulnerable to cladoceran grazing), and lower protist-to-bacteria ratio (#11) also support this suggestion. The latter ratio also indicates strong bacterial grazing pressure at 'river' and 'main lake', particularly if only suspended cells are considered at 'river' (ratio increases to 0.29). Bacteria may dominate plankton biomass in Meiliang Bay due to sediment resuspension introducing benthic bacteria into the water column, indicated by unusually large bacterial cells and balanced dark NH_4^+ regeneration and uptake.

Conclusions

Nitrogen dynamics relative to MFW composition were characterized along the Liangxihe River discharge gradient in and around Meiliang Bay, Lake Taihu, during late summer/early fall cyanobacteria bloom conditions. The observations provide insights into ecological characteristics of a subtropical, shallow, well-mixed, eutrophic lake. The most interesting finding is that N limits or co-limits primary production in and near central Lake Taihu, contrary to the previous paradigm of exclusive P limitation. A gradient was observed from strong P limitation at the river discharge to N limitation or co-limitation in the main lake. Denitrification in Meiliang Bay may drive main lake N limitation by removing excess N before physical forces transport it to the main lake. This result also exemplifies the importance of characterizing N cycling in freshwater systems, where most studies have focused on P dynamics.

Another important finding is the importance of water column N recycling relative to sediment processes. Nutrient flux from Lake Taihu sediments do not appear to drive primary production during late summer/early fall when cyanobacteria bloom conditions are beginning to wane. However, sediments have an important role in nutrient and MFW dynamics by providing a site for denitrification, which, in conjunction with N fixation and other processes, can regulate available nutrient ratios. Reflecting N cycling patterns, MFW structure shifted from autotrophic (phytoplankton dominated) at the river discharge to heterotrophic (bacteria dominated) in and near the main lake. More studies are needed to evaluate seasonal and spatial variability in nutrient limitation status, N transformation rates, and phytoplankton and MFW structure and function in Lake Taihu.

Acknowledgements This project was supported in part by the Chinese Academy of Sciences (contract No. KZCX1-SW-12), the National Science Foundation of China (Grant #40371104), and the Nancy Lee and Perry R. Bass Regents Chair in Marine Science held by Gardner. Several members of the faculty, staff, and graduate students at the Taihu Laboratory for Lake Ecosystem Research contributed field assistance and discussion during this project. Special assistance was provided by Dr. Zhu Guangwei. The authors thank Dr. Dmitri Sobolev for comments on the manuscript and Dr. Hilairy Hartnett for discussion on dissolved gas sample preservation techniques.

References

An, S. & W. S. Gardner, 2002. Dissimilatory nitrate reduction to ammonium (DNRA) as a nitrogen link, versus denitrification as a sink in a shallow estuary (Laguna Madre/Baffin Bay, Texas). Marine Ecology Progress Series 237: 41–50.

An, S., W. S. Gardner & T. M. Kana, 2001. Simultaneous measurement of denitrification and nitrogen fixation using isotope pairing with membrane inlet mass spectrometry analysis. Applied and Environmental Microbiology 67: 1171–1178.

Blackburn, T. H., 1979. Method for measuring rates of NH_4^+ turnover in anoxic marine sediments, using a ^{15}N-NH_4^+ dilution technique. Applied and Environmental Microbiology 37: 760–765.

Blomqvist, P., A. Pettersson & P. Hyenstrand, 1994. Ammonium–nitrogen: a key regulatory factor causing dominance of non-nitrogen-fixing cyanobacteria in aquatic systems. Archiv fur Hydrobiologie 132: 141–164.

Bode, A. & Q. Dortch, 1996. Uptake and regeneration of inorganic nitrogen in coastal waters influenced by the Mississippi River: spatial and seasonal variations. Journal of Plankton Research 18: 2251–2268.

Cai, Q., X. Gao, Y. Chen, S. Ma & M. Dokulil, 1997. Dynamic variations of water quality in Lake Taihu and multivariate analysis of its influential factors. Journal of Chinese Geography 7: 72–82.

Caperon, J., D. Schell, J. Hirota & E. Laws, 1979. Ammonium excretion rates in Kaneohe Bay, Hawaii, measured by a ^{15}N isotope dilution technique. Marine Biology 54: 33–40.

Chen, Y., B. Qin, K. Teubner & M. Dokulil, 2003. Long-term dynamics of phytoplankton assemblages: *Microcystis*-domination in Lake Taihu, a large shallow lake in China. Journal of Plankton Research 25: 445–453.

Cochlan, W. P., N. M. Price & P. J. Harrison, 1991. Effects of irradiance on nitrogen uptake by phytoplankton: comparison of frontal and stratified communities. Marine Ecology Progress Series 69: 103–116.

Colt, J., 1984. Computation of dissolved gas concentrations in water as functions of temperature, salinity, and pressure. American Fisheries Society Special Publication 14.

Dickman, M., P. Pu & C. Zheng, 1998. Taihu Lake: past, present and future. Journal of Lake Sciences (China) 10(suppl.): 75–83.

Dokulil, M. T. & K. Teubner, 2000. Cyanobacterial dominance in lakes. Hydrobiologia 438: 1–12.

Downing, J. A., S. B. Watson & E. McCauley, 2001. Predicting cyanobacteria dominance in lakes. Canadian Journal of Fisheries and Aquatic Sciences 58: 1905–1908.

Edwards, C. A., T. A. Powell & H. P. Batchelder, 2000. The stability of an NZP model subject to realistic levels of vertical mixing. Journal of Marine Research 58: 37–60.

Elser, J. J., 1999. The pathway to noxious cyanobacteria blooms in lakes: the food web as the final turn. Freshwater Biology 42: 537–543.

Eyre, B. D., S. Rysgaard, T. Dalsgaard & P. B. Christensen, 2002. Comparison of isotope pairing and N_2:Ar methods for measuring sediment denitrification—assumptions, modifications, and implications. Estuaries 25: 1077–1087.

Fahnenstiel, G. L., A. E. Krause, M. J. McCormick, H. J. Carrick & C. L. Schelske, 1998. The structure of the planktonic food web in the St. Lawrence Great Lakes. Journal of Great Lakes Research 24: 531–554.

Gardner, W. S., H. A. Bootsma, C. Evans & P. A. St. John, 1995a. Improved chromatographic analysis of ^{15}N:^{14}N ratios in ammonium or nitrate for isotopic addition experiments. Marine Chemistry 48: 271–282.

Gardner, W. S., J. F. Cavaletto, T. H. Johengen, J. R. Johnson, R. T. Heath & J. B. Cotner, 1995b. Effects of the zebra mussel, *Dreissena polymorpha*, on community nitrogen dynamics in Saginaw Bay, Lake Huron. Journal of Great Lakes Research 21: 529–544.

Gardner, W. S., P. J. Lavrentyev, H. A. Bootsma, J. F. Cavaletto, F. Troncone & J. B. Cotner, 2000. Effects of natural light on nitrogen dynamics in diverse aquatic environments. Internationale Vereinigung Fur Theoretische Und Angewandte Limnologie 27: 64–73.

Gu, B., K. E. Havens, C. L. Schelske & B. H. Rosen, 1997. Uptake of dissolved nitrogen by phytoplankton in a eutrophic subtropical lake. Journal of Plankton Research 19: 759–770.

Havens, K. E., T. Fukushima, P. Xie, T. Iwakuma, R. T. James, N. Takamura, T. Hanazato & T. Yamamoto,

2001. Nutrient dynamics and the eutrophication of shallow lakes Kasumigaura (Japan), Donghu (PR China), and Okeechobee (USA). Environmental Pollution 111: 263–272.

Hewson, I., J. M. O'Neil, C. A. Heil, G. Bratbak & W. C. Dennison, 2001. Effects of concentrated viral communities on photosynthesis and community composition of co-occurring benthic microalgae and phytoplankton. Aquatic Microbial Ecology 25: 1–10.

Huang, Y., 2000. The water quality of Lake Taihu and its protection, pp. 239–246. In Nanjing Institute of Geography and Limnology (eds), Research on Lake Sciences. Chinese Academy of Sciences.

Hyenstrand, P., P. Blomqvist & A. Pettersson, 1998. Factors determining cyanobacterial success in aquatic systems: a literature review. Ergebnisse der Limnologie 51: 41–62.

Jacoby, J. M., D. C. Collier, E. B. Welch, F. J. Hardy & M. Crayton, 2000. Environmental factors associated with a toxic bloom of *Microcystis aeruginosa*. Canadian Journal of Fisheries and Aquatic Sciences 57: 231–240.

Jin, X. & Q. Tu (eds), 1990. Standard Methods for Observation and Analysis of Lake Eutrophication (2nd edn.). Chinese Environmental Science Press, Beijing, 240 pp (in Chinese).

Kana, T. M., C. Darkangelo, M. D. Hunt, J. B. Oldham, G. E. Bennett & J. C. Cornwell, 1994. Membrane inlet mass spectrometer for rapid high-precision determination of N_2, O_2, and Ar in environmental water samples. Analytical Chemistry 66: 4166–4170.

Lavrentyev, P. J., H. A. Bootsma, T. H. Johengen, J. F. Cavaletto & W. S. Gardner, 1998. Microbial plankton response to resource limitation: insights from the community structure and seston stoichiometry in Florida Bay, USA. Marine Ecology Progress Series 165: 45–57.

Lavrentyev, P. J., W. S. Gardner & L. Yang, 2000. Effects of the zebra mussel on nitrogen dynamics and the microbial community at the sediment–water interface. Aquatic Microbial Ecology 21: 187–194.

Loferer-Krossbacher, M., J. Klima & R. Psenner, 1998. Determination of bacterial cell dry mass by transmission electron microscopy and densitometric image analysis. Applied and Environmental Microbiology 64: 688–694.

McCarthy, M. J. & W. S. Gardner, 2003. An application of membrane inlet mass spectrometry to measure denitrification in a recirculating mariculture system. Aquaculture 218: 341–355.

Menden-Deuer, S. & E. J. Lessard, 2000. Carbon to volume relationships for dinoflagellates, diatoms, and other protist plankton. Limnology and Oceanography 54: 569–579.

Nedwell, D. B. & L. F. Dong, 2002. Authors' reply to letters to the editor: nitrous oxide formation in the Colne estuary in England: the central role of nitrite. Applied and Environmental Microbiology 68: 5202–5204.

Nielsen, L. P., 1992. Denitrification in sediment determined from nitrogen isotope pairing. FEMS Microbiology Ecology 86: 357–362.

Porter, K. G. & Y. C. Feig, 1980. The use of DAPI for identifying and counting aquatic microflora. Limnology and Oceanography 25: 943–948.

Pu, P. & J. Yan, 1998. Taihu Lake—a large shallow lake in the East China Plain. Journal of Lake Sciences (China) 10(suppl.): 1–12.

Putt, M. & D. K. Stoecker, 1989. An experimentally determined carbon: volume ratio for marine oligotrichous ciliates from estuarine and coastal waters. Limnology and Oceanography 34: 177–183.

Qin, B., P. Xu, Q. Wu, L. Luo & Y. Zhang, 2007. Environmental issues of Lake Taihu, China. Hydrobiologia 581: 3–14.

Seitzinger, S. P., 1988. Denitrification in freshwater and coastal marine ecosystems: ecological and geochemical significance. Limnology and Oceanography 33: 702–724.

Shapiro, J., 1990. Current beliefs regarding dominance by blue-greens: the case for the importance of CO_2 and pH. Internationale Vereinigung Fur Theoretische Und Angewandte Limnologie 24: 38–54.

Sherr, E. B., D. A. Caron & B. F. Sherr, 1993. Staining of heterotrophic protists for visualization via epifluorescence microscopy. In Kemp, P. F., B. F. Sherr, E. B. Sherr & J. J. Cole (eds), Handbook of Methods in Aquatic Microbial Ecology. Lewis, London: 213–228.

Tiedje, J. M., 1988. Ecology of denitrification and dissimilatory nitrate reduction to Ammonium. In Zehnder, A. J. B. (ed.), Biology of Anaerobic Microorganisms. Wiley, New York, 179–244.

Tobias, C. R., I. C. Anderson, E. A. Canuel & S. A. Macko, 2001. Nitrogen cycling through a fringing marsh-aquifer ecotone. Marine Ecology Progress Series 210: 25–39.

Velji, M. I. & L. J. Albright, 1993. Improved sample preparation for enumeration of aggregated aquatic substrate bacteria. In Kemp, P. F., B. F. Sherr, E. B. Sherr & J. J. Cole (eds), Handbook of Methods in Aquatic Microbial Ecology. Lewis, London, 139–142.

Zumft, W. G., 1997. Cell biology and molecular basis of denitrification. Microbiology and Molecular Biology Reviews 61: 533–616.

Hydrobiologia (2007) 581:209–215
DOI 10.1007/s10750-006-0493-5

EUTROPHICATION IN LAKES

Different competitive outcomes among four species of cladocerans under different alga combinations of colonial *Microcystis* spp. and green alga *Scenedesmus obliquus*

Feizhou Chen · Ping Xie · Boqiang Qin

Abstract Cyanobacteria blooms (especially *Microcystis* spp.) are thought to alter dominance of large-sized daphnids into small-sized metazoan zooplankton. However, several field investigations show different phenomena. Laboratory experiments were conducted based on the hypothesis that different *Microcystis* spp. concentrations would influence competitive outcomes using two algal combinations of different concentrations and four species of cladocerans. In the algal combination of 50 mg l^{-1} colonial *Microcystis* spp. + 1 mg l^{-1} *Scenedesmus obliquus* (fresh weight), *Daphnia carinata* was absent during the experiment in competition with other cladocerans. Decreasing colonial *Microcystis* spp. concentration (10 mg l^{-1}) resulted in a shift from dominance by small-sized cladocerans to dominance by *D. carinata*. No significant effects of different concentrations of colonial *Microcystis* spp. on competitive outcomes were shown among three small-sized cladocerans. These results support the idea that cyanobacteria concentration affects the dominance status of large-bodied daphnid.

Keywords cladoceran · *Daphnia* · competition · *Microcystis* spp.

Guest editors: B. Qin, Z. Liu & K. Havens
Eutrophication of shallow lakes with special reference to Lake Taihu, China

F. Chen (✉) · P. Xie
Donghu Experimental Station of Lake Ecosystems, State Key Laboratory for Freshwater Ecology and Biotechnology of China, Institute of Hydrobiology, Chinese Academy of Sciences, Wuhan 430072
P.R. China
e-mail: feizhch@niglas.ac.cn

F. Chen · B. Qin
Nanjing Institute of Geography and Limnology, Chinese Academy of Sciences, Nanjing 210008
P.R. China

Introduction

Cyanobacteria blooms, often occurring in eutrophic systems, can alter zooplankton community structure by disturbing interspecific actions. Generally, large-sized species often are replaced by small-sized species (e.g. Infante & Riehl, 1984). For example, *Ceriodaphnia* and *Bosmina* replaced the large-sized cladoceran *Daphnia ambigua* during a cyanobacteria bloom (Allan, 1977). Other field investigations showed that large-sized cladocerans dominated during blooms (George & Edwards, 1974; Kim et al., 2001). In another case, colonial *Microcystis* did not favor rotifers (*Brachionus calyciflorus*) in competition with *D. ambigua* (Fulton & Paerl, 1988).

Many factors, including concentration, size, *Microcystis* toxin and alternative food abundance during blooms may result in zooplankton shifts (Jarvis et al., 1987; Fulton & Paerl, 1988; Sartonov, 1995; Chen & Xie, 2003, 2004). Different responses of species to these factors result in observed zooplankton community structure.

Since there is little or no evidence for lethal effects of toxic cyanobacteria on crustacean plankton (Haney, 1987) and disregarding effects of planktivorous fish, zooplankton shifts during cyanobacteria blooms may be determined, to a certain extent, by biomass/concentration of cyanobacteria and alternative food. Low concentrations of edible algae favors small-sized cladocerans, while high concentrations favor large-sized cladocerans, such as *Daphnia* (Goulden et al., 1982; Tessier & Goulden, 1987). However, the opposite also has been reported (Nandini et al., 2002). During cyanobacteria blooms, alternative food availability may influence competition outcomes, except feeding inhibition. Although colonial *Microcystis* spp. did not favor *D. ambigua* in competition with *Diaphanosoma brachyurum* and *Bosmina longirostris* in a laboratory experiment (Fulton & Paerl, 1988), field investigation showed daphnid dominance in the cladoceran community. In the present study, we hypothesized that different colonial *Microcystis* spp. concentrations could affect competition outcomes among cladocerans, expecially large-sized daphnids and small-sized cladocerans. Two concentration combinations of colonial *Microcystis* spp. and green alga *S. obliquus* were used to test the hypothesis.

Materials and methods

All stock culture and experimental procedures were conducted at 25°C ± 0.5°C and a 12:12 (hours light:dark) photoperiod. Wild *Daphnia carinata*, *Moina micrura*, *D. brachyurum* and *Ceriodaphnia cornuta* were collected from Lake Donghu and reared through many generations in the laboratory. *S. obliquus*, a green alga, was sourced from cultures maintained by the Institute of Hydrobiology of the Chinese Academy of Sciences. *Sceneesmus obliquus* was cultured in Shuisheng VI medium (Li et al., 1959) with aeration. Colonial *Microcystis* spp. (90–95% *Microcystis aeruginosa*) was collected from a dense bloom in a nearby fishpond, and previous work indicated that this population produced microcystin that was lethal to *Daphnia magna* and *D. carinata* (Carmichael et al., 1988; He et al., 1997; Chen & Xie, 2004). Microcystin concentration was 4.5×10^{-6} μg cell^{-1}. Colonial *Microcystis* between 64–112 μm were used for competitive experiments. A low concentration of this size range of *Microcystis* improves development of *D. carinata* (Chen & Xie, 2003). When colonies were destroyed by ultrasonic waves, toxicity to *D. carinata* increased (Chen & Xie, 2004). Thus, toxicity to cladocerans was less in this study and can be neglected. Colonies were held at 4–6°C in the dark for 24–48 h prior to experiments.

During the cyanobacteria bloom in Lake Donghu, *Microcystis* spp. biomass ranged from 0 to 46 mg l^{-1} (fresh weight) (Tang, 2002). Two algal biomass combinations were used: 10 mg l^{-1} colonial *Microcystis* spp. + 1 mg l^{-1} *S. obliquus* (hereafter 10 M:1 S), and 50 mg l^{-1} colonial *Microcystis* spp. + 1 mg l^{-1} *S. obliquus* (hereafter 50 M:1 S). Algal biomass was fresh weight. Each combination had two replicates, and each replicate included 10 neonates (<12 h old) for *D. carinata*, *M. micrura*, and *D. brachyurum* and 15 neonates for *C. cornuta* considering body size difference. Experiments were conducted in a 250 ml beaker. Experiment length was determined by competitive results, and the maximum was four weeks. The culture medium was changed every two days, and the number of cladocerans was counted.

Maximal population growth rate was calculated as:

$$g = (\ln N_t - \ln N_0)/t$$

where N_0 and N_t are the mean individual number initially and reaching highest density after t days, respectively. For two peaks in certain growth curves, we selected the first peak for calculation of population growth rate. Competition comparison among *M. micrura*, *D. brachyurum* and *C. cornuta* was made using 2-way ANOVA with the index of the maximal population growth rate. All statistical analyses were conducted in Statistica 5.1.

Results

Competitive outcomes were different between the two alga combinations. In the 50 M:1 S combination, *D. carinata* could not compete with *M. micrura, C. cornuta*, and *D. brachyurum*. *D. carinata* was absent within 18 days. Time to extinction of *D. carinata* was shortest (8 days) in the presence of *M. micrura*. However, at 10 M:1 S, an equilibrium was reached in the presence of *M. micrura* and *C. cornuta* without any extinction (Fig. 1). For *D. brachyurum*, extinction occurred within two weeks in competition with *D. carinata* in the 10 M:1 S combination.

Competitions among *M. micrura, C. cornuta* and *D. brachyurum* were less intense than between *D. carinata* and other cladocerans (Fig. 2). No species disappeared within four weeks in the 50 M:1 S combination, while only *C. cornuta* was absent in competition with *D. brachyurum* in the 10 M:1 S concentration. *Moina micrura* and *C. cornuta* had stable density at the end of 50 M:1 S experiments. In 10 M:1 S trials, *M. micrura* dominated over *C. cornuta*. At both *Microcystis* spp. concentrations, *M. micrura* had a higher population growth rate than *D. brachyurum*, but, after 10 days, its density declined gradually along with increasing density of *D. brachyurum*. *C. coruta* dominated in all 50 M:1 S experiments. ANOVA analysis indicted that algal mixtures had no significant effect on the maximal population growth rate of *M. micrura* while species and interaction had. For *D. brachyurum* and *C. cornuta*, algal mixtures, species and interactions had significant

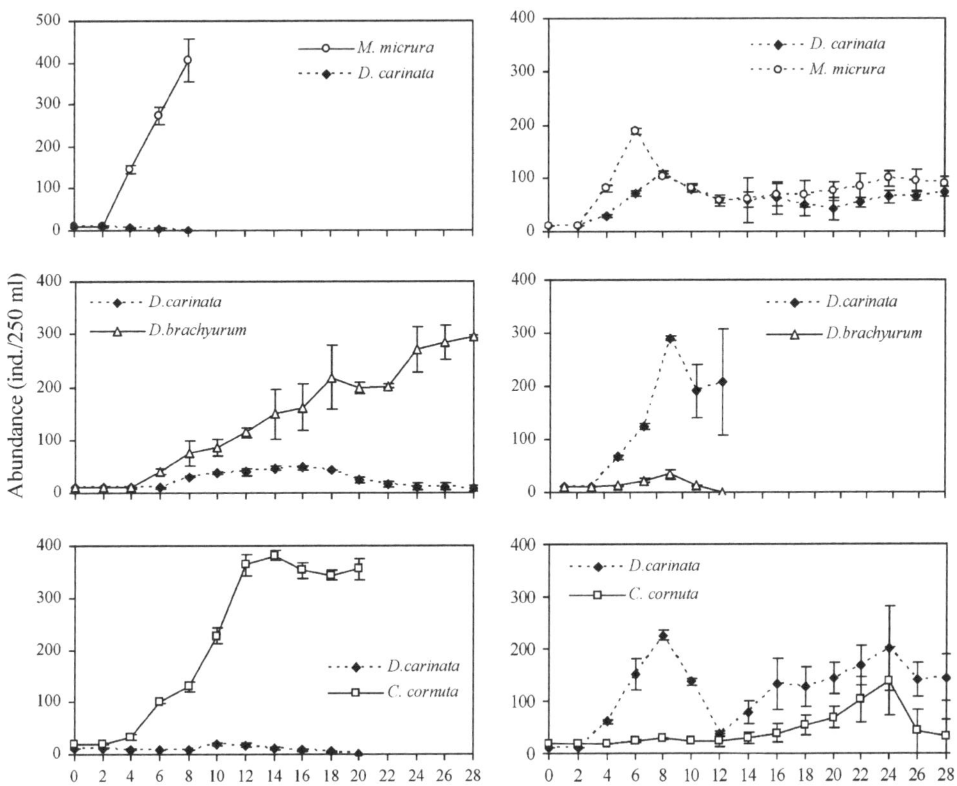

Fig. 1 Cladoceran abundances (±SD) in laboratory competition between *Daphnia carinata* and three small-sized cladoceran at different alga combinations. Ultimate time of experiments is four weeks. Left shows the results at the alga combination of 50 mg l^{-1} colonial *Microcystis* spp. + 1 mg l^{-1} *Scenedesmus obliquus*. Right at the alga combination of 10 mg l^{-1} colonial *Microcystis* spp. + 1 mg l^{-1} *S. obliquus*. Algal biomass is fresh weight. Abundance showed in Y-axis is number in 250 ml

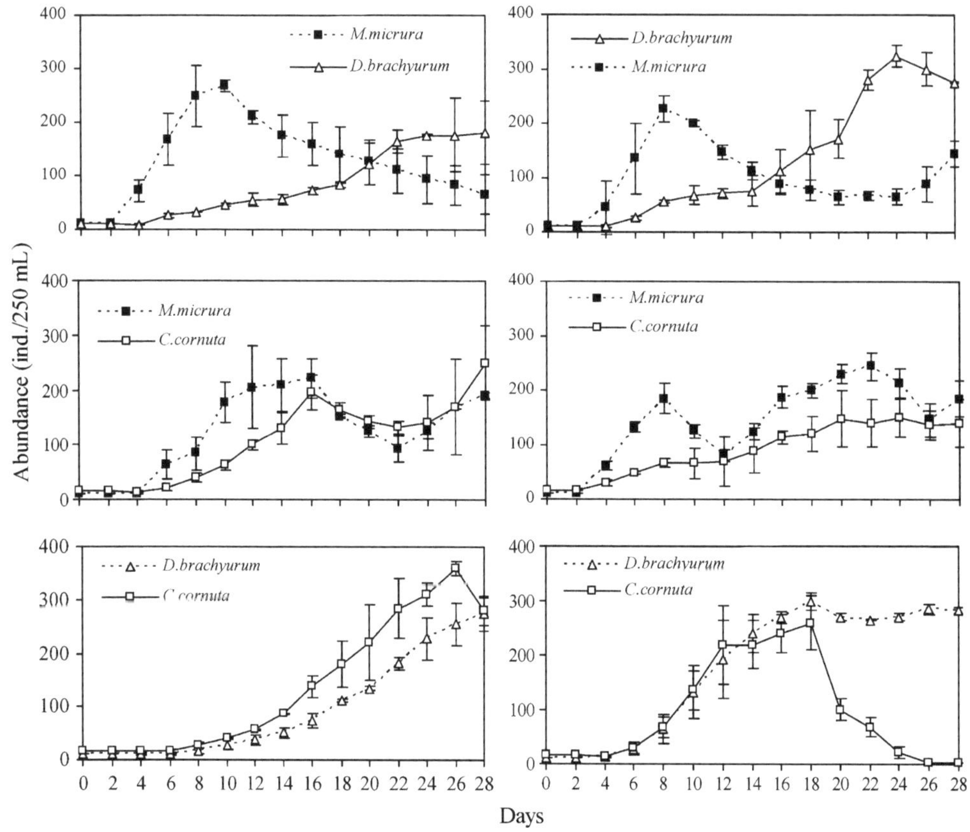

Fig. 2 Cladoceran abundances (±SD) in laboratory competition among *Moina micrura, Diaphanosoma brachyurum* and *Ceriodaphnia cornuta* at different alga combinations. Ultimate time of experiments is four weeks. Left shows results at the alga combination of 50 mg l^{-1} colonial *Microcystis* spp. + 1 mg l^{-1} *Scenedesmus obliquus*. Right at the alga combination of 10 mg l^{-1} colonial *Microcystis* spp. + 1 mg l^{-1} *S. obliquus*. Algal biomass is fresh weight. Abundance in Y-axis is number in 250 ml

effect on their maximal population growth rate (Table 1).

Table 2 shows maximum population growth rate (MGR) and time needed to reach MGR (days) for three cladocerans in competition under two algal combinations. Except for the competition between *M. micrura* and *C. cornuta*, MGR were higher in the 10 M:1 S trials and time reaching MGR shortened correspondingly.

Discussion

Results show that competitive outcomes among cladocerans, especially between large-sized *D. carinata* and small-sized cladocerans, are related to colonial *Microcystis* spp. concentration. Low concentrations of colonial *Microcystis* spp. may not result in competitive extinction of large-sized *D. carinata*, while high *Microcystis* spp. concentration brought on its extinction.

During a cyanobacteria bloom, it is thought that community succession is from larger-sized cladocerans to copepods, rotifers and small-sized cladocerans (e.g. Infante & Riehl, 1984). This has not always been the case in laboratory experiments and field investigations (Mesfin et al., 1988). In fact, concentration, size, *Microcystis* toxin, and alternative food abundance affect zooplankton shifts. Large-sized phytoplankton and high concentrations of colonial algae often inhibit *Daphnia* sp. Feeding and interfere with food availability (Hessen & Donk, 1993). *Daphnia ambigua* dominated without *M. aeruginosa*

Table 1 Results of variance (ANOVA) analysis performed on maximal population growth rate of *M. micrura*, *D. brachyurum* and *C. cornuta* in relation to food mixture and competitive interaction

Source	df	SS	MS	F
M. micrura				
Food combination (A)	1	0.0002	0.0002	7.35^{ns}
With species (B)	1	0.0733	0.0733	2444.68**
Interaction of A·B	1	0.0076	0.0076	252.14**
Error	7	0.0812	0.0116	
D.brachyurum				
Food combination (A)	1	0.0062	0.0062	842.59**
With species (B)	1	0.0018	0.0018	239.94**
Interaction of A·B	1	0.0004	0.0004	58.98**
Error	7	0.0085	0.0012	
C. cornuta				
Food combination (A)	1	0.0008	0.0008	127.99**
With species (B)	1	0.0046	0.0046	737.24**
Interaction of A·B	1	0.0005	0.0005	76.88**
Error	7	0.0059	0.0008	

Ns and **represent significant level of $p > 0.05$ and $p < 0.001$, respectively

and approached extinction with 50% and 90% *M. aeruginosa* (Fulton & Paerl, 1988). Our results show a different competitive shift for *D. carinata* under different concentrations of *Microcystis*. For small-sized cladocerans in the present study, growth rate was related negatively to colonial *M. aeruginosa* concentration with fixed concentration of *S. obliquus*. *D. carinata* obtained the highest growth rate at a biomass level of 10 mg l^{-1} colonial *M. aeruginosa*, indicating its ability to utilize colonial forms (Chen & Xie, 2003). The competitive shift in the present study was from small-sized cladocerans to *D. carinata* when concentration of colonial *Microcystis* spp. was low. It appears that a low concentration of colonial *Microcystis* does not lead to the general successive pattern for zooplankton.

Colony size also affects cladoceran population and their interactions. Only *Daphnia pulex* could feed on 60–100 μm *Microcystis* colonies, while *Ceriodaphnia reticulata*, *M. micrura* and *Diaphanosoma excisum* could not (Jarvis et al., 1987). Previous work showed similar results (Chen & Xie, 2003). The inability of the other three cladocerans to feed on 64–112 μm *Microcystis* colonies determined the status of *D. carinata* in competition. In the present study, larger-sized *M. aeruginosa* (>112 μm) colonies that may have a more negative effect on *D. carinata* were not evaluated. For the selected size of *Microcystis* spp., inhibition of *Daphnia* from low *Microcystis* concentration was weak, and *Daphnia* maintained dominance.

Microcystin and other compounds in *Microcystis* can be lethal to cladocerans (e.g. Jungamnn, 1992; DeMott, 1999). *Keratella cochlearis* was superior in competition with *D. pulex* under toxic *Microcystis* (Sartonov, 1995). However, there is little or no evidence for lethal effects of toxic cyanobacteria on crustacean plankton in the field (Haney, 1987). In fact, many cladocerans can coexist well with *Microcystis* in nature (McNaught et al., 1997; Nandini et al., 2000). Colonial *Microcystis* spp. used in the present study had no obvious toxicity to cladocerans based on the previous study, but it exhibited toxicity when colonies were destroyed (Chen & Xie, 2004). Microcystin released into the water is thought to be inhibited by the mucilaginous structure of the colony (Stangenberg, 1968; Goarant et al., 1994). Once colonial *Microcystis* decomposes, microcystin and other toxic compounds are released (Watanabe et al., 1992). Although toxicity is possible, alternative food availability may weaken toxic effects of microcystin (Ferrão-Filho et al., 2000; Chen & Xie, 2004).

During the bloom, alternative food, including bacteria attached to colonial *Microcystis*, detritus, algae and protozoans also affect population sizes

Table 2 Maximum population growth rate (MGR) and time to reach MGR (days) of three cladocerans among competitions under two types of mixtures of *Microcystis* spp. (M) and *Scenedesmus obliquus* (S)

	50 mg l^{-1} M + 1 mg l^{-1} S			10 mg l^{-1} M + 1 mg l^{-1} S		
	M. micrura	*D. brachyurum*	*C. cornuta*	*M. micrura*	*D. brachyurum*	*C. cornuta*
M. micrura		0.328(10)	0.194(16)		0.390(8)	0.145(22)
D. brachyurum	0.103(28)		0.118(28)	0.145(24)		0.189(18)
C. cornuta	0.089(28)	0.122(26)		0.096(24)	0.158(18)	

of cladocerans. Laboratory experiments showed that growth rates of the four cladocerans used in the present study were correlated positively with *S. obliquus* biomass (Chen & Xie, 2003). Many species, including *Daphnia* sp., can consume colonial *Microcystis* sp. of edible size (Matveev et al., 1994; Nandini et al., 2000). *Daphnia* sp. can consume the mucilaginous part of colonies and associated bacteria and may be more resistant to toxin (Goarant et al., 1994). In the present study, *D. carinata* was not inhibited by low colonial *Microcystis* spp. concentration. *Scenedesmus obliquus*, the mucilaginous part of colonies and associated bacteria may be used by *D. carinata*, which may improve population growth. *Moina micrura* may have a competitive advantage over rotifers in exploiting *Microcystis* as a food source or use other food sources (e.g. bacteria) when *Microcystis* are under stress (Kim et al., 2001). Several genera of *Moina, Diaphanosoma, Bosmina* and *Daphnia* can use cyanobacteria detritus (Hanazato & Yasuno, 1985, 1987; Gulati et al., 2001; Chen & Xie, 2003). This may promote populations of these species in the lake and affect their dominant status in the cladoceran community through interactions.

However, other research shows that low concentrations of edible algae favors small-sized cladocerans, while high concentrations favor large-sized cladocerans, such as *Daphnia* (Goulden et al., 1982; Tessier & Goulden, 1987). During a *Microcystis* bloom, other edible algae may have actual or relative low concentration. Resultant food scarcity may favor small-sized cladocerans. If *Daphnia* are inhibited by colonial *Microcystis* and less alternative food, its population may decline.

No obvious competition outcomes between *M. micrura* and *D. brachyurum* or *C. cornuta* under two alga combinations were observed in the present study. *Diaphanosoma brachyurum* outweighs *M. micrura* later in the experiment. Similar trends appear in the competition between *D. brachyurum* and *C. cornuta*. *D. brachyurum* may utilize bacteria around *Microcystis* spp. and detritus in the beaker more than *M. micrura* and *C. cornuta* (Champ & Pourriot, 1977; Taira, 1989). For these three cladocerans, competition may affect competitive outcomes (Chaneton & Bonsall, 2000; Martínez & Medel, 2002). In nature, this effect should be considered.

In conclusion, our experiments reveal that in 50 mg l^{-1} colonial *Microcystis* spp. + 1 mg l^{-1} *S. obliquus* combinations, *D. carinata* was extinct during the experiment in competition with other cladocerans. Decreasing colonial *Microcystis* spp. concentration (10 mg l^{-1}) resulted in a shift from small-sized cladocerans to *D. carinata*. No obvious effect on competitive outcomes was shown among three small-sized cladocerans. These results suggest that cyanobacteria concentration may influence dominance status of large-bodied daphnids.

Acknowledgements This work was supported by National Natural Science Foundation of China (No. 30600069), a key project of Chinese Academy of Sciences (No. KZCX1-SW-12) and National Natural Science Foundation of China (No. 40501078).

References

Allan, J. D., 1977. An analysis of seasonal dynamics of a mixed population of *Daphnia* and associated cladoceran community. Freshwater Biology 7: 505–512.

Carmichael, W. W., M. J. Yu, Z. R. He, J. W. He & J. L. Yu, 1988. Occurrence of the toxic cyanobacterium (blue-green alga) *Microcystis aeruginosa* in central China. Archiv für Hydrobiologie 114: 21–30.

Champ, P. & R. Pourriot, 1977. The food of freshwater planktonic cladocera (Crustacea). Annee Biology 16: 317–342.

Chaneton, E. J. & M. B. Bonsall, 2000. Enemy-mediated apparent competition: empirical patterns and the evidence. Oikos 88: 380–394.

Chen, F. Z. & P. Xie, 2003. The effects of fresh and decomposed Microcystis aeruginosa on cladocerans from a subtropic Chinese lake. Journal of Freshwater Ecology 18: 97v–104v.

Chen, F. Z. & P. Xie, 2004. The toxicities of single-celled *Microcystis aeruginosa* PCC7820 and liberated *M. aeruginsa* to *Daphnia carinata* in the absence and present of the green alga *Scenedesmus obliquus*. Journal of Freshwater Ecology 19: 539–545.

DeMott, W. R., 1999. Foraging strategies and growth inhibition in five daphnids feeding on mixtures of a toxic cyanobacterium and a green alga. Freshwater Biology 42: 263–274.

Ferrão-Filho, A. S., S. M. F. O. Azevedo & W. R. DeMott, 2000. Effects of toxic and non-toxic cyanbacteria on the life history of tropical and temperature cladocerans. Freshwater Biology 45: 1–19.

Fulton, III R. S. & H. W. Paerl, 1988. Effects of the blue-green alga *Microcystis aeruginosa* on zooplankton competitive relations. Oecologia 76: 383–389.

George, D. G. & R. W. Edwards, 1974. Population dynamics and production of *Daphnia hyalina* in a eutrophic reservoir. Freshwater Biology 4: 445–465.

Goarant, E., G. Prensier & N. Lair, 1994. Specific immunological probes for the identification and tracing of prey in crustacean gut contents. The example of cyanobacteria. Archiv für Hydrobiologie 131: 243–252.

Goulden, C. E., L. L. Henry & A. J. Tessier, 1982. Body size, energy reserves, and competitive ability in three species of cladocera. Ecology 63: 1780–1789.

Gulati, R. D., M. Bronkhorst & E. Van Donk, 2001. Feeding in *Daphnia galeata* on *Oscillatoria limnetica* and on detritus derived from it. Journal of Plankton Research 23: 705–718.

Hanazato, T. & M. Yasuno, 1985. Population dynamics and production of cladoceran zooplankton in the highly eutrophic Lake Kasumigaura. Hydrobiologia 124: 13–22.

Hanazato, T. & M. Yasuno, 1987. Experimental studies on competition between *Bosmina longirotris* and *Bosmina fatalis*. Hydrobiologia 154: 189–199.

Haney, J. F., 1987. Field studies on zooplankton-cyanobacteria interactions. New Zealand Journal of Marine and Freshwater Research 21: 467–475.

He, J. W., Z. R. He & Q. L. Guo, 1997. The toxicity of *Microcystis aeruginosa* to fishes and *Daphnia*. Journal of Lake Sciences 9: 49–56 (in Chinese).

Hessen, D. O. & E. Van Donk, 1993. Morphological changes in *Scenedesmus* induced by substances release from *Daphnia*. Archiv für Hydrobiologie, 127: 129–140.

Infante, A. & W. Riehl, 1984. The effect of cyanophyta upon zooplankton in a eutrophic tropical lake (Lake Valencia, Venezuela). Hydrobiologia 113: 293–298.

Jarvis, A. C., R. C. Hart & S. Combrink, 1987. Zooplankton feeding on size fractionated *Microcystis* colonies and *Chlorella* in a hypertrophic lake (Hartbeespoort Dam, South Africa): Implications to resource utilization and zooplankton succession. Journal of Plankton Research 9: 1231–1249.

Jungmann, D., 1992. Toxic compounds isolated from *Microcystis* PCC7806 that are more active against *Daphnia* than two microcystins. Limnology and Oceanography 37: 1777–1783.

Kim, H. W., G. J. Joo, K. H. Chang & S. J. Hwang, 2001. Zooplankton community dynamics during the summer Mycrocystis bloom in the lower part of the Nakdong River, South Korea. Verhandlungen Internationale Vereinigung für theoretische und angewandte Limnologie 27: 1044–1049.

Li, S. H., H. Zhu, J. Z. Xia, K. E. Lin, K. S. Liu, Z. Y. Le & Y. X. Chen, 1959. Experiment on mass culture of single-celled green alga. Acta Hydrobiological Sinica 4: 462–472 (in Chinese).

Martínez, G. & R. Medel, 2002. Indirect interactions in a microcosm-assembled cladoceran community: implications for apparent competition. Oikos 97: 111–115.

Matveev, V., L. Matveeva & G. Jones, 1994. Study of the ability of *Daphnia carinata* King to control phytoplankton and resist cyanobacterial toxicity: implications of biomanipulation in Australia. Australia Journal of Marine and Freshwater Research 45: 889–904.

McNaught, D. C., W. Chen & H. J. Cai, 1997. Grazing on a *Microcystis* bloom by cladocerans and copepods in Tai Lake. Verhandlungen Internationale Vereinigung für theoretische und angewandte Limnologie 26: 550–552.

Mesfin, M., C. Tudorancea & R. M. Baxter, 1988. Some limnological observations on two Ethiopian hydroelectric reservoirs: Koka (Shewa administrative district) and Finchaa. Hydrobiologia 157: 47–55.

Nandini, S., S. S. S. Sarma & P. Ramirez-Garcia, 2000. Life table demography and population growth of *Daphnia laevisi* (Cladocera, Anomopoda) under different densities of *Chlorella vulgaris* and *Microcystis aeruginosa*. Crustaceana 73: 1273–1286.

Sartonov, A., 1995. Effects of *Microcysits aeruginosa* on interference competition between *Daphnia pulex* and *Keratella cochlearis*. Hydrobiologia 307: 117–126.

Stangenberg, M., 1968. Toxic effects of *Microcystis aeruginosa* Kg. Extracts on *Daphnia longispina* O. F. Müller and *Eucypris virens* Jurine. Hydrobiologia 32: 81–87.

Taira, M., 1989. Species composition and feeding habits of crustacean plankters in the pools of high moor. Japanese Journal of Limnology 50: 333–339.

Tang, H. J., 2002. Ecological studies on phytoplankton of the shallow, eutrophic Lake Donghu. Dotoral dissertation thesis, Institute of Hydrobiology, Chinese Academy of Sciences (in Chinese).

Tessier, A. J. & C. E. Goulden, 1987. Cladoceran juvenile growth. Limnology and Oceanography 32: 680–686.

Watanabe, M. M., K. Kaya & N. Takamura, 1992. Fate of the toxic cyclic heptapeptides, the microcystins, from blooms of *Microcystis* (Cyanobacteria) in a hypertrophic lake. Journal of Phycology 28: 761–767.

Hydrobiologia (2007) 581:217–223
DOI 10.1007/s10750-006-0506-4

EUTROPHICATION IN LAKES

The effect of temperature on growth characteristics and competitions of *Microcystis aeruginosa* and *Oscillatoria mougeotii* in a shallow, eutrophic lake simulator system

Zhaosheng Chu · Xiangcan Jin · Norio Iwami · Yuhei Inamori

Abstract Blue-green algal blooms formed by *Microcystis* and *Oscillatoria* often occur in shallow eutrophic lakes, such as Lake Taihu (China) and Lake Kasumigaura (Japan). Growth characteristics and competitions between *Microcystis aeruginosa* and *Oscillatoria mougeotii* were investigated using lake simulator systems (microcosms) at various temperatures. *Oscillatoria* was the superior competitor, which suppressed *Microcystis*, when temperature was <20°C, whereas the opposite phenomenon occurred at 30°C. *Oscillatoria* had a long exponential phase (20 day) and a low growth rate of 0.22 day^{-1} and 0.20 day^{-1} at 15°C and 20°C, respectively, whereas *Microcystis* had a shorter exponential phase (2–3 days) at 30°C and a higher growth rate (0.86 day^{-1}). Interactions between the algae were stronger and more complex in the lake simulator system than flask systems. Algal growth in the lake simulator system was susceptible to light attenuation and pH change, and algae biomasses were lower than those in flasks. The outcome of competition between *Microcystis* and *Oscillatoria* at different temperatures agrees with field observations of algal communities in Lake Taihu, indicating that temperature is a significant factor affecting competition between *Microcystis* and *Oscillatoria* in shallow, eutrophic lakes.

Keywords *Microcystis* · *Oscillatoria* · Lake simulator · Competition · Lake Taihu

Guest editors: B. Qin, Z. Liu & K. Havens
Eutrophication of shallow lakes with special reference to Lake Taihu, China

Z. Chu · X. Jin (✉) · N. Iwami
Research Center for Lake Ecology & Environments, Chinese Research Academy of Environmental Sciences (CRAES), Beijing 100012, China
e-mail: jinxc@craes.org.cn

Y. Inamori
National Institute for Environmental Studies (NIES), 16-2 Onogawa, Ibaraki 305-0053, Japan

Introduction

Blue-green algal blooms often cause significant problems, such as odors, water deoxygenation and clogging of water supply systems. Furthermore, some blue-green algae produce toxins, which are dangerous to aquatic organisms, domestic animals and human beings. *Microcystis* and *Oscillatoria* are two genera of bloom-forming blue-green algae often present in shallow eutrophic lakes (Reynolds & Walsby, 1975). *Microcystis* usually forms colonies, and *Oscillatoria* grows in filamentous form. In Lake Taihu (China), *Microcystis* often dominates in summer and autumn (Fan, 1996; Dokulil et al., 2000; Yang et al., 2003) whereas *Oscillatoria* often dominates in early spring and later autumn. The climate of Lake Kasumigaura (Japan) is similar to Lake Taihu, and *Microcystis* dominance often occurs in summer (Wei et al., 2001). In recent years, *Microcystis* has been declining and was replaced by filamentous blue-green algae, such as *Oscillatoria* and *Phormidium* (Takamura et al., 1992; Ohkubo et al., 1993; Wei et al., 2001). Despite considerable research on photosynthetic and growth responses of *Microcystis* and *Oscillatoria* in laboratory and field studies (Mur & Schreurs, 1995; Zhang & Prepas, 1996; Fujimoto et al., 1997; Coles and Jones 2000), there is limited information about mechanisms of *Microcystis* and *Oscillatoria* dominance in seasonal dynamics and year-by-year changes in shallow eutrophic lakes.

Generally, there are three categories of studies on competition of bloom-forming algae. The first is large-scale investigations based on field surveys; second, small-scale studies based on laboratory experiments; and, third, intermediate-scale studies applying 'microcosms'. Natural systems are often too complicated to obtain detailed information in a short time or only provide insight into lake-specific circumstances. The second method often oversimplifies natural systems, and information only can be applied in a limited sense. Two kinds of microcosms being used are the 'in-situ microcosm', placed directly into lakes, and the 'laboratory microcosms', constructed in the laboratory. Construction of in situ microcosms depends on weather and space (Xie et al., 2003a, b). Here, a laboratory microcosm, called the lake simulator system, was used. It is a large-scale simulation system where light, temperature, nutrients, water and turbulence can be controlled (Okada & Sudo, 1980). Conveniently, chemical components, such as pH, dissolved oxygen (DO) and oxidation–reduction potential (ORP) can be monitored. Furthermore, the system can be sterilized by steam.

In this study, the lake simulator and a flask system were used to investigate growth and competition of *Microcystis* and *Oscillatoria* at different temperatures, thus increasing understanding of shifts in bloom-forming blue-green algae during seasonal dynamics in shallow eutrophic lakes.

Materials and methods

Algae. Microcystis aeruginosa Kützing and *Oscillatoria mougeotii* Kützing *ex* Lemmermann were isolated from Lake Kasumigaura in 1996 and maintained at 20°C under a 12 h:12 h light:dark cycle in M_{11} medium, which consists of 100 mg $NaNO_3$, 10 mg K_2HPO_4, 75 mg $MgSO_4 \cdot 7H_2O$, 40 mg $CaCl_2 \cdot 2H_2O$, 20 mg Na_2CO_3, 6 mg Fe-citrate and 1 mg $Na_2EDTA \cdot 2H_2O$ in 1:l deionized water. Both algal cultures were axenic.

Lake Simulator System. The lake simulator system consists of two artificial lakes (tanks, 4 m high and 1 m diameter), two artificial suns (Xenon lamp, 5 kW), a sterilization tank, monitoring system, temperature control system, sterilized air supply system and a steam system (an artificial lake and artificial sun is omitted in Fig. 1).

Growth and competition experiments

The lake simulator and small conical flasks (500 ml) were used to investigate growth and competition between *Microcystis* and *Oscillatoria.* M_{11} culture medium was used to cultivate the two algae. Before culture experiments, the simulator was sterilized with steam. The culture medium was sterilized in the sterilization tank and transported to the artificial lake by the pressure of

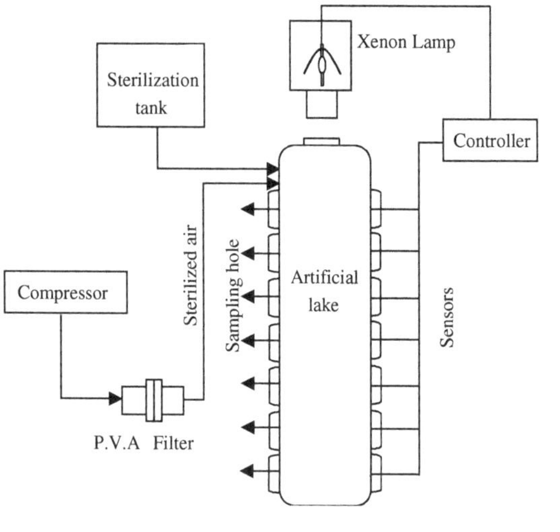

Fig. 1 Schematic diagram of the lake simulator system

sterilized air. Each tank contained 2400 l of culture medium with 3.4 m in depth. To simulate lake turbulence, sterilized air was supplied via the sampling hole, which was 0.7 m above the bottom of the artificial tank, producing 5–10 cm S^{-1} turbulence at the water surface. pH was adjusted to 8.0–8.5 using concentrated HCl. When the medium was saturated with DO, *Microcystis* (in single cell) and *Oscillatoria* were inoculated to the artificial lake resulting in an initial population density of 2.5×10^4 cells ml^{-1} and 5.0 cm ml^{-1} (the length of filaments), respectively. These numbers correspond to 0.40 and 0.20 mg DW (dry weight) l^{-1}. Illumination at the surface water was 100,000 lux (12 h light:12 h dark cycle), similar to natural sunlight at noon at the lake surface (Lake Taihu). 200 ml water was sampled from seven sampling holes and pooled for analysis. Flask cultures (containing 150 ml M_{11} medium) were incubated in light (2,000 lux). The culture medium, the light:dark cycle, temperatures and initial cell densities were the same as those in the lake simulator.

Chlorophyll *a* was extracted with acetone (90%) and determined spectrophotometrically (EPA Method, 1997). pH was determined potentiometrically. Algal density was counted with a haemacytometer (TATAI, Japan) under a microscope. 30–300 cells were counted for *Microcystis*, and 10 μl of algal solution determined for *Oscillatoria*. Counting was carried out three times for each sample. If *Oscillatoria* filaments were longer than 0.25 mm, they were disrupted by ultrasonication before counting. Cell counts were converted to dry weight to determine dominance. The sampling culture medium was filtered through glass fiber filter (Whatman/C) and dried at 90°C for 12 h to determine DW. Conversion factors for DW and cell count were 2.3×10^{-11}, 1.6×10^{-11}, and 1.7×10^{-11} g $cell^{-1}$ for *Microcystis* and 4.1×10^{-8}, 4.0×10^{-8}, and 6.0×10^{-8} g cm^{-1} for *Oscillatoria*, at 15°C, 20°C and 30°C, respectively. Specific growth rates were determined by linear regression of logarithm of biomass and time (Eq. 1) on data from exponential growth.

$$\ln B = \mu * t + b \quad (1)$$

where μ is the growth rate, B is dry weight of biomass (mg DW l^{-1}), t is time (day) and b is the ln of inoculation density.

Results

Flask system

Growth curves of uni-algal and mixed cultures of *Microcystis* and *Oscillatoria* in flask experiments consisted of exponential and stationary phases without a lag phase (Figs. 2, 3; since experiments were stopped at the end of the saturation phase, the death phase was not included in the curves). *Oscillatoria* grew very well at 15–30°C (Fig. 3), whereas *Microcystis* did not grow well at 15°C (Fig. 2), whether in uni- or mixed cultures. Growth curves of uni-algal cultures of both algae were similar to those of mixed cultures in flask experiments from 15°C to 30°C (Figs. 2, 3). They had similar growth rates in mixed and uni-algal cultures (Table 1), but maximum biomass in mixed cultures was lower than uni-algal cultures (Table 2).

Growth rates of the two algae increased with temperature (Table 1), and the increase was more pronounced for *Microcystis* than *Oscillatoria.* At 15°C, the growth rate of *Oscillatoria* was higher than *Microcystis*, at 20°C the growth rates were comparable, and at 30°C the growth rate of

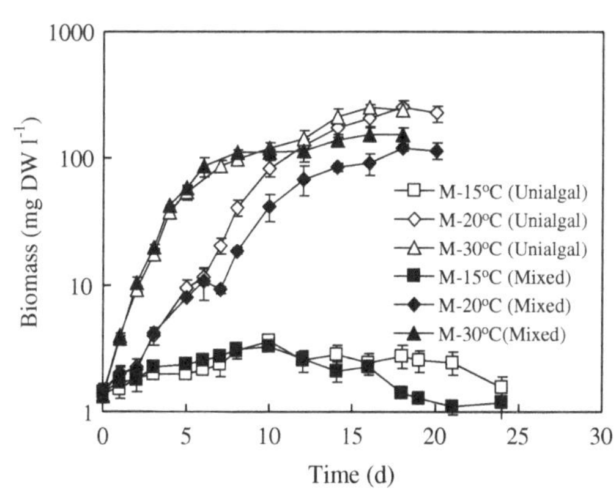

Fig. 2 The growth curves of uni-algal cultures and mixed cultures of *Microcystis* at different temperature in flask experiments

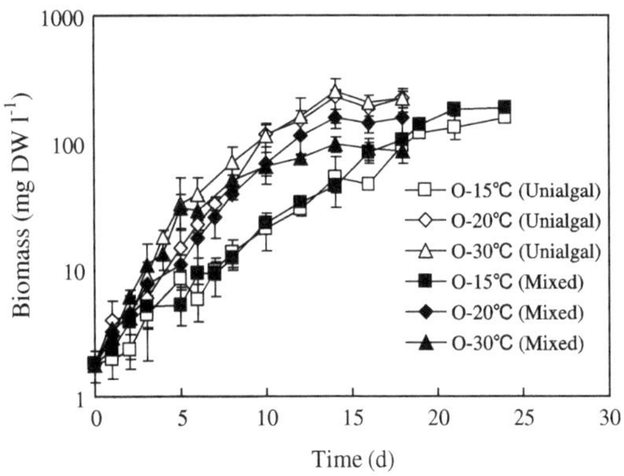

Fig. 3 The growth curves of uni-algal cultures and mixed cultures of *Oscillatoria* at different temperature in flask experiments

Microcystis was higher. Flask experiments showed that *Oscillatoria* was the superior competitor at 15°C, whereas *Microcystis* was superior at 30°C, and the competition was about equal at 20°C (Tables 1, 2).

Lake simulator system

Growth and competition of *Microcystis and Oscillatoria* in the lake simulator system are shown in Fig. 4, revealing that *Oscillatoria* dominated at 15°C and 20°C, whereas *Microcystis* dominated at 30°C. At 15°C, *Oscillatoria* had a long exponential phase (20 day) and a low growth rate, but *Microcystis* grew little, and, therefore, *Oscillatoria* contributed 91% to the total biomass on day 20. When temperature was increased to 20°C, growth and competition of the two algae changed little, and *Oscillatoria* accounted for 98% of total mass on day 20. When temperature was 30°C, *Microcystis* doubled quickly in the exponential phase, but this phase was very short (about 2–3 days). Afterwards, *Microcystis* biomass increased with time, *Oscillatoria* grew very slowly, and *Microcystis* accounted for 99.7% of total biomass on day 29.

Changes in growth rates of *Microcystis* and *Oscillatoria* in lake simulator experiments did not show the same trends. With increased temperature, *Microcystis* growth rate increased, while that of *Oscillatoria* decreased. Moreover, total maximum biomass increased with temperature: 9.13, 19.9 and 28.1 mg DW l^{-1} at 15°C, 20°C and 30°C, respectively. These biomasses accounted for 4.8, 5.0 and 7.4% of those reached in flasks experiments.

Table 2 Maximum biomass (B_{max}) of *Microcystis* and *Oscillatoria* at different temperature in the lake simulator and flasks experiments

		B_{max} (mg DW l^{-1})		
		Flask uni-algal cultures	Flask mixed cultures	Lake simulator cultures
Microcystis	15°C	3.56	3.23	1.14
	20°C	345	155	0.60
	30°C	250	155	28.1
Oscillatoria	15°C	156	188	8.35
	20°C	220	121	19.9
	30°C	254	97.2	0.50

Table 1 Growth rates ($\mu \pm$ SE) of *Microcystis* and *Oscillatoria* at different temperature in the lake simulator and flasks experiments

		Growth rate μ (day^{-1})		
		Flask uni-algal cultures	Flask mixed cultures	Lake simulator cultures
Microcystis	15°C	0.083 ± 0.009 (0–10 day)	0.079 ± 0.007 (0–10 day)	–
	20°C	0.42 ± 0.02 (0–10 day)	0.33 ± 0.02 (0–10 day)	–
	30°C	0.81 ± 0.04 (0–4 day)	0.85 ± 0.05 (0–4 day)	0.86 (0–2 day)
Oscillatoria	15°C	0.22 ± 0.01 (0–19 day)	0.22 ± 0.01 (0–19 day)	0.22 ± 0.02 (2–20 day)
	20°C	0.40 ± 0.02 (0–10 day)	0.35 ± 0.02 (0–10 day)	0.20 ± 0.01 (0–20 day)
	30°C	0.56 ± 0.05 (0–4 day)	0.55 ± 0.04 (0–5 day)	–

Period of exponential phase is in parentheses

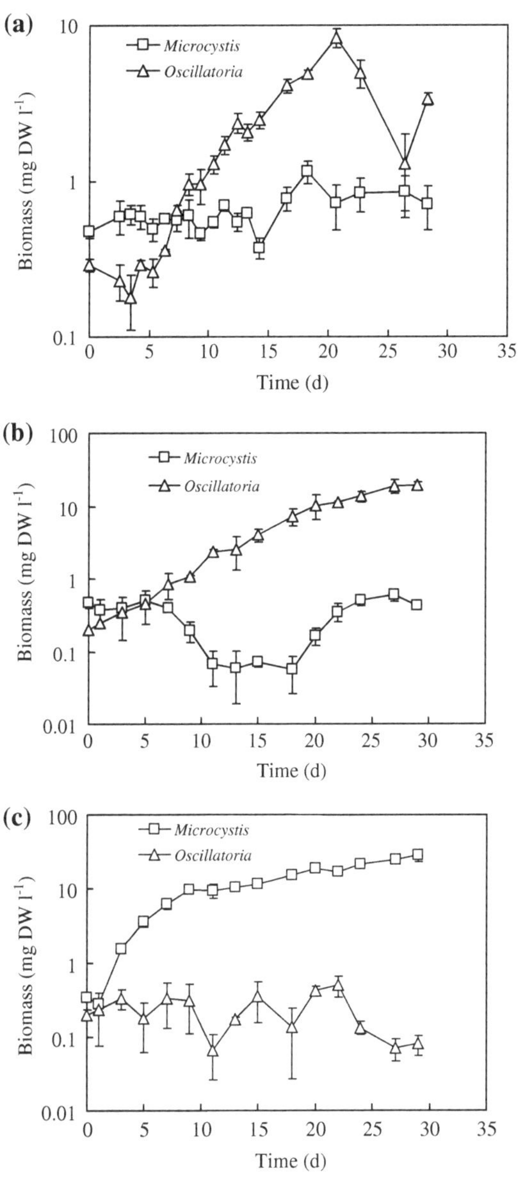

Fig. 4 Competition between *Microcystis* and *Oscillatoria* in lake simulator experiments at 15°C (**a**), 20°C (**b**) and 30°C (**c**)

Discussion

Uni-algal cultures of *Microcystis and Oscillatoria* in flasks showed increased growth rates with increasing temperature. Flasks can be considered a homogeneous system due to small volume, and they are useful for determination of specific growth rates under different conditions. When grown together in flasks, the two algal species grew independently and reached high biomasses. However, the flasks represent a simple system and cannot simulate a lake.

Competition between *Microcystis and Oscillatoria* in the lake simulator experiments showed that *Oscillatoria* dominated when temperature was <20°C, whereas *Microcystis* dominated when temperature was 30°C. Maximum biomasses reached in lake simulator experiments were <10% of those reached in corresponding flasks at all temperatures. Since interactions between the two algae in the lake simulator experiments were stronger than those in flasks, these results show that flask experiments cannot predict competition between algae in a lake system.

Changes in nutrients, light, pH and other factors during algal growth may be important in algal competition. Due to the low biomass in lake simulator experiments, all nutrients except carbon were considered sufficient and could not have limited algal growth rates. Since speciation of total inorganic carbon (TIC) depends on pH, the available inorganic carbon resource is determined by both TIC and pH (Stumm & Morgan, 1996). An increase in pH results in a decrease in the fraction of free CO_2, which is assimilated easily by algae. Low CO_2 conditions decrease growth rate (Thyssen et al., 2001). Indeed, growth rates of the dominant algae decreased when pH increased to 9–9.5 (Fig. 5). This suggests that changes in the growth rate of the dominant alga may be a result of pH changes. Furthermore, the light climate

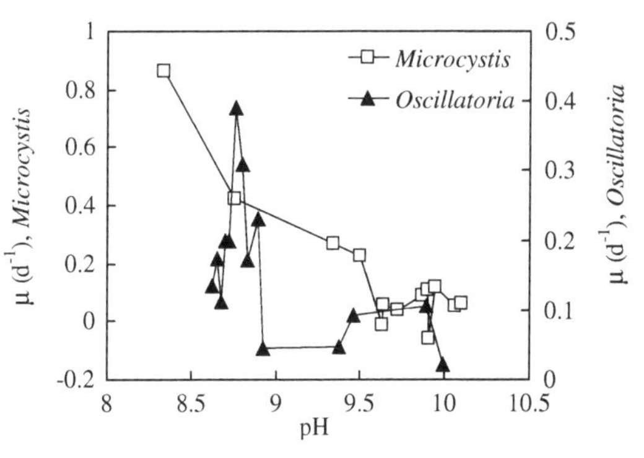

Fig. 5 The effect of pH on the growth rate on the dominant alga (*Microcystis* is the dominant alga at 30°C and *Oscillatoria* is the dominant alga at 20°C)

underwater changed during algal growth. If the wall effect on light is neglected, Eq. 2 (Huisman et al., 2002) can be used to describe changes of light intensity in the lake simulator system.

$$I(z,t) = I_{\text{in}} \exp[-\int_0^z k\varpi(\sigma,t)\mathrm{d}\sigma - K_{\text{bg}}z] \quad (2)$$

where I_{in} is incident light intensity, $\varpi(\sigma,t)$ is population density at depth z and time t, σ is an integration variable, k is the specific light attenuation coefficient of algae, and K_{bg} is the background attenuation. As shown by Eq. 2, light intensity attenuates exponentially with biomass and depth. Due to high incident light intensity, increasing algal biomass may alleviate high light stress. However, growth rate did not increase with biomass, suggesting high light stress was not evident in the lake simulator system. Furthermore, increasing algal biomass leads to decreased transparency and development of a steep light gradient in the upper water column. Light limitation caused by self-shading of algae may limit growth when the biomass of dominant alga is significant. At 30°C, a long linear phase following the exponential phase may be the result of light limiting growth of *Microcystis*. At 15°C and 20°C, *Oscillatoria* died quickly after the exponential phase, which may result from increased pH. Due the rapid growth of the superior alga, elevated pH and light shading also may influence growth of inferior algae in the lake simulator system.

As the lake simulator system simulated sunlight, water turbulence, nutrient concentrations and depth in natural, shallow, eutrophic lakes and controlled temperature precisely, results might reflect temperature influences on growth and competition of these two algae in natural lakes. In some areas of Lake Taihu (China), such as Gonghu Bay in the northeast, *Oscillatoria* dominates in spring and late autumn, whereas *Microcystis* dominates in summer and early autumn. Temperatures in natural systems often are between 15 and 25°C, and, therefore, results from the lake simulator experiments can be used to predict cyanobacterial dominance in nature. Although there are many environmental factors affecting the algal community, this study suggests that water temperature is an important factor in the shift of *Microcystis* and *Oscillatoria* dominance in a simulated, shallow, eutrophic lake.

Acknowledgements This work was supported by the Major State Basic Research Development Program of China (No. 2002CB412301), JICA Taihu Water Environment Restoration Model Project and National Natural Science Foundation of China (No. 40501078). We thank Dr. Fengchang Wu, Dr. Shao Yang and three anonymous reviewers for their useful comments and suggestions.

References

Coles, J. F. & R. C. Jones, 2000. Effect of temperature on photosynthesis-light response and growth of four phytoplankton species isolated from a tidal freshwater river. Journal of Phycology 36: 7–16.

Dokulil, M., W. Chen & Q. Cai, 2000. Anthropogenic impacts to large lakes in China: the Tai Hu example. Aquatic Ecosystem Health and Management 3: 81–94.

EPA Method 446.0, 1997. In vitro Determination of Chlorophylls *a*, *b*, $c_1 + c_2$ and Pheopigments in Marine and Freshwater Algae by Visible Spectrophotometry.

Fan, C. X., 1996. Historical evolution of water ecological setting in Taihu Lake. Journal of Lake Sciences 8: 297–304 (in Chinese).

Fujimoto, N., R. Sudo, N. Sugiura & Y. Inamori, 1997. Nutrient-limited growth of *Microcystis aeruginosa* and *Phormidium tenue* and competition under various N:P supply ratios and temperatures. Limnology and Oceanography 42: 250–256.

Huisman, J., M. Arrayas, U. Ebert & B. Sommeijer, 2002. How do sinking phytoplankton species manage to persist? The American Naturalist 159: 245–254.

Mur, L. R. & H. Schreurs, 1995. Light as a selective factor in the distribution of phytoplankton species. Water Science and Technology 32: 25–34.

Ohkubo, N., O. Yagi & M. Okada, 1993. Studies on the succession of blue-green-algae, *Microcystis*, anabaena, *Oscillatoria* and *Phormidium* in lake Kasumigaura. Environmental Technology 14: 433–442.

Okada, M. & R. Sudo, 1980. Microcosm system study of algal bloom formation. Journal WPCF 52: 1029–1033.

Reynolds, C. S & A. E. Walsby, 1975. Water-blooms. Biology Review 50: 437–481.

Stumm, W. & J. J. Morgan, 1996. Aquatic Chemistry: Chemical Equilibria and Rates in Natural Waters, 3rd edn. John Wiley & Sons, New York, 148–205.

Takamura, N., A. Otsuki, M. Aizaki & Y. Nojiri, 1992. Phytoplankton species shift accompanied by transition from nitrogen dependence to phosphorus dependence of primary production in lake Kasumigaura, Japan. Archiv für Hydrobiologie 124: 129–148.

Thyssen, C., R. Schlichting & C. Giersch, 2001. The CO_2-concentrating mechanism in the physiological context: lowering the CO_2 supply diminishes culture growth

and economises starch utilisation in Chlamydomonas reinhardtii. Planta 213: 629–639.
Wei, B., N. Sugiura & T. Maekawa, 2001. Use of artificial neural network in the prediction of algal blooms. Water Research 35: 2022–2028.
Xie, L., P. Xie, L. Sixin, H. Tang & H. Liu, 2003a. The low TN:TP ratio, a cause or a result of *Microcystis* blooms? Water Research 37: 2073–2080.
Xie, L., P. Xie & H. J. Tang, 2003b. Enhancement of dissolved phosphorus release from sediment to lake water by *Microcystis* blooms – an enclosure experiment in a hyper-eutrophic, subtropical Chinese lake. Environmental Pollution 122: 391–399.
Yang, Z., W. Shi, L. Chen & Y. Chen, 2003. Ecological environment succession and countermeasure of East Taihu. China Environmental Science 23: 64–68 (in Chinese).
Zhang, Y. & E. E. Prepas, 1996. Regulation of the dominance of planktonic diatoms and cyanobacteria in four eutrophic hardwater lakes by nutrients, water column stability, and temperature. Canadian Journal of Fisheries and Aquatic Sciences 53: 621–633.

Hydrobiologia (2007) 581:225–240
DOI 10.1007/s10750-006-0508-2

EUTROPHICATION IN LAKES

Stability and change of phytoplankton communities in a highly dynamic environment—the case of large, shallow Lake Balaton (Hungary)

Márk Honti · Vera Istvánovics · András Osztoics

Abstract Time series data of key environmental variables (water temperature, global radiation, vertical light attenuation, internal P load) and biomass of four colour classes of photosynthetically active algae were collected during 2003 and 2004 with daily resolution. Using these data, seasonal patterns of phytoplankton were analyzed as a function of the dynamic environment. Abstraction of the environmental state as a point in multi-dimensional space was used to identify habitat templates of bloom-forming groups and derive an indicator of environmental stability/physical disturbance. These templates were synthesized into a simple threshold model that sufficiently simulated development and collapse of various blooms. Blooms were, however, rare events related to specific environments with strong, unidirectional forcing. Tentative quantification of disturbance and compositional stability/community change allowed discriminating disturbance-driven changes and autogenic succession with reasonable success. The two processes were found to be equally important in shaping the composition and biomass of phytoplankton.

Keywords Daily time series · Environmental stability · Physical disturbance · Autogenic succession · Predictability · Ground state

Introduction

Phytoplankton communities undergo a seasonal change comparable to the decade-scale succession in terrestrial ecosystems (Reynolds, 1997). Succession is, in a strict sense, an autogenic process during which species selection is driven by environmental changes resulting from metabolism of the community itself. External forcing represents the rest of environmental variability, which phytoplankton cannot influence. Below certain thresholds, external forcing is absorbed efficiently by the community without leaving its recognizable fingerprint on composition or abundance. Beyond these thresholds, it acts as a disturbance upon phytoplankton development. Such events alter community structure, repeatedly distorting the succession pattern and propagating it into altered directions of development. The unpredictable frequency of disturbance sets the time span allowed for autogenic succession and, thus, decreases the chance of attaining stable states.

The intermediate disturbance hypothesis (IDH) of Connell (1978) recognizes that disturbance

Guest editors: B. Qin, Z. Liu & K. Havens
Eutrophication of shallow lakes with special reference to Lake Taihu, China

M. Honti (✉) · V. Istvánovics · A. Osztoics
Department of Civil and Environmental Engineering, Budapest University of Technology and Economics, Műegyetem rkp. 3., Budapest 1111, Hungary
e-mail: mark@vkkt.bme.hu

in a specific range of frequency is a prime factor maintaining species diversity of ecosystems. This range of frequencies must be scaled to the characteristic doubling time of biota making up the examined communities (Reynolds, 1997). Fulfilling this criterion, the IDH recently has been shown to be a successful approach in phytoplankton ecology (Reynolds, 1988; Padisák et al., 1993; Padisák, 1993; Sommer, 1995). However, disturbance is not measured when studying natural phytoplankton but inferred from specific compositional changes (Juhász-Nagy, 1993). Nevertheless, few theoretical studies aimed at resolving the logical problem of this circular argumentation (Reynolds, 1997; Elliott et al., 2001). This is not surprising if one considers the difficulties faced when characterizing environmental stability/disturbance. One point is a priori clear: such a trial will not succeed unless environmental dynamics is captured at a temporal scale matching characteristic period time of phytoplankton. As shown by Fourier spectrum analysis of daily biomass time series from Lake Balaton, the characteristic period time is in the range of a few days (Istvánovics et al., 2005). Consequently, environmental dynamics is to be described at least at a daily scale in this lake.

Large, shallow lakes represent a dynamic environment for phytoplankton, where it is difficult to separate various pathways of community development due to frequent resuspension of sediment. Field work and theoretical analysis revealed the decisive impact of sediment resuspension acting as disturbance on phytoplankton dynamics in Lake Balaton (Padisák et al., 1988, 1990; Padisák, 1993) as well as in a broad assortment of shallow lakes (Padisák et al., 2003).

A high-resolution monitoring program in Lake Balaton yielded time series of key environmental variables and phytoplankton at a daily scale (Istvánovics et al., 2005). The objectives of this work are to identify patterns of community development, which are predictable using available environmental variables, and distinguish compositional changes of phytoplankton driven by disturbance and succession.

Materials and methods

Description of study site

Lake Balaton is a large (593 km^2), shallow ($z_{mean} = 3.2$ m) lake in Western Hungary (Fig. 1). The long wind fetch from all directions results in almost permanent mixing of the whole water column. Due to frequent resuspension of highly calcareous (50–60% carbonates) sediments and slow sedimentation of precipitating carbonates, the water is always turbid. In the shallowest, westernmost area ($z_{mean} = 2.3$ m), the coefficient of vertical light attenuation (K_d) averages 4 m^{-1} (Istvánovics et al., 2004). As a consequence, mean euphotic depth is shallower (1.2 m) than mixing depth. Phosphorus desorption from resuspended sediments results in a pulsed internal load (Istvánovics et al., 2004). Mean suspended solids concentration approaches 36 g m^{-3}, and the number of inorganic particles in the ingestible size range exceeds food items of filter-feeding zooplankton by one to three orders of magnitude (G.-Tóth, 1992). Due to strong physical constraints acting upon non-selective cladocerans, the copepod *Eudiaptomus gracilis* (Kiefer, 1932) makes up 80–100% of crustacean plankton. Thus, community grazing rate rarely exceeds 8–10% (G.-Tóth, 2000), and phytoplankton live in a dynamic environment where light, temperature and nutrient supply fluctuate rapidly, but sedimentation and grazing are negligible to moderate.

Data and analysis

An online monitoring station has been maintained since 2001 in the westernmost area of the lake (Fig. 1). The station is equipped with meteorological and turbidity sensors and an online delayed fluorescence (DF) spectrometer that automatically registers composition and biomass of photosynthetically active phytoplankton. DF spectroscopy is a new method (Gerhardt & Bodemer, 2000; Istvánovics et al., 2005) that distinguishes four colour classes of algae: cyanobacteria, chlorophytes (Chlorophyceae, Euglenophyceae, Conjugatophyceae), chromophytes (Bacillariophyceae, Chrysophyceae, Dinophyta,

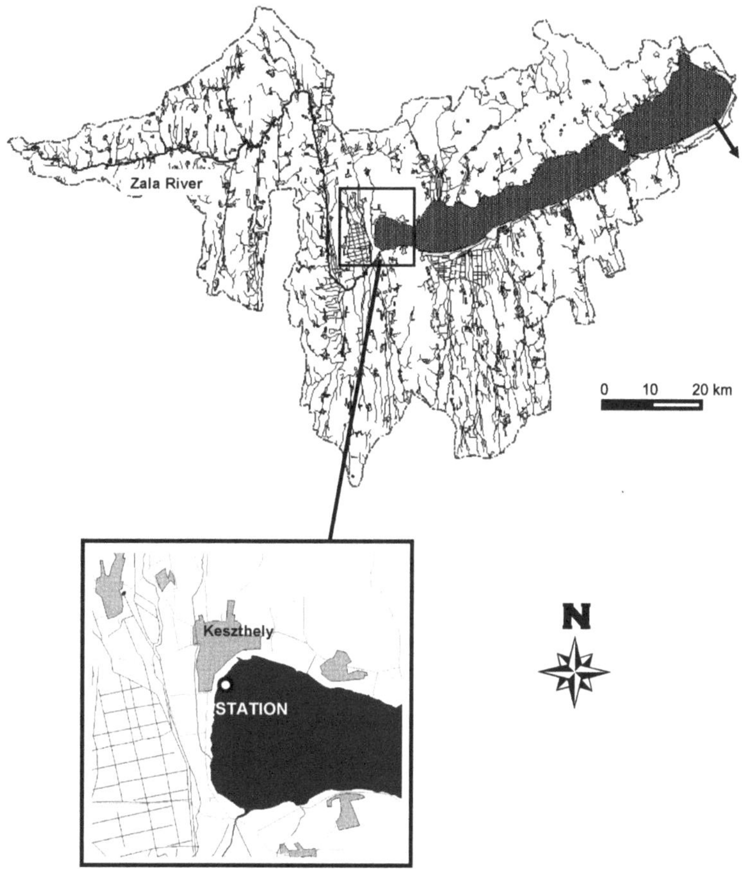

Fig. 1 Lake Balaton and its catchment. (Insertion shows location of the sampling station. Zala River is the largest tributary. Arrow indicates the only outflow)

Xanthophyceae, Haptophyta) and cryptophytes. Phytoplankton data were recorded every 20 min at four depths in a single profile. Meteorological and turbidity data were obtained every 10 and 6 sec, respectively. These data were averaged in 15 min intervals. Turbidity was calibrated against coefficient of vertical light attenuation in the laboratory and using weekly K_d values determined manually. For a more detailed description of the monitoring program and a critical evaluation of the DF method, see Istvánovics et al. (2004, 2005). At the monitoring site, weekly phytoplankton samples were taken and analysed by microscopic counting. (Padisák, pers. comm.)

A database was created consisting of K_d, underwater light, temperature, phytoplankton composition and biomass. Each variable was averaged by day and depth. Time series of dominance was defined with five categories: one for each colour class of algae and an 'uncertain' category. A colour class was considered dominant if its biomass exceeded that of any other class by more than 10%. The uncertain category was reserved for rare cases when this criterion was not satisfied. Since time series collected in 2001 and 2002 were fragmented, we rely on data from 2003 and 2004. Our unique dataset has very high temporal resolution compared to that allowed by traditional methods. Obviously, resolution of phytoplankton along the diversity axis is poor relative to microscopic enumeration.

Daily external loads of various P and N forms are monitored daily at the mouth section of the largest tributary of the lake (database of the West Transdanubian Authority of Water and Environment). This load is estimated to represent 90–95% of total external load of the westernmost area of Lake Balaton (Clement, 2000).

Results and discussion

In 2003, the seasonal pattern of phytoplankton dynamics was typical for shallow, eutrophic lakes (Fig. 2). Three major biomass peaks developed: (i) a spring diatom bloom in late March, (ii) a mixed bloom of *Aphanizomenon flos-aquae* (Lyngbye) Ralfs, *A. klebahnii* Elenkin and *Planktothrix agardhii* Gomont in early July and (iii) a *Cylindrospermopsis raciborskii* (Woloszynska) Seenayya et Subba Raju bloom in mid-August. Development of two successive cyanobacteria blooms is common in Lake Balaton. The high temperature requirement of subtropical *C. raciborskii* is the prime reason to account for the phenomenon (Padisák & Istvánovics, 1987; Padisák & Reynolds, 1998). In addition to the

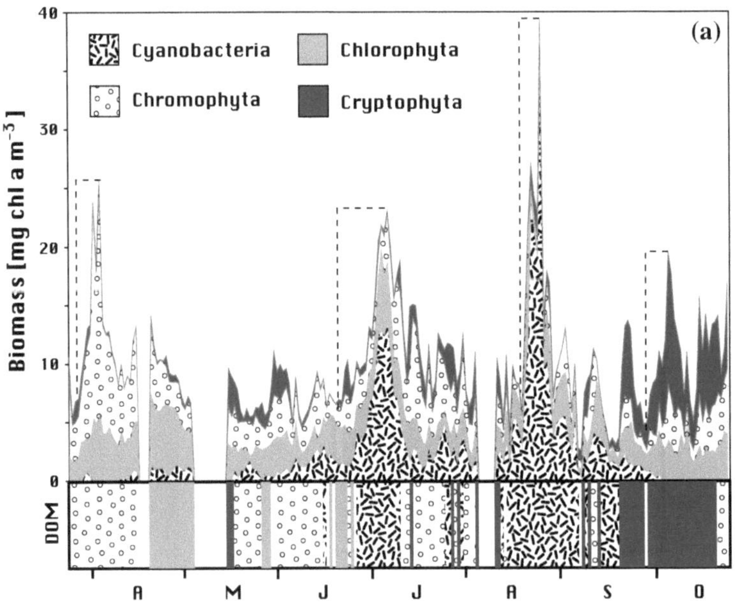

Fig. 2 Biomass and dominance (DOM) of four colour classes of phytoplankton in 2003 (**a**) and 2004 (**b**). Gaps indicate data deficiency due to technical reasons. (Dashed boxes border major blooms used to derive habitat templates for functionally distinct groups; for details see text)

major blooms, smaller peaks of chromo- and cryptophytes occurred during autumn. Since our dataset was recorded at a single point, biomass variations originating from water motion could not be resolved. However, Balaton is well-mixed (Luettich et al., 1990), which allows neglecting spatial heterogeneities.

Environmental conditions were different in 2004 from those in 2003 (Table 1). Higher water discharge resulted in higher external nutrient loads, particularly during spring. Nutrient supply also was more favorable during summer. Water column stability was high throughout the summer and autumn 2004 due to shading and turbulence dampening effects of macrophytes. The latter appeared in moderate densities in 2003, but surface coverage reached about 85% by early summer 2004. Mean water temperature was significantly lower in summer 2004. As a result of the strong influence of macrophytes, phytoplankton dynamics was atypical for Lake Balaton (Fig. 2). A sustained spring bloom of diatoms and the lack of a well-defined summer bloom resembled the seasonal pattern characteristic of mesotrophic areas of the lake. At the same time, there was a long-lasting and virtually unbroken dominance of cryptophytes from early July to the end of the study period (cf. Padisák, 1994).

A few characteristic patterns evolved in dominancy and biomass of phytoplankton: (i) major blooms were separated by periods of low biomass; (ii) each bloom was dominated by species of a single colour class; (iii) between blooms, small peaks of one or the other colour class started but collapsed before attaining significant biomass levels and produced capricious variability in dominance. This pattern suggests that separated blooms may be bound to specific circumstances, while, between blooms, there are chaotic periods of low biomass with various species struggling to outgrow each other.

As a first step to identify environmental conditions selecting for certain species during bloom periods, physical forcing was characterized by two indicators: water temperature and vertical light attenuation. The choice of day- and depth-averaged water temperature is self-explanatory. Daily mean coefficient of vertical light attenuation (K_d) was selected as an indirect indicator of underwater light climate. Although global radiation was measured, shading effect of macrophytes was not quantified. Therefore, time series of incident photon flux density (PFD) was heterogeneous. K_d showed large, irregular fluctuations between about 1 and 15 m^{-1} due to vast sediment resuspension. The corresponding variability in

Table 1 Mean ± standard deviation values of environmental variables (daily data) and nutrient status of phytoplankton (weekly data) in 2003 and 2004

Variable	2003	2004
Spring (March–May)		
Coefficient of vertical light attenuation (m^{-1})	2.9 ± 1.9	2.2 ± 1.8
Water temperature (°C)	14.0 ± 6.3	15.5 ± 2.9
Water discharge ($m^3 s^{-1}$)	4.8 ± 2.4	11.6 ± 4.9
External load of nitrate (mg N m^{-2} day^{-1})	8.3 ± 10.0	10.2 ± 7.4
External load of phosphate (mg P m^{-2} day^{-1})	0.57 ± 0.43	0.70 ± 0.62
Internal load of phosphate (mg P m^{-2} day^{-1})	1.32 ± 0.99	0.78 ± 0.63
Specific surplus P content (μg P [μg chl *a*]$^{-1}$)	0.70 ± 0.23	0.98 ± 0.61
Specific N content (μg N [μg chl *a*]$^{-1}$)	19.1 ± 4.2	22.7 ± 2.9
Late summer (July–August)		
Coefficient of vertical light attenuation (m^{-1})	2.5 ± 1.2	1.1 ± 1.0
Water temperature (°C)	25.9 ± 1.9	23.8 ± 2.3
Water discharge ($m^3 s^{-1}$)	0.4 ± 0.2	2.8 ± 1.5
External load of nitrate (mg N m^{-2} day^{-1})	0.08 ± 0.05	0.80 ± 0.58
External load of phosphate (mg P m^{-2} day^{-1})	0.15 ± 0.10	1.78 ± 0.79
Internal load of phosphate (mg P m^{-2} day^{-1})	1.11 ± 0.50	0.44 ± 0.47
Specific surplus P content (μg P [μg chl *a*]$^{-1}$)	0.53 ± 0.15	1.30 ± 0.36
Specific N content (μg N [μg chl *a*]$^{-1}$)	13.0 ± 4.9	18.6 ± 6.3

underwater light—proportionate to exp(-K_d)—was in the range of 10^6. In comparison, daily variability in incident light (maximum 25 fold) was negligible. Thus, K_d is a prime determinant of underwater PFD and an indicator of sediment resuspension.

Plotting daily values of T and -K_d on a two dimensional graph, the path followed by physical constraints emerged (Fig. 3). Noticeably, seasonality of temperature governed this path into well-expressed directions in early spring and late autumn. In summer, a complex knot pattern was observed with no detectable progress towards any direction.

Dominant colour classes of phytoplankton were inserted into the T versus K_d space (Fig. 3). Their positions agreed with the known physiological requirements of characteristic member species and clearly reflected their different sensitivity to stability of the physical environment (Padisák

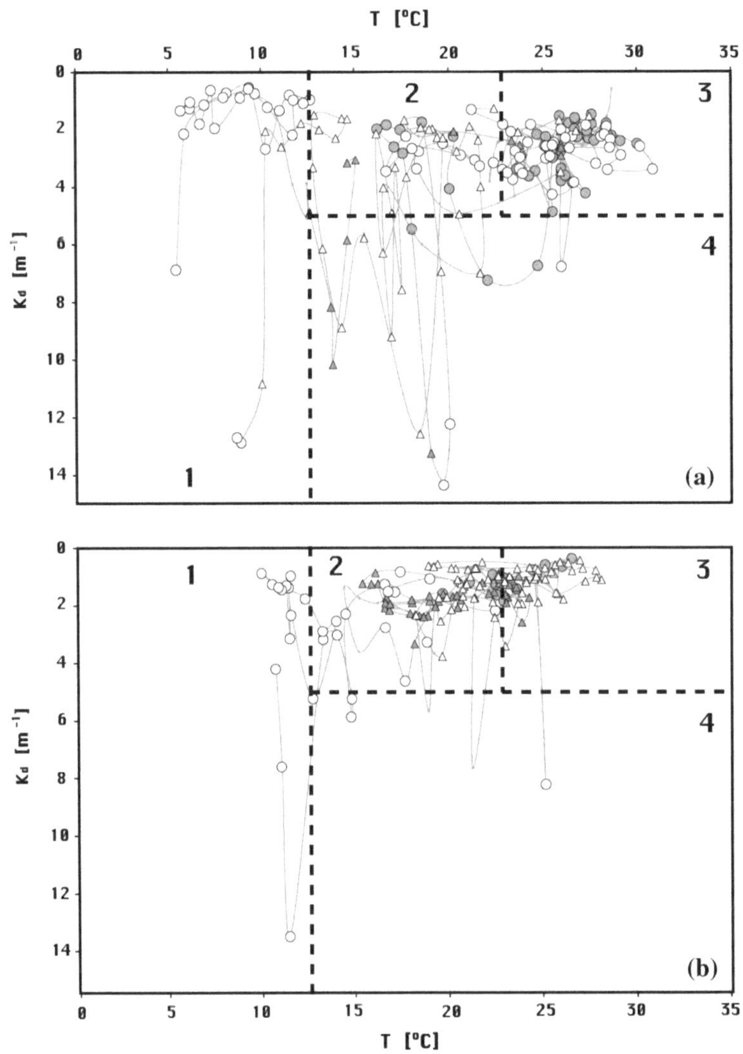

Fig. 3 The path followed by the environment in the space of physical constraints in 2003 (**a**) and 2004 (**b**) and dominant colour classes of phytoplankton placed into this space. (T—vertically averaged daily mean water temperature; K_d—daily mean coefficient of vertical light attenuation. Cyanobacteria—closed circle; chlorophytes—closed triangle; chromophytes—open circle; cryptophytes—open triangle. Numbered areas separated by dashed lines indicate characteristic habitat ranges of various functional groups. 1—acclimating R species, 2—early successional phase, 3—stress tolerant S species, 4—meroplankton). Seasonality means advancing from left to right in spring, staying on the right side in summer and returning to left in autumn

et al., 1988, 1990; Padisák, 1994; Reynolds, 1997). Spring chromophytes (diatoms) are typical acclimating strategists (R species) adapted to low temperature and requiring sufficient turbulence to keep entrained in the water column. Assemblages of early successional stages can be recognized by concentrated occurrence of chlorophytes and chromophytes in the moderate T versus low to moderate K_d sub-space of physical constraints. Summer blooms of cyanobacteria, particularly those of subtropical *C. raciborskii*, are associated with high temperature. Low turbulence allows this group to exploit the advantage of buoyancy regulation. Besides cyanobacteria, stress tolerant chromophytes (*Ceratium hirundinella* (O.F. Müller) Schrank) are typical members of late-summer quasi-equilibria in Lake Balaton. The unusual summer dominance of cryptophytes in 2004 may be explained by metaphytic affiliation of certain species within this group (Borics et al., 2003). Small cryptophytes and several meroplanktonic diatoms benefit from intense wind-induced sediment resuspension, which temporarily extends their habitat into the water column.

Major phytoplankton blooms were identified manually in 2003 (Fig. 2) and placed into the space of physical constraints (Fig. 4). Habitat templates of various colour classes were determined by selecting sequences of days when the bloom-forming class achieved significant and undisturbed growth. The sub-space determined by these days (13 days in 2003) represents physical conditions where the given class has obvious advantage over the others. These regions were separated from each other and occupied margins of the environmental space observed during our measurements (Fig. 4). The marginal position of blooming sub-spaces indicated that strong, unidirectional selective pressure was a prerequisite for bloom formation. Conclusively, blooms are rare events bound to a specific environmental state that provides a few species of a single colour class the opportunity to grow faster than others. Such states are, however, transient and represent the extreme.

A simple threshold model was developed to test the consistency of our observations and hypotheses about seasonal dynamics of phytoplankton. Obviously, the two physical constraints (Fig. 4) did not discriminate habitat templates of various blooms. For example, the path of environmental changes in the T versus K_d space was almost symmetrical during the growth and collapse of the spring diatom bloom in 2003. Thus, additional variables had to be included.

Minimal water temperature in the last two weeks (T_{min}) was chosen because it reflects

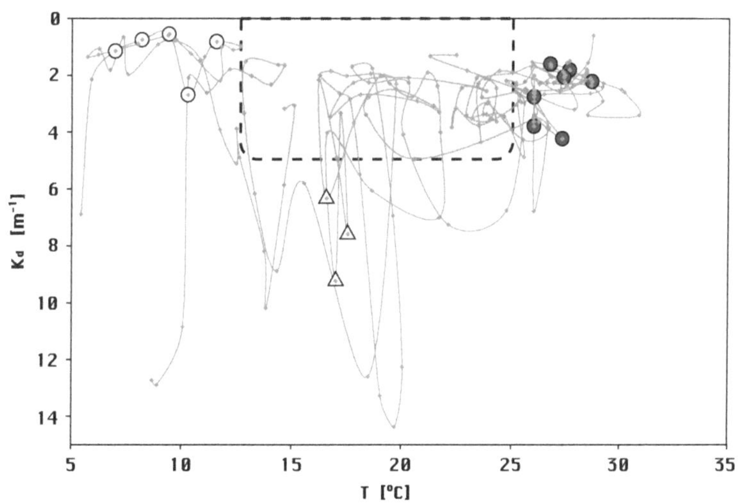

Fig. 4 The path followed by the environment in the space of physical constraints in 2003 and four major biomass peaks placed into this space. (T—vertically averaged daily mean water temperature; K_d—daily mean coefficient of vertical light attenuation. Cyanobacteria—closed circle; chromophytes—open circle; cryptophytes—open triangle. The four blooms are those shown in Fig. 2; the two cyanobacterial blooms do not separate in this space. Shaded area indicates the region of "mean" environmental states emphasizing marginal positioning of blooms)

sensitivity of N_2-fixing cyanobacteria to abrupt temperature changes. At the same time, it incorporates the experience that cyanobacteria blooms are preceded regularly by two to three weeks of calm, warm weather (Gorzó, 1987; Padisák, 1994). Compared to *Aphanizomenon* and *Anabaena* species, subtropical *Cylindrospermopsis raciborskii* has an unusually narrow temperature range (22–24°C), where akinetes germinate (Gorzó, 1987), and a high threshold temperature (18°C), where the species starts to grow (Shafik et al., 1997). These differences were expressed in the model by determining different T_{min} values (Table 2).

Nutrient supply of phytoplankton is a discriminative factor during growth periods of various blooms. In Lake Balaton, concentration of silica is always high enough to satisfy requirements of even large diatom blooms (V. Balogh, pers. comm.). Physiological and chemical data (Istvánovics et al., unpublished) showed that carrying capacity might be N determined during spring in dry years without typical spring floods, including 2003. Previous observations also suggest that N deficiency was likely to put an end to spring diatom blooms (Istvánovics & Herodek, 1985, 1995). Thus, area specific external nitrate load was chosen to indicate N supply.

In summer, algal growth can be limited by either N or P, but extreme P deficiency prevails and the carrying capacity usually is P determined (Istvánovics & Herodek, 1985, 1995; Istvánovics & Somlyódy, 2001). This also was the situation in 2003. However, no signs of serious P deficiency were detected in summer 2004, when light availability probably limited algal growth (Table 1). Although daily loads of external and internal P were available (Istvánovics et al., 2004), differential ability of various algal groups to grow on previously acquired and stored P during low external supply prevented incorporation of these variables into the simple threshold model.

Table 2 Discriminative rules for habitat templates and growth rate of the four functionally different blooms, derived from 13 days in 2003

Bloom	Criteria	Growth rate (d^{-1})
Spring chromophytes (diatoms)	$6.5 \le T \le 12.5$, $L_{nitrate\text{-}N} \ge 13$	0.74
Early summer cyanobacteria	$T_{min} \ge 23.7$	0.15
Late summer cyanobacteria	$T_{min} \ge 25.5$	0.40
Meroplankton	$T \ge 10$ $K_d \ge 5.6$	–

T, °C is the actual water temperature; T_{min}, °C is the minimum water temperature during the last two weeks; $L_{nitrate}$, mg N m^{-2} d^{-1} is the external load of nitrate; K_d, m^{-1} is the coefficient of vertical light attenuation. Biomass changes of meroplankton were related to frequency of resuspension; see in text

In this way, a four dimensional environmental space was established, including T, K_d, T_{min} and external nitrate load. Actual environmental state is indicated by a point in the space. Considering the seasonal biomass pattern in 2003 (Fig. 2), habitat templates of four functionally distinct blooms could be quantified for modeling purposes: (i) spring diatoms, (ii) early summer cyanobacteria, (iii) late summer cyanobacteria and (iv) meroplankton associated with resuspension. A few simple binary rules were assumed to delimit habitat templates for each functional group. Rules were set up according to ecological knowledge and refined to fit observed data (Table 2). Growth of spring diatoms was dependent on external N supply and water temperature. Early and late summer cyanobacteria grew above a T_{min} of 23.7°C and 25.5°C, respectively. Meroplanktonic biomass peaks were associated with intense sediment resuspension (indicated by high absolute value of K_d) in a certain temperature range. Each resuspension event exceeding the preset criterion was assumed to increase the biomass of this group by 1 mg chl *a* m^{-1}, and the rate of decay was 0.1 d^{-1} in the absence of critical resuspension.

The community most often existed in a "ground state" corresponding to periods between blooms characterized by low biomass and capricious compositional changes (cf. Fig. 2). To ease mathematics, an equal share of biomass was assigned to each group in ground state. As the environment entered a blooming subspace, or fulfilled the preset rules of a habitat template, the appropriate group began to grow at "full speed". Truly planktonic groups (i–iii) were assumed to grow at a fixed exponential rate, while meroplankton (iv) growth was set proportional to frequency of resuspension. As soon as the environment left the template, growth stopped and the bloom started to decay towards ground state. Since growth and collapse of each bloom was roughly symmetrical, loss rates were assumed to equal growth rates, an assumption supported by Sommer (1981).

Biomass simulations fit reasonably to measurements (Fig. 5). Although model parameters were tuned using data from 2003, the model reproduced the lack of cyanobacterial blooms in 2004. This suggested that low mean water temperature during summer 2004 (Table 1) may have been a more influential factor in preventing blooming of cyanobacteria than other differences experienced in this year.

Our simple model reduced the influence of environmental dynamics to a narrow range of conspicuous events: crossing the borders of habitat templates. This oversimplification was useful to demonstrate predictability of bloom formation but could not explain the complexity of phytoplankton dynamics in the ground state, which accounted for about 80% of the study period (Figs. 2, 3, 5). The surprising length of the ground state interestingly demonstrated Reynold's (1997) contention that phytoplankton inhabited an extremely hostile habitat. To get an insight into the effect of environmental dynamics on community composition, the compositional stability was examined as a function of environmental dynamics. For this purpose, appropriate indicators of

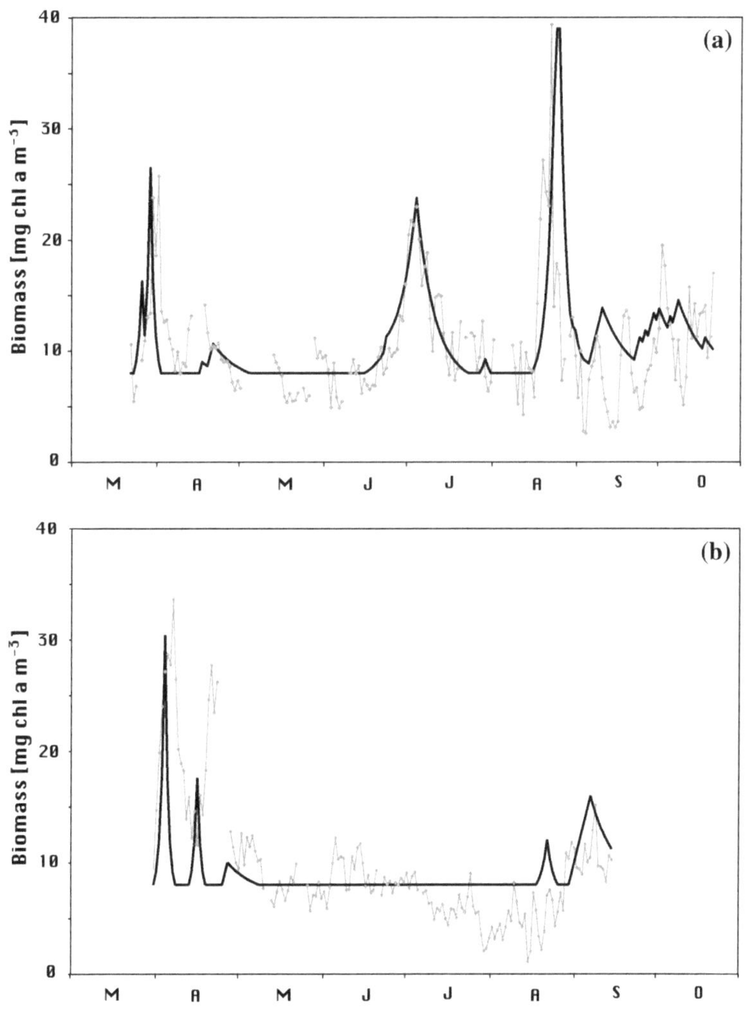

Fig. 5 Observed (thin line) and simulated (thick line) biomass of phytoplankton in 2003 (**a**) and 2004 (**b**)

environmental stability/physical disturbance and community stability/change were identified.

Environmental instability (IS_e) was defined as the step length between the states (e) of the environment in two subsequent days in the T versus K_d space. It can be computed as Euclidean distance:

$$IS_e = \|e_t - e_{t-1}\| = \sqrt{(T_t - T_{t-1})^2 + (Kd_t - Kd_{t-1})^2} \quad (1)$$

where t is the current day. The smaller the step is between two days, the more stable the environment is. Large IS_e is indicative of high instability, which, above some level, can act as physical disturbance. Based on the distribution of step lengths, the upper 15% of events was designated as disturbance (Fig. 6). According to this definition, an environmental change qualifies as a disturbance when $IS_e > 2.5$. As the dimensions of the environmental space have different units

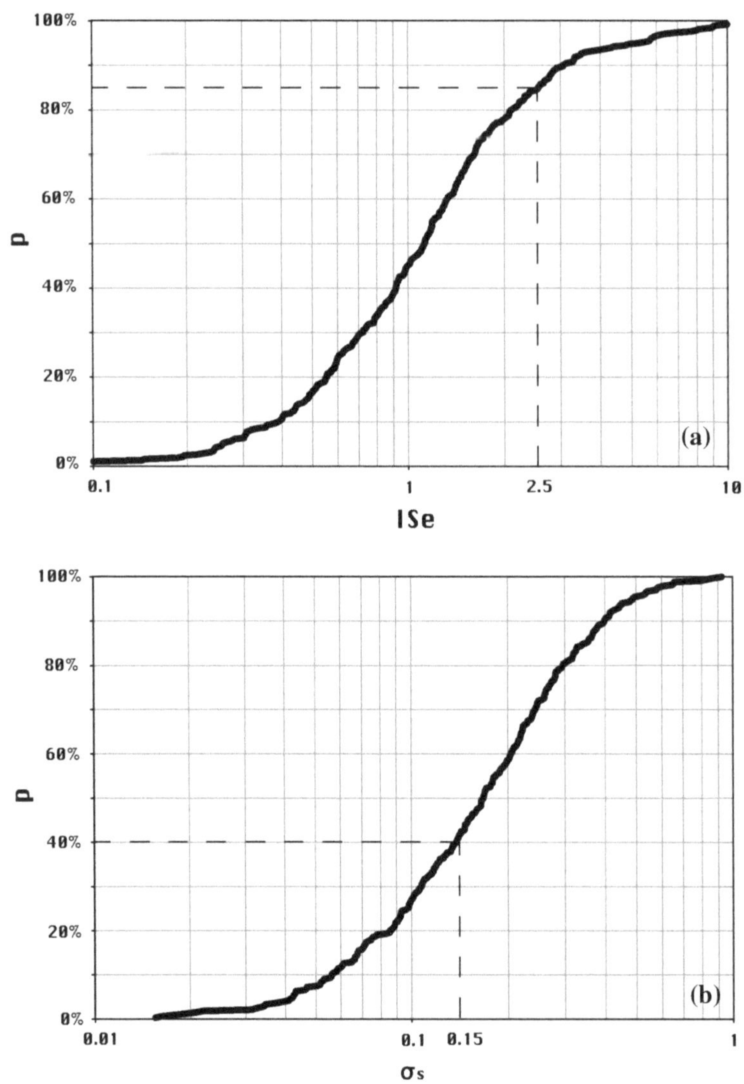

Fig. 6 Probability distribution (P) of environmental instability (IS_e; **a**) and that of the rate of community change (σ_s; **b**). (Dashed lines in **a** and **b** indicate threshold values beyond which instability is assumed to act as a disturbance and σ_s is assumed to represent a real community change, respectively)

(m^{-1} for K_d and °C for T), distances are subject to the weighting used between dimensions. Noticeably, Equation 1 gives equal weights to T and K_d, but the change in the latter variable by 2.5 units is equivalent to a twelve-fold resuspension induced change in underwater PFD. Since, however, a temperature change of 2.5°C represents about 10% of the temperature range covered by the study period, whereas the respective value is 0.001% for the change in underwater PFD, IS_e is far more sensitive to changes in resuspension and PFD than temperature. Disturbances happened throughout 2003 and were less frequent in 2004 (Fig. 7). These events almost each time were coupled with an instantaneous or slightly delayed flip in dominance (Fig. 7).

Compositional changes were characterized using a highly sensitive and more robust indicator

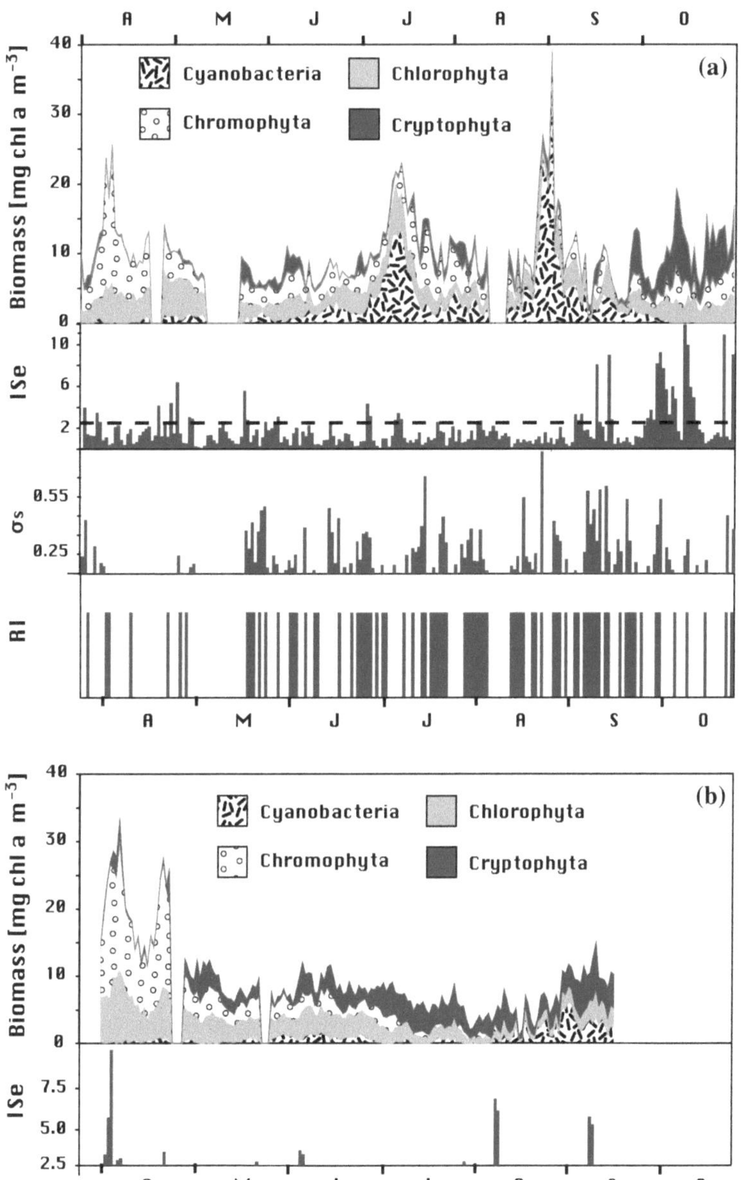

Fig. 7 Seasonal pattern of phytoplankton related to disturbance ($IS_e > 2.5$) in 2003 (**a**) and 2004 (**b**). (In 2003 the rate of community change, σ_s, and community stability, RI, are also indicated)

(Fig. 7). In the former case, the rate of community change (σ_s, d^{-1}) was calculated as the sum of daily change in relative biomass of various colour classes (Reynolds, 1997):

$$\sigma_s = \sum_{i=1}^{4} \left| \frac{b_{t+1}^i}{B_{t+1}} - \frac{b_t^i}{B_t} \right| \quad (2)$$

where B is total biomass (mg chl a m^{-3}), b^i is fractional biomass of the four colour classes (mg chl a m^{-3}), and t is the current day. Taking into account the distribution of σ_s (Fig. 6) and uncertainty of the DF method (Istvánovics et al., 2005), σ_s values below 0.15 d^{-1} were assigned to measurement error.

To derive a less sensitive indicator, the four colour classes were ordered on the basis of their fractional biomass. A positional change was assumed to occur whenever a class jumped one to three ranks up or down compared to the preceding day. The rank indicator (RI) was a simple binary value with RI = 0 when no jump occurred and RI = 1 when any kind of jump was observed. The rate of community change is sensitive to differences in net growth and/or loss rates of various colour classes. These differences, however, are not necessarily large enough to provoke positional changes ensuring lower RI sensitivity. In this way, RI can be considered a measure of community stability rather than an indicator of community change.

To examine the effect of physical disturbance on phytoplankton composition, community change ($\sigma_s > 0.15$ d^{-1}) and stability (RI) were related to IS_e (Figs. 7, 8). We assumed that compositional changes were driven by disturbance when an $IS_e > 2.5$ event was followed by compositional changes within a three day period, including the day of disturbance. Both σ_s and RI showed that disturbance most often induced changes in community composition. In a somewhat larger number of cases, community changes were not related to physical disturbance (Fig. 8). Most of the latter cases occurred during summer (Fig. 7). Although a partial failure of the disturbance indicator could not be excluded, timing of disturbance-independent compositional changes suggested that such changes were driven by autogenic succession of phytoplankton. Certainly, some of the disturbance-independent changes may be associated with selective grazing of zooplankton or other selective loss processes, like viral infection. Although community grazing rate of zooplankton is typically low in Lake Balaton (G.-Tóth, 1992), two pieces of indirect evidence suggested that the situation might be different in 2004. First, dense macrophyte stands provide refuge for zooplankton (Sheffer et al., 1993), where restricted turbulence might favor passive feeders. Second, prolonged dominance of cryptophytes was indicative of sufficient food supply. In spite of these uncertainties, succession and disturbance were equally important in shaping the structure of phytoplankton.

Both σ_s and RI indicated that the most frequent case was communities with stable composition (Fig. 8). Stable composition did not imply that the environment was stable. On the contrary, the environment changed all the time (Figs. 3, 4), but variability usually stayed within the limits where phytoplankton were capable of absorbing external fluctuations. This was indicative of a considerable structural resilience. At the same time, invoked stress usually prevented significant growth.

Cases of stable composition rarely concentrated into long-lasting intervals (Fig. 7). Surprisingly, the longest unbroken period of stable community structure was observed during spring in 2003 and 2004 (Fig. 7). The explanation was twofold. (i) Phytoplankton experienced the most severe and frequent disturbance during autumn, resulting in a number of small biomass peaks with divergent compositions. (ii) The rate of community change increased with water temperature (Fig. 9), indicating that more species were released from stress caused by a sub-optimal level of this variable. Internally driven compositional changes thus were slow in spring compared to the period of autogenic succession.

Fragmented occurrence of stable communities suggested that it was difficult for any colour class to preserve its position for any significant period of time in the dynamic environment of Lake Balaton. This is an important difference compared to deep lakes, where environmental stability associated with thermal stratification supports less disturbed successional sequences, which frequently result in prolonged dominance of one or another K-selected stress tolerant species (Reynolds, 1997). This difference between deep and shallow lakes explain the observation of Padisák et al. (2003) that equilibrium phases sensu Sommer et al. (1993) were relatively infrequent (17 out of 80) in shallow lakes during summer.

Cases when disturbance failed to influence phytoplankton composition (Fig. 8) may reflect measuring errors and/or unsatisfactory definition of disturbance. The number of such cases, however, was low, indicating the merits of the approach. Refinement of this approach deserves attention considering the limited efforts taken to characterize disturbance using independent methods (cf. Juhász-Nagy, 1993; Reynolds, 1997).

Compositional changes were related to fluctuations in the T versus K_d space, considering that stress caused by low nutrient supply was likely a key factor determining phytoplankton dynamics. On one hand, due to lack of top-down control of zooplankton (G.-Tóth, 1992, 2000), autogenic succession is driven by competition for gradually depleting resources, including nutrients. On the other hand, resuspension related disturbance is coupled to increased nutrients, primarily P availability from desorption from sediment particles. The numerous storage and delay factors hidden in the system make it impossible to characterize the

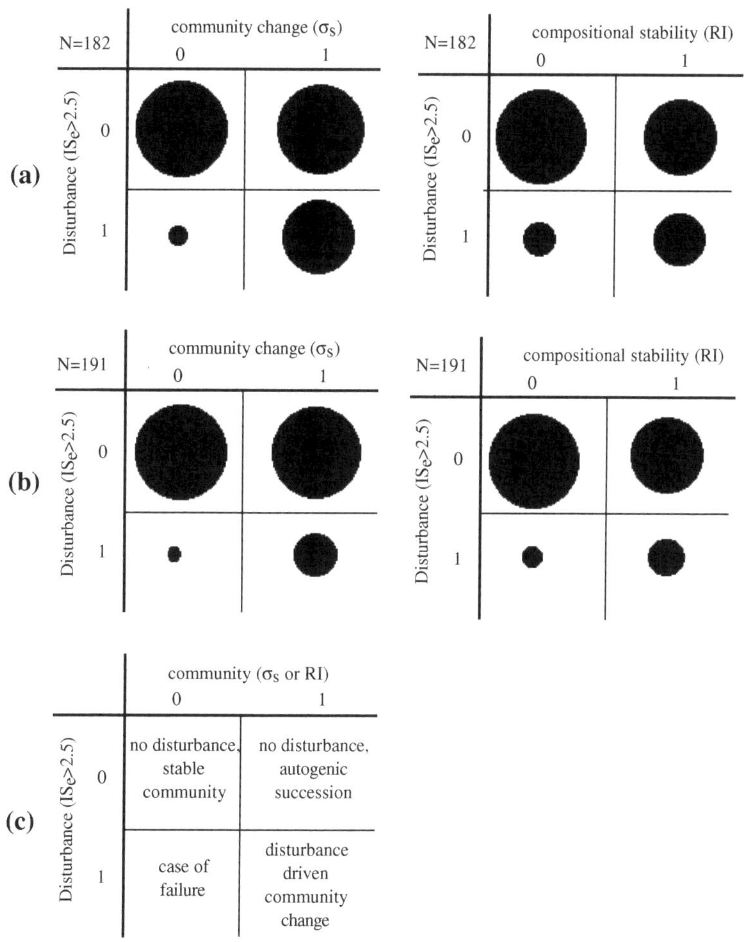

Fig. 8 Compositional stability expressed as either the rate of community change or community stability as a function of disturbance in 2003 (**a**) and in 2004 (**b**). (Disturbance was identified as high environmental instability indicated by $IS_e > 2.5$ values. N—total number of cases. Area of symbols is proportionate to number of cases. **c**—interpretation of matrix areas)

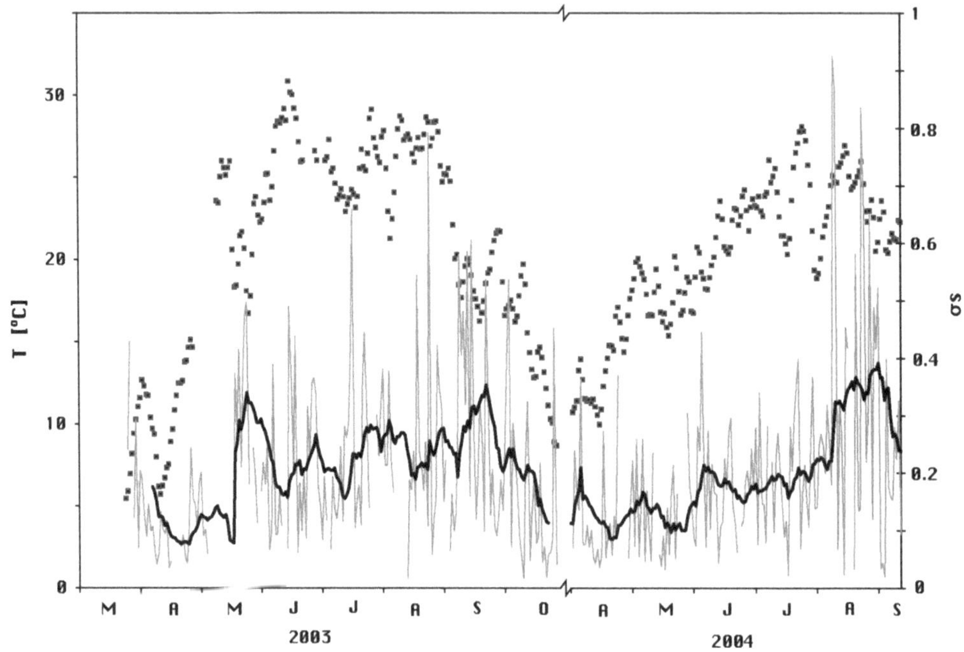

Fig. 9 The rate of community change (σ_s) compared to vertically averaged daily mean water temperature (T). (Temperature—closed symbol; observed σ_s—thin line, moving average on σ_s-thick line)

role of nutrient stress on a daily basis. One should, however, recognize that any change in the phytoplankton community examined in the two dimensional space of physical constraints implicitly reflects variability of nutrient stress.

Conclusions

(1) High-resolution online monitoring systems are required to follow dynamics of the environment and phytoplankton in sufficient detail. Such time series reveal the enormous complexity of ecosystems.

(2) Environmental states providing the opportunity for certain species or functional groups to develop blooms represent extreme margins in the space of environmental constraints with strong unidirectional selective pressure. Correspondingly, blooms are rare, exceptional events and phytoplankton spend nearly 80% of the time in a ground state with low biomass and high diversity.

(3) Growth and collapse of blooms could be described by a simple threshold model based on habitat templates of the bloom-forming functional groups. In contrast to the predictability of major blooms, dominance and biomass was unpredictable during the ground state.

(4) Disturbances could be identified with success using a simple indicator of environmental instability developed here. Most of the identified disturbance events provoked detectable compositional changes in the phytoplankton community, even though time series of phytoplankton consisted of only four colour classes. Autogenic succession and disturbance were found to be equally important in shaping the phytoplankton community.

(5) The highly dynamic environment of shallow lakes is predicted to restrict the occurrence and duration of equilibrium phases.

Acknowledgements This study was supported financially by the "Phytoplankton-on-line" (No. EVK1-CT-1999-00037) project and by the Balaton project of the Office of the Prime Minister (Hungary). We are greatly indebted to Dr. Judit Padisák for microscopic counting results and reading the manuscript as well as to Mr. Norbert Turay, Mr. Gábor Poór, and Mrs. Rozália Kelemen for their invaluable technical assistance with the monitoring program. We thank Dr. János Korponai, Mr. György Süle and Mr. István Nádor from the West Transdanubian Authority of Water and Environment for allowing access to their load data and three anonymous referees for their useful comments.

References

Borics, G., B. Tóthmérész, I. Grigorszky, J. Padisák, G. Várbíró & S. Szabó, 2003. Algal assemblage types of bog-lakes in Hungary and their relation to chemistry, hydrological conditions and habitat diversity. Hydrobiologia 50: 145–155.

Clement, A., 2000. Improving uncertain nutrient load estimates for Lake Balaton. Water Science and Technology 43: 279–286.

Connell, J. H., 1978. Diversity in tropical rain forests and coral reefs. Science 199: 1302–1310.

Elliott, J. A., A. E. Irish & C. S. Reynolds, 2001. The effects of vertical mixing on a phytoplankton community: a modelling approach to the intermediate disturbance hypothesis. Freshwater Biology 46: 1291–1297.

Gerhardt, V. & U. Bodemer, 2000. Delayed fluorescence excitation spectroscopy: a method for determining phytoplankton composition. Archiv für Hydrobiologie Special Issues Advances in Limnology 55: 101–120.

Gorzó, Gy., 1987. The influence of physical and chemical factors on akinete germination of heterocytic cyanobacteria from Lake Balaton. Hidrológiai Közlöny 67: 127–133 (in Hungarian).

G.-Tóth, L., 1992. Limiting effect of abioseston on food ingestion, postembryonic development time and fecundity of daphnids in Lake Balaton (Hungary). Journal of Plankton Research 14: 435–446.

G.-Tóth, L., 2000. Feeding biology of planktonic crustaceans and their role in elimination of phytoplankton in Lake Balaton. D.Sc. Thesis, Balaton Limnological Research Institute, Tihany (in Hungarian).

Istvánovics, V. & S. Herodek, 1985. Orthophosphate uptake of planktonic microorganisms in Lake Balaton. Hydrobiologia 122: 159–166.

Istvánovics, V. & S. Herodek, 1995. Estimation of net uptake and leakage rates of orthophosphate from ^{32}P uptake kinetics by a force-flow model. Limnology and Oceanography 40: 17–32.

Istvánovics, V., M. Honti, A. Osztoics, H. M. Shafik, J. Padisák, Y. Yacobi & W. Eckert, 2005. Continuous monitoring of phytoplankton dynamics in Lake Balaton (Hungary) using on-line delayed fluorescence excitation spectroscopy. Freshwater Biology 50: 1950–1970.

Istvánovics, V., A. Osztoics & M. Honti, 2004. Dynamics and ecological significance of daily internal load of phosphorus in shallow Lake Balaton, Hungary. Freshwater Biology 49: 232–252.

Istvánovics, V. & L. Somlyódy, 2001. Factors influencing lake recovery from eutrophication-the case of Basin 1 of Lake Balaton. Water Research 35: 729–735.

Juhász-Nagy, P., 1993. Notes on compositional diversity. Hydrobiologia 249: 173–182.

Luettich, R. A., D. R. F. Harleman & L. Somlyódy, 1990. Dynamic behavior of suspended sediment concentrations in a shallow lake perturbed by episodic wind events. Limnology and Oceanography 35: 1050–1067.

Padisák, J., 1993. The influence of different disturbance frequencies on the species richness, diversity and equitability of phytoplankton in shallow lakes. Hydrobiologia 249: 135–156.

Padisák, J., 1994. Relationships between short-term and long-term responses of phytoplankton to eutrophication of the largest shallow lake in Central Europe (Balaton, Hungary). In Sund, H., H.-H. Stabel, W. Geller, Y. Xiaogan, Y. Kechang & S. Fengning (eds), Environmental Protection and Lake Ecosystem. Science & Technology Press, Beijing, 419–437.

Padisák, J., G. Borics, G. Fehér, I. Grigorszky, I. Oldal, A. Schmidt & Zs. Zámbóné-Doma, 2003. Dominant species, functional assemblages and frequency of equilibrium phases in late summer phytoplankton assemblages in Hungarian small shallow lakes. Hydrobiologia 502: 157–168.

Padisák, J., L. G.-Tóth & M. Rajczy, 1988. The role of storms in the summer succession of phytoplankton in a shallow lake (Lake Balaton, Hungary). Journal of Plankton Research 10: 249–265.

Padisák, J., L. G.-Tóth & M. Rajczy, 1990. Stir-up effect of wind on a more-or-less stratified shallow lake phytoplankton community, Lake Balaton, Hungary. Hydrobiologia 191: 249–254.

Padisák, J. & V. Istvánovics, 1997. Differential response of blue-green algal groups to phosphorus load reduction in a large shallow lake: Balaton, Hungary. Verhandlungen der Internationale Vereiningung für theoretische und angewandte Limnologie 26: 574–580.

Padisák, J. & C. S. Reynolds, 1998. Selection of phytoplankton associations in Lake Balaton, Hungary, in response to eutrophication and restoration measures, with special reference to the cyanoprokaryotes. Hydrobiologia 384: 41–53.

Padisák, J., C. S. Reynolds & U. Sommer (eds), 1993. The intermediate disturbance hypothesis in phytoplankton ecology. Hydrobiologia 249: 1–199.

Reynolds, C. S., 1988. The concept of ecological succession applied to seasonal periodicity of freshwater phytoplankton. Verhandlungen der Internationale Vereiningung für theoretische und angewandte Limnologie 23: 683–691.

Reynolds, C. S., 1997. Vegetation Processes in the Pelagic: A Model for Ecosystem Theory. Ecology Institute, Oldendorf/Luhe.

Shafik, H. M., L. Vörös, M. Présing, A. Kovács & I. Kóbor, 1997. Growth of Cylindrospermopsis raciborskii in batch and continuous cultures. Hidrológiai Közlöny 77: 17–18 (in Hungarian).

Scheffer, M., S. H. Hosper, M.-L. Meijer, B. Moss & E. Jeppesen, 1993. Alternative equilibria in shallow lakes. Trends in Ecology & Evolution 8: 275–279.
Sommer, U., 1981. The role of r- and K-selection in the succession of phytoplankton in Lake Constance. Acta Oecologica 2: 327–342.
Sommer, U., 1995. An experimental test of the intermediate disturbance hypothesis using cultures of marine phytoplankton. Limnology and Oceanography 40: 1271–1277.
Sommer, U., J. Padisák, C. S. Reynolds & P. Juhász-Nagy, 1993. Hutchinson's heritage: the diversity-disturbance relationship in phytoplankton. Hydrobiologia 249: 1–8.

Hydrobiologia (2007) 581:241–254
DOI 10.1007/s10750-006-0500-x

EUTROPHICATION IN LAKES

Intra-habitat heterogeneity of microbial food web structure under the regime of eutrophication and sediment resuspension in the large subtropical shallow Lake Taihu, China

Qinglong L. Wu · Yuwei Chen · Kuidong Xu · Zhengwen Liu · Martin W. Hahn

Abstract Planktonic microbial community structure and classical food web were investigated in the large shallow eutrophic Lake Taihu ($2338\ km^2$, mean depth 1.9 m) located in subtropical Southeast China. The water column of the lake was sampled biweekly at two sites located 22 km apart over a period of twelve month. Site 1 is under the regime of heavy eutrophication while Site 2 is governed by wind-driven sediment resuspension. Within-lake comparison indicates that phosphorus enrichment resulted in increased abundance of microbial components. However, the coupling between total phosphorus and abundance of microbial components was different between the two sites. Much stronger coupling was observed at Site 1 than at Site 2. The weak coupling at Site 2 was mainly caused by strong sediment resuspension, which limited growth of phytoplankton and, consequently, growth of bacterioplankton and other microbial components. High percentages of attached bacteria, which were strongly correlated with the biomass of phytoplankton, especially *Microcystis* spp., were found at Site 1 during summer and early autumn, but no such correlation was observed at Site 2. This potentially leads to differences in carbon flow through microbial food web at different locations. Overall, significant heterogeneity of microbial food web structure between the two sites was observed. Site-specific differences in nutrient enrichment (i.e. nitrogen and phosphorus) and sediment resuspension were identified as driving forces of the observed intra-habitat differences in food web structure.

Guest editors: B. Qin, Z. Liu & K. Havens
Eutrophication of shallow lakes with special reference to Lake Taihu, China

Q. L. Wu · Y. Chen · Z. Liu
Nanjing Institute of Geography & Limnology,
Chinese Academy of Sciences,
Nanjing 210008, P.R. China

Q. L. Wu · M. W. Hahn
Institute for Limnology,
Austrian Academy of Sciences,
A-5310 Mondsee, Austria

K. Xu
Institute of Oceanography,
Chinese Academy of Sciences, Qingdao
266071, P.R. China

Z. Liu
Institute of Hydrobiology, Jinan University,
Guangzhou 510632, P.R. China

Present Address:
Q. L. Wu (✉)
Division of Environmental Science, Engineering,
Faculty of Engineering, National University
of Singapore, 1 Engineering Drive 2,
Singapore 117576, Singapore
e-mail: wuqinglong@hotmail.com

Keywords Microbial food web · Eutrophication · Resuspension · Shallow lake · Heterogeneity

Introduction

Eutrophication in lakes associated with cyanobacterial blooms, notably the cyanobacteria *Microcystis* spp., is becoming common in many parts of the world as a consequence of excessive input of nutrients and water pollution (Dokulil & Teubner, 2000). Although many studies on pelagic classical food webs have been performed, the microbial food webs of eutrophic lakes in tropical or subtropical zone have received only little attention (Robarts & Wicks, 1990; Sommaruga, 1995). The currently available information on microbial food webs in the pelagic of freshwater systems is mainly about temperate lakes. In these lakes nutrient has significant impacts on microbial food webs (e.g. Weisse, 1991; Berninger et al., 1993). The nutrients enrichment in lakes resulted in increased abundance and biomass of different microbial components (e.g. Beaver & Crisman, 1989; Christoffersen et al., 1990; Berninger et al., 1991). Also under eutrophic and hypertrophic conditions, the microbial food webs may play an important role in the carbon flow and nutrient cycling as they do under oligotrophic conditions (Weisse & Stockner, 1993). On the other hand, the energy transfer efficiency is assumed to decrease with increasing trophic status of lakes (Weisse & Stockner, 1993; Sommaruga, 1995). Along a gradient towards eutrophy, 'bottom-up' control (i.e. supply of nutrients) of the microbial components is increasingly shifted to 'top-down' interactions (Weisse, 1991; Weisse & Stockner, 1993). In eutrophic lakes with high abundance of colonial *Microcystis* spp., the dominance of *Microcystis*-attached or *Microcystis*-associated bacteria may occur (Worm & Søndergaard, 1998). Due to the decreased sensitivity of attached bacteria to grazing by most bacterivorous protists (Jürgens & Güde, 1994), the increased proportion of attached bacteria may have consequences for the structure and energy transfer efficiency of microbial food web in *Microcystis* spp. dominated habitats.

Physical forces may influence the microbial food webs of shallow water habitats. One of the most important processes is wind-driven sediment resuspension and water column mixing. It has been indicated that sediment resuspension may increase the total abundances and may also change the community composition of protists due to passive dispersal of protists from sediment (Rogerson & Laybourn-Parry, 1992; Shimeta & Sisson, 1999; Garstecki et al., 2000) and due to species-specific differences in susceptibility to resuspension (Shimeta & Sisson, 1999; Garstecki et al., 2002). Previous experiments demonstrated that at least some protists are generally well adapted to environmental situations of high sediment load (Boenigk & Novarino, 2004). Field and experimental studies indicate that sediment resuspension may cause release of nutrients and organic matter into the water column (Arfi & Bouvy, 1995). This can result in increased bacterial abundance and biomass (Wainright, 1987, 1990; Ritzrau & Graf, 1992), as well as in increased numbers of bacterivorous protists benefiting from the increased prey availability (Garstecki et al., 2002). The currently available information on the ecological impacts of sediment resuspension was mainly obtained in studies on estuarine and costal systems.

In this study, the structure and dynamics of the planktonic microbial food web in the large subtropical shallow eutrophic Lake Taihu (Qin et al., 2007) were investigated. This lake has been the subject of many limnological studies during the last 50 years (Qin et al., 2004), however, detailed studies on the microbial food web were lacking thus far. For comparison of intra-habitat differences of microbial food web structure two sites located 22 km apart were selected for investigation over a period of twelve month. The two sites differ strongly in their degree of sediment resuspension and eutrophication. In order to reveal possible site-specific differences in the interactions between classical and microbial food web, the dynamics and composition of phytoplankton and zooplankton were also investigated. Attempts to analyse the potential effects of eutrophication and sediment resuspension on microbial food web in this large and shallow lake were made.

Methods

Study sites and sampling scheme

General information of Lake Taihu can be found in the introductory paper (Qin et al., 2007) or elsewhere (Chen et al., 1997; Qin et al., 2004). The lake can be divided into several different regions according to the spatial differences of physical and chemical conditions and phytoplankton community structure (Chen et al., 1997, 2003). One sampling station (Site 1) is located in Meiliang Bay, which was highly enriched with nitrogen and phosphorus (Fig. 1). The bay is located in the northern part of Lake Taihu, has a surface area of ca. 100 km^2, and an average depth of ca. 1.8 m. The eutrophication of the bay was mainly caused by pollution with domestic wastewater discharged from Lujiang and Liangxi River (Fig. 1). The second sampling station (Site 2) is located in the open lake close to its centre (Fig. 1). The open lake is less enriched with nitrogen and phosphorus than most of the bays. The open lake is on average only a few centimetres deeper than Meiliang Bay, but much more exposed to wind.

Twenty-four samplings were performed at both sampling sites from September 30, 2002 to September 30, 2003, at intervals of ca. 15 days. All samples were collected from surface waters (top 50 cm) with a 5-litre Schindler sampler. Sampling was undertaken between 9 and 11 a.m. Fifty ml water samples for determination of numbers of freely suspended bacteria, attached and aggregated bacteria, and heterotrophic nanoflagellates (HNF) were fixed with 2% formalin (final concentration). 250 ml water samples for determination of phytoplankton, ciliates, and nauplii were preserved with 1.5% Lugol's Iodine. Macrozooplankton was collected by filtering 5 litre lake water through a nylon mesh of 30 μm. Samples were fixed with formalin (final concentration 4%). All samples were stored at 4°C in dark until further analysis within 5 months.

At each second sampling, secchi-depth (SD), water temperature (WT), pH, conductivity (Cond), dissolved oxygen (DO), total suspended solid (TSS), total nitrogen (TN), total dissolved nitrogen (TDN), ammonium (NH_4^+), nitrite (NO_2^-), nitrate (NO_3^-), total phosphorus (TP), total dissolved phosphorus (TDP), ortho-phosphorus (PO_4^{3-}-P), and chlorophyll *a* (Chl *a*) were measured according to standard methods (Jin & Tu, 1990).

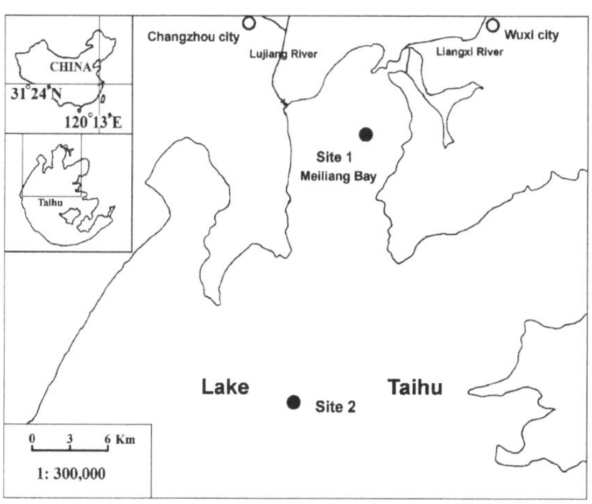

Fig. 1 Map of the northern part of Lake Taihu showing the locations of the sampling sites in Meiliang Bay (Site 1) and the open lake (Site 2)

Abundance and biomass determination

Phytoplankton samples were counted using a sedimentation chamber of 1–5 ml at 200× or 400× magnification on an inverted microscope. Phytoplankton species were identified according to Hu et al. (1980). Algal biovolumes were calculated from cell numbers and cell size measurements using appropriate geometric morphs (Hillebrand et al., 1999). Conversion to biomass assumed that 1 mm^3 of biovolume is equivalent to 1 mg of fresh-weight biomass.

Ciliates and nauplii were counted in sedimentation chambers of 5–50 ml at 200× magnification on an inverted microscope. Ciliates were identified mainly according to the keys of Foissner and Berger (1996), and Foissner et al. (1999). Linear dimensions of about 30 individuals of abundant taxa were measured by using the LUCIA G image analysis system (Lucia 4.51, Laboratory Imaging, Prague, Czech Republic), which was connected to the inverted microscope via a video camera. The biovolumes were calculated using appropriate geometric shapes and converted into biomass using a conversion factor of 110 fg C μm^{-3} (Turley et al., 1986). Macrozooplankton were counted in a Bogorov chamber at 45× magnifications, and identified according to Zhang & Huang (1991). For counting of rotifers fixed samples of 0.1 to 0.5 litre were filtered through a 30 μm mesh and poured into a chamber, and counted at 100× magnification following the same approach as for ciliates and nauplii.

For determination of numbers of freely suspended bacteria and numbers of attached bacteria untreated and sonicated subsamples were counted (Velji & Albright, 1993). Subsamples (0.2–1 ml) were diluted to 1 or 2 ml with autoclaved inorganic medium, stained with 4,6-diamidino-2-phenylindole (DAPI) (0.1% wt/vol, final concentration) for 10 min (Porter & Feig, 1980), filtered onto black 0.2 μm pore-size membrane filters (Millipore), and counted at 1250× magnifications by using an epifluorescence microscope (Zeiss, Germany) equipped with a BP 365, FT 395, and LP 397 filter set. Attached bacterial abundances were determined as the differences between total cell numbers (sonicated samples) and total numbers of freely suspended cells (non-sonicated samples). Autotrophic picoplankton cells were identified by checking for the presence of autofluorescence under green excitation and excluded from bacterial counts. Length and width of bacteria were determined with LUCIA G image analysis system as described in Psenner (1993). Cell volumes were calculated with the formula: $V = (w^2 \times \pi/4) \times (l-w) + (\pi \times w^3/6)$, where V = Volume (μm^3), w = width (μm) and l = length (μm). The biovolumes were converted into biomass using allometric relationship as suggested by Norland (1993).

Heterotrophic nanoflagellates (HNF) were counted and measured in the same way as for bacteria except sonication of samples. Individual autofluorescence was measured under green excitation using a BP 510–560, FT 580, and LP 590 filter set. Biovolumes were calculated using the formula: $V = (\pi \times w^2/6) \times L$, where V = Volume (μm^3), w = width (μm) and L = length (μm). The biomass of HNF was estimated from mean cell volume using a conversion factor of 220 fg C μm^3 (Borsheim & Bratbak, 1987).

Statistics

Redundancy analysis (RDA) was applied for revealing relationship between the members of planktonic microbial and classical food web and the major environmental variables controlling the dynamics of different members. The biotic variables include the abundances of total bacteria, attached bacteria, HNF, ciliates, rotifers, Cladocerans, *Daphnia* sp., total phytoplankton, and *Microcystis* spp. The tested abiotic environmental variables were water temperature, TP, TN, TSS and COD. All data were log $(x + 1)$ transformed. The RDA was performed with the software CANOCO 4.5 (SCIENTIA Software) by assuming linear species-environment relationships because Detrended Correspondence Analysis (DCA) run on biotic variables indicated that the length of first axis was <2. The significance of first ordination and canonical axes together was assessed in permutation tests with 499 unrestricted Monte Carlo permutations. The statistical analyses were performed separately for data of Site 1 and Site 2. *T*-test was used to test the differences of parameters between the two sites. Log-transformation was used to normalize the data if necessary.

Results

Major ecological differences at the two study sites

Main ecological parameters measured at the two sites of Lake Taihu during the one-year sampling period were summarized in Table 1. Major between-sites differences in chemo-physical parameters were observed for conductivity, secchi depth, concentration of total suspended solid, and concentrations of major algal nutrients (Table 1). No anoxic condition were observed at both sites over the water column and near the bottom, but much lower dissolved oxygen was found at the sediment water interface at Site 1 (unpublished data from Fan et al.). Minimum water temperature (3.8°C) was measured in January and maximum water temperature occurred in July (30°C) (Fig. 2a and b). The concentrations of TSS, TN, NH_4^+, NO_3^-, TP, TDP, and Chl *a* were higher at Site 1 than those at Site 2. Abundance and biomass of components of planktonic community were higher at Site 1 than at Site 2. Little fluctuation of chl *a* value was observed at Site 2

(Fig. 2b). But high seasonal variation of TSS was found at Site 2 relative to Site 1 (Fig. 2e, f). Results of RDA indicated that the most important factors structuring the food web at Site 1 were TP and water temperature (Fig. 3a), while at Site 2 it was only the water temperature (Fig. 3b).

Seasonal changes of microbial components

At Site 1, total bacterial abundance ranged from 2.9×10^6 cells ml^{-1} to 14.5×10^6 cells ml^{-1} with a mean of 7.4×10^6 cells ml^{-1} (Fig. 4a). Bacterial numbers peaked in the middle of summer and showed the lowest values in late winter. The mean of bacterial cell volume ranged from 0.03 to 0.13 μm^3, while the median ranged from 0.02 μm^3 to 0.09 μm^3 (Fig. 4e). Total bacterial biomass ranged from 34.8 μg C l^{-1} to 299.5 μg C l^{-1} with a mean of 98.3 μg C l^{-1} (Fig. 4a). Fluctuation of bacterial biomass was generally in accordance with bacterial abundance. Attached bacterial abundance oscillated between 0 (not measurable) and 6.1×10^6 cells ml^{-1} with a mean of 1.4×10^6 cells ml^{-1} accounting for 0–51.8 % and 14.4% of the total bacterial abundance (Fig. 4c). High numbers of attached bacteria were observed from May to October.

At Site 2, the mean bacterial abundance was 5.7×10^6 cells ml^{-1} with a highest value of 9.2×10^6 cells ml^{-1} in middle of summer and a lowest value of 2.0×10^6 ml^{-1} in early winter (Fig. 4b). The bacterial biomass ranged from 28.9 μg C l^{-1} to 128.4 μg C l^{-1} with an average of 64.9 μg C l^{-1}. Little fluctuation of bacterial abundance and biomass was observed at Site 2. Attached bacteria ranged from 0 (not measurable) to 4.3×10^6 cells ml^{-1} with a mean value of 1.4×10^6 cells ml^{-1}, which account for 0–51.2% and 19.0% of the total bacterial abundance

Table 1 Physical, chemical, and biological parameters determined for the two sampling sites of Lake Taihu in the period from September 2002 to September 2003. The data of the two sampling sites were compared by *t*-test

Parameters	Site 1 Mean (range)	Site 2 Mean (range)	*p* value (*t*-test)
Water depth (m)	2.0–2.7	2.4–3.2	<0.01
Water temperature (°C)	17.6 (4.3–30.0)	17.2 (3.8–30.0)	0.91
Dissolved oxygen (g m^{-3})	9.28 (6.38–12.31)	9.01 (6.82–12.51)	0.91
Conductivity (μs cm^{-1})	457 (380–570)	388 (320–440)	<0.01
PH	8.43 (8.01–8.98)	8.38 (8.19–8.66)	0.59
Secchi-depth (cm)	50 (10–100)	36 (15–50)	0.08
Total suspended substances (g m^{-3})	32.31 (9.56–51.09)	68.18 (20.24–170.12)	0.02
Chemically oxygen demanded (mg l^{-1})	7.6 (4.6–34.2)	5.9 (3.6–5.9)	0.04
Total nitrogen (μmol l^{-1})	282.9 (108.6–780.0)	154.8 (61.4–252.9)	0.03
Total dissolved nitrogen (μmol l^{-1})	182.9 (32.1–311.4)	117.9 (21.4–234.9)	0.09
NH_4^+-N (μmol l^{-1})	40.0 (2.14–110.0)	6.4 (1.4–17.9)	<0.01
NO_2^--N (μmol l^{-1})	5.0 (0.7–19.3)	0.7 (0.1–1.4)	0.03
NO_3^--N (μmol l^{-1})	70 (3.6–132.1)	59.3 (4.3–124.3)	0.58
TP (μmol l^{-1})	3.39 (1.26–6.87)	2.36 (1.45–3.81)	0.06
Total dissolved phosphorus (μmol l^{-1})	1.07 (0.52–2.32)	0.65 (0.32–1.84)	0.04
PO_4^{3-}P (μmol l^{-1})	0.19 (0–1.0)	0.13 (0–0.39)	0.19
Chlorophyll a (μg l^{-1})	20.75 (1.70–52.01)	8.88 (3.83–16.12)	0.04
Total nitrogen: total phosphorus	44.95 (8.82–103.08)	32.35 (12.2–75.8)	0.40
Total bacterial abundance (10^6 cells ml^{-1})	7.4 (2.9–14.5)	5.7 (2.0–9.2)	<0.01
Total bacterial biomass (μg C l^{-1})	98.3 (34.8–299.5)	64.9 (28.9–128.4)	<0.01
HNF abundance (10^7 cells l^{-1})	0.8 (0.1–3.5)	0.5 (0.1–1.6)	0.03
HNF biomass (μg C l^{-1})	127.1 (4.9–699.7)	84.7 (1.6–454.3)	0.02
Ciliate abundance (10^3 cells l^{-1})	9.1 (0.5–29)	8.7 (1.8–62.6)	0.89
Ciliate biomass (μg C l^{-1})	16.2 (0.7–68.6)	10.4 (1.4–37.4)	0.21
Phytoplankton biomass (mg l^{-1})	2.38 (0.06–9.02)	0.28 (0.01–0.86)	0.02
Rotifer abundance (ind. l^{-1})	640 (1–4650)	156 (1–696)	0.02
Cladocerans abundance (ind. l^{-1})	183 (0.2–740)	87.5 (1.4–496)	0.03

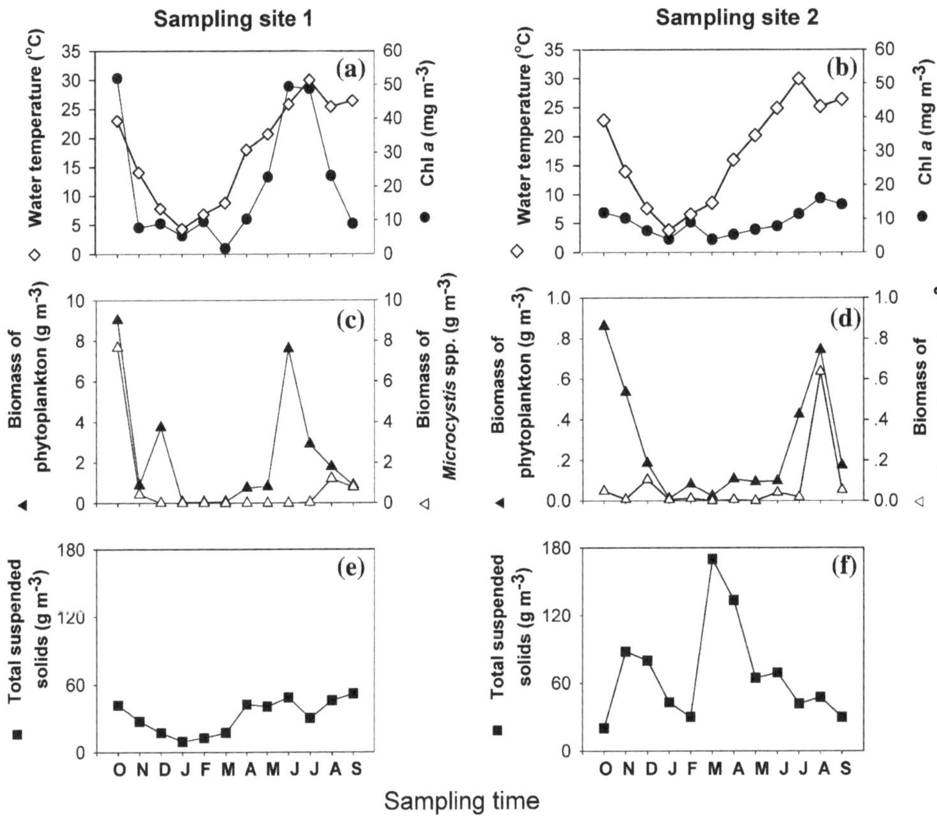

Fig. 2 Seasonal development of Chlorophyll a and water temperature (**a** and **b**), biomass of phytoplankton and biomass of colonial *Microcystis* spp. (**c** and **d**, and total suspended solids (TSS) (**e** and **f**) in the water column at the two sampling stations in Lake Taihu

respectively (Fig. 4d). High abundance and percentage of attached bacteria were found from early spring to late autumn. The mean and median of cell volume fluctuated from 0.03 to 0.07 μm^3 and 0.02 μm^3 to 0.03 μm^3 respectively (Fig. 4f). No correlations between the cell volumes of freely suspended bacteria or attached bacteria, and temperature were found at both sampling stations (linear regression, $P > 0.05$).

The abundance of HNF at Site 1 ranged from 0.1×10^7 cells l^{-1} to 3.5×10^7 cells l^{-1} with a median and mean density of 0.6×10^7 cells l^{-1} and 0.8×10^7 cells l^{-1} respectively (Fig. 5a). HNF in the size class from 5 to 10 μm were the most abundant. HNF biomass ranged from 4.9 μg C l^{-1} to 699.7 μg C l^{-1} with a median and mean biomass of 56.6 μg C l^{-1} and 127.1 μg C l^{-1}, respectively (Fig. 5a). Single peak of HNF abundance and biomass was observed in late autumn, while no strong oscillation was observed during other months (Fig. 5a). At Site 2, the median and average density of HNF was 0.4×10^7 cells l^{-1} and 0.5×10^7 cells l^{-1} with a range from 0.1×10^7 cells l^{-1} to 1.6×10^7 cells l^{-1}, respectively (Fig. 5h). HNF biomass ranged from 1.6 μg C l^{-1} to 454.3 μg C l^{-1} with a median and mean biomass of 57.3 μg C l^{-1} and 84.7 μg C l^{-1} (Fig. 5h). HNF biomass formed one peak in late fall (Fig. 5h). Both HNF abundance and HNF biomass were not correlated with the mean size of total bacteria and freely suspended bacteria at both sites (linear regression, $P > 0.05$).

At Site 1, the abundance of ciliates ranged from 470 cells l^{-1} to 29,000 cells l^{-1} with a mean and median density of 9100 cells l^{-1} and 4000 cells l^{-1} (Fig. 5b). Ciliates in the size class of 20–40 μm were the most abundant in most time of the year (Fig. 5b). Dominant taxa in this size

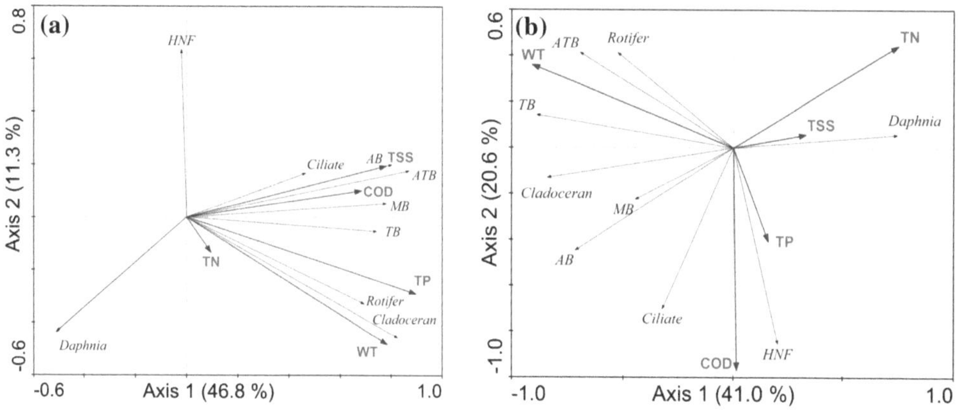

Fig. 3 RDA biplots show different planktonic groups in relation to main environmental factors at Site 1 (**a**) and Site 2 (**b**) in Lake Taihu. TN, TP, TSS, WT and COD refer to total nitrogen, total phosphorus, total suspended substances, water temperature, and chemically oxygen demanded respectively. AB, MB, TB, and ATB refer to abundance of phytoplankton, *Microcystis* spp., total bacteria and attached bacteria, respectively

class were *Strobilidium*-like and unidentified oligotrich small ciliates. The lowest abundance occurred in winter and early spring with exception of 15th of January, while the highest density was observed in summer. Ciliates <20 μm were abundant only from July to September (Fig. 5b)

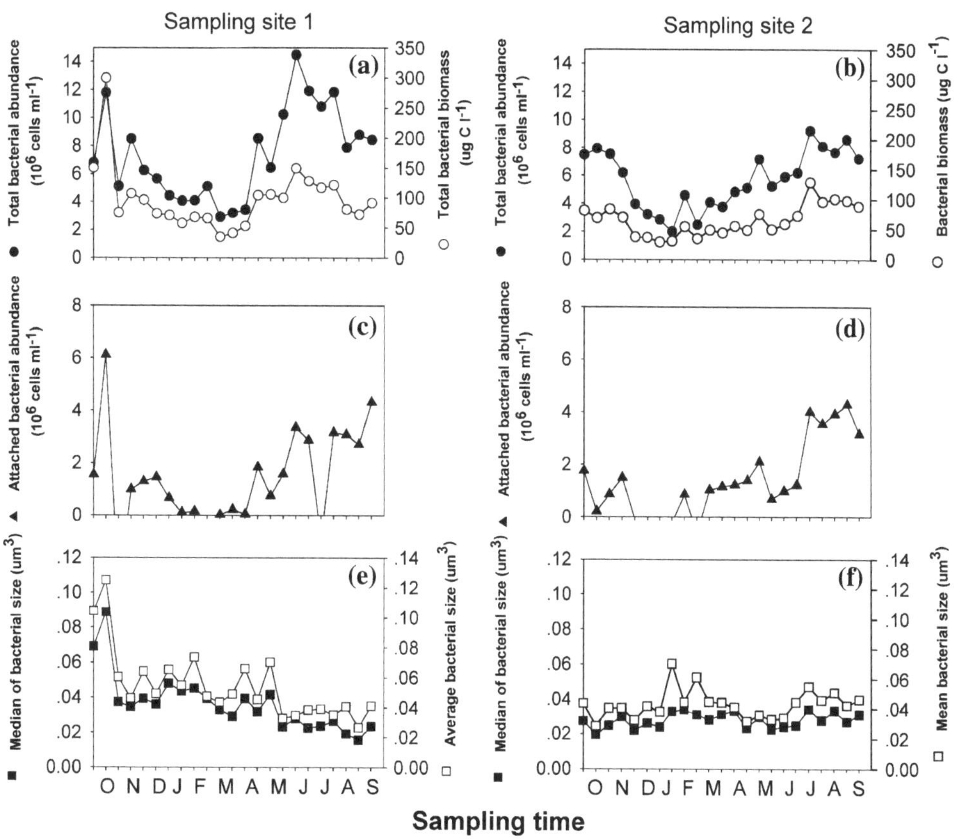

Fig. 4 Seasonal development of total bacterial abundance and total bacterial biomass (**a** and **b**), attached bacterial abundance (**c** and **d**), mean and median of bacterial size (**e** and **f**) at the two sampling sites in Lake Taihu

and dominated by *Urotricha*-like species. In January and June, ciliates >40 μm contributed 50% of total ciliate abundance (Fig. 5b). They were represented by *Vorticella* spp., *Epistylis* spp., *Codenella* sp., and *Tintinnidium* spp. Total ciliate biomass ranged from 0.7 μg C l^{-1} to 68.6 μg C l^{-1} with a mean and median of 16.2 μg C l^{-1} and 10.3 μg C l^{-1} (Fig. 5b). High biomass was observed in January and June, which was mainly contributed by *Vorticella* spp., and *Epistylis* spp. At Site 2, total ciliate abundance averaged 8700 cells l^{-1}, showing two higher values at the end of January with density of 26,100 cells l^{-1} and in midsummer with density of 62,600 cells l^{-1}, while showing lower values during most time of the year (Fig. 5i). During the winter peak, ciliates in the size class of <40 μm were the dominant groups (Fig. 5i). They were represented by an unidentified small ciliate and *Pelagovorticella*-like species, respectively. During the summer peak, *Urotricha*-like ciliates were dominating. During other periods, ciliates were mainly composed of *Codenella* sp., *Tintinnidium* spp., *Stenosemella* sp., and *Vorticella* spp. Total biomass of ciliates ranged from 1.4 μg C l^{-1} to 37.4 μg C l^{-1} with a mean and median of 10.4 μg C l^{-1} and 7.5 μg C l^{-1} (Fig. 5i). Higher biomass was observed from May to August, which was mainly contributed by *Vorticella* spp. and *Codenella* sp.

Community dynamics of phytoplankton and zooplankton

At Site 1, the phytoplankton biomass ranged from 0.06 g m^{-3} in January to 9.02 g m^{-3} in October. The cyanobacteria *Microcystis* spp. dominated in terms of biomass among the cyanobacteria, as well as among the whole phytoplankton assemblages from July to November (Fig. 2a). At Site 2, low phytoplankton biomass was observed with a range from 0.01 g m^{-3} in December to 0.86 g m^{-3} in October. *Microcystis* spp. dominated phytoplankton communities in August and September (Fig. 2b).

At Site 1, rotifer abundance averaged 640 ind. l^{-1} showing the highest value of 4650 ind. l^{-1} in the middle of July and the lowest value of 1 ind. l^{-1} at the end of December (Fig. 5c). The abundant rotifers include *Branchionus* sp., *Keratella valga*, and *Polyarthra* sp., while *Asplanchna priodonta* only contributed to more than 50% of rotifer biomass in middle of July. Nauplii abundance ranged from 6 to 1200 ind. l^{-1} with an average and median of 144 and 27 ind. l^{-1} (Fig. 5d). The highest abundance occurred at the end of August, while much lower abundance was observed during most of other period (Fig. 5d). Abundance of calanoid copepods ranged from 0 to 22 ind. l^{-1} with maximum abundances occurring in middle of fall and spring, and minimum in summer (Fig. 5e). Abundance of cyclopoid copepods averaged 49 ind. l^{-1}, showing highest value of 408 ind. l^{-1} in middle of July and much lower values during most of the other periods (Fig. 5f). The succession of cladocerans can be divided into two main phases (Fig. 5g). In the first phase, *Daphnia* sp. dominated the plankton community from end of February to end of May, with maximum abundance of 96 ind. l^{-1} at the end of March (Fig. 5g). In the second phase, from June to November, *Bosmina* sp. was the most abundant species, showing the highest abundance of 722 ind. l^{-1} at the end of June (Fig. 5g). *Moina* sp. only appeared in considerable numbers in middle of September (Fig. 5g). At Site 2, low rotifer abundances were observed all over the year, ranging from 1 to 696 ind. l^{-1} with an average of 156 ind. l^{-1} (Fig. 5j). Little fluctuation of rotifer abundance was found (Fig. 5j). The abundant rotifers include *Branchionus* sp., *Keratella cochlearis*, and *Polyarthra* sp. Nauplii abundance ranged from 0.6 to 2830 ind. l^{-1} with an average and median of 226 and 17 ind. l^{-1} (Fig. 5k). One nauplii peak appeared at the end of June, while much lower abundance was observed during most of the other periods (Fig. 5k). Higher abundance of calanoid copepods and cyclopoid copepods was observed in later summer while lower abundance appeared during most of other period (Fig. 5l, m). The succession of cladocerans could again be divided into two main phases (Fig. 5n). In the first phase, *Daphnia* sp. also dominated the plankton community at this site from middle of January to

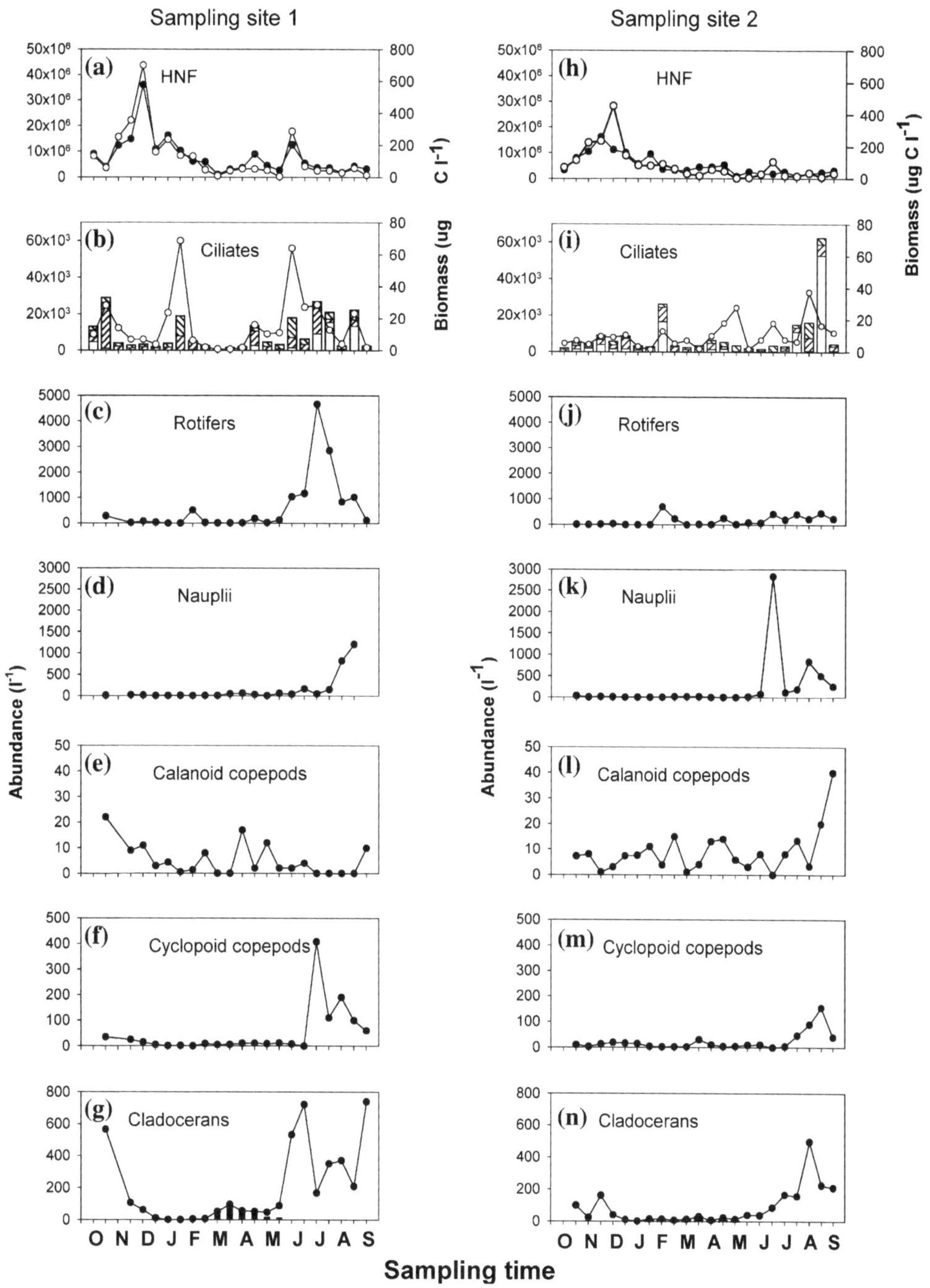

Fig. 5 Seasonal changes in abundance (•) and biomass (○) of HNF (**a** and **h**), ciliates (**b** and **i**), rotifers (**c** and **j**), nauplii (**d** and **k**), calanoid copepods (**e** and **l**), cyclopoid copepods (**f** and **m**), and cladocerans (**g** and **n**). In Fig. **b** and **i**, the vertical bars show abundances of ciliates in three size classes: <20 μm (open part of the bars); 20–40 μm (upward diagonally hatched part of the bars); and > 40 μm (downward diagonally hatched part of the bars). In figure g and n, the vertical black bars show the abundance of *Daphnia*

middle of May, with maximum abundance of 29 ind. l^{-1} at the end of March (Fig. 5n). In the second phase from May to middle of November, *Bosmina* sp. and *Moina* sp. were the most abundant species, showing the highest abundance of 496 ind. l^{-1} in the middle of August (Fig. 5n).

Statistics

RDA model calculated with the data of Site 1 and Site 2 significantly (for first canonical axis, $P < 0.05$) explained the variability of the abundance of members of planktonic food webs at both sites (Figs. 3a and b). At Site 1 (Fig. 3a), the first axis showed high canonical correlation with TP, WT, COD, and TSS, which were positively related to planktonic components except for HNF and *Daphnia*. HNF was negatively related to such micrograzers like rotifer and *Bosmina*-dominated zooplankton. *Daphnia* abundance was negatively related with all microbial components. There was a significant correlation between attached bacteria and TSS and abundance of *Microsystis* spp. At site 2 (Fig. 3b), the first axis showed only high canonical correlation with water temperature. In contrast to Site 1, TN, TSS and TP were even negatively related to abundance of phytoplankton and microbial communities. *Daphnia* abundance was negatively related with all microbial components as the same at Site 1.

Discussion

Influence of sediment resuspension on microbial components

Based on the concentrations of TP and TN (Table 1), Site 1 was more enriched with phosphorus and nitrogen than Site 2. On the other hand, Site 2 showed a higher intensity of sediment resuspension than Site 1, as indicated by higher amount of TSS (Table 1). Furthermore, most planktonic organisms showed on average lower biomass at Site 2.

Sediment resuspension, as a common feature of large shallow lakes, may have positive and negative impacts on microbial food webs. Positive effects on abundance of microbial components could include (i) direct transportation of benthic bacteria and protozoan into the water column (Rogerson & Laybourn-Parry, 1992; Shimeta & Sisson, 1999; Garstecki et al., 2000, 2002); (ii) release of dissolved nutrient and organic matter into water column (Arfi & Bouvy, 1995), and thereby increasing the abundance of bacteria (Wainright, 1987, 1990; Ritzrau & Graf, 1992) and some bacterivorous protists (Garstecki et al., 2002). However, no correlations between the abundance of microbial components (i.e., bacteria, HNF and ciliates) and TSS, the indicator of sediment resuspension, were found at Site 2. The same result was found for Site 1 except the bacterial abundance. Possible explanation for these findings is the different intensity of sediment resuspension, which is dependent on wind speed, water depth, and the size of a lake. Site 1 is located in the Meiliang bay, where only very strong wind can cause a significant resuspension of sediment (Qin et al., 2004). Therefore, concentration of suspended solids in the water column at Site 1 was lower during periods of weak wind intensity or at calm situations. This is confirmed by little variations of TSS and high contribution of phytoplankton biomass to the TSS fraction. This contribution was indicated by the strong positive correlations between phytoplankton biomass and TSS. The microbial organisms in the upper sediment may only contribute to the microbial components in the water column of the bay during periods with high wind intensity. Site 2 is located in the centre of the lake, where moderate wind (3.3–5.0 m s^{-1}) caused strong mixing of the water column and sediment resuspension during almost two thirds of a year (Qin et al., 2004; Luo et al., 2004). Long-lasting sediment accumulation at the lake bottom was not observed at Site 2 (Qin et al., 2004). The bottom (hard clay) water interface at Site 2 can only allow biofilm-forming microorganisms to inhabit this surface, and these biofilm organisms cannot contribute much to the planktonic microbial community in the water column. For the same reason, not much organic carbon can be deposited and later introduced into the water column by resuspension events. Therefore, the bacterial community at Site 2 could, in contrast to Site 1, not benefit from release of organic material from the sediment during resuspension events. This fits to the observation, that higher amount of resuspended sediment did not lead to higher abundances of total bacteria and freely suspended bacteria at Site 2. To reveal the exact contribution of benthic microbial organism to planktonic microbial food web and its controlling factors, more detailed temporal in situ studies are needed at different locations of the lake.

At Site 2 a negative correlation between TSS and phytoplankton biomass and a relatively low phytoplankton biomass despite relatively high concentration of total phosphorous were observed. Obviously, the phytoplankton was negatively impacted by decrease in light intensities as a result of the frequent resuspension of sediment and the high concentration of TSS (Chen et al., 2003). A negative indirect impact of sediment resuspension on bacterial growth conditions could be caused by the limited growth and production of phytoplankton. Decrease in primary production may lead to a lower supply of dissolved organic carbon for bacterial growth (Chen & Wangersky, 1996). At Site 1, the positive correlations between TSS and the bacterial abundances or phytoplankton biomass suggest no apparent negative impacts.

Impacts of eutrophication on microbial components

Higher abundances of microbial components (i.e. bacteria, HNF) were observed at site 1 than those at Site 2. This within-lake comparison supports the hypothesis that abundances of microbial components are positively related to phosphorus enrichment of a lake (e.g. Weisse, 2003). However, the coupling between TP and abundance of microbial components was much different at the two sites (Fig. 3). Indeed no obvious correlation between TP and abundance of microbial components was observed at Site 2 (Fig. 3b). This weak coupling could be duo to the strong sediment resuspension at Site 2, which limited growth of phytoplankton and, consequently, growth of bacterioplankton and other microbial components (i.e. HNF, ciliate). This was supported by the fact that no correlation between TP and phytoplankton abundance were observed at Site 2 (Fig. 3b), while good correlation between abundance of phytoplankton and bacterioplankton was found at both sites (Figs. 3a and b). On the other hand, on average higher biomass ratios of phytoplankton to bacteria were observed at Site 1 than at Site 2 (Fig. 6). This may indicate different rates of bacterial production at two different locations.

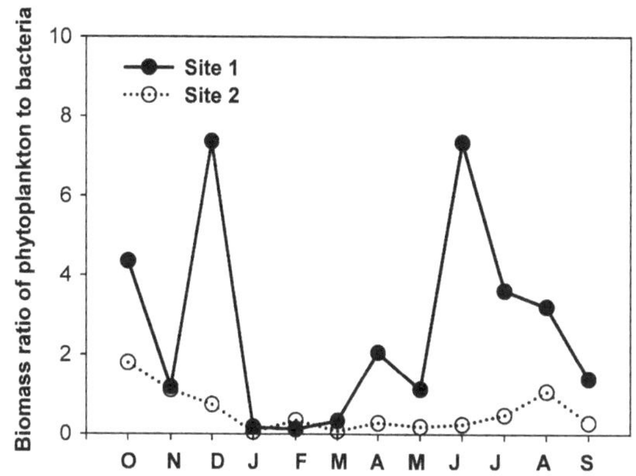

Fig. 6 Biomass ratios of phytoplankton to bacteria at the two sampling sites in Lake Taihu

Along a gradient towards eutrophy, microbial food webs are progressively controlled by 'top-down' forces (Weisse & Stockner, 1993), and zooplankton predation on components of microbial food webs is gaining an increasing importance in linking of planktonic microbial food webs and the classical food chains. Among the zooplankton, *Daphnia*, can strongly suppress the microbial food web by their size-selective feeding abilities (Weisse & Stockner, 1993; Jürgens, 1994, Jürgens & Stolpe, 1995). The data on the seasonal succession of cladocerans, flagellates, bacteria and ciliates at both sites allow insights into the trophic interactions between metazoans and the microbial loop (Figs. 3, 5), and indicate the similar 'top-down' forces of *Daphnia* during spring and early summer in such a subtropical lake. Following the collapse of *Daphnia* population in mid-summer, possibly due to fish predation and blooming of toxic colonial *Microcystis* spp., *Bosmina* sp. and *Moina* sp., dominated the crustacean community.

Significance of attached bacteria

Particles from sediment resuspension, terrestrial input, and large phytoplankton are important for colonization of attached bacteria (Simon et al., 2002). Lake Taihu received a large amount of particles (organic and inorganic) from its watershed due to intensive agriculture and soil erosion (Dokulil et al., 2000). Sediment resuspension especially in the open lake is very significant due to the lake's shallowness (Qin et al., 2004).

Under the process of eutrophication, phytoplankton blooms, notably the colonial *Microcystis* spp. fuel Lake Taihu with high amount of particulate organic matter (Chen et al., 2003). Our investigations indicated that high abundances of attached bacteria occurred from summer to autumn at both sites in this lake. At Site 1, the strong correlations between attached bacterial abundance and biomass of *Microcystis* spp. suggests that much of the attached bacteria were associated with *Microcystis* spp. colonies (Fig. 3a). This fits to the findings in a eutrophic Danish lake (Worm & Søndergaad, 1988). Worm et al. (2001) demonstrated that bacteria attached to phytoplankton blooms hold up to nearly 80% of total bacterial numbers in an enclosure experiment. At Site 2, however, no strong correlations between attached bacterial abundance and phytoplankton biomass, *Microcystis* spp. biomass, and TSS were observed. This might suggest that the communities of attached bacteria differed in their composition at the two investigated sites.

In comparison to freely suspended bacteria, attached bacteria contributed higher proportion to bacterial production (Simon et al., 2002). The dominance of attached bacteria in Lake Taihu from summer to autumn may have consequences for the structure of the microbial food web, as well as for the carbon flow through the entire food web. Since attached bacteria could be less susceptible to predation by protozoan (Jürgens & Güde, 1994), the production by attached bacteria might not be channeled efficiently through the microbial food web to higher trophic levels. On the other hand, planktivorous fish (e.g. *Hypophthalmichthys molitrix*, *Aristichthys nobilis* commonly found in Lake Taihu), which are able to directly graze upon phytoplankton dominated by *Microcystis* spp. (Xie & Liu, 2001), may directly crop a part of the production of attached bacteria. This may lead to a direct channeling of a part of the bacterial production to higher trophic levels.

Conclusion

Significant intra-habitat heterogeneity of microbial food web structure and dynamics was observed in Lake Taihu. Site-specific differences in wind-driven sediment resuspension, as well as differences in intensity of eutrophication were identified as major causes of the observed within habitat heterogeneity. In contrast to previous observations in coastal regions, our investigations indicate that frequent sediment resuspension in large lakes does not contribute to growth of planktonic microbial organisms.

Acknowledgements This work was supported by Chinese Academy of Sciences (No. KZCX1–SW–12), NSFC (No. 30370278, 30200037), and Austrian Academy of Sciences. Jiang Ji and some other staff members at the Taihu Laboratory for Lake Ecosystem Research contributed field assistance. Q. L. Wu was supported by a scholarship from North-South dialogue Program, Ministry of Foreign Affairs, Austria (EZA project 894/01). We thank Thomas Weisse for providing comments on an early version of the manuscript.

References

Alfri, R. & M. Bouvy, 1995. Size, composition and distribution of particles related to wind-induced resuspension in a shallow tropical lagoon. Journal of Plankton Research 17: 557–574.

Beaver, J. R. & T. L. Crisman, 1989. The role of ciliated protozoa in pelagic freshwater ecosystems. Microbial Ecology 17: 111–136.

Berninger, U. G., B. J. Finlay & P. Kuuppo-Leinikki, 1991. Protozoan control of bacterial abundances in freshwater. Limnology and Oceanography 36: 139–146.

Berninger, U. G., S. A. Wickham & B. J. Finlay, 1993. Trophic coupling within the microbial food web: a study with fine temporal resolution in a eutrophic freshwater ecosystem. Freshwater Biology 30: 419–432.

Boenigk, J. & G. Novarino, 2004. Effect of suspended clay on the feeding and growth of bacterivorous flagellates and ciliates. Aquatic Microbial Ecology 34: 181–192.

Borsheim, K. Y. & G. Bratbak, 1987. Cell volume to carbon conversion factors for a bacterivorous Monas sp. enriched from sea water. Marine Ecology Progress Series 36: 171–175.

Chen, W., Y. Chen, X. Gao & I. Yoshida, 1997. Eutrophication of Taihu and its control. Agriculture Engineering Journal 6: 109–120.

Chen, Y., B. Qin, K. Teubner & M. T. Dokulil, 2003. Long-term dynamics of phytoplankton assemblages: *Microcystis*-domination in Lake Taihu, a large shallow lake in China. Journal of Plankton Research 25: 445–453.

Chen, W. & P. Wangersky, 1996. Rates of microbial degradation of dissolved organic carbon from phytoplankton cultures. Journal of Plankton Research 18: 1521–1533.

Christoffersen, K., B. Riemann, L. R. Hansen, A. Klysner & H. B. Sorensen, 1990. Qualitative importance of the microbial food loop and plankton community structure in a eutrophic lake during a bloom of cyanobacteria. Microbial Ecology 20: 253–272.

Dokulil, M. T. & K. Teubner, 2000. Cyanobacterial dominance in lakes. Hydrobiologia 438: 1–12.

Dokulil, M. T., W. Chen & Q. Cai, 2000. Anthropogenic impacts to large lakes in China: the Taihu example. Aquatic Ecosystem Health and Management 3: 81–94.

Foissner, W. & H. Berger, 1996. A user-friendly guide to the ciliates (Protozoa Ciliophora) commonly used by hydrobiologist as bioindicators in rivers, lakes, and waste waters, with notes on their ecology. Freshwater Biology 35: 375–482.

Foissner, W., H. Berger & J. Schaumberg, 1999. Identification and Ecology of Limnetic Plankton Ciliates. Bayer. Landesamt für Wasserwirtschaft, München.

Garstecki, T., R. Verhoeven, S. A. Wickham & H. Arndt, 2000. Benthic-pelagic coupling: a comparison of the community structure of benthic and pelagic heterotrophic protists in shallow inlets of the southern Baltic. Freshwater Biology 45: 147–168.

Garstecki, T., S. A. Wickham & H. Arndt, 2002. Effects of experimental sediment resuspension on a coastal planktonic microbial food web. Estuarine, Coastal and Shelf Science 55: 751–762.

Hillebrand, H., C. D. Dürselen, D. Kirschtel, T. Zohary & U. Pollingher, 1999. Biovolume calculation for pelagic and benthic microalgae. Journal of Phycology 35: 403–424.

Hu, H., Y. Li, Y. Wei, H. Zhu, J. Chen & Z. Shi, 1980. Freshwater Algae in China. Shanghai Science and Technology Press, Shanghai (in Chinese).

Jin, X. & Q. Tu, 1990. The Standard Methods for Observation and Analysis of Lake Eutrophication, 2nd edn. China Environmental Science Press, Beijing (in Chinese).

Jürgens, K. & G. Stolpe, 1995. Seasonal dynamics of crustacean zooplankton, heterotrophic nanoflagellates and bacteria in a shallow eutrophic lake. Freshwater Biology 33: 27–38.

Jürgens, K. & H. Güde, 1994. The potential importance of grazing-resistant bacteria in planktonic systems. Marine Ecology Progress Series 112: 169–188.

Jürgens, K., 1994. The impact of Daphnia on microbial food webs-A review. Marine Microbial Food Webs 8: 295–324.

Luo, L., B. Qin, W. Hu & F. Zhang, 2004. Sediment re-suspension under different hydrodynamic distribution in Lake Taihu. Journal of Lake Science 16: 273–276.

Norland, S., 1993. The relationship between biomass and volume of bacteria. In Kemp, P., B. Sherr, E. Sherr & J. J. Cole (eds), Handbook of Methods in Aquatic Microbial Ecology. Lewis, 303–308.

Porter, K. G. & Y. S. Feig, 1980. The use of DAPI for identifying and counting aquatic microflora. Limnology and Oceanography 25: 943–948.

Psenner, R., 1993. Determination of size and morphology of aquatic bacteria by automated image analysis. In Kemp, P., B. Sherr, E. Sherr & J. J. Cole (eds), Handbook of Methods in Aquatic Microbial Ecology. Lewis, pp. 339–346.

Qin , B., W. Chen & W. Hu, 2004. Succession of Ecological Environment and Its Mechanism in Lake Taihu. China Science Press, Beijing (in Chinese).

Qin, B., P. Z. Xu, Q. L. Wu, L. C. Luo & Y. L. Zhang, 2007. Environmental issues of Lake Taihu, China. Hydrobiologia 581: 3–14.

Ritzrau, W. & G. Graf, 1992. Increase of microbial biomass in the benthic turbidity zone of Kiel Bight after resuspension by a storm event. Limnology and Oceanography 37: 1081–1086.

Robarts, R. D. & R. J. Wickham, 1990. Heterotrophic bacterial production and its dependence on autotrophic production in a hypertrophic African reservoir. Canadian Journal of Fisheries and Aquatic Science 47: 1027–1037.

Rogerson, A. & J. Laybourn-Parry, 1992. The abundance of marine naked amoebae in the water column of the Clyde Estuary. Estuarine, Coastal and Shelf Science 33: 187–196.

Shimeta, J. & J. D. Sisson, 1999. Taxon-specific tidal resuspension of protists into the subtidal benthic boundary layer of a coastal embayment. Marine Ecology Progress Series 177: 51–62.

Simon, M., H. P. Grossart, B. Schweitzer & H. Ploug, 2002. Microbial ecology of organic aggregates in aquatic ecosystems. Aquatic Microbial Ecology 28: 175–211.

Sommaruga, R., 1995. Microbial and classical food webs: A visit to hypertrophic lake. FEMS Microbiology Ecology 17: 257–270.

Turley, C. M., R. C. Newell & D. B. Robins, 1986. Survival strategise of two small marine ciliates and their role on regulating bacterial community structure under experimental conditions. Marine Ecology Progress Series 33: 59-70.

Velji, M. I. & L. J. Albright, 1993. Improved sample preparation for enumeration of aggregated aquatic substrate bacteria. In Kemp, P., B. Sherr, E. Sherr & J. J. Cole (eds), Handbook of Methods in Aquatic Microbial Ecology. Lewis, 139–142.

Wainright, S. C., 1987. Stimulation of heterotrophic microplankton production by resuspended marine sediments. Science 238: 1710–1711.

Wainright, S. C., 1990. Sediment-to-water fluxes of particulate material and microbes by resuspension and their contribution to the planktonic food web. Marine Ecology Progress Series 52: 271–281.

Weisse, T. & J. G. Stockner, 1993. Eutrophication: the role of microbial food webs. Memorie dell'Istituto Italiano di Idrobiologia 52: 133–150.

Weisse, T., 1991. The microbial food web and its sensitivity to eutrophication and contaminant enrichment: a cross-system overview. International Review of Hydrobiology 76: 327–337.

Weisse, T., 2003. Pelagic microbes-Protozoa and the microbial food web. In O'Sulliva, P. E. & C. S. Reynolds (eds), The Lakes Handbook. Blackwell Science Ltd, 417–460.

Worm, J. & M. Søndergaad, 1998. Dynamics of heterotrophic bacteria attached to *Microcystis* spp. (Cyanobacteria). Aquatic Microbial Ecology 14: 19–28.
Worm, J., K. Gustavson, K. Garde, N. H. Borch & M. Søndergaard, 2001. Functional similarity of attached and free-living bacteria during freshwater phytoplankton blooms. Aquatic Microbial Ecology 25: 103–111.
Xie, P. & J. Liu, 2001. Practical success of biomanipulation using filter-feeding fish to control cyanobacteria blooms: a synthesis of decades of research and application in a subtropical hypereutrophic lake. The Scientific World 1: 337–356.
Zhang, Z. S. & Q. F. Huang, 1991. Methods in Freshwater Plankton Study. China Science Press, Beijing (in Chinese).

Hydrobiologia (2007) 581:255–267
DOI 10.1007/s10750-006-0501-9

Effects of limiting nutrients and N:P ratios on the phytoplankton growth in a shallow hypertrophic reservoir

Ho-Sub Kim · Soon-Jin Hwang · Jae-Ki Shin · Kwang-Guk An · Chun Gyeong Yoon

Abstract The purpose of this study was to evaluate the effects of limiting nutrients and the N:P ratios on the growth of phytoplankton (mainly cyanobacteria) in a shallow hypertrophic reservoir between November 2002 and December 2003. Nutrient enrichment bioassays (NEBs) were conducted, along with analyses of seasonal ambient nutrients and phytoplankton taxa, in the reservoir. The average DIN:TDP and TN:TP mass ratios in the ambient water were 90 (range: 17–187) and 34 (13–60), respectively, during the study period. The dissolved inorganic phosphorus showed seasonal variation, but less than that of inorganic nitrogen. The TN:TP ratios ranged from 13 to 46 (mean: 27 ± 6) during June–December when the cyanobacteria, *Microcystis*, dominated the phytoplankton composition. The NEBs showed that phytoplankton growth was mainly stimulated by the phosphorus (all of total 17 cases), rather than the nitrogen concentration (8 of 17 cases). The rapid growth rate of cyanobacteria was evident with TN:TP ratios less than 30. According to the results of the NEBs with different N concentrations (0.07, 0.7 and 3.5 mg l^{-1}), but the same N:P ratios and when the nitrogen concentration was highest, the cyanobacterial growth reached a maximum at N:P ratios <1. Overall, the response of cyanobacterial growth was a direct function of added phosphorus in the NEBs, and was greater with increased N concentrations. Thus, cyanobacterial blooms favored relatively low N:P ratios in this hypertrophic reservoir system.

Keywords N:P ratio · P-limitation · Phytoplankton growth · Cyanobacteria · Hypertrophic reservoir

Guest editors: B. Qin, Z. Liu & K. Havens
Eutrophication of shallow lakes with special reference to Lake Taihu, China

H.-S. Kim · S.-J. Hwang (✉) · C. G. Yoon
Han River Environmental Research Laboratory,
National Institute of Environmental Research,
Gyeonggi 476-823, Korea
e-mail: sjhwang@konkuk.ac.kr

J.-K. Shin
Korea Institute of Water and Environment,
Korea Water Resources Corporation,
Daejon 305-730, Korea

K.-G. An
School of Bioscience and Biotechnology,
Chungnam National University, Daejon 305-764,
Korea

Present Address:
H.-S. Kim
Watershed Management Research Division,
National Institute of Environmental Research, Inchon
404-170, Korea

Introduction

The nitrogen-to-phosphorus ratio (N:P) in lentic systems has been frequently used as a key indicator in predicting algal biomass and compositions, and its seasonal succession (Tilman, 1982; Kilham, 1990). It can also serve as an index that represents the nutrient limitation for the algal growth (Smith, 1983; Fugimoto & Sudo, 1997). The atomic ratio, 16N:1P, which is known as the Redfield ratio, has been used for generally describing the average elemental composition of phytoplankton. Many investigations; however, have shown this ratio to vary considerably, depending on the algal species and surrounding environmental conditions (Rhee, 1978; Forsberg & Ryding, 1980; Geider & La Roche, 2002; Ho et al., 2003), and that optimum nutrient ratios vary interspecifically within the diatoms due to growth competitions (Tilman, 1977, 1978).

A considerable variation in the nutrient stoichiometry of the natural phytoplankton community is a reflection of the type and extent of nutrient limitation and availability. In turn, this is likely related to the supply ratios of the elements loaded to the lake affecting the development of particular phytoplankton communities (Smith, 1983; Smith et al., 1987; Sommer, 1989; Fujimoto & Sudo, 1997; Xie et al., 2003). The hypothesis of "the dominance of cyanobacteria at low N:P ratios" is one of these examples. Smith (1983) pointed out that bloom-forming cyanobacteria had a tendency to dominate the lake when the N:P ratio was less than 29. The significance of the N:P ratio as a critical standard, however, is still controversial, due to variabilities in the other chemical characteristics and phytoplankton compositions within a geographic region.

Unlike Smith's 'N:P ratio rule', which stated that the relative proportion of cyanobacteria to the total algal biomass might decrease when the N:P ratio exceeded 29 (Smith et al., 1987), the bloom of cyanobacteria was affected by increases in the phosphorus concentration rather than simply by a decrease in the N:P ratio (Trimbee & Prepas, 1987; Sheffer et al., 1997; Xie et al., 2003). Moreover, the possibility that the growth response of algae to the limiting nutrient would be inconsistent with the N:P ratio rule in eutrophic systems that had greater amounts of nitrogen and phosphorus than are required for algal growths (Paerl et al., 2001). These results suggest that the nutrient stoichiometry that algae require for growth would vary with the extent of nutrient deficiency and availability.

In this study, we evaluated the response of phytoplankton growth in relation to different ambient concentrations of N and P, and N:P ratios, in a hypertrophic reservoir where the nitrogen background concentrations were high. For these experiments, we conducted nutrient enrichment bioassays (NEBs) and compared the growth rates of phytoplankton among several treatments of the NEBs, along with seasonal analyses of the ambient nutrients and phytoplankton compositions.

Materials and methods

The study system

Singu Reservoir is a small agricultural reservoir, with surface area of 0.1 km^2 and maximum depth of 7.0 m, which has two inflowing streams with channel length of 1.3 and 3.4 km, respectively. The total watershed area is 0.255 km^2, with the water quality mainly influenced by the point source of livestock wastes (KARICO, 2001). The watershed of the reservoir is located in Monsoon climate region, and precipitation is concentrated during June–September. The use for irrigation affects the water level of the reservoir (usually between May and June). Thus, both irrigation and Monsoon rainfall affect both water quality and water level of the reservoir.

Limnological variables

Various limnological variables including temperature, dissolved oxygen (DO) concentration, pH, and electric conductivity (EC) were measured by Hydrolab data logger (RE-232/SDI-12) in situ. Temperature and DO were measured vertically at 20–50 cm intervals. Transparency was measured with a white disk with 20 cm diameter. Water level was measured as the maximum depth at each time. Rainfall data were obtained from the nearest meteorological observation center (Boryung Meteorological Observation Center, Chungnam Province, S. Korea).

Chemical analyses

The surface water was collected biweekly from the reservoir between November 2002 and December 2003. Stream water samples were also collected from an inflowing location prior to entering the reservoir. The discharge volume from the reservoir was estimated through the linear multiplication of the water velocity and the channel area.

Five liter water samples were collected for analyses of nutrients and phytoplankton taxa, using polyethylene bottles, which were kept in a dark ice-box during transportation. On arrival at the laboratory (within 5 h), the water samples were filtered through GF/F filters, and the filtrate used to measure the concentration of chlorophyll *a* (Chl *a*). Phosphorus fractions, such as dissolved inorganic phosphorus (DIP), total dissolved phosphorus (TDP) and total phosphorus (TP), were analyzed in triplicate using the ascorbic acid method from the standard method of the APHA (1995). The NH_4^+–N concentrations were measured using an indolphenol reagent, and those of nitrate–nitrogen (NO_3^-–N) and nitrite–nitrogen (NO_2^-–N) using the cadmium reduction method (APHA, 1995). The total nitrogen (TN) was measured using the cadmium reduction method, following persulfate digestion. Dissolve inorganic nitrogen (DIN) was obtained from the sum of NO_3^-–N, NO_2^-–N and NH_4^+–N. The Chl *a* concentration was analyzed using the methanol extraction method of Maker et al. (1980).

Phytoplankton analysis

Water samples for the identification of phytoplankton species and the measurement of cell densities were collected from a 0.5–1 m depth from the reservoir, stored in Whirl-Pak bags and fixed with Lugol's solution (2%, final conc.). The cell density was measured using the Sedgwick-Rafter counting chamber under microscopic magnification of 200–400×. The phytoplankton composition was identified as containing three major and two minor taxa; Bacillariophyceae, Cyanophyceae and Chlorophyceae, and Dinophyceae and Cryptophyceae, respectively.

Nutrient enrichment bioassays (NEBs)

In addition to the regular sampling, nutrient enrichment bioassays (NEBs) were conducted for the evaluations of the limiting nutrients that regulate phytoplankton growth. The experiments for the evaluation of NEBs were conducted in a shaking incubator at a temperature of 20–25°C and light intensity of 100 ± 5 μE m^{-2} s^{-1}. For the experiments, the water was filtered using 100 μm mesh to minimize the effects associated with zooplankton grazing. Aliquots of filtrate (50 ml) were placed in 250 ml flasks, followed by the addition of KNO_3 and KH_2PO_4 as nutrient sources, to final concentrations of 1 mg l^{-1} for both nutrients. The final volume was made up to 100 ml with the addition of GF/F filtrate. A set of algal bioassay was conducted in triplicate sets for the control (no addition), +N (addition of nitrogen) and N + P (addition of nitrogen and phosphorus), and the responses measured as the difference between the final and initial Chl *a* at 0, 1, 3, 5 and 7 days after incubation. The growth rate (μ: day^{-1}) under each set of treatment conditions was calculated using the equation of the APHA (1995), as follows:

$$\mu\ (\text{day}^{-1}) = \ln\ (X_2/X_1)/(T_2 - T_1)$$

where, X_1 = the concentration of Chl *a* at the initial incubation stage (T_1 h);
X_2 = the concentration T_2 h

Other triplicate experiments to measure the phytoplankton growth rate were conducted with five different P concentrations (+0.05, 0.1, 0.5, 1.0 and 1.5 mg l^{-1}, as KH_2PO_4 spiking), with nitrogen fixed at 1 mg l^{-1}. The medium into which different P concentrations were added was the same GF/F filtered lake water as above. The nitrogen concentrations of the medium used in the NEBs were similar between the treatments categorized with different N:P ratios. The spiking concentration of phosphorus, however, varied within the range 0.05–1.5 mg l^{-1}.

The DIN:TDP mass ratio was calculated using the concentration measurements from the subsamples taken from each treatment prior to the starting of the NEBs. The maximum growth rate (μ_{max}) and half saturation constant (K_u) were calculated as values to the phosphorus concentrations, with the corresponding growth rate (μ) obtained from the model of Monod (1950) employing the Sigma plot program (Version 7.0, SPSS Inc.).

Analysis of cyanobacterial growth under different N:P ratios

Using the same N:P ratio, the cyanobacterial growth rates were determined for different N and P concentrations. The incubation conditions were maintained at 20°C, with a light intensity of 53 ± 2 μE m^{-2} s^{-1} (L:D cycle = 12:12). The light intensities used were values observed during the study period at the water depths from which the samples were taken. The cyanobacteria were collected by vertically pulling up a 30 μm pore size plankton net, with the collected sample then filtered through a 100 μm net to remove the large zooplankton. In order to isolate the cyanobacteria from composite phytoplankton, samples were placed in a large-mouth container for 24 h, and the cyanobacteria that rose to the surface due to their buoyancy were transferred by pipette to a GF/F pre-filtered lake water.

Cyanobacterial cells were starved in MA2 medium with phosphorus and nitrogen excluded. After 2-week incubation, a 2 ml aliquot of the cyanobacteria sample was placed in an acid-washed 250 ml flask, along with MA2 medium that had been modified with different phosphorus and nitrogen concentrations. The N:P mass ratio was controlled by changing the amount of phosphorus with respect to fixed portions of nitrogen (0.07, 0.7 and 3.5 mg l^{-1}). Triplicate treatment sets were incubated for 5 days, and the cell growth determined as the difference between initial Chl *a*, prior to the incubation, and that 5 days later.

Statistical analysis

Statistical analyses were conducted between the variables using Pearson's correlation (SPSS 10.0). The differences in the growth response between the various NEB treatments were analyzed by an ANOVA, and the level of significance identified at $P < 0.05$.

Results

Variation of limnological variables

Rainfall during the study period was concentrated between June and September (41.5 mm in June, 391 mm in July, 79.5 mm in August, and 393 mm in September) (Fig. 1). The use for irrigation between the end of May and mid June decreased the reservoir water level from 7.1 to 5 m. The water level returned to 7 m due to seasonal monsoon rainfall during June–July. Transparency of the reservoir varied seasonally with the peak during May. The time of low transparency was

Fig. 1 Seasonal changes of precipitation, Secchi transparency, pH and electric conductivity in Shingu Reservoir during November 2002–February 2004

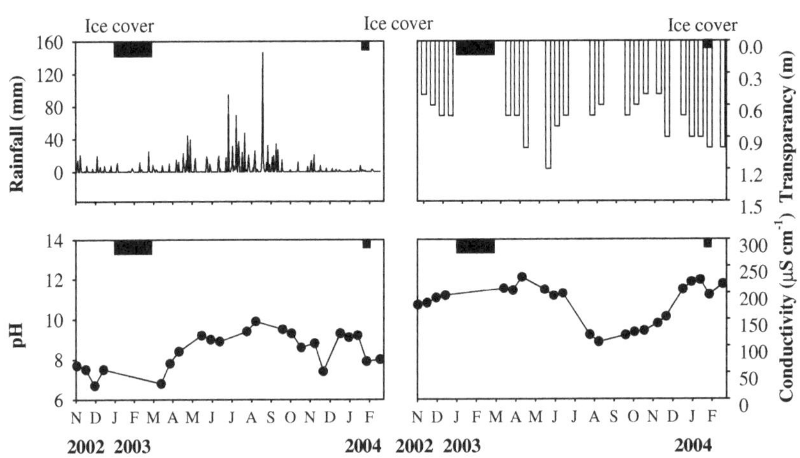

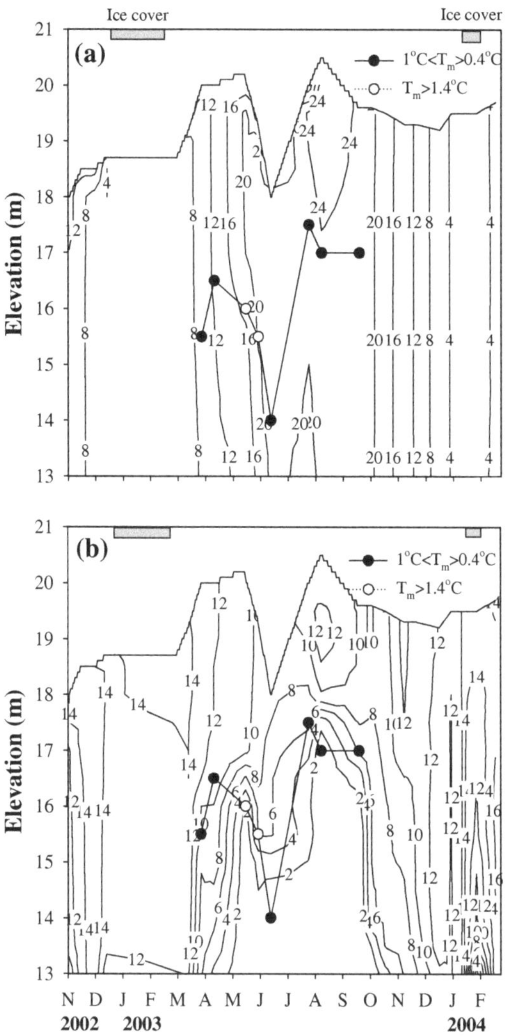

Fig. 2 Temporal and spatial changes of temperature (**a**) and dissolved oxygen concentration (**b**) in Shingu Reservoir during November 2002–February 2004. The line with *black circles* indicates the vertical temperature difference between 0.4 to 1.0°C, and the *open-circled line* indicates the vertical temperature difference greater than 1.0°C

coincided with the period of high pH, but EC variation was almost mirrored with pH. The reservoir had a weak thermal stratification near the hypolimnion, which persisted annually from mid-June to October (Fig. 2). Hypoxia <2 mg O_2 l^{-1} in the hypolimnion was frequently observed from the end of May, and was most intense during mid-August. Hypolimnetic hypoxia was maintained until late October, the beginning of fall.

Variation of nutrients and phytoplankton

During the study period, the TN concentrations in the reservoir were >1.2 mg l^{-1} (1.6–3.2 mg l^{-1}) (Fig. 3). The TN was mainly composed of NO_3^-–N, rather than NH_4-N, with NO_2-N + NO_3-N and NH_4-N ranging from 0.4 to1.9 and 0.01 to 0.92 mg N l^{-1}, respectively (Fig. 3). Dissolved inorganic N peaked in early spring (March–April), then abruptly decreased during May–August. Thus, the variation in the dissolved inorganic N was greater than that of the TN. DIP concentration ranged from 0.6 to 8.8 μg P l^{-1} and varied not much during the study period. DTP ranged from 10.1 to 25.6 μg P l^{-1}, with the peak concentration during August and minimum during December. The average TP was 76.4 μg P l^{-1} and showed large seasonal fluctuations (48.0–126.6 μg P l^{-1}). The TP increased from June, reaching a maximum in October. This reservoir was adjudged to be hypertrophic, based on the conventional criteria of nitrogen and phosphorus (Forsberg & Ryding, 1980). The TN:TP mass ratios in this hypertrophic reservoir were <60, whereas those of DIN:TDP increased to >150 during March–June.

The concentration of Chl *a* varied from 28.8 to 109.7 μg l^{-1} during the study period, and showed a similar seasonal pattern to that of the algal cell density (Fig. 4). The Chl *a* was at a minimum during early spring of 2003, but began to increase from May when cyanobacteria dominated the phytoplankton community. Thus, Chl *a* maxima >90 μg l^{-1} were observed between July and November, which coincided with the peaks in the cell density (Fig. 4). The Chl *a*:TP ratios, the Chl *a* yield for a given phosphorus concentration, were >1.0, indicative of a massive algal bloom. Pearson's correlation analyses showed the Chl *a* concentrations to be positively correlated ($r = 0.66$, $P = 0.002$, $n = 19$) with the TP, but negatively ($r = -0.48$, $P = 0.042$, $n = 18$) with the TN. This result coincided with the inverse relations between Chl *a* and the TN:TP ratio ($r = -0.70$, $P < 0.001$, $n = 18$).

The cell densities (cell number ml^{-1}) in the reservoir showed high seasonal fluctuations. The community structure was characterized by the dominance of cyanobacteria during most of

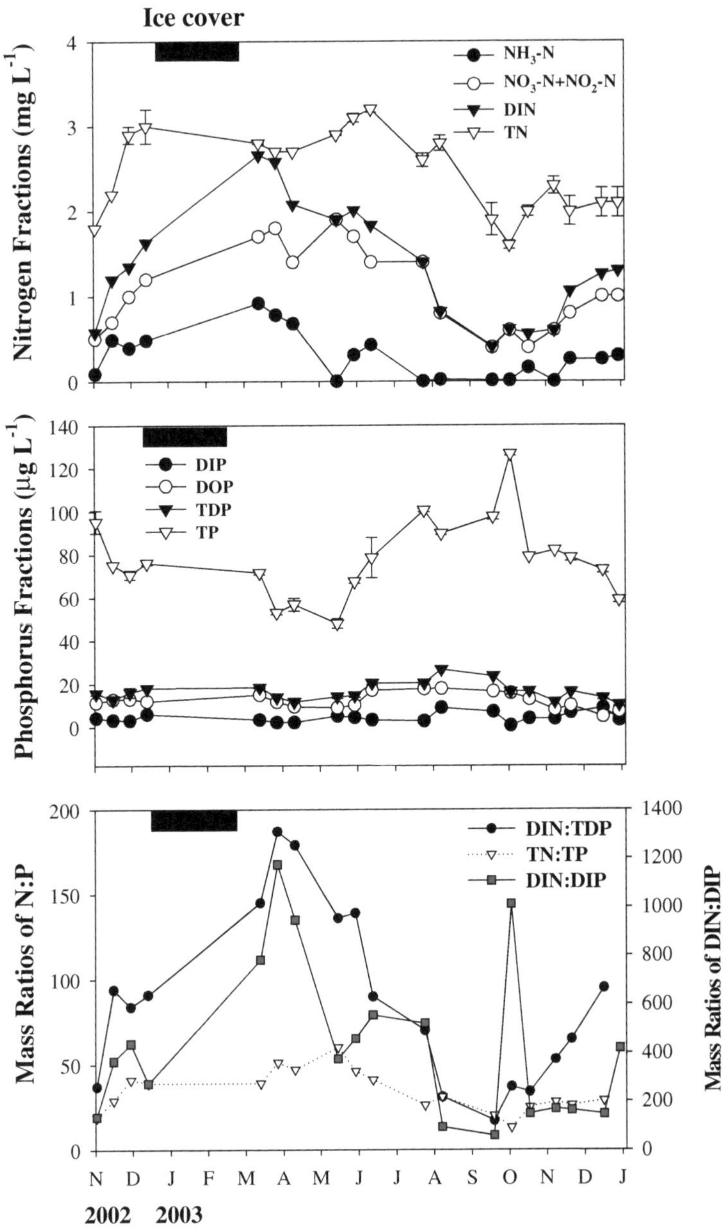

Fig. 3 Seasonal changes in the nitrogen and phosphorus fractions and N:P ratios in Shingu Reservoir during November 2002–December 2003

the year, with the exception of a short period during spring, which was dominated by a few other phytoplankton, such as a diatom (*Melosira varians*) and a green alga (*Dictyosphaerium puchellum*), and cryptomonads (*Rohdomonas* spp.) (Fig. 5). The dominant taxa of cyanobacteria were *Oscillatoria* spp., *Microcystis* spp. and *Aphanizomenon* sp., which dominated from May until the winter ice season. Seasonal succession of the phytoplankton community showed that *Oscillatoria* spp. dominated at the beginning of the study period, which was then followed by *Aphanizomenon* sp. in late spring when the water level of the reservoir abruptly decreased. During the study, the major species causing the algal bloom was identified as *Microcystis* spp. (Fig. 5), and the high Chl *a* >60 μg l^{-1}, continued during July–December 2003 (Fig. 4).

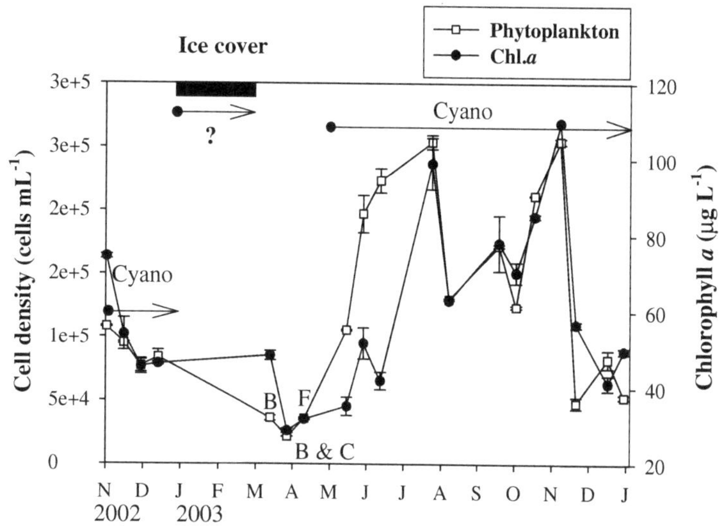

Fig. 4 Seasonal changes in the phytoplankton cell density and chlorophyll-*a* in Shingu Reservoir during November 2002–January 2004. Cyano, B, C and F denote cyanophytes, bacillariophytes, chlorophytes and flagellated algae, respectively

Phytoplankton growth under limiting nutrients and various N:P ratios

Overall, the results of the NEBs experiments showed phosphorus as the key nutrient regulating the phytoplankton growth in the system (Fig. 6, lower panel), but nitrogen was also considered an important nutrient during a short period (Fig. 6, upper panel). Phosphorus enrichment caused a significantly ($P < 0.001$, $n = 17$, ANOVA) greater growth rate than that seen in the control and nitrogen treatments (Fig. 6), indicating P-limitation of the algal growth in the system. No difference was found in the NEBs between the P and N + P additions ($P = 0.229$, $n = 17$, ANOVA). Six of the 17 NEBs experiments (20 July, 8 August, 21 September, 2 October, 16 October, and 1 November) showed significantly ($P < 0.05$, $n = 6$, ANOVA) higher growth rates with the N + P treatments than the P, N treatments and control. This outcome suggests that simultaneous limitation of nitrogen and phosphorus occurred between July and early November 2003. In the mean time, only 3 cases (November 2002, March and April 2003) of the NEBs had greater growth rates in the +P than the N + P treatments (Fig. 6).

The phytoplankton growth was largely determined by the magnitude of P, although the ambient N and P concentrations were different for the same N:P ratio (Fig. 7). The growth rates over a wide range of DIN:TDP ratios decreased with increasing ratio, but the most noticeable decrease was observed at ratios above 30 (Fig. 7). The growth rate consistently increased until the P concentration reached 50 $\mu g\ l^{-1}$ (Fig. 8). The results plotted using the equation of Monod (1950) indicated a maximum growth rate of 0.54 day^{-1} and with a half saturation concentration of 33.1 $\mu g\ l^{-1}$ for phosphorus. The maximum growth rate occurred in those treatments enriched with phosphorus alone or with both phosphorus and nitrogen. The treatments enriched with only nitrogen (+N) had lower growth rate, with the exception of one case during December 2002.

Cyanobacterial growth variation under different N:P ratios

The magnitude of the cyanobacterial growth responses showed differences with respect to the N:P ratios and N concentrations for the same N:P level. The N:P ratios under 20 displayed large differences in the growth rate with different nitrogen concentration (Fig. 9a), with the growth level tending to increase with increasing nitrogen concentration (Fig. 9b). The difference in the growth rates between 0.7 and 3.5 mg N l^{-1} were smaller ($P = 0.86$, ANOVA) than that of the lowest concentration, 0.07 mg N l^{-1} ($P = 0.001$, ANOVA).

At a constant N:P ratio, but with variable concentrations of N, the cyanobacteria showed

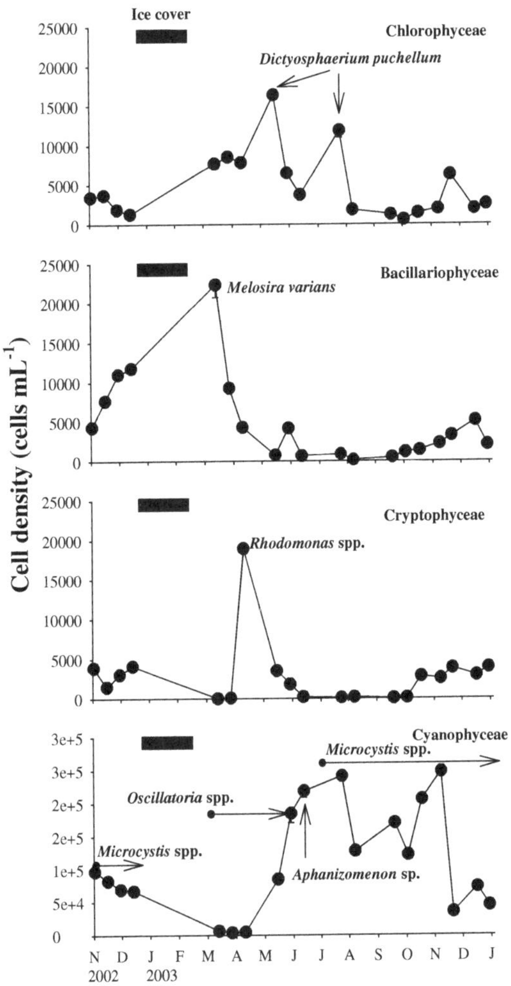

Fig. 5 The seasonal succession of dominant phytoplankton taxa in Shingu Reservoir during November 2002–December 2003

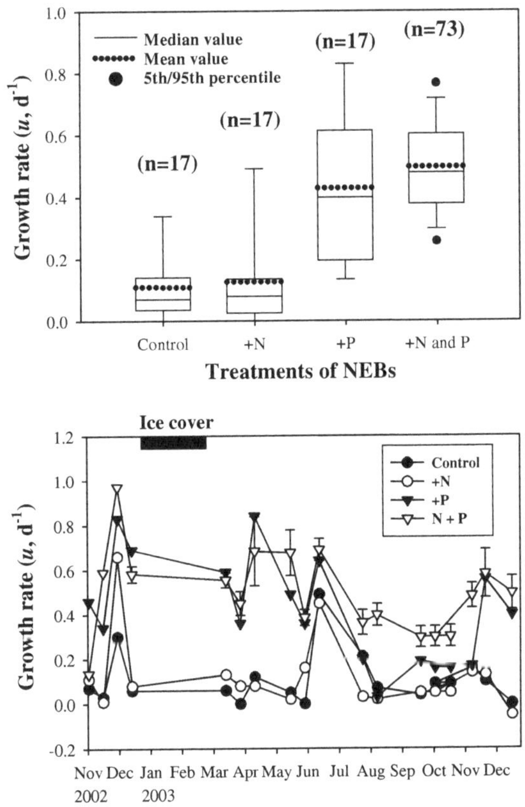

Fig. 6 The growth rates of phytoplankton toward various treatments (control, +N, +P, and N + P) in the NEBs, during November 2002–December 2003 (upper panel), and the seasonal growth response to nutrient additions in the NEBs (lower panel)

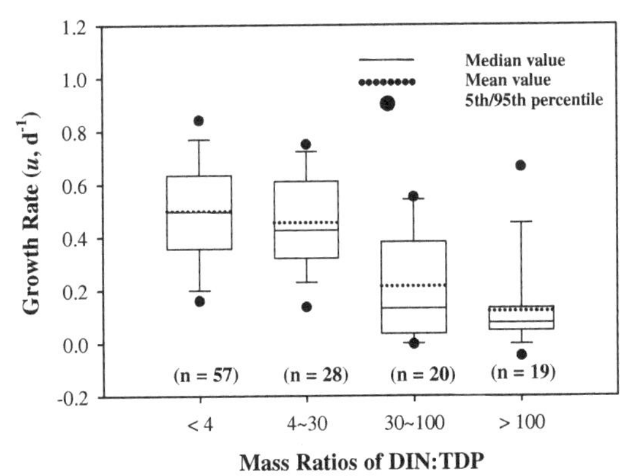

Fig. 7 The growth rates of phytoplankton at various DIN:TDP ratios

different growth patterns in the NEBs (Fig. 9c). The growth response reached a maximum at an N concentration of 3.5 mg l^{-1} when the N:P ratio in the NEBs was 1.0. The growth tended to decrease as the N:P ratio increased ($P < 0.01$, ANOVA) (Fig. 9c). A nitrogen concentration at 0.7 mg N l^{-1} caused the highest algal growth when the N:P ratio was 0.7 ($P < 0.05$, ANOVA). The growth rates were similar within the N:P ratio range 1–10 ($P > 0.5$, ANOVA), while the level of growth tended to decrease for N:P ratios above 10 ($P = 0.009$, ANOVA) (Fig. 9a). Conversely, the lowest nitrogen concentration, 0.07 mg N l^{-1}, showed a much lower level of growth, especially when the N:P ratio was smaller than 1. Ratios >7 tended to give greater phytoplankton growth ($P < 0.004$, ANOVA) (Fig. 9c), compared to ratios <1.

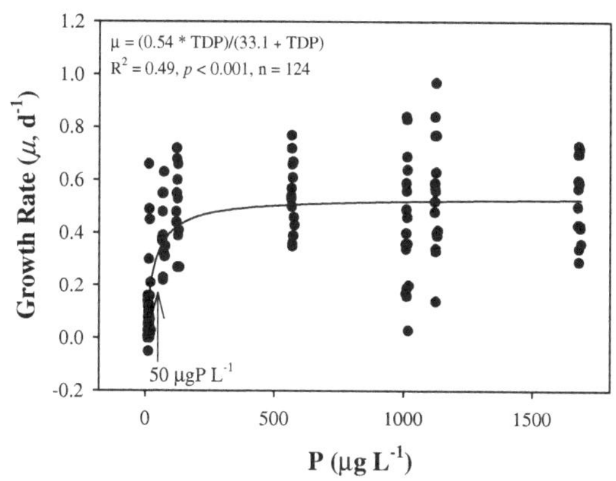

Fig. 8 Phytoplankton growth rate at various P (TDP = ambient + addition) concentrations using natural phytoplankton assemblages during November 2002–December 2003

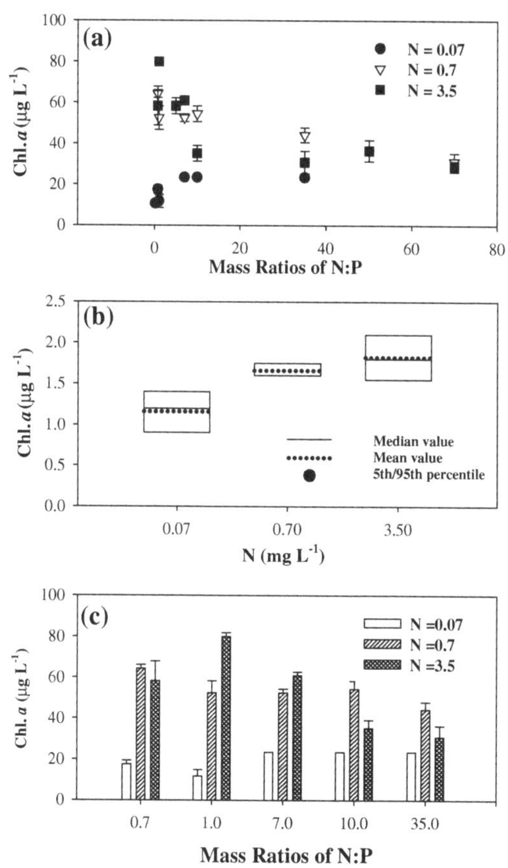

Fig. 9 Response of chlorophyll *a* to the N:P mass ratios (upper panel), and cyanobacterial responses to nitrogen (mid panel) and the N:P ratios (range: 0.7–35.0) at various nitrogen concentrations (lower panel)

Discussion

As important factors for determining the growth of phytoplankton, the ambient nutrient concentrations and nutrient ratios, known as stoichiometry (e.g., Sterner & Elser, 2002), are being widely used to estimate the physico-ecological consequences, such as the algal growth and succession of specific phytoplankton compositions (Goldman et al., 1979; Tilman, 1982; Sommer, 1989). An explanation for the mass conservation in aquatic organisms is that the external nutrient supply determines the cellular element composition of algae, which was supported by the earlier seminal work on stoichiometry (Redfield et al., 1963; Rhee, 1973) in an oligotrophic ocean and culture medium. The rules determining the constraints on stoichiometry; however, have been understood to vary depending on the organism, feeding mode and metabolic process (Geider & La Roche, 2002; Sterner & Elser, 2002; Ho et al., 2003).

A degree of nutrient deficiency and differential nutrients loading to a lake are also constraints on the stoichiometric variation and; thus, cause not only an imbalanced algal growth, but also lead the development of certain algal populations that have adapted to that particular situation (Smith et al., 1987). This implication calls attention to the need for eutrophic systems to be seen as having more variable and complicated limiting nutrients and critical nutrient ratio dynamics (Terry et al., 1985; Leonardos & Geider, 2004). Baring this in mind, our main concern in this study was to elucidate how algal growth (usually cyanobacteria) in a hypertrophic reservoir responded to both variations in ambient nutrient regimes and the N:P ratio in the NEBs in relation to the ambient N:P ratio.

Our major finding was that the phytoplankton (mostly cyanobacteria) in our hypertrophic study reservoir were P-limited and the degree of their growth in P addition increased with higher N concentrations added, indicating that at high nitrogen level, low N:P ratios were favorable for cyanobacterial development. These results suggest that phytoplankton in an N-rich environment appeared to demand a large amount of P to enable the use of a large supply of both N and P simultaneously to meet the great biomass building. These results also suggest that higher nitrogen concentration is likely to induce stronger P-limitation on the phytoplankton growth, and are supported by the inverse relationship between the TN:TP ratio and Chl *a* concentration ($r = -0.7$, $P < 0.001$: Figs. 3, 4). An evident increase in the growth rate was observed with TN:TP ratios <30, which was sustained with increases in TDP up to 50 $\mu g\ l^{-1}$, but reached a plateau thereafter (Figs. 7, 8).

With respect to the seasonal succession of phytoplankton, the variation in the N:P ratio appeared to be related with the cyanobacterial predominance in the reservoir. Due to the characteristic blooms in lentic systems with human-induced eutrophication, there have been numerous studies regarding the development of cyanobacteria and the interspecific changes of phytoplankton within eutrophic systems (Smith, 1983; Smith et al., 1987; Xie et al., 2003). The TN:TP ratios during the cyanobacterial blooms were 27 ± 6, almost coinciding with the criteria of Smith (1983), who suggested a high possibility of cyanobacterial dominance at TN:TP ratios under 29. The phytoplankton growth rate to specific ambient nutrient concentrations and N:P ratios, via the nutrient enrichment experiments, suggests that the level of the nutrient limiting ratios was frequently found near 30 (Fig. 7), even if the criteria are still controversial. P limitation was observed with ratios exceeding this level, but no N-limitation was detected at ratios below 30. The possibility of N-limitation was only detectable under low N concentration conditions, even in the experiment subjected to the cyanobacteria (Fig. 9).

The causes of these results were suspected to be due to the special characteristics of our reservoir, which maintained much higher nitrogen than phosphorus levels, as well as DIN:TN ratios (53%) much higher than those for TDP:TP (23%). The N:P ratio standard for evaluating the possibility of potential P and N limitation would vary in cases where the absolute concentrations were different, even though the N:P ratio would be maintained at the same level. Also, the eutrophic system where the concentration of N is much higher than that of P might create the possibility of N limitation being relatively lower than that of phosphorus, even with a decreasing N:P ratio. This result suggests that any standard for evaluating the limiting nutrient should consider the factors, not only the characteristics of phytoplankton species (Tilman 1977, 1978; Rhee, 1978), but also the absolute concentration of the nutrients existing within the aquatic system. A relative difference in the concentrations of the nutrients is necessary to endorse the growth of phytoplankton, rather than simply by their determination using the relative ratios of the nutrients in various aquatic systems with various nutrient stati (Leonardos & Geider, 2004).

Even though a number of studies have suggested the possibility of nitrogen-fixing cyanobacteria as the dominant species during periods of decreased N:P ratio (Horne, 1979; Tilman, 1982; Howarth et al., 1988; Paerl et al., 2001), our result indicated there was a dominant species shift, from *Aphanozomenon* to *Microsystis*, during periods when the N concentration decreased, and after completing the stratification, it has been suggested that the nitrogen concentration may not be the cause of the cyanobacterial species succession. The level of inorganic N tends to decrease at a higher rate than that of inorganic P during the algal growth period in both oligotrophic and eutrophic lakes and; thus, the growth limiting factor could change from P to N during the stratification period (Hendrey & Welch, 1974). Under this condition, dominant cyanobacteria species could succeed to taxa armed with the ability to fix nitrogen (Horne, 1979; Tilman, 1982; Howarth et al., 1988; Paerl et al., 2001). However, the nitrogen fixing *Apanizomenon* which was observed before establishing the stratification had shifted to non-N fixing *Microcystis* during the period of N:P ratio decrease, which was caused by the drop in the N concentration and the rise in the P concentration after stratification had been established. Therefore, phosphorus seemed to be one of the most substantial factors influencing the dominance of *Microsystis* (Xie et al., 2003). Our result is also supported by Fujimoto & Sudo (1997), who showed that *Microcystis aeruginosa* was a superior competitor under low N:P supply ratio conditions than the filamentous cyanobacterium, *Phormidium tenue*.

A low phytoplankton growth rate was demonstrated, even in the experiments with P or N + P additions between July and October (Fig. 6), which suggests the physiological property of phytoplankton needs to be considered when predicting their growth response, rather than simply relying on the nutrient concentration and N:P ratio (Droop, 1968; Rhee & Gotham, 1981). Due to the luxury of P consumption, cyanobacteria could store enough P in the cell, so their growth would not necessarily rely on the ambient phosphate concentration (Goldman et al., 1987). The DIN:DTP ratio in our reservoir between July and October was in the range 17–37, with the exception of July (DIN:DTP = 70), and this range seemed to be comparable to the optimum DIN:DTP ratio (<30) (Fig. 7) for phytoplankton growth and; thus, both P and N could function as potential limiting nutrients. On the other hand, there could have been an excess amount of external phosphorus supplied to the phytoplankton with the inflowing discharge around this period (Table 1). Moreover, after the end of May, the phosphorus released from the sediment under hypo-oxic conditions (Kim, 2004) could have been supplied to the surface by wind disturbances. Therefore, these phosphorus additions could have resulted in the low growth rate. During August and September especially, when DIP concentration in the water column was at the maximum seasonal concentration, there had been a lot of standard precipitation prior to sampling. The precipitation from June to September accounted for 66% of the total annual precipitation (1,359 mm) and; thus, this concentrated precipitation may have been responsible for increasing the discharge and P loading from the watershed. The increase in the TP concentration with both decreasing N and Chl *a* concentrations during this period may also be reflected by the increase of inorganic particulate matter.

Considering the mass conservation between an algal cell and the external nutrient supply, there should be no problem in using any type of nutrient for evaluating nutrient limitation (e.g., Sterner & Elser, 2002); for example, cellular or external and inorganic or organic forms. However, the factors causing variations in the cellular stoichiometry, such as the physiological status and the property of nutrient storage in phytoplankton species, need to be carefully considered to understand the ecological stoichiometry. The TN:TP ratios have been most frequently used as a measure, or criterion, to evaluate cyanobacterial blooms and limiting nutrients (Forsberg & Ryding, 1980; Smith, 1983; Trimbee & Prepas, 1987; Fujimoto & Sudo, 1997; Sheffer et al., 1997; Xie et al., 2003). Also, the DIN:DIP (Rhee, 1978; Patricia et al., 2000), DIN:OP or DIN:TP (Perkins & Underwood, 2000) ratios have been employed as surrogates of the TN:TP ratio for the assessment of lake conditions. In this study, the TN:TP ratios were closely related to the cyanobacterial density and nutrient ratios, which agreed with the growth response, as based on the soluble fractions of the DIN:TDP ratios. Even though organic N could be used for phytoplankton growth in the long run, the DIN content of our reservoir was 45 times higher than that of the DIP (less than 8.0 $\mu g\ l^{-1}$), and the average DIP

Table 1 Ambient nutrients, chlorophyll-*a* (Chl *a*) and discharge in two inflow streams and the outflow of Shingu Reservoir between May and October, 2003

Date	Inflow 1			Inflow 2			Outflow		
	Discharge $m^3\ day^{-1}$	TN $mg\ l^{-1}$	TP $\mu g\ l^{-1}$	Discharge $m^3\ day^{-1}$	TN $mg\ l^{-1}$	TP $\mu g\ l^{-1}$	TN $mg\ l^{-1}$	TP $\mu g\ l^{-1}$	Chl *a* $\mu g\ l^{-1}$
30 May	–	2.1 ± 0.03	197.7 ± 0.8	–	3.5 ± 0.10	315.8 ± 11.2	–	–	–
13 Jun.	–	6.2 ± 0.17	143.2 ± 2.3	–	7.6 ± 0.08	125.3 ± 9.3	–	–	–
25 Jul.	3,606	5.8 ± 0.04	94.4 ± 0.8	31,548	2.9 ± 0.17	89.8 ± 0.8	3.2 ± 0.10	134.4 ± 3.0	132.0 ± 4.6
8 Aug.	384	1.6 ± 0.04	74.9 ± 0.7	6,150	2.4 ± 0.12	91.2 ± 0.7	2.9 ± 0.13	86.8 ± 0.7	55.6 ± 2.7
19 Sep.	1,051	1.7 ± 0.25	96.7 ± 0.7	7,833	2.5 ± 0.17	105.3 ± 0.7	–	–	–
3 Oct.	303	1.4 ± 0.01	94.6 ± 2.8	2,378	2.5 ± 0.06	68.2 ± 2.1	–	–	–
18 Oct.	370	1.8 ± 0.10	29.1 ± 0.7	2,229	2.5 ± 0.00	84.2 ± 0.7	–	–	–

–: No inflow or outflow water

proportion of the total P was 5.8%. Therefore, we applied the soluble fraction of the TDP, which included the organic form of phosphorus, for phytoplankton to use under the environment with low DIP concentration levels (Islam & Whitton, 1992; Yelloy & Whitton, 1996). The phytoplankton growth in our system had a positive relationship with the dissolved P in the NEBs, which increased as both the TN:TP ($r = 0.596$, $P = 0.012$, $n = 17$) and DIN:DTP ($r = 0.575$, $P = 0.016$, $n = 17$) ratios increased. Thus, our experimental approach of NEBs, based on the algal growth response to the nutrient ratio and ambient nutrients, may be used as a useful tool for the diagnosis of lake eutrophication and efficient managements.

To conclude, cyanobacteria dominated in our hypertrophic study reservoir were P-limited in most times, and thus their growth rate increased with P addition. Also the degree of the cyanobacterial growth in P addition was magnified with more N supplied, reflecting that at high nitrogen level, low N:P ratios were favorable for cyanobacterial dominance. Our results suggest that limiting nutrients and N:P ratio are critical factors to evaluate the cynaobacterial bloom in eutrophic systems. However, cyanobacterial bloom events would be complicated in the ambient condition, so that some other factors may also be important to affect it such as temperature, light, and grazing (Hutchinson, 1957; Reynolds, 1984; Carpenter & Kitchell, 1993). In the companion study (Kim, 2004), we found that temperature had significant effect on the maximum growth rate (U_{max}) of cyanobacteria between 10 and 25°C ($Q_{10} = 2$). Light was also detrimental to their growth rate, but the degree of the growth rate by any single factor of either temperature or light varied with nutrients. As such, we do not deny the potentials of other factors to the cyanobacterial bloom in our reservoir, but, as we found in this study, limiting nutrients and their stochiometry are important factors to understand cyanobacterial dominance in the eutrophic system.

Acknowledgement This paper was supported by Konkuk University in 2000.

References

APHA, 1995. Standard Methods for the Examination of Water and Wastewater, 19th edn. APHA-AWWA-WEF, Washington, DC, USA.

Carpenter, S. R. & J. R. Kitchell, 1993. Cascading trophic interactions and lake productivity. Bioscience 35: 634–639.

Droop, M. R., 1968. Vitamin B_{12} and marine ecology. 4. The kinetics of uptake, growth and inhibition of *Monochrysis lutheri*. Journal of Marine Biology Association, UK 48: 689–733.

Forsberg, C. & S. O. Ryding, 1980. Eutrophication parameters and trophic state indices in 30 Swedish waste-receiving lakes. Archive fur Hydrobiologie 89: 189–207.

Fujimoto, N. & R. Sudo, 1997. Nutrient-limited growth of *Microcystis aeruginosa* and *Phormidium tenue* and competition under various N:P supply ratios and temperatures. Limnology and Oceanography 42: 250–256.

Geider, R. J. & J. La Roche, 2002. Redfield revisited: Variability of C:N:P in marine microalgae and its biochemical basis. European Journal of Phycology 37: 1–17.

Goldman, J. C., J. J. McCarthy & D. G. Peavey, 1979. Growth rate influence on the chemical composition of phytoplankton in oceanic waters. Nature 279: 210–215.

Goldman, J., D. A. Caron & M. R. Dennet, 1987. Nutrient cycling in a microflagellate food chain, IV. Phytoplankton–microflagellate interactions. Marine Ecology Progress Series 38: 75–87.

Hendrey, G. R. & E. B. Welch, 1974. Phytoplankton productivity in Findley Lake. Hydrobiology 45: 45–63.

Ho, T. Y., A. Quigg, Z. V. Finkel, A. J. Milligan, K. Wyman, P. G. Falkowski & F. M. M. Morel, 2003. The elemental composition of some marine phytoplankton. Journal of Phycology 39: 1145–1159.

Horne, A. J., 1979. Management of lakes containing N-fixing blue-green algae. Archive fur Hydrobiologie 13: 133–144.

Howarth, R. W., R. Marino & J. J. Cole, 1988. Nitrogen fixation in freshwater, esturine, and marine ecosystems. 2. Biogeochemical controls. Limnology and Oceanography 33: 669–687.

Hutchinson, G. E., 1957. A Treatise on Limnology. Geography Physics and Chemistry, Vol. 1. Wiley, New York: 1015 pp.

Islam, M. R. & B. A. Whitton, 1992. Retention of P-nitrophenol and 4-methylumbelliferone by marine macroalgae and implications for measurement of alkaline phosphatase activity. Journal of Phycology 32: 819–825.

KARICO, 2001. Report of Water Monitoring in Agricultural Reservoirs, S. Korea. Korea Agriculture and Rural Infrastructure Cooperation (in Korean).

Kilham, P., 1990. Mechanisms controlling the chemical composition of lakes and rivers: Data from Africa. Limnology and Oceanography 35: 80–93.
Kim, H. S., 2004. Study on the growth dynamics and ecotechnological control of algae in reservoirs. Ph. D dissertation thesis. Konkuk University, Seoul, South Korea (in Korean).
Leonardos, N. & R. J. Geider, 2004. Responses of elemental and biochemical composition of *Chaetoceros muelleri* to growth under varying light and nitrate: Phosphate supply ratios and their influence on critical N:P. Limnology and Oceanography 49: 2105–2114.
Marker, A. F. H., E. A. Nusch, I. Rai & B. Riemann, 1980. The measurement of photosynthetic pigments in freshwaters and standardization of methods: Conclusions and recommendations. Archive fur Hydrobiologie 14: 91–106.
Monod, J., 1950. La technique de culture continue: theorie at applications. Annales de l'Institut Pasteur de Lille 79: 390–410.
Paerl, H. W., R. S. Fulton, P. H. Moisander & J. Dyble, 2001. Harmful freshwater algal blooms, with an emphasis on cyanobacteria. The Scientific World Journal 1: 76–113.
Patricia, A. S., W. H. Shaaw & P. A. Bukaveckss, 2000. Differences in nutrient limitation and grazer suppression of phytoplankton in seepage and drainage lakes of the Adirondack region, NY, USA. Freshwater Biology 43: 391–407.
Perkins, R. G. & G. J. C. Underwood, 2000. Gradients of chlorophyll *a* and water chemistry along an eutrophic reservoir with determination of the limiting nutrient by In Situ nutrient addition. Water Research 34: 713–724.
Redfield, A. C., B. H. Ketchum & F. A. Richards, 1963. The influence of organisms on the composition of sea water. In Hill, M. N. (ed.), The Sea. Interscience, New York, 26–27.
Reynolds, C. S., 1984. Phytoplankton periodicity: The interactions of form, function and environmental variability. Freshwater Biology 14: 111–142.
Rhee, G. Y., 1973. A continuous culture study of phosphate uptake, growth rate and polyphosphate in *Scenedesmus* sp. Journal of Phycology 9: 495–506.
Rhee, G. Y., 1978. Effects of N:P atomic ratios and nitrate limitation on algal growth, cell composition, and nitrate uptake. Limnology and Oceanography 23: 10–24.
Rhee, G. Y. & I. J. Gotham, 1981. The effect of environmental factors on phytoplankton growth: Light and the interactions of light with nitrate limitation. Limnology and Oceanography 26: 649–650.
Sheffer, M., S. Rinaldi, A. Grangnani, L. R. Mur & E. H. Nes, 1997. On the dominance of filamentous cyanobacteria in shallow, turbid lakes. Ecology 78: 272–282.
Smith, V. H., 1983. Low nitrogen to phosphorus ratios favors dominance by blue-green algae in lake phytoplankton. Science 221: 669–671.
Smith, V. H., E. Willen & B. Karlsson, 1987. Predicting the summer peak biomass of four species of blue-green algae (cyanphyta/cyanobacteria) in Swedish lakes. Water Research Bulletin 23: 397–402.
Sommer, U., 1989. The role of competition for resources in phytoplankton succession. In Sommer, U. (ed.), Plankton Ecology: Succession in Plankton Communities. Springer-Verlag, Berlin, 57–106.
Sterner, R. W. & J. J. Elser, 2002. Ecological Stoichiometry: The Biology of Elements from Molecules to the Biosphere. Princeton University Press. New Jersey, 439.
Terry, K. L., E. A. Laws & D. J. Burns, 1985. Growth rate variation in the N:P requirement ratio of phytoplankton. Journal of Phycology 21: 323–329.
Tilman, D. 1977. Resource competition between planktonic algae: An experimental and theoretical approach. Ecology 58: 338–348.
Tilman, D. 1978. Ecological competition between algae: Experimental confirmation of resource based competition theory. Science 192: 463–465.
Tilman, D. 1982. Resource competition and community structure. Princeton Monographs in Population Biology 17. Princeton University Press.
Trimbee, A. M. & E. E. Prepas, 1987. Evaluation of total phosphorus as a predictor of the relative biomass of blue-green algae with emphasis on Alberta lakes. Canadian Journal of Fisheries and Aquatic Science 44: 1337–1342.
Xie, L., P. Xie, S. Li, H. Tang & H. Liu, 2003. The low TN:TP ratio, a cause or a result of *Microcystis* blooms. Water Ressearch 37: 2073–2080.
Yelloly, J. M. & B. A. Whitton, 1996. Seasonal changes in ambient phosphate and phosphatase activities of the cyanobacterium *Rivularia atra* in intertidal pools at Tyne Sands, Scotland. Hydrobiologia 325: 201–212.

Hydrobiologia (2007) 581:269–285
DOI 10.1007/s10750-006-0507-3

EUTROPHICATION IN LAKES

Restoration of shallow lakes by nutrient control and biomanipulation—the successful strategy varies with lake size and climate

E. Jeppesen · M. Meerhoff · B. A. Jacobsen · R. S. Hansen · M. Søndergaard · J. P. Jensen · T. L. Lauridsen · N. Mazzeo · C. W. C. Branco

Abstract Major efforts have been made worldwide to improve the ecological quality of shallow lakes by reducing external nutrient loading. These have often resulted in lower in-lake total phosphorus (TP) and decreased chlorophyll *a* levels in surface water, reduced phytoplankton biomass and higher Secchi depth. Internal loading delays recovery, but in north temperate lakes a new equilibrium with respect to TP often is reached after <10–15 years. In comparison, the response time to reduced nitrogen (N) loading is typically <5 years. Also increased top-down control may be important. Fish biomass often declines, and the percentage of piscivores, the zooplankton:phytoplankton biomass ratio, the contribution of *Daphnia* to zooplankton biomass and the cladoceran size all tend to increase. This holds for both small and relatively large lakes, for example, the largest lake in Denmark (40 km^2), shallow Lake Arresø, has responded relatively rapidly to a ca. 76% loading reduction arising from nutrient reduction and top-down control. Some lakes, however, have proven resistant to loading reductions. To accelerate recovery several physico-chemical and biological restoration methods have been developed for north temperate lakes and used with varying degrees of success. Biological measures, such as selective removal of planktivorous fish, stocking of piscivorous fish and implantation or protection of submerged plants, often are cheap versus traditional physico-chemical methods and are therefore attractive. However, their long-term effectiveness is uncertain. It is argued that additional measures beyond loading reduction are less cost-efficient and often not needed in very large lakes.

J. P. Jensen was deceased.

Guest editors: B. Qui, Z. Liu & K. Havens
Eutrophication of shallow lakes with special reference to Lake Taihu, China

E. Jeppesen (✉) · M. Meerhoff · M. Søndergaard · J. P. Jensen · T. L. Lauridsen
Department of Freshwater Ecology, National Environmental Research Institute, Vejlsøvej 25, 8600 Silkeborg, Denmark
e-mail: ej@dmu.dk

E. Jeppesen · M. Meerhoff
Department of Plant Biology, University of Aarhus, Ole Worms Allé 135, 8000 Aarhus, Denmark

M. Meerhoff · N. Mazzeo
Departamento de Ecología, Facultad de Ciencias, Iguá 4225, Montevideo, Uruguay

B. A. Jacobsen · R. S. Hansen
Frederiksborg Amt, Technical Department, Amtsgården, Kongens Vænge 2, 3400 Hillerod, Denmark

C. W. C. Branco
Departamento de Ciências Naturais, Universidade Federal do Estado do Rio de Janeiro, Av. Pasteur 458, Urca, Rio de Janeiro, Brazil 22290-040

Although fewer data are available on tropical lakes these seem to respond to external loading reductions, an example being Lake Paranoá, Brazil (38 km^2). However, differences in biological interactions between cold temperate versus warm temperate-subtropical-tropical lakes make transfer of existing biological restoration methods to warm lakes difficult. Warm lakes often have prolonged growth seasons with a higher risk of long-lasting algal blooms and dense floating plant communities, smaller fish, higher aggregation of fish in vegetation (leading to loss of zooplankton refuge), more annual fish cohorts, more omnivorous feeding by fish and less specialist piscivory. The trophic structures of warm lakes vary markedly, depending on precipitation, continental or coastal regions locations, lake age and temperature. Unfortunately, little is known about trophic dynamics and the role of fish in warm lakes. Since many warm lakes suffer from eutrophication, new insights are needed into trophic interactions and potential lake restoration methods, especially since eutrophication is expected to increase in the future owing to economic development and global warming.

Keywords Lake restoration · Large lakes · Recovery · Nutrient loading reduction · Climate

Introduction

Efforts in the last 20–30 years have attempted to combat eutrophication by reducing external phosphorus (P) loading in Western Europe and North America. Sewage and industrial loading has declined, whereas nutrient input from diffuse sources often remains high, particularly in countries with intensive agriculture (Van der Molen & Portielje, 1999; Kronvang et al., 2005).

Many lakes have responded positively to nutrient loading reductions (Sas, 1989; Marsden, 1989; Jeppesen et al., 2005c). An analysis of European and North American lakes based on 35 case studies, including long time-series data, revealed that, although internal P loading does delay recovery, new equilibria with respect to (total phosphorus) TP were reached in most lakes after <10–15 years, marginally dependent on lake retention time and depth (Jeppesen et al., 2005c). In comparison, response times to reduced N loading are typically <5 years. These findings also apply to shallow Danish lakes after a reduction in mainly external P loading (Søndergaard et al., 2002a, 2005; Jeppesen et al., 2002, 2005a, b). In these lakes, TP has declined during all seasons, particularly in winter, followed by spring and autumn, while weaker effects in late summer can be ascribed to internal P loading. Changes in chlorophyll *a* follow TP, indicating that nutrient constraints are important in determining seasonal behaviour. Diatoms declined precipitously in winter and spring. Non-heterocystous cyanobacteria declined throughout the summer, primarily in early summer, while heterocystous cyanobacteria increased in late summer and autumn. Green algae declined throughout the year but least so in mid-summer. In Barton Broad, a shallow lake in the UK, Phillips et al. (2005) observed a similar pattern with strong effects on diatoms in spring and a decline first in non-heterocystous cyanobacteria and secondly, albeit slower, in heterocystous genera after 15–20 years. Thus, there is clear evidence for enhanced resource control of phytoplankton after an external loading reduction.

However, enhanced top-down control by zooplankton may also play a role. Fish biomass often declined in the Danish lakes and the percentage of piscivores often increased. Similar observations were made in the study of 35 North American and European lakes in recovery (Jeppesen et al., 2005c); in 82% of the lakes for which data were available, fish biomass declined with TP, while the piscivore proportion increased in 80% of the lakes. This is surprising given the high longevity of the dominant species. While there were no changes in zooplankton biomass, the contribution of *Daphnia* to the biomass and mean individual biomass of cladocerans increased most notably during summer. Moreover, the zooplankton:phytoplankton biomass ratio, and probably phytoplankton grazing, rose. This is suggested also by the cladoceran size increase. Higher abundance of large-bodied zooplankton, particularly in summer, may reflect that increased piscivory leads to stronger predation control of young-of-the year (YOY), primarily planktivorous, fish. This reduces predation on zooplankton during summer, when YOY fish are abundant and forage in the pelagic zone. Enhanced abundance of submerged macrophytes or filamentous algae in some lakes may also enhance survival of large-bodied zooplankton as they can serve as a daytime refuge from predation (Burks et al., 2002).

A north temperate example: recovery of Lake Arresø, Denmark

The positive response to external nutrient loading reduction is not restricted to small lakes. Here we elaborate in more detail on the response of Arresø to nutrient loading reduction. Arresø (coordinates) is the largest lake in Denmark (40 km^2). It is shallow (mean depth 3.1 m, max depth 5.9 m) and has a 215 km^2 catchment area. The retention time is long (1.3–10.3 years, mean 1989–2003: 3.5 years) compared with most Danish shallow lakes. The lake is near the coast and exposed to prevailing north-westerly winds and, thus, wave-induced resuspension (Kristensen et al., 1992). Based on intensive sampling (every 2–8 h) during different seasons in 1991, and modelling, resuspension was calculated to occur 50% of the time and to reduce Secchi depth to <0.5 m on average (Kristensen et al., 1992). The lake was loaded heavily with P and N from the catchment, particularly P from the upstream city of Hillerød. TP loading decreased from 0.68 g P m^{-2} $year^{-1}$ in 1989 to 0.16 g P m^{-2} $year^{-1}$ for 1996–2003, and TN decreased from 13 to 11 g N m^{-2} $year^{-1}$ (Jacobsen et al., 2004). The reduction mostly was due to improved sewage treatment, establishment of artificial lakes on the main inlet stream, and possibly to reduced fertilization in the catchment. Nutrient load reduction led to substantial improvements in the lake. TN, TP, chlorophyll *a* and the total biomass (and especially of green algae) of phytoplankton have declined markedly (Fig. 1). Diatom biomass has tended to increase with associated effects on silicate concentration, whereas the trend for cyanobacteria is ambiguous. Algal biomass changes cannot be ascribed entirely to enhanced nutrient control, since there have been marked zooplankton community changes and potential grazing, most likely due to fish community changes. Biomasses of the dominant plankti-benthivorous bream (*Abramis brama*) (Fig. 2) and benthic ruffe (*Gymnocephalus cernua*) are decreasing, whereas zander (*Sander lucioperca*), a potential predator, is increasing, implying enhanced top-down control on prey fish. Accordingly, zooplankton biomass and the share of large-sized cladocerans have increased substantially. This can be explained by reduced predation pressure on zooplankton, since the algal biomass was dominated by edible green algae when biomass was low and the zooplankton were dominated by small forms. With increased zooplankton biomass the grazing pressure on phytoplankton has intensified, and the phytoplankton face simultaneous, enhanced resource control due to lower P and N concentrations. Grazing control is strengthened by the shift to large-sized cladocerans *Daphnia hyalina* and *D. galeata* (Fig. 1), which have broader feeding capabilities than previously dominant *Bosmina* spp. and *Chydorus sphaericus*. Following the increase in zooplankton biomass and cladoceran size, and the zooplankton:phytoplankton ratio, water transparency has improved (Fig. 1), emphasizing the role of enhanced grazer control of phytoplankton.

Seasonal phytoplankton biomass reduction and Secchi depth increases are larger in spring, autumn and winter than in summer (Fig. 3). This also is true for the zooplankton:phytoplankton biomass ratio and cladoceran body mass, suggesting that changes in grazing pressure are important. The minor effect in summer coincides with fish larvae exploiting the pelagic in Danish lakes (Jeppesen et al., 2004). Moreover, during this period, nitrate is low (in all years) (Fig. 3), which likely may have dampened P loading reduction effects.

However, loading still remains too high to promote a clearwater state with macrophytes, which prevailed in the 1900's per paleolimnological analyses (Klein, 1989).

Accelerating the recovery of northern temperate lakes

Positive effects may not occur after nutrient input reduction. Several lakes have been resistant to loading reductions and shown slow recovery (Sas,

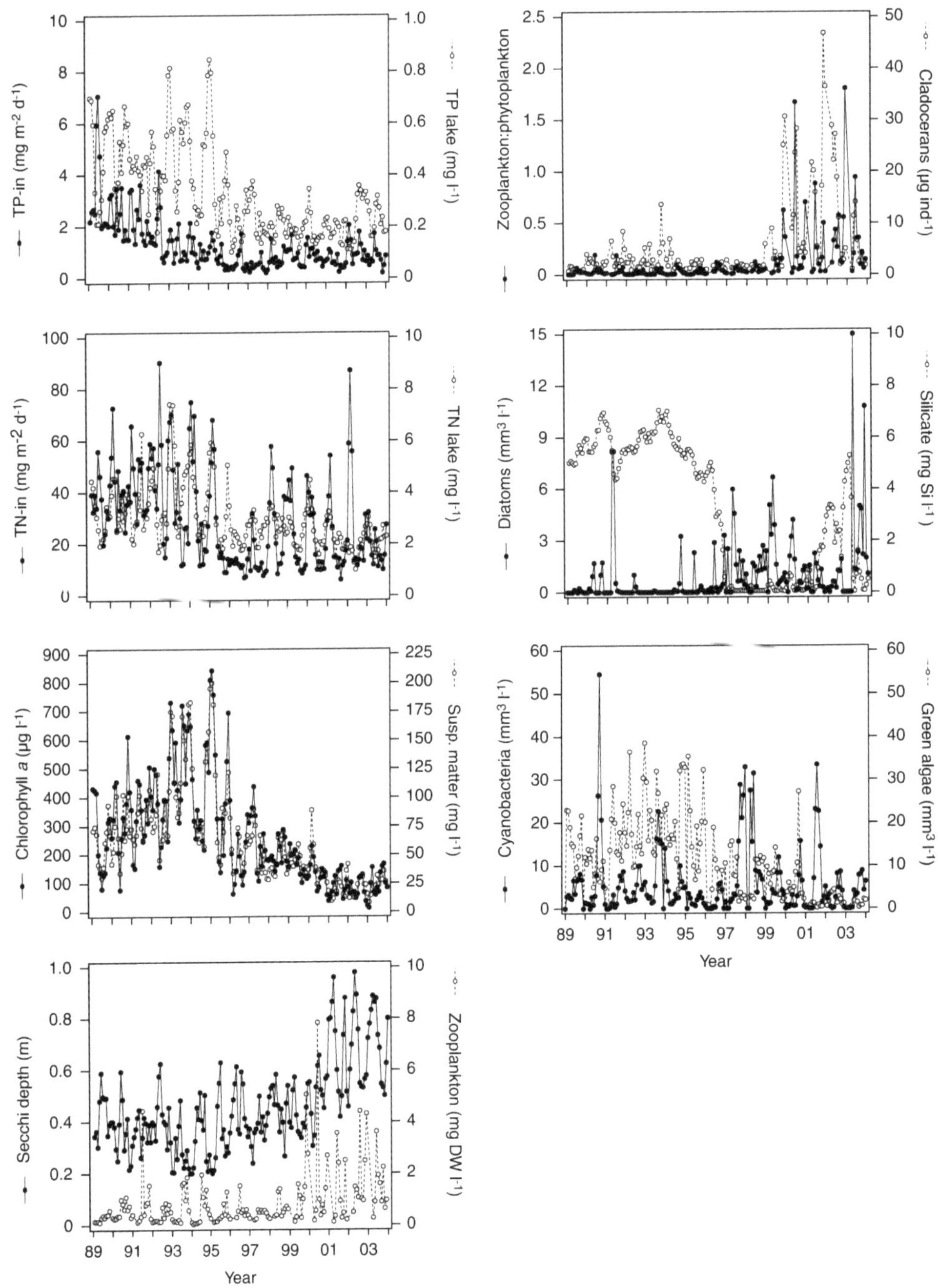

Fig. 1 Monthly mean data on various environmental variables in Arresø. TP_{in} and TN_{in} are the external loading of the lake. Zooplankton:phytoplankton is the biomass (mg DW) ratio

1989; Marsden, 1989). For some lakes, this reflects insufficient nutrient input reduction to promote a clearwater state. For example, significant and sustained changes in the biological community and water transparency of shallow, temperate, freshwater lakes may not appear unless TP is reduced below 0.05–0.1 mg P l^{-1} (Jeppesen et al., 2000) or, for deep lakes, 0.02–0.03 mg P l^{-1} (Sas, 1989). Even when P loading has been reduced sufficiently, resistance to improvement often is observed. This resistance may be "chemical", P concentrations remain high due to release from

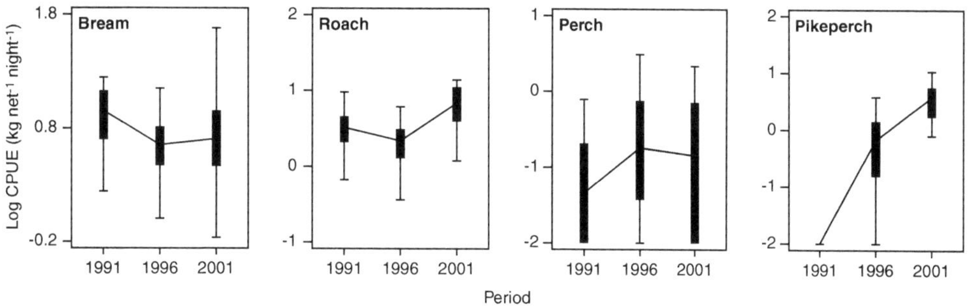

Fig. 2 Box plot (10, 25, 75, 90% percentiles) of log(base e) transformed catch per net per night in Arresø using 52 nets distributed in the lake according to Jeppesen et al. (2004)

the sediment pool accumulated when loading was high (Søndergaard et al., 2002a). Various methods have been used to reduce internal P loading (Cooke et al., 1993; Søndergaard et al., 2002a, b), including sediment removal and chemical treatment with aluminum or iron salts. In stratified lakes, oxygen or nitrate injections to the bottom layer or destabilization of the thermocline have been used (Cooke, 1993). For large, shallow lakes, it is debatable whether sediment removal and chemical treatment are feasible methods. Sediment removal is expensive, particularly for large lakes due to transport and disposal. Moreover, sediment resuspension and redistribution may hamper removal of the upper, nutrient-rich sediment layers. Resuspension also is a problem with chemical methods. P bound to aluminum or iron may return to the water on windy days and loosely bound P made available to phytoplankton. Finally, the sediment P pool often is small in large lakes because resuspension leads to washout of particulate TP (unless hydraulic retention time is long) and organic net sedimentation is low, the latter due to high mineralization. Thus, the usefulness of sediment removal or treatment is more limited in large than small lakes. The fast response of Arresø was attained without sediment removal.

The resistance also may be "biological". Planktivorous and benthivorous fish contribute to biological resistance in shallow eutrophic lakes (Moss, 1990; Scheffer et al., 1993). High fish predation prevents the appearance of large herbivorous zooplankton, which would otherwise clear the water, and diminishes benthic invertebrates, which stabilize and oxidize the sediment. Moreover, nutrient excretion to overlaying waters by benthic-feeding fish or fish bioturbation of surface sediment may play a role (Breukelaar et al., 1994; Persson, 1997). To overcome biological resistance, various fish manipulation methods have been developed (Benndorf, 1995; Drenner & Hambright, 1999; Søndergaard et al., 2001). One method is enhancement of top-down control of phytoplankton by selective removal of planktivorous fish; a method employed extensively in the temperate zone, but only just beginning in the subtropics (Scasso et al., 2001) and other warmer climate zones, such as Mediterranean arid lakes (Beklioglu & Tan, personal communication). Removal of 75–80% of the planktivorous and benthivorous fish stock over 1–2 years is recommended to avoid regrowth and stimulate growth of potentially piscivorous perch (Perrow et al., 1997; Hansson et al., 1998; Meijer et al., 1999). An alternative or supplementary method to fish removal is stocking of 0^+ pike to control newly hatched plankti-benthivorous roach and bream (Prejs et al., 1994; Berg et al., 1997; Skov & Berg, 1999) though the results are ambiguous (C. Skov et al., unpublished data). Others have stocked pikeperch, walleye (*Stizostedion vitreum*) and largemouth bass (*Micropterus salmoides*) (Benndorf, 1995; Lathrop et al., 2002). In contrast to chemical engineering methods, fish manipulation often is cheap (Jeppesen & Sammalkorpi, 2002) and therefore attractive, though its long-term stability is uncertain as a return to a more turbid state has occurred in many of the lakes. The reasons for this remain to be elucidated, but

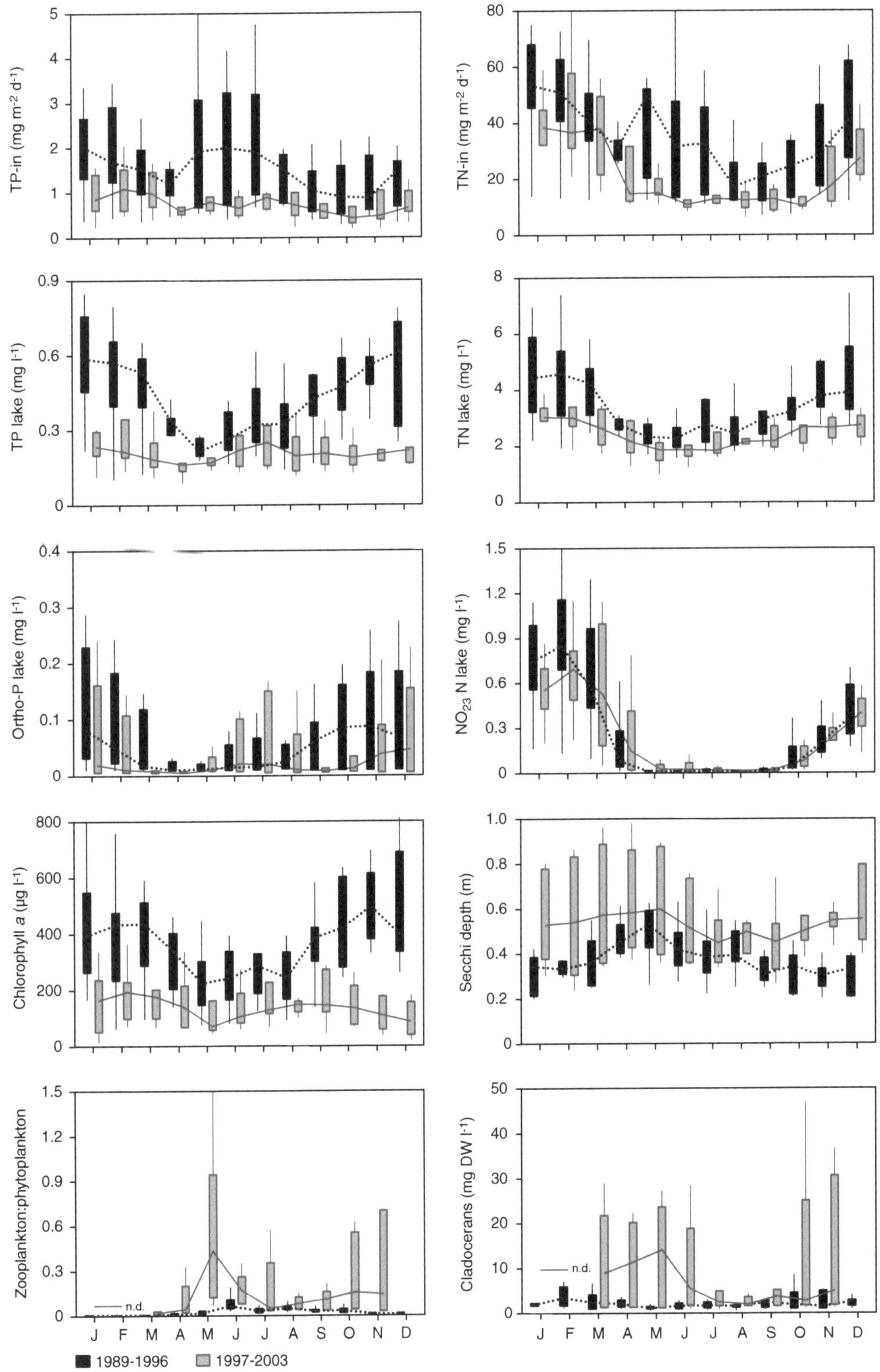

Fig. 3 Box plot (10, 25, 75, 90% percentiles) of selected environmental variables in Arresø during the first (1989–1996) and the last part of the study period (1997–2003). n.d. means no data in Jan–Feb for 1997–2003. See also legend of Fig. 1

factors such as reinforced internal loading of P, return of planktivorous dominance (particularly roach in Danish lakes) and large year-to-year variations in the coverage of the submerged macrophyte community may play a role. However, in Danish lakes suspended matter remains lower and Secchi depth higher likely reflecting an apparent permanent lower bream population (authors unpublished data).

It is unlikely that fish manipulation will be feasible for very large lakes (say >500 km^2). First, it is difficult to obtain a 'sledgehammer' effect on the target fish population in very large lakes. The fish cannot easily be removed fast enough to compensate for their annual reproductive rate. Changes in the fish stock may simply become difficult to discern unless the fishing effort is substantial. Second, recent studies indicate that the fish population responds quickly to a nutrient loading reduction (Jeppesen et al., 2005a, b, c). This may render fish stock intervention less attractive in very large lakes as the natural changes may occur at higher speed than those invoked by biomanipulation in such lakes. Third, fish stocking is expensive for large lakes. However, biomanipulation combined with nutrient loading reduction has yielded a fast response (higher transparency, fewer cyanobacteria and in some cases extensive growth of submerged macrophytes) in medium large lakes, examples being Lake Vesijärvi, Finland (109 km^2), Lake Wolderwijd, Holland (27 km^2), and Lake Christina, USA (16 km^2) (Hanson & Butler, 1990; Peltonen et al., 1999; Meijer et al., 1999).

Delay in recovery and re-establishment of submerged vegetation after nutrient loading reduction have often been noted (Sas, 1989; Cooke et al., 1993; Jeppesen & Sammalkorpi, 2002; Jeppesen et al., 2005c). The delay in plant appearance may be due to lack of seed banks, waterfowl grazing, limited grazing on periphyton or too high turbidity (Søndergaard et al., 1996; Mitchell & Perrow, 1997; Lauridsen et al., 2003). Construction of exclosures to protect macrophytes against waterfowl grazing to improve plant re-establishment has been used as an alternative or supplementary restoration tool to fish manipulation (Cooke, 1993; Jeppesen & Sammalkorpi, 2002). The exclosures enable macrophytes to grow in a grazer-free environment where they can spread seeds, turions or plant fragments and thereby augment colonization. Moreover, they serve as a daytime refuge for zooplankton. Plant refuges as a restoration tool is probably most useful in small lakes and in the littoral zone of other lakes where colonization is restricted to the near shore areas due to light limitation in the deeper main lake. Waterfowl aggregate in littoral emergent zones, which provide shelter and nest sites, and may impart higher grazing pressure on plants. Plants as a refuge have greater effect in nutrient-rich lakes where plant density often is highest and prey fish are confined to the pelagial (Jeppesen et al., 1997; Blindow et al., 2000; Burks et al., 2002). Transplantation or protection of plants (in sheltered beds) and seeds may be useful in large shallow lakes if pre-studies reveal that (re)-colonization is seed-limited (Weisner & Strand, 2002; Ke & Li, 2006). However, it is important to evaluate whether wind (wave) disturbance will prevent colonization. In large lakes effort should be concentrated on sheltered areas where plants can colonize gradually, as seen in large (27 km^2) Lake Veleuwe (Meijer et al., 1999).

Warm lakes

Few long-term lake recovery studies after reduced external loading exist for tropical and subtropical lakes: here we present an example from Brazil. Lake Paranoá (15°48′S, 47°47′W) is a shallow, urban, man-made lake (area 38 km^2, mean depth: 13 m, retention time: 0.9 years). It was created in 1959 for recreational purposes, hydroelectric power production, to increase the humidity of the region, as well as to receive wastewater and run-off from the city of Brasilia. The first plankton study recorded dominance of desmids and large-bodied cladocerans (*Daphnia* and *Ceriodaphnia*) (Oliveira & Krau, 1970). Later, phytoplankton biomass was dominated by *Cylindrospermospsis raciborskii*, which remained dominant in subsequent decades (Cronberg, 1976; Pinto-Coelho & Giani, 1985; Branco & Senna, 1991). *Microcystis aeruginosa* blooms occurred periodically during the end of the dry season

(Branco & Senna, 1994). Besides the permanent bloom of filamentous cyanobacteria (*C. raciborskii*), eutrophic conditions were also indicated by small-bodied zooplankton (mainly rotifers and small cladocerans, such as *Diaphanosoma birgei* and *Bosmina* spp.), low Secchi depth (<50 cm) and high levels of nutrients and chlorophyll *a* (Branco & Senna, 1996; Branco & Cavalcanti, 1999, Fig. 4). Aquatic macrophytes, such as free-floating *Eichhornia crassipes* and *Pistia stratiotes*, also spread in some areas. Eleven fish species colonized the lake from the catchment. In addition, the lake was stocked with exotic fish species, such as bluegill sunfish (*Lepomis macrochirus*), black-bass (*Micropterus salmoides*), carp (*Cyprinus carpio*), Congo (*Tilapia rendalli*) and Nile (*Oreochromis niloticus*) tilapias, and the piscivorous tucunaré (*Cichla ocellaris*) from the Amazon Basin (Ribeiro et al., 2001). In the late 1980s, tilapia dominated and accounted for 50% of total fish biomass. In subsequent years, the lake was stocked with tambaqui (*Colossoma macropomum*) and tamoatá (*Callychthys callychthys*), both from the Amazon Basin and with high commercial value.

In 1993–1994, two tertiary sewage treatment plants capable of processing 2,400 l s^{-1} were constructed in addition to a complementary sewage collecting system. Resultant decrease in external point-source of TP loading, from 6.3 g m^{-2} $year^{-1}$ in 1989 to 0.48 g m^{-2} $year^{-1}$ in 1998 (Pereira & Cavalcanti, 1997; Burnett et al., 2001), led to decreased mean TP, and also total Kjeldahl nitrogen (TKN) loading and concentration decreased (Fig. 4). Chlorophyll *a* values decreased (Fig. 4), and Secchi depth attained 2.4 m in 1999. Important planktonic community level changes were observed from 1996 to 1999: (1) decreased cyanobacteria dominance and replacement of *C. raciborskii* by green algae and diatoms by late 1998; (2) short-term dominance of colonial floating green algae (*Botryococcus braunii*), replacing *M. aeruginosa* in some hypertrophic areas in 1996; (3) re-appearance of *Ceriodaphnia cornuta*, *Daphnia gessneri*, other small cladocerans, such as *Moina micrura* and *Bosminopsis deitersi* (Elmoor-Loureiro et al., 2004), and *Notodiaptomus cearensis*, an oligotrophic calanoid copepod species (Padovesi-

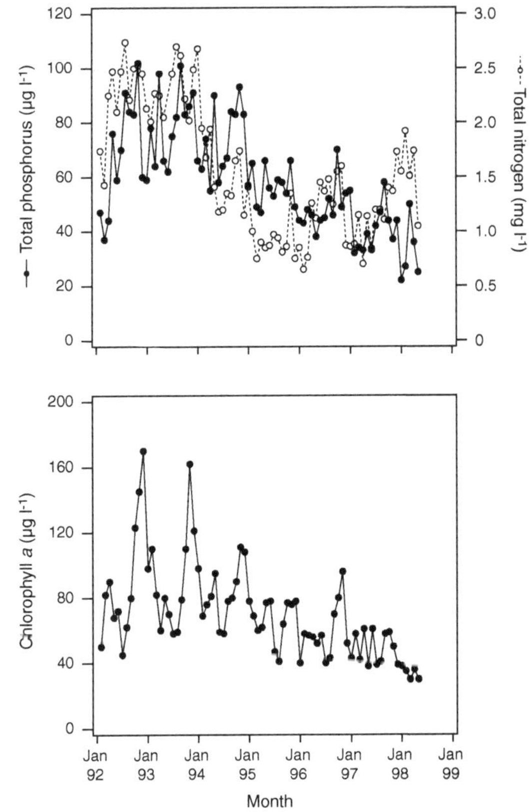

Fig. 4 Total phosphorus, total Kjeldahl nitrogen (TKN) and chlorophyll *a* in Lake Paranoá, Brazil, before and after (1993–1994) a major reduction of nutrient loading

Fonseca et al., 2001). Other strategies, such as reducing the retention time of the lake and fish population management, were used to generate further improvements. An echo-sounding campaign in 1998 revealed a fish stock of 1,500 tons, with tilapia constituting up to 90% of fish biomass. Laboratory studies on fish feeding rates on lake plankton and P excretion rates, plus an assessment of planktivorous fish impacts in enclosures, indicated that tilapia overpopulation control would prevent accumulated P from being available to primary production, especially due to bottom feeding (Starling, 1993a, b; Starling & Rocha, 1990). Tilapia control was initiated in the most eutrophicated region by legalizing commercial cast-net fisheries. Continuous improvement of lake environmental conditions was confirmed in early 2000 by lower chlorophyll *a* values, enhanced water transparency, and decreased fish biomass to 800 tons. Tilapia dominance declined, allowing recovery of a heterogeneous fish population, including native omnivores and piscivores, and some exotic species, all of which contributed to enhancement of large zooplankton populations.

Trophic dynamics and biomanipulation in warm lakes

While results from Lake Paranoá illustrate the importance of reducing external loading to restore subtropical and tropical lakes, it is difficult to evaluate effects of fish manipulations in the lake. It is debatable whether fish manipulation methods used in cold temperate lakes can be employed in warm temperate, subtropical and tropical lakes. Several factors indicate that fish stock manipulation may not have the same positive effects in warm lakes (Moss et al., 2004; Jeppesen et al., 2005d): (a) fish species richness often is higher and many fish show partial niche overlap, which should increase predator control of prey (Lazzaro, 1997; Aguiaro & Caramaschi, 1998); (b) fish stocks often are dominated by omnivores independent of trophic state (Lazzaro, 1997; Branco et al., 1997; Yafe et al., 2002; Blanco et al., 2003; Mazzeo et al., unpublished data) and generally few piscivores are present (Quirós, 1998); (c) fish density, but not necessarily biomass, is higher (1–2 orders of magnitude) (Scasso et al., 2001; Aguiaro et al., 2003; Mazzeo et al., 2003; Meerhoff et al., 2003); (d) fish reproduction occurs throughout the year (Fernando, 1994; Paugy & Lévêque, 1999) and many species are viviparous and/or show parental care (Lorier & Berois, 1995), assuring higher juvenile survival rates. Therefore, top-down control by piscivores probably is weaker in warm lakes. Since small fish are more zooplanktivorous and have higher energy demand per unit biomass (Kalff, 2002), small fish in high abundance leads to higher predation pressure on zooplankton.

Accordingly, zooplankton communities in tropical and subtropical lakes often are dominated by small cladocerans (i.e. *Diaphanosoma*, *Ceriodaphnia* and *Bosmina*), rotifers, and copepod juveniles and nauplii (Dumont, 1994; Lewis, 1996; Branco et al., 2002; Garcia et al., 2002; Lacerot et al., unpublished data). Large *Daphnia* spp. may appear in early spring and autumn in warm-temperate lakes (Romo et al., 2005) or, if fish are absent, for longer periods in subtropical lakes (Mazzeo et al., unpublished data). Omnivorous copepods usually dominate zooplankton biomass in oligo-mesotrophic systems, whereas microzooplankton prevail in eutrophic systems. Therefore, classic control of phytoplankton by large zooplankton usually is not found in warm lakes. However, the numerous microzooplankton in these systems may control nanophytoplankton, thus favouring larger cyanobacteria. Some works demonstrate that cyclopoid copepods can control diatoms, rotifers can graze small green algae, and particulate matter and associated bacteria is an important food resource for zooplankters in tropical lakes (Infante, 1978; González et al., 2005). The zooplanktivorous phantom midge *Chaoborus* also seems to be more abundant in tropical lakes (e.g. Aguilera & Goitia, 1999), perhaps because bottom water anoxia is more common, which may provide *Chaoborus* with a fish predation refuge (Lewis, 1996) and allow strong predation pressure on zooplankton (Arcifa et al., 1992). Finally, shrimps (e.g. *Palaemonetes argentinus* or *Macrobrachium borelli*) also can be important predators on zooplankton considering their high abundance, food preferences and several reproductive events from spring to summer (Boschi, 1981; Collins, 1999; Collins & Paggi, 1998).

Aquatic plants play an important structural role in most freshwater ecosystems (Moss, 1990; Scheffer et al., 1993), also in the tropics (Thomaz & Bini, 2003). In temperate nutrient-rich lakes, submerged plants act as daytime refuges for zooplankton against fish predators (Timms & Moss, 1984; Lauridsen et al., 1996; Burks et al., 2002). At night, large zooplankton move out to the open water to feed, thereby contributing to clearwater conditions in lakes with high macrophyte coverage (Jeppesen et al., 1997). However, in the tropics and subtropics, effects of macrophytes on trophic interactions are more complex, since all forms (i.e. emergent, submerged, floating-leaved and large free-floating species) can be prominent. Macrophytes are extremely important to fish in the tropics and subtropics. Removal of *Chara* spp. beds in an Itaipú reservoir resort

(Santa Helena, Brazil) with the purpose to decrease piranha (*Serrassalmus marginatus*) population led to decreased fish abundance and species richness from 25 to 10 species (Agostinho et al., 2003; Fig. 5). Studies comparing different habitats in tropical and subtropical lakes indicate that the smallest fish species and individuals aggregate in all vegetation forms (Conrow et al., 1990; Meschiatti et al., 2000; Meerhoff et al., 2003; Meerhoff et al. unpublished data). Densities above 140 ind m^{-2} of a single, small-sized species were found in submerged plant beds in a subtropical lake (Meerhoff et al., 2003). Therefore, submerged vegetation in warm lakes may be a poor refuge for large-bodied zooplankton, particularly cladocerans (Meerhoff et al., 2003, 2006). An inverse diel horizontal migration pattern (aggregating in the pelagial during day and among plants at night) was found for *Bosmina longirostris* and *Diaphanosoma birgei* during spring and autumn in a subtropical lake without piscivores (Lake Blanca, Uruguay; Iglesias et al., 2007). The preference of small omnivorous fish for emergent and submerged plant beds during day and the occurrence of *Chaoborus* in the pelagic at night determined the spatial distribution observed. In accordance with the suggestion of a poor refuge effect for zooplankton in warm lakes, no positive effect of plants on water clarity was found in nutrient-rich subtropical lakes in Florida, and nutrient rich lakes with high plant biomass often were turbid (Bachmann et al., 2002).

Considering the higher fish density in tropical systems and lack of refuges for zooplankton among macrophytes, fish biomass thresholds allowing major grazing pressure by zooplankton may be lower and fish reductions therefore have to be very large compared to temperate lakes to allow large-bodied grazers, like *Daphnia* spp, to dominate (Scasso et al., 2001; Mazzeo et al., unpublished data). The efficiency of decreasing planktivorous fish biomass by fish removal is further hampered by continuous and/or viviparous reproduction in many (sub)tropical species (Scasso et al., 2001 and Fig. 6). Therefore, it may be more difficult to provoke and maintain a trophic cascade effect in subtropical and tropical lakes than in temperate lakes. Accordingly, positive, but only short-term, cascading effects of a massive (>80%) fish kill (due to fungal infection) of the dominant planktivorous mosquito fish (*Gambusia affinis*) were found in Lake Naini Tal, India (Nagdali & Gupta, 2002). Zooplankton abundance increased, phytoplankton biomass and productivity declined, as did

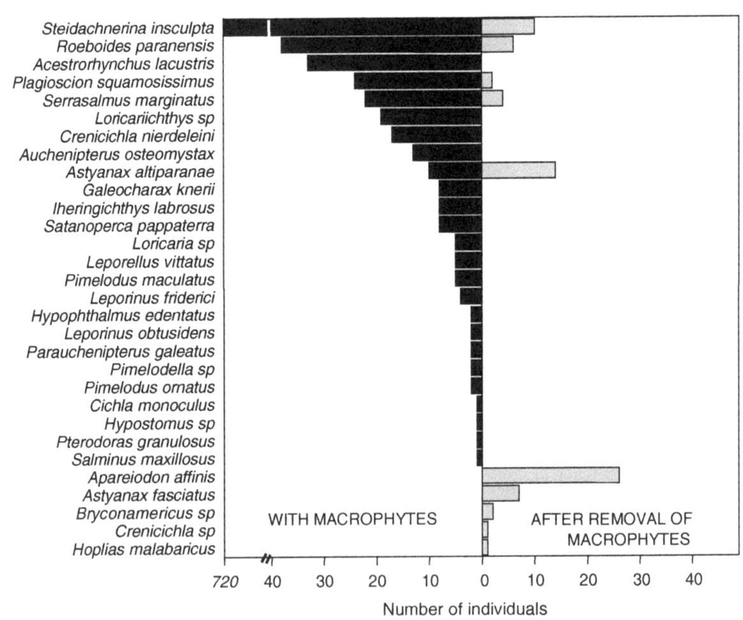

Fig. 5 Specific composition of fish assemblages in Santa Helena, Itaipú Reservoir (Paraná, Brazil) before (1987) and after (1988) a substantial removal of submerged macrophytes. Reproduced with kind permission from Agostinho et al. (2003)

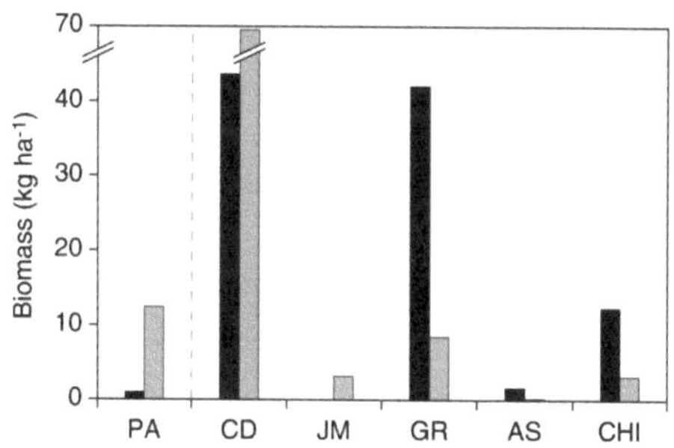

Fig. 6 Changes in the biomass of nekton at the beginning (black) and end (white) of summer 2003 in biomanipulated Lake Rodó (Uruguay) after an intensive programme of planktivorous fish removal. The biomanipulation had no effect on shrimps (*Paleomonetes argentinus*, PA) and viviparous fish, such as *Cnesterodon decemmaculatus* (CD) and *Jenynsia multidentata* (JM), while effective on the short-term for oviparous species: *Gymnogeophagus rhabdotus* (GR), *Astyanas* sp (AS) and *Cheirodon interruptus* (CHI). Moreover, no effect was observed on total fish biomass (from F. Quintans, unpublished data)

nutrient concentrations, resulting in higher water transparency. However, four months later, mosquito fish, plankton and nutrients had returned to previous levels.

In subtropical and tropical lakes, stocking of herbivorous silver carp (*Hypophthalmus molitrix*) has, with some success, been used to combat potentially toxic cyanobacteria (Starling et al., 1998; Datta & Jana, 1998). However, silver carp also consume zooplankton (Wu et al., 1997; Starling et al., 1998; Xu & Zie, 2004), which reduces grazing on small-sized phytoplankton. Therefore, it is unlikely that silver carp stocking, which is a symptom treatment rather than a restoration, will lead to a shift to a clear, macrophyte dominated stage (Matyas et al., 2003). Introduction of strictly piscivorous fish might be the only efficient biomanipulation strategy in (sub)tropical regions to control planktivorous fish. Ongoing mesocosm experiments using the sit-and-wait piscivore tararira (*Hoplias malabaricus*) support this hypothesis (Mazzeo et al., unpublished), but clearly more experiments are needed in this field before any firm conclusions can be drawn.

However, it is important to notice that trophic structure varies among warm lakes, depending on whether they are situated in wet or dry regions and temporal variations in temperature and salinity, and this may have implications on the restoration strategy as well. For example, tropical lakes are particularly sensitive to hydrological changes, which strongly affect their structure and functioning. In Cochabamba, Bolivia (17°S), switches from phytoplankton to macrophyte dominance in Lake Alalay have been associated to an increase in water level (Cadima, 1997). Higher densities and richness of zooplankton and fish were found during the low water season in several shallow lakes in Bolivia (Pouilly et al., 1999; Rejas & Maldonado, 2000). Water level changes also have strong effects on macrophyte abundance and water clarity in Turkish lakes (Beklioglu et al., 2003). These lakes, as well as northern Greek lakes, are, though warm in summer, subjected to cold winters, leading to a mixed fish community, often with low species richness compared to the subtropical and tropical lakes, and dominance of northern species (Beklioglu et al., 2003; Zalidis et al., 2002). These form an intermediate state between the north temperate lakes and winter warm temperate lakes and the subtropical/tropical lakes, which have to be taken into account when planning lake restoration (Beklioglu et al., 2003; Beklioglu & Tan, personal communication). Likewise, saline lakes are typically species poor (Williams, 1998) and, like in the subtropical lakes, submerged macrophytes appear to have little effect on water clarity in saline lakes (Moss, 1994a; Jeppesen et al., 1994) and form special cases.

In many experiments in temperate lakes, improvement in environmental state was observed without a trophic cascade after biomanipulation, i.e. without zooplankton species composition changes towards dominance of large-sized individuals and, consequently, higher phytoplankton grazing (Horppila et al., 1998). However, reductions in cyanobacteria and TP were observed, and water clarity increased. These phenomena may be due to reduced sediment P release and fish foraging following biomanipulation (Horppila et al., 1998). Biomanipulation also may reduce sediment nutrient release in tropical lakes, but the dominance of small fish species and better growth conditions for cyanobacteria suggest that effects may be temporary. Other methods, such as the combined use of aquatic plant harvests (mostly free-floating, if present) and hydraulic management, can be useful, though insufficient, for restoring small, shallow lakes in warm regions by decreasing internal nutrient loads (Rodríguez-Gallego et al., 2004). Sediment removal might be more attractive in such lakes, since P-content per unit volume of sediment exceeds that in the water column and plants (Søndergaard et al., 2003). However, drastic reduction of external nutrient loading seems to be the best method for restoring small or large subtropical and tropical lakes, but the scientific basis (e.g. nutrient threshold levels) to make decisions remains limited. Focus may well have to be on reducing external N loading as nitrogen has often been found to be the limiting nutrient for phytoplankton growth in tropical lakes (Lewis, 1996). Focus on nitrogen, in addition to phosphorus, is also relevant for temperate shallow lakes, as recent studies indicate that submerged macrophyte recovery after loading reduction is more likely to occur at moderately high TP when TN loading is low (Moss et al., 1994b; González Sagrario et al., 2005; James et al., 2005; Jeppesen et al., 2005b).

Conclusions

For small and large shallow lakes, the best strategy for improving the environmental state is external nutrient loading reduction. Like small lakes, large shallow lakes also may respond strongly and quickly to loading reduction. For smaller lakes resistant to nutrient loading reduction, physico-chemical and biological methods may enhance recovery. However, these methods are unlikely to be feasible in large shallow lakes. Sediment removal and chemical treatment, in particular, are less attractive because: (a) sediment transport and storage on land (if removed) will be expensive; (b) there is a large risk of sediment redistribution from resuspension; (c) the sediment P pool often is relatively small in large lakes. Fish manipulation also may be impractical because: (a) the large size makes it difficult to obtain a strong effect on the target fish population; and (b) fish population is among the first to respond to nutrient loading reduction, and natural changes in fish stock become difficult to discern in large lakes. If needed, transplantation/protection of aquatic plants (in sheltered beds) and seeds may be useful in large shallow lakes, the latter if pre-studies reveal that (re)-colonization is seed-limited. However, it is important first to evaluate whether wind (wave) disturbance will prevent colonization where seeds or plants are added.

Differences in biological interactions in cold temperate versus warm temperate-subtropical-tropical lakes make it difficult to apply biological restoration methods to warm lakes. Warm lakes often have prolonged growth seasons with a larger risk of long-lasting algal blooming and dense floating plant communities, higher dominance and abundance of small fish, higher fish aggregation in vegetation (loss of zooplankton refuge), more fish cohorts per year, and more omnivory by fish and less specialist piscivory. However, trophic structure varies among these lakes, depending on whether they are situated in wet or dry regions and temporal variations in temperature and salinity and this have implication for the restoration strategy.

Acknowledgements We thank the Danish counties for access to data from the Survey Programme of Danish Lakes. The work was supported by Eurolimpacs (GOCE-CT-2003-505540) and the Danish Natural Science Research Council funded research project "Consequences of weather and climate changes for marine and freshwater ecosystems. Conceptual and operational forecasting of the aquatic environment – CONWOY" (SWF: 2052-01-0034) and CLEAR (a Villum Kann Rasmussen centre of excellence project). We acknowledge the main Uruguayan Scientific Agencies for funding some of the research projects: CSIC, DINACYT, and PDT. Thanks to Sidinei M. Thomaz and Angelo A. Agostinho (Universidade Estadual de Maringá – Nupélia, Brazil) for their kind permission to use data and figures, and to Federico Quintans for data used in Fig. 6. We also thank Kathe Møgelvang and Anne Mette Poulsen for technical assistance. The two unknown reviewers are thanked for most valuable scientific comments and editorial suggestions and the organisers of the conference about Lake Taihu for inviting this paper.

References

Agostinho, A. A., L. C. Gomes & H. J. Ferreira Jr. 2003. Relações entre macrófitas aquáticas e fauna de peixes. In Thomaz, S. M. & L. M. Bini (eds), Ecologia e manejo de macrófitas aquáticas. EDUEM, Maringá, 261–279.

Aguiaro, T. & E. P. Caramaschi, 1998. Trophic guilds in fish assemblages in three coastal lagoons of Rio de Janeiro State (Brazil). Verhandlungen der Internationale Vereinigung der Limnologie 26: 2166–2169.

Aguiaro, T., C. W. C. Branco, J. R. Verani & E. P. Caramaschi, 2003. Diet of the clupeid fish *Platanichthys platana* (Regan, 1917) in two different Brazilian coastal lagoons. Brazilian Archives of Biology and Technology 46: 215–222.

Aguilera, X. & E. Goitia, 1999. Structure of the zoobenthic community of the Laguna Bufeos (Cochabamba, Bolivia). Revista Boliviana de Ecología 6: 55–64.

Arcifa, M. S., E. A. T. Gomes & A. J. Meschiatti, 1992. Composition and fluctuations of the zooplankton of a tropical Brazilian reservoir. Archiv für Hydrobiologie 123: 479–495.

Bachmann, R. W., C. A. Horsburgh, M. V. Hoyer, L. K. Mataraza & D. E. Canfield Jr. 2002. Relations between trophic state indicators and plant biomass in Florida lakes. Hydrobiologia 470: 219–234.

Beklioglu, M., Ö. Ince & I. Tüzün, 2003. Restoration of eutrophic Lake Eymir, Turkey, by biomanipulation undertaken following a major external nutrient control I. Hydrobiologia 489: 93–105.

Benndorf, J., 1995. Possibilities and limits for controlling eutrophication by biomanipulation. International Revue der Gesamten Hydrobiologie 80: 519–534.

Berg, S., E. Jeppesen, M. Søndergaard, 1997. Pike (*Esox lucius* L.) stocking as a biomanipulation tool. 1. Effects on the fish population in Lake Lyng (Denmark). Hydrobiologia 342/343: 311–318.

Blanco, R., S. Romo, M. J. Villena & S. Martinez, 2003. Fish communities and food web interaction in some shallow Mediterranean lakes. Hydrobiologia 506/509: 473–480.

Blindow, I., A. Hargeby, B. Wagner & G. Andersson, 2000. How important is the crustacean plankton for the maintenance of water clarity in shallow lakes with abundant submerged vegetation? Freshwater Biology 44: 185–197.

Boschi, E., 1981. Fauna de agua dulce de la República Argentina. Vol. XXVI, Decapoda Natantia. FECIC, Buenos Aires, 61 pp.

Branco, C. W. C. & P. A. C. Senna, 1991. The taxonomic elucidation of the Paranoá Lake (Brasília, Brazil) problem: *Cylindrospermopsis raciborskii*. Bulletin du Jardin Botanique National de Belgique 61: 85–91.

Branco, C. W. C. & P. A. C. Senna, 1994. Factors influencing the development of *Cylindrospermopsis raciborskii* and *Microcystis aeruginosa* in Paranoá Reservoir, Brasilia, Brazil. Algological Studies 75: 85–96.

Branco, C. W. C. & P. A. C. Senna, 1996. Plankton studies in Paranoá Reservoir, Brasília, Brazil. I. Relations among heterotrophic bacteria, chlorophyll *a*, total phytoplankton, total zooplankton and physichochemical factors. Hydrobiologia 337: 171–181.

Branco, C. W. C., T. Aguiaro, F. A. Esteves & E. P. Caramaschi, 1997. Food sources of the Teleost *Eucinostomus argenteus* in two coastal lagoons of Brazil. Studies on Neotropical Fauna & Environment 32: 33–40.

Branco, C. W. C. & C. B. Cavalcanti, 1999. A ecologia das Comunidades Planctônicas no Lago Paranoá. Chapter 19. In Henry, R. (ed.), Ecologia de reservatórios: estrutura, função e aspectos sociais. Editora da UNESP, Botucatu, 573–595.

Branco, C. W. C., M.-I. Rocha, G. F. S. Pinto, G. A. Gômara & R. D. Filippo, 2002. Limnological features of Funil Reservoir (Brazil, RJ) and indicator properties of rotifers and cladocerans of the zooplankton community. Lakes and Reservoirs: Research and Management 7: 87–92.

Breukelaar, A. W., E. H. R. R. Lammens, J. P. G. Klein Breteler & I. Tatrai, 1994. Effects of benthivorous bream (*Abramis brama* L.) and carp (*Cyprinus caprio* L.) on sediment resuspension and concentration of nutrients and chlorophyll *a*. Freshwater Biology 32: 113–121.

Burks, R. L., E. Jeppesen, D. M. Lodge & T. Lauridsen, 2002. Diel horizontal migration of zooplankton: costs and benefits of inhabiting littoral zones. Freshwater Biology 47: 343–365.

Burnett, J. A. B., S. P. Mattos & N. M. P. Azzolin, 2001. Intervenções da Companhia de Saneamento. In Olhares sobre o Lago Paranoá. Secretaria de Meio Ambiente e Recursos Hídricos do DF, Brasília, Capítulo IX: 199–211.

Cadima, M. M., 1997. Algae and macrophytes of the lake Alalay (Cochabamba, Bolivia). Revista Boliviana de Ecologia 3: 35–46.

Collins, P, 1999. Feeding of *Palaemonetes argentinus* (Decapoda: Palaemonidae) from an oxbow lake of the Paraná River, Argentina. Journal of Crustacean Biology 19: 485–492.

Collins, P. & J. C. Paggi, 1998. Feeding ecology of *Macrobrachium borelli* (Nobili) (Decapoda: Palaemonidae) in the food valley of the River Paraná, Argentina. Hydrobiologia 363: 21–20.

Conrow, R., A. V. Zale & R. W. Gregory, 1990. Distributions and abundances of early stages of fishes in a Florida lake dominated by aquatic macrophytes. Transactions of the American Fisheries Society 119: 521–528.

Cooke, G. D., 1993. Phosphorus inactivation and sediment oxidation. In Cooke, G. D., E. B. Welch, S. A. Peterson & P. R. Newroth (eds), Restoration and Management of Lakes and Reservoirs, 2nd ed. Lewis Publishers, Boca Raton: 161–209.

Cooke, G. D., E. B. Welch, S. A. Peterson & P. R. Newroth, 1993. Restoration and Management of Lakes and Reservoirs. Boca Raton, Lewis Publishers, Florida.

Cronberg, G., 1976. The Lago Paranoá Restoration Project: Phytoplankton ecology and taxonomy. A preliminary report. Brasília (Project FAHO – World Health Organization, 76/PW/BRA/2000, 24 pp.

Datta, S. & B. B. Jana, 1998. Control of blooom in a tropical lake: grazing efficiency of some herbivorous fishes. Journal of Fish Biology 53: 12–24.

Drenner, R. & D. Hambright, 1999. Review: biomanipulation of fish assemblages as a lake restoration technique. Archiv für Hydrobiologie 146: 129–165.

Dumont, H. J., 1994. On the diversity of the Cladocera in the tropics. Hydrobiologia 272: 27–38.

Elmoor-Loureiro, L. M. A., L. Mendonça-Galvão & C. Padovesi-Fonseca, 2004. New cladocerans from Lake Paranoá, Central Brazil. Brazilian Journal of Biology 64: 415–422.

Fernando, C. H., 1994. Zooplankton, fish and fisheries in tropical freshwaters. Hydrobiologia 272: 105–123.

Garcia, P. R., S. Nandini, S. S. S. Sarma, E. R. Valderrama, I. Cuesta & M. D. Hurtado, 2002. Seasonal variations of zooplankton abundance in the freshwater reservoir Valle de Bravo (Mexico). Hydrobiologia 467: 99–108.

González, Sagrario, M. A., E. Jeppesen, J. Gomà, M. Søndergaard, T. Lauridsen & F. Landkildehus, 2005 Does high nitrogen loading prevent clear-water conditions in shallow lakes at moderately high phosphorus concentrations? Freshwater Biology 50: 27–41.

González, E. J., G. Bernal, H. Hernández, M. L. Matos & C. Penaherrera, 2005. De qué se alimenta el zooplancton de los embalses venezolanos? Memorias del Instituto de Biología Experimental 4: 137–141.

Hanson, M. A. & M. G. Butler, 1990. Early responses of plankton and turbidity to biomanipulation in a shallow prairie lake. Hydrobiologia 200/201: 317–327.

Hansson, L.-A., H. Annadotter, E. Bergman, S. F. Hamrin, E. Jeppesen, T. Kairesalo, E. Luokkanen, P.-Å. Nilsson, M. Søndergaard & J. Strand, 1998. Biomanipulation as an application of food chain theory: constraints, synthesis and recommendations for temperate lakes. Ecosystems 1: 558–574.

Horppila, J., H. Peltonen, T. Malinen, E. Juokkanen & T. Kairesalo, 1998. Top-down or bottom-up effects by fish – issues of concern in biomanipulation of lakes. Restoration Ecology 6: 1–10.

Iglesias, C., G. Goyenola, N. Mazzeo, M. Meerhoff, E. Rodó & E. Jeppesen, 2007. Horizontal dynamics of zooplankton in subtropical Lake Blanca (Uruguay) hosting multiple zooplankton predators and aquatic plant refuges. Hydrobiologia DOI 10.1007/s10750-007-0599-4 (in press).

Infante, A., 1978. Natural food of herbivorous zooplankton of lake Valencia (Venezuela). Archiv Hydrobiologia 82: 347–358.

Jacobsen, B. A., R. S. Hansen, B. Moeslund, E. Jørgensen & G. Jensen, 2004. Arresø-tilstand og udvikling 2003. Report from Frederiksborg Amt, Teknik og Miljø, Denmark, 115 pp.

James, C., J. Fisher, V. Russell, S. Collings & B. Moss, 2005. Nitrate availability and plant species richness: implications for management of freshwater lakes. Freshwater Biology 50: 1049–1063.

Jeppesen, E., M. Søndergaard, E. Kanstrup, B. Petersen, R. B. Henriksen, M. Hammershøj, E. Mortensen, J. P. Jensen & A. Have, 1994. Does the impact of nutrients on the biological structure and function of brackish and freshwater lakes differ? Hydrobiologia 275/276: 15–30.

Jeppesen, E., T. L. Lauridsen, T. Kairesalo & M. Perrow, 1997. Impact of submerged macrophytes on fish-zooplankton relationships in lakes. In Jeppesen, E., Ma. Søndergaard, Mo. Søndergaard & K. Christoffersen (eds), The Structuring Role of Submerged Macrophytes in Lakes. Ecological Studies, Vol. 131, Springer Verlag, 91–115.

Jeppesen, E., J. P. Jensen, M. Søndergaard, T. Lauridsen & F. Landkildehus, 2000. Trophic structure, species richness and biodiversity in Danish lakes: changes along a phosphorus gradient. Freshwater Biology 45: 201–213.

Jeppesen, E. & I. Sammalkorpi, 2002. Lakes. In Perrow, M. & T. Davy (eds), Handbook of Restoration Ecology, Chapter 14, Cambridge University Press, 297–324.

Jeppesen, E., J. P. Jensen & M. Søndergaard, 2002. Response of phytoplankton, zooplankton and fish to re-oligotrophication: an 11-year study of 23 Danish lakes. Aquatic Ecosystem Health & Management 5: 31–43.

Jeppesen, E., J. P. Jensen, M. Søndergaard, M. Fenger-Grøn, K. Sandby, P. Hald, Møller & U. H. Rasmussen, 2004. Does fish predation influence zooplankton community structure and grazing during winter in north-temperate lakes? Freshwater Biology 49: 432–447.

Jeppesen, E., J. P. Jensen, M. Søndergaard & T. Lauridsen, 2005a. Response of fish and plankton to nutrient loading reduction in 8 shallow Danish lakes with special emphasis on seasonal dynamics. Freshwater Biology 50: 1616–1627.

Jeppesen, E., M. Søndergaard, J. P. Jensen, T. Lauridsen, L. Liboriussen, R. Bjerring Hansen, L. S. Johansson & F. Landkildehus, 2005b. The response of north temperate lakes to reduced nutrient loading with special emphasis on shallow Danish lakes. Verhandlungen der Internationale Vereinigung der Limnologie 29: 115–122.

Jeppesen, E., M. Søndergaard, J. P. Jensen, K. Havens, O. Anneville, L. Carvalho, M. F. Coveney, R. Deneke, M. T. Dokulil, B. Foy, D. Gerdeaux, S. E. Hampton, S. Hilt, K. Kangur, J. Köhler, E. Lammens, T. L. Lauridsen, M. Manca, R. Miracle, B. Moss, P. Nõges, G. Persson, G. Phillips, R. Portielje, S. Romo, C. L. Schelske, D. Straile, I. Tatrai, E. Willén & M. Winder, 2005c. Lakes' response to reduced nutrient loading–an analysis of contemporary data from 35 European and North American long term studies. Freshwater Biology 50: 1747–1771.

Jeppesen, E., M. Søndergaard, N. Mazzeo, M. Meerhoff, C. Branco, V. Huszar & F. Scasso, 2005d. Lake restoration and biomanipulation in temperate lakes: relevance for subtropical and tropical lakes. Chapter 11. In Reddy, M. V. (ed.), Tropical Eutrophic Lakes: Their Restoration and Management. Science Publishers, Enfield, 331–359.

Kalff, J., 2002. Limnology – Inland Water Ecosystems. Prentice-Hall Inc.

Ke, X. S. & W. Li, 2006. Germination requirement of *Vallisneria natans* seeds: implications for lake restoration in Chinese Lakes. Hydrobiologia 559: 357–362.

Klein, T., 1989. Søerne i de gode gamle dage-om den "Økologiske baggrundstilstand". Vand & Miljø 5: 211–215.

Kristensen, P., M. Søndergaard & E. Jeppesen, 1992. Resuspension in a shallow hypertrophic lake. Hydrobiologia 228: 101–109.

Kronvang, B., E. Jeppesen, D. J. Conley, M. Søndergaard, S. E. Larsen, N. B. Ovesen & J. Carstensen, 2005. Nutrient pressures and ecological responses to nutrient loading reductions in Danish streams, lakes and coastal waters. Journal of Hydrology 304: 274–288.

Lathrop, R.C., B. M. Johnson, T. B. Johnson, M. T. Vogelsang, S. R. Carpenter, T. R. Hrabik, J. F. Kitchell, J. J. Magnuson, L. G. Rudstam & R. S. Stewart, 2002. Stocking piscivores to improve fishing and water clarity: a synthesis of the Lake Mendota biomanipulation project. Freshwater Biology 47: 2410–2424.

Lauridsen, T. L., L. J. Pedersen, E. Jeppesen & M. Søndergaard, 1996. The importance of macrophyte bed size for cladoceran composition and horizontal migration in a shallow lake. Journal of Plankton Research 18: 2283–2294.

Lauridsen, T., H. Sandsten & P. H. Møller, 2003. The restoration of a shallow lake by introducing *Potamogeton* spp.: the impact of waterfowl grazing. Lakes & Reservoirs: Research & Management 8: 177–187.

Lazzaro, X., 1997. Do the trophic cascade hypothesis and classical biomanipulation approaches apply to tropical lakes and reservoirs? Verhandlungen der Internationale Vereinigung der Limnologie 26: 719–730.

Lewis, W. M. Jr. 1996. Tropical lakes: how latitude makes a different. In Schiemer, F. & K. T. Boland (eds), Perspectives in Tropical Limnology. SPB Academic Publishing B.V., Amsterdam, The Netherlands, 43–64.

Lorier, E. & N. Berois, 1995. Reproducción y nutrición embrionaria en *Cnesterodon decemmaculatus* (Teleoste: Poeciliidae). Revista Brasileira de Biologia 55: 27–44.

Marsden, S., 1989. Lake restoration by reducing external phosphorus loading: the influence of sediment phosphorus release. Freshwater Biology 21: 139–162.

Matyas, K, I. Oldal, J. Korponai, I. Tatrai & G. Paúlovits, 2003. Indirect effect of different fish communities in nutrient chlorophyll relationship in shallow hypertrophic water quality reservoir. Hydrobiologia 504: 231–239.

Mazzeo, N., L. Rodríguez-Gallego, G. Kruk, M. Meerhoff, J. Gorga, G. Lacerot, F. Quintans, M. Loureiro, D. Larrea & F. García-Rodrígues, 2003. Effect of *Egeria densa* Planch. beds on a shallow lake without piscivorous fish. Hydrobiologia 506/509: 591–602.

Meerhoff, M., N. Mazzeo, B. Moss & L. Rodríguez-Gallego, 2003. The structuring role of free-floating versus submerged plants in a subtropical shallow lake. Aquatic Ecology 37: 377–391.

Meerhoff, M., C. Fosalba, C. Bruzzone, N. Mazzeo, W. Noordoven & E. Jeppesen, 2005. An experimental study of habitat choice of Daphnia: plants signal danger more than refuge in subtropical lakes. Freshwater Biology 51: 1320–1330.

Meijer, M.-L., I. de Boois, M. Scheffer, R. Portielje & H. Hosper, 1999. Biomanipulation in the Netherlands: an evaluation of 18 case studies in shallow lakes. Hydrobiologia 408/409: 13–30.

Meschiatti, A. J., M. S. Arcifa & N. Fenerich-Verani, 2000. Fish communities associated with macrophytes in Brazilian floodplain lakes. Environmental Biology of Fishes 58: 133–143.

Mitchell, S. F. & M. Perrow, 1997. Interactons between grazing birds and macrophytes. In Jeppesen, E., Ma. Søndergaard, Mo. Søndergaard, K. Christoffersen (eds), The Structuring Role of Submerged Macrophytes in Lakes. Springer-Verlag, New York, 175–196.

Moss, B., 1990. Engineering and biological approaches to the restoration from eutrophication of shallow lakes in which aquatic plant communities are important components. Hydrobiologia 200/201: 367–377.

Moss, B., 1994a. Brackish and fresh-water shallow lakes – different systems or variations on the same theme? Hydrobiologia 276: 1–14.

Moss, B., S. McGowan & L. Carvalho, 1994b. Determination of phytoplankton crops by top-down and bottom-up mechanisms in a group of English lakes, the West Midland Meres. Limnology and Oceanography 39: 1020–1030.

Moss, B., D. Stephen, D. M. Balayla, E. Bécares, S. E. Collings, C. Fernández-Aláez, M. Fernández-Aláez, C. Ferriol, P. García, J. Gomá, M. Gyllström, L.-A. Hansson, J. Hietala, T. Kairesalo, M. R. Miracle, S. Romo, J. Rueda, V. Russell, A. Ståhl-Delbanco, M. Svensson, K. Vakkilainen, M. Valentín, W. J. Van de Bund, E. Van Donk, E. Vicente & M. J. Villena, 2004. Continental-scale patterns of nutrient and fish effects on shallow lakes: synthesis of a pan-European mesocosm experiment. Freshwater Biology 49: 1633–1649.

Nagdali, S. S. & P. K. Gupta, 2002. Impact of mass mortality of a mosquito fish, *Gambusia affinis* on the ecology of a fresh water eutrophic lake (Lake Naini Tal, India). Hydrobiologia 468: 45–52.

Oliveira, L. P. H. & L. Krau, 1970. Hidrobiologia geral aplicada particularmente a veiculadores de esquistossomos: hipereutrofia, mal moderno das águas. Memórias do Instituto Oswaldo Cruz 68: 89–118.

Padovesi-Fonseca, C., L. Mendonça-Galvão, D. F. Pereira, M. G. Philomeno & D. L. P. Rocha, 2001. O zooplâncton do Lago Paranoá. In Olhares sobre o Lago Paranoá. Secretaria de Meio Ambiente e Recursos Hídricos do DF, Brasília, 115–117.

Paugy, D. & C. Lévêque, 1999. La Reproduction. In Lévêque, C. & D. Paugy (eds), Les poissons des eaux continentales africaines: diversité, écologie, utilisation par ĺhomme. IRD Editions, Paris, 129–152.

Peltonen, H., J. Ruuhijärvi, M. Olin, T. Malinen, J. Horppila & J. Keto, 1999. The effects of food-web management and environmental variations on fish assemblage dynamics in a north-temperate lake. Journal of Fish Biology 55: 54–67.

Pereira, C. E. B. & C. G. P. Cavalcanti, 1997. Lago Paranoá – Rumo à Recuperação. Technical Report, CAESB, 15 pp.

Persson, A., 1997. Phosphorus release by fish in relation to external and internal load in a eutrophic lake. Limnology & Oceanography 43: 577–583.

Perrow, M. R., M.-L. Meijer, P. Dawidowicz & H. Coops, 1997. Biomanipulation in shallow lakes: state of the art. Hydrobiologia 342/343: 355–365.

Phillips, G., A. Kelly, J-A. Pitt, R. Sanderson & E. Taylor, 2005. The recovery of Barton Broad, a very shallow eutrophic lake, 20 years after the control of effluent derived phosphorus. Freshwater Biology 50: 1628–1638.

Pinto-Coelho, R. M. & A. Giani, 1985. Variações sazonais do fitoplâncton e fatores físico-químicos no Reservatório do Paranoá, Brasília. Ciência e Cultura 37: 2000–2006.

Pouilly, M., C. Ibáñez, M. Gutiérrez & T. Yunoki, 1999. Ecological functioning of floodplain lakes of Mamoré river (Beni, Bolivia). Revista Boliviana de Ecologia 6: 41–54.

Prejs, A., A. Martyniak, S. Boroń, P. Hliwa & P. Koperski, 1994. Food web manipulation in a small eutrophic Lake Wirbel, Poland: effect of stocking with juvenile pike on planktivorous fish. Hydrobiologia 275/276: 65–70.

Rejas, D. & M. Maldonado, 2000. Spatial-temporal variations in the composition of the fish community of a shallow lake in the floodplain of the Ichilo river (Cochabamba, Bolivia). Revista Boliviana de Ecologia 7: 37–46.

Quirós, R., 1998. Fish effects on trophic relationships in the pelagic zone of lakes. Hydrobiologia 361: 101–111.

Ribeiro, M. C. L. B., F. L. R. M. Starling, T. Walter & E. M. Farah, 2001. Peixes. In Olhares sobre o Lago Paranoá (ed.), Secretaria de Meio Ambiente e Recursos Hídricos do DF, Brasília, 121–128.

Rodríguez-Gallego, L., N. Mazzeo, M. Meerhoff, J. Clemente, C. Kruk, F. Scasso, G. Lacerot, J. García & F. Quintans, 2004. Effects of a water recirculation system covered by free-floating plants on the restoration of a hypertrophic subtropical lake. Lakes and Reservoirs: Research and Management 9: 205–213.

Romo, S., M.-J. Villena, M. Sahuquillo, J. M. Soria, T. Alfonso, E. Vicente & M. R. Miracle, 2005. Response of a shallow Mediterranean lake to nutrient diversion: does it follow similar patterns as northern shallow lakes? Freshwater Biology 50: 1706–1717.

Sas, H. (ed.), 1989. Lake restoration by reduction of nutrient loading. Expectation, experiences, extrapolation. Acad. Ver. Richardz Gmbh.

Scasso, F., N. Mazzeo, J. Gorga, C. Kruk, G. Lacerot, J. Clemente, D. Fabián & S. Bonilla, 2001. Limnological changes of a subtropical shallow hypertrophic lake during its restoration. Two years of whole-lake experiments. Aquatic Conserversation: Marine Freshwater Ecosystems 11: 31–44.

Scheffer, M., S. H. Hosper, M.-L. Meijer, B. Moss & E. Jeppesen, 1993. Alternative equilibria in shallow lakes. Trends in Ecology and Evolution 8: 275–279.

Skov, C., & S. Berg, 1999. Utilization of natural and artificial habitats by YOY pike in a biomanipulated lake. Hydrobiologia 408/409: 115–122.

Starling, F. L. R. M., 1993a. Control of eutrophication by silver carp (*Hypophthalmus molitrix*) in the tropical Paranoá Reservoir (Brasília, Brazil): a mesocosm experiment. Hydrobiologia 257: 143–152.

Starling, F. L. R. M., 1993b. Análise experimental dos efeitos da tilápia do Congo (*Tilapia rendalli*) e carpa prateada (*Hypophthalmus molitrix*) sobre a estrutura da comunidade planctônica do Lago Paranoá, Brasília (DF). Acta Limnologica Brasileira 4: 144–156.

Starling, F. L. R. M. & A. J. A. Rocha, 1990. Experimental study of the impacts of planktivorous fishes on plankton community and eutrophication of a tropical Brazilian reservoir. In Gulati, R. D., E. H. R. R. Lammens, M. L. Meijer & E. van Donk (eds), Biomanipulation – Tool for Water Management. Developments in Hydrobiology 61. Kluwer Academic Publishers, Dordrecht, 581–591.

Starling, F. L. R. M., M. Beveridge, X. Lazzaro & D. Baird, 1998. Silver carp biomass effectsd on the plankton community in Paranoa Reservoir (Brazil) and an assessment of its potential for improving water quality in lacustrine environments. Internationale Revue der Hydrobiologie 83: 499–507.

Søndergaard, M., L. Olufsen, T. L. Lauridsen, E. Jeppesen, & T. Vindbæk Madsen, 1996. The impact of grazing waterfowl on submerged macrophytes: in situ experiments in a shallow eutrophic lake. Aquatic Botany 53: 73–84.

Søndergaard, M., J. P. Jensen & E. Jeppesen, 2001. Retention and internal loading of phosphorus in shallow, eutrophic lakes. The Scientific World 1: 427–442.

Søndergaard, M., J. P. Jensen, E. Jeppesen & P. H. Møller, 2002a. Seasonal dynamics in the concentrations and retention of phosphorus in shallow Danish lakes after reduced loading. Aquatic Ecosystem Health & Management 5: 19–23.

Søndergaard, M., K.-D. Wolter & W. Ripl, 2002b. Chemical treatment of water and sediments with special reference to lakes. In Perrow, M. & T. Davy (eds), Handbook of Restoration Ecology, Chapter 10. Cambridge University Press, 184–205.

Søndergaard, M., J. P. Jensen & E. Jeppesen, 2005. Seasonal response of nutrients to reduced phosphorus loading in 12 Danish lakes. Freshwater Biology 50: 1605–1615.

Søndergaard, M., J. P. Jensen & E. Jeppesen, 2003. Role of sediment and internal loading of phosphorus in shallow lakes. Hydrobiologia 506/509: 135–145.

Thomaz, S. M. & L. M. Bini, 2003. Ecologia e manejo de macrófitas aquáticas. EDUEM, Maringá, 341 pp.

Timms, R. M. & B. Moss, 1984. Prevention of growth of potentially dense phytoplankton populations by zooplankton grazing in the presence of zooplanktivorous fish, in a shallow wetland ecosystem. Limnology & Oceanography 29: 472–486.

Van der Molen, D.T. R. Portielje, 1999. Multi-lake studies in The Netherlands: trends in eutrophication. Hydrobiologia 409: 359–365.

Weisner, S. E. B. & J. A., Strand, 2002. Ecology and management of plants in aquatic systems. In Perrow,

M. R., & A. J. Davy (eds), Handbook of Ecological Restoration. Vol. 1. Principles of restoration. Cambridge University Press, Cambridge, 242–256.

Williams, W. D., 1998. Salinity as determinant of the structure of biological communities in salt lakes. Hydrobiologia 381: 191–201.

Wu, J., P. Xie, M. Dai & J. Wang, 1997. Effects of silver carp density on zooplankton and water quality: implications for eutrophic lakes in China. Journal of Freshwater Ecology 12: 437–444.

Xu, J. & P. Xie, 2004. Studies on the food web structure of Lake Donghu using stable carbon and nitrogen isotope ratios. Journal of Freshwater Ecology 19: 645–650.

Yafe, A., M. Loureiro, F. Scasso & F. Quintans, 2002. Feeding of two Cichlidae species in a hypertrophic urban lake. Iheringia Série Zoologica, Porto Alegre 92: 73–79.

Zalidis, G. C., T. L. Crisman, & P. A. Gerakis, 2002. Restoration of Mediteranean wetlands. Hellnic Ministry of Environment. Physical Planning and Public Works, Athens and Greek Biotope/Wetland Centre, Thermi, Greece, 237 pp.

Hydrobiologia (2007) 581:287–295
DOI 10.1007/s10750-006-0516-2

EUTROPHICATION IN LAKES

Dredging effects on P status and phytoplankton density and composition during winter and spring in Lake Taihu, China

Xiuyun Cao · Chunlei Song · Qingman Li · Yiyong Zhou

Abstract Phytoplankton density and composition, together with phosphorus (P) concentrations and size-fractionated alkaline phosphatase activity (APA), were investigated in dredged and undredged zones in Lake Taihu from January to April 2004. P concentrations were also determined in the corresponding interstitial water. Enzyme Labeled Fluorescence (ELF) was used for localizing extracellular phosphatase on phytoplankton cell membranes in April. The increase in phytoplankton density was paralleled by a significant increase in soluble reactive P (SRP) concentrations in the water column and interstitial water at all sites from January to April, with chlorophyte gradually becoming dominant. In February, at the undredged site, more algae dominated by chlorophyte occurred in overlying water, rather than in the surface, coinciding with higher SRP concentrations in overlying and interstitial water. Therefore, P status in the bottom is important to phytoplankton development in terms of density and composition. Undredged sites had higher SRP concentrations in interstitial water than dredged sites. Furthermore, Higher APA was observed, accompanied by higher dissolved organic P (DOP) and lower total P at the undredged site in February. Enzymatic hydrolysis of DOP may have been an additional source of P for phytoplankton. In April, *Schroederia* sp. was ELF labeled in surface water at the dredged site, which showed markedly lower SRP concentration, but not at the undredged site with higher SRP concentration. Thus, the dredging might regulate algal density and composition in water column by reducing P bioavailability.

Keywords Phytoplankton · Phosphorus · Alkaline phosphatase · ELF · Sediment dredging · Shallow lakes · Eutrophication

Introduction

One symptom of lake eutrophication is an excess phytoplankton growth, which is controlled by various factors. Besides light and temperature, phytoplankton requires a wide range of inorganic nutrients for growth. These include potentially growth-limiting nutrients, such as carbon dioxide, nitrogen, phosphorus (P), and silicon. Phosphorus is the most commonly limiting nutrient in oligotrophic freshwaters (Graham & Wilcox, 2000). However, high phosphorus loading complicates the situation in eutrophicated waters. Lake

Guest editors: B. Qin, Z. Liu & K. Havens
Eutrophication of shallow lakes with special reference to Lake Taihu, China

X. Cao · C. Song · Q. Li · Y. Zhou (✉)
Institute of Hydrobiology, The Chinese Academy of Sciences, Wuhan 430072, China
e-mail: zhouyy@ihb.ac.cn

recovery often is delayed after external loading reductions due to internal P load (Jeppesen et al., 1990). Phosphorus release from the sediment into the lake water may be so intense and persistent that it may substantially increases the bioavailable P pool and consequently also the algal biomass (Knuttila et al., 1994). Dredging to remove surface sediment is widely used to reduce the internal loading, but its effects on phytoplankton are still equivocal. Cell counts and chlorophyll *a* concentrations suggest that after sediment removal, phytoplankton quantity may remain virtually unchanged compared with the period before sediment removal (Poulickova et al., 1998), while the eutrophic Lake Trummen (Sweden) was restored by removal of the upper most 0.5 m nutrient-rich sediment layer by suction dredging, which drastically reduced nutrient concentrations and phytoplankton biomass (Cronberg, 1982).

Phytoplankton can take up only inorganic orthophosphate (Overbeck, 1991). However, an important indirect source is dissolved organic phosphorus (DOP) after hydrolysis with extracellular phosphatase. Phytoplankton may adapt to P limitation by changing species composition and/or increasing alkaline phosphatase activity (APA; Vrba et al., 1993). A method known as ELF (enzyme labeled fluorescence; Gonzáles-Gil et al., 1998), allows for phosphatase labeling and direct microscopic detection. The product of ELF-phosphate hydrolysis can fluorescently tag individual algal cells at or near the fixed enzymatic activity sites (Rengefors et al., 2001).

A comparative study was conducted at dredged and undredged sites in Lake Taihu from January to April 2004. The objective of this study was to investigate the dredging effects on P status and phytoplankton density and composition concerning a large, eutrophic, shallow Chinese lake. We investigated the fluctuations of phytoplankton in terms of density and composition, coupled with phosphorus concentrations and phosphatase activities in the water column. ELF was used to localize extracellular phosphatase on algal cells, and P concentrations were determined in interstitial water.

Materials and methods

Study site and sampling

The common description of Lake Taihu was given elsewhere (Qin et al., 2006). Dredging was carried

Fig. 1 The map of sampling sites

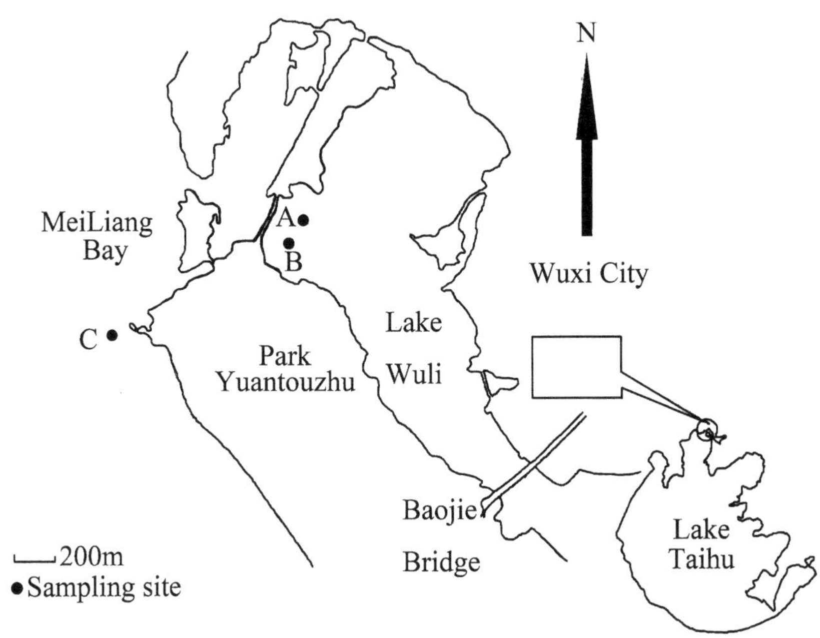

out to remove the surface sediment in order to improve the water quality in 2003–2004. Three sampling sites (Fig. 1), Site A (31°32′4.6″ N, 120°13′22.6″ E) and Site B (31°32′1.7″ N, 120°13′18.7″ E) located in Wuli Bay area and Site C (31°31′32.3″ N, 120°12′37.6″ E) located in the main embayment (Meiliang Bay) were sampled in January, 2004. Site A was dredged before sampling, Site B was sampled before dredging in January and after dredging in February and April, and Site C was undredged. Water samples were collected with a Friedinger sampler at the surface and in the water layer overlying the sediment. Phytoplankton samples (1 l) were preserved immediately with Lugol's iodine solution. Another liter of water was taken for P and phosphatase analysis. Sediment was collected using a Peterson dredge. Interstitial water was extracted from the sediment by centrifugation at 3,000 rpm for 30 min.

Phytoplankton density

Preserved phytoplankton samples in the surface and overlying water were concentrated to 30 ml after sedimentation for one week. After mixing, 0.1 ml concentrated samples were counted directly using a Zeiss microscope at ×640 magnification. Phytoplankton species were identified according to Hu et al. (1979).

P concentration

Total P (TP), dissolved total P (DTP) and soluble reactive phosphorus (SRP) were determined in the surface, overlying and sediment interstitial water. TP was determined following digestion according to Golterman et al. (1978). Phosphorus compounds were hydrolysed by H_2SO_4 and $K_2S_2O_8$ to orthophosphate, which was then determined after Murphy and Riley (1962). 0.5 ml H_2SO_4 and 1 g $K_2S_2O_8$ were added to 50 ml samples. Flasks were covered with an inverted beaker and autoclaved at 120°C for 2 h. DTP was determined using water samples filtered through a 0.45 μm membrane filters (Millipore) using the same procedure. SRP was determined according to Murphy & Riley (1962). DOP was calculated as follows: DOP = DTP-SRP.

Phosphatase activity and microscopic phosphatase detection

Extracellular alkaline phosphatase activity (APA) in the water samples was determined using a procedure modified from Gage and Gorham (1985) and Boon (1989). Triplicate 5 ml water samples were supplied with Tris–HCl buffer (pH = 8.5, final concentration 13 mmol l^{-1}), Na_3N (final concentration 5 mmol l^{-1}), and *p*-nitrophenyl phosphate (*p*NPP, final concentration 0.3 mmol l^{-1}), and incubated at 37°C for 4 h. Absorbency of *p*-nitrophenol was measured spectrophotometrically at 410 nm. APA was determined in unfiltered samples (APA_T) and in samples filtered through 3 μm and 0.45 μm membrane filters ($APA_{<3}$ and $APA_{<0.45}$, respectively). The contribution of APA to the coarser ($APA_{>3}$) and finer ($APA_{0.45-3}$) fractions were calculated as follows: $APA_{>3} = APA_T - APA_{<3}$, $APA_{0.45-3} = APA_{<3} - APA_{<0.45}$ (Chróst et al., 1984).

ELF 97 phosphate (ELFP, Molecular Probes) used for microscopic detection of extracellular PA in the phytoplankton according to the protocol by Štrojsová et al. (2003). Incubations started by adding the ELFP solution (final concentration 27 μmol l^{-1}), and samples were incubated at room temperature for 2–4 h. Each incubation was terminated by transferring the sample to a filter holder (diameter 7 or 10 mm, depending on phytoplankton density) with a membrane filter (Millipore; 0.45 μm pore size) and immediate filtering with a mild vacuum (<20 kPa). The filter with retained plankton was placed on a microscope slide, embedded with anti-fading reagent (Citifluor AF1, London, UK), and covered with a cover slip. To inspect phytoplankton for presence of ELFA precipitates, an epifluorescence microscopy (Carl Zeiss Axioskop 40FL; Olympus AX-70) with UV-excitation was used. For more details, see Štrojsová et al. (2003) and Nedoma et al. (2003).

Results

Variations in P forms

SRP concentration gradually increased both in surface and overlying water from January to April. The same trend was found in interstitial water in February and April (Fig. 2). TP concentration in interstitial water was significantly higher than that in the water column, and, for SRP, interstitial water always showed higher values than overlying water (Figs. 2, 3). In January, Site B (undredged) had significantly ($p < 0.01$) higher SRP concentrations in interstitial water than Site A (dredged). In February and April, the undredged site (Site C) also showed significantly ($p < 0.05$ in February and $p < 0.01$ in April) higher SRP values in both interstitial and overlying waters than dredged sites (Site A and B, which was dredged after January). In February, Site C had the highest TP concentration in interstitial water and DOP concentration in

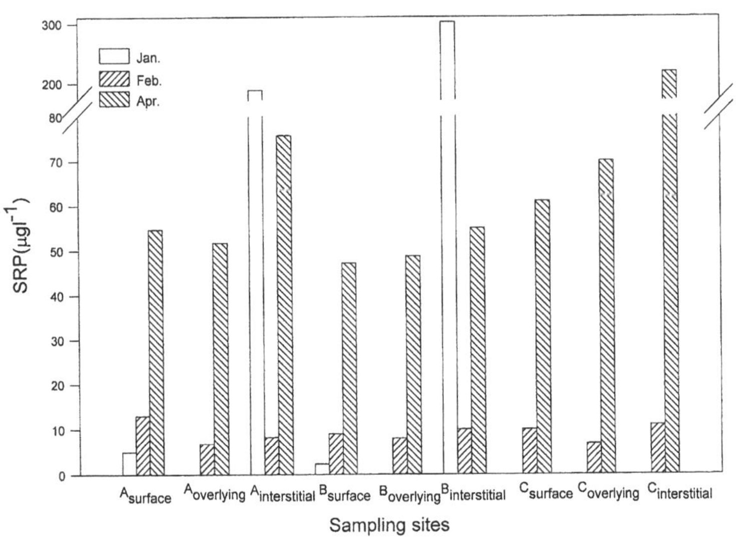

Fig. 2 Seasonal fluctuations of SRP concentrations in surface, overlying and interstitial water of all sampling sites

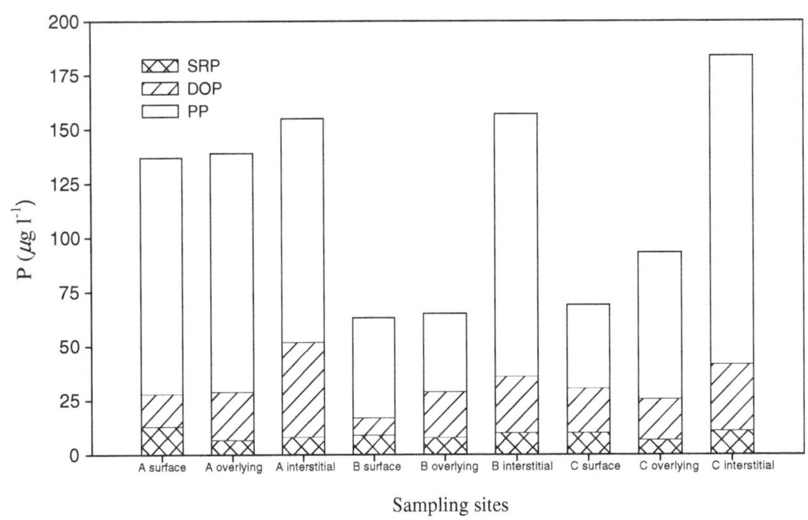

Fig. 3 Concentrations of different forms of P at all sampling sites in February

surface water among all the sites (Fig. 3). Furthermore, Site A showed the highest SRP among the sampling sites in January and February and Site B the highest in April in surface water. For overlying water, Site C had the highest SRP in February and April.

Variations in phytoplankton density and composition

Phytoplankton density gradually increased in surface waters during study period at all sampling sites (Table 1). The same trend was observed in overlying water except at Site C. Algal density was higher in surface water than in overlying water except at Sites B and C in February. Overlying water at Site C had higher phytoplankton density, especially of green algal cells, than surface water in February. Surface water was richer in algal species in April, while it showed similar species numbers to those in overlying water in February. In surface water, Site A showed the highest algal density among the sampling sites in January and February, and Site B had the highest in April. For overlying water, Site C showed the highest value in February and Site B in April, especially in green algae both in density and diversity.

Cryptophytes dominated the phytoplankton assemblage from January to April at Sites A and B, but not at Site C, in surface water. Diatoms were present but were not abundant. Green algae began to appear in February and increased markedly in April at all sites. Few cyanobacteria were found in April at Sites B and C. Similar changes were found in overlying water in February, while more green algae were observed in surface water than in overlying water in April. In overlying water at Site C, green algae (*Coelastrum* sp.) were very abundant.

Extracellular phosphatase activity

The contribution of the coarser ($APA_{>3.0}$) fraction, conventionally taken as "algal" APA to total APA was always over 49% (Fig. 4). No clear trend was found in the relationship between total APA and P concentrations, but an inverse relationship was observed in February. With the highest TP, Site A showed the lowest total APA in surface and overlying water, while the highest total APA was associated with the lowest TP in surface water at Site C. Algae excreted membrane bound phosphatase, as evidenced by ELFA at Sites A and C in surface water in April. A green alga *Schroederia* sp. present, but not dominant at Sites A and C in February and April was ELFA-labeled at Site A. Even so, not all cells were ELF positive. For all the labeled cells, ELFA precipitated on the membrane of the algae in a linear pattern around the whole cell. No labeled cells of this same species were observed at Site C, where SRP concentration in surface water was higher relative to that at Site A.

Discussion

Phytoplankton development is influenced by several factors, e.g., light intensity, water temperature, nutrient concentrations and herbivore grazing (Graham & Wilcox, 2000). Others have reported green algae becoming abundant with increasing water temperature (Robert et al., 2003). In Lake Taihu, algal density increased with temperature, and green algae were abundant in April at all sites. Major shifts in species composition of benthic algae did not result from nutrient additions but seemed to be related to light penetration and *Dreissena* herbivory in Saginaw Bay, Lake Huron (Pillsbury et al., 2002). At all sites in Lake Taihu, algal density

Table 1 Algal density in surface and overlying water during the study period for all sampling sites (10^6 ind l^{-1})

	A surface			A overlying		B surface			B overlying		C surface		C overlying	
	Jan.	Feb.	Apr.	Feb.	Apr.	Jan.	Feb.	Apr.	Feb.	Apr.	Feb.	Apr.	Feb.	Apr.
Density	4.25	5.55	7.3	2.13	4.2	2.65	3.69	15.6	4.09	6.43	4.27	8.06	5.64	3.41

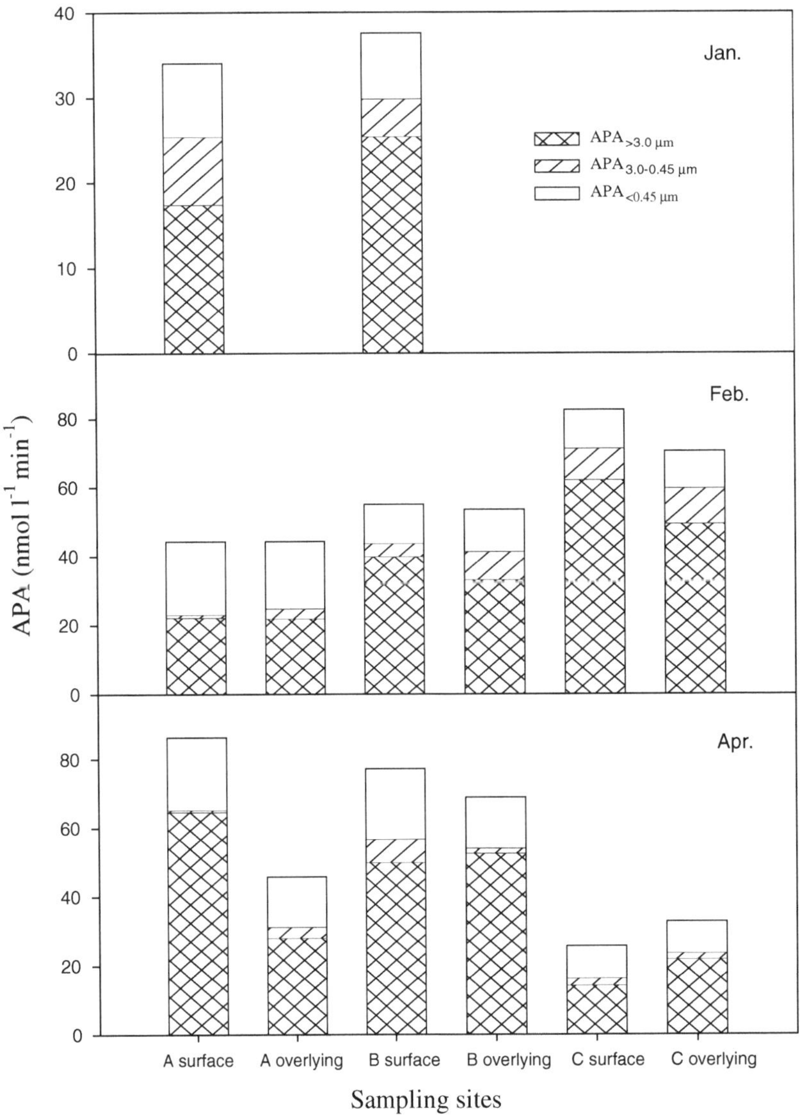

Fig. 4 Size-fractionated APA at all sites studied during experiment period

was higher in surface versus overlying water indicating that light might act as an important factor affecting growth.

SRP appears to have been important in phytoplankton development. From January to April, an increase in bioavailable phosphorus (SRP) was observed at all sites. Following the increase in SRP, water column algae cell counts increased. Higher SRP was accompanied by higher algal density at Site A in January and February and Site B in April. PO_4-P is a major factor in phytoplankton biomass and development elsewhere (Karjalainen et al., 1997–1998; Levine & Whalen, 2001). In a subtropical wetland, algal species composition was more sensitive to P concentration than algal biomass (Pan et al., 2000). In the present study, SRP increased from January to April, and algal composition changed. Green algae occurred in February and became prevalent in April, with cyanobacteria occurring as well. Algal biomass increased after nutrient addition and chlorophytes replaced chrysophytes as the dominant phytoplankton group in another study (Lessmann et al., 2003). In Lake Greifensee

(Switzerland), although total biomass values of phytoplankton did not show any correlation with the long-term decline of phosphorus concentrations, the composition of phytoplankton showed clear trends. During the period with Total P concentrations of >0.25 mg l^{-1}, phytoplankton biomass was dominated by chlorophytes and cryptomonads. With declining P load, chlorophytes and dinoflagellates decreased (Buergi et al., 2003). Overall, in Lake Taihu, higher SRP concentrations in water column was accompanied by higher phytoplankton density, dominated by green algae.

Extracellular phosphatase is important in algal development. For green algae, APA was responsible for making P available, at least partially. For example, *Schroederia* sp. at Site A, responding to low SRP, excreted extracellular phosphatase in April, as illustrated by ELFA-labeling. Total and "algal" APA at Site A was higher than those at Site C (Fig. 4). In a eutrophic Czech reservoir, ELFA-labeling indicated that several green algae produced extracellular phosphatase whenever present (Štrojsová et al., 2003). Green algae can lead to increased phosphatase activity (Sala & Guede, 1995). APA in *Selenastrum capricornutum* was not correlated to molybdate reactive P over a concentration range. The P fraction made available by alkaline phosphatases, enzyme hydrolyzable P, increased the P supply to organisms with high APA (Klotz, 1985). Thus, P fractions mediated by APA, in addition to SRP, are important for P supply to algae. In the present study, higher APA (total or "algal") was accompanied by lower TP and higher DOP in surface water at Site C in February, where green algae were abundant. The alternative method for phytoplankton to obtain P would be use of DOP by APA (Labry et al., 2002). These facts likely reflected the key role of bioavailable phosphorus in the regulation of phytoplankton composition.

From February to April, SRP concentrations in interstitial water paralleled with those in water column. Among all sampling sites, Site C had the highest SRP in interstitial and overlying waters in February. Highest water column phytoplankton density occurred in April, with abundant green algae including *Coelastrum* sp., responding to highest SRP concentrations in interstitial water. In wetland, sediment is of paramount importance, promoting a more diverse phytoplankton assemblage. When sediment or nutrient pulses are absent, phytoplankton is dominated by slow-growing algae (Ortega-Mayagoitial et al., 2003). In Lake Kinneret (Israel), from the bottom to 3 m, green algae (*Pediastrum, Scenedesmus, Coelastrum*, etc.) were observed and changed with increased nutrients from cultural disturbance (Pollingher, 1986). In Canadian lakes, increases at fall turnover consisted mainly of the previous hypolimnetic P and were responsible for fall and winter blooms. These estimates of P were similar in lakes with little internal load, but could differ up to 700% in lakes with high internal load (Nuernberg et al., 1995). In our results, highest algal density was found in overlying water, with green algae (*Coelastrum* sp.) being dominant, which is contrary to other results showing more algae occurring in surface waters. Therefore, the upward flux of phosphorus could support the excess growth of phytoplankton, whereas phytoplankton could adapt for P nutrient by concentrating in bottom as well, which provided further evidence that P in the overlying and interstitial waters is an important factor affecting phytoplankton density and composition.

Compared to the dredged sites, the undredged controls of Lake Taihu showed significantly higher concentration of bioavailable phosphorus (SRP) in interstitial and overlying waters, and considerably larger amounts of algae occurred in bottom water, rather than at the surface, with green algae being dominant. Therefore, in addition to temperature and light, nutrient P may serve as a driving force for development of phytoplankton. A significantly higher density of phytoplankton, dominated by green algae responded to the greatly increased SRP concentrations in interstitial water at all stations in April, strengthening the relation between the density and composition of phytoplankton and P loading from sediments. In general, the undredged sites in Lake Taihu exhibited significantly higher SRP concentrations in interstitial water than the dredged sites. Furthermore, in February, higher total and "algal" APA was accompanied by significantly lower TP and higher DOP concentrations at the undredged sites, where green algae dominated. In

April, *Schroederia* sp. (green algae) excreted extracellular phosphatase as shown by ELF-labeling at a dredged site with relative lower SRP concentration, while, at the dredged site with higher SRP concentration, no such labeling was observed. These facts suggested that organic phosphorus, and its regeneration mediated by phosphatase, constituted an additional way of supplying P for some algae to compensate ambient P deficiency. This was also related to internal loading, and consequently to the sediment dredging.

In summary, the undredged sites in Lake Taihu had significantly higher bioavailable P (SRP) in the interstitial and overlying waters, coinciding with larger amounts of algae in the overlying water, rather than in the surface water, with green algae being dominant. Moreover, higher phytoplankton density, dominated by green algae, responded to higher SRP in the interstitial water at all sites in April. Therefore, in addition to temperature and light, P status in the bottom is important to phytoplankton development in terms of density and composition in Lake Taihu. Additionally, the undredged sites generally had higher SRP in the interstitial water than the dredged sites. Furthermore, in February, higher total and "algal" APA was accompanied by lower TP and higher DOP at the undredged sites, where green algae dominated, suggesting that extracellular phosphatase of chlorophyte mainly mediate P regeneration to improve its bioavailability. In April, *Schroederia* sp. (green algae) excreted extracellular phosphatase as shown by ELF-labeling at a dredged site with lower SRP, while, at the undredged site with higher SRP, no such a labeling was observed. Based on results in an urban lake, Half Moon Lake in Wisconsin, James et al. (2002) found that reduction of internal P sources could substantially reduce by greater than 70% the high concentrations of algae in the lake. Suggested internal P control measures included a sediment chemical treatment to bind P, greater harvesting of *P. crispus* to reduce the macrophyte P pool at the time of senescence, and limiting motor boat activity when the lake is weakly stratified. Our results indicated that the dredging might regulate algal density and composition in water column by reducing P bioavailability in the bottom of eutrophic lakes.

Acknowledgements This work was supported by the National Key Basic Research and Development Program (2002CB412304) and the grant (2002AA601013). We also obliged to the funds from the National Science Foundation of China (20177033). This work was partly supported by Sino-Czech Scientific and Technological Cooperation—project No. 36–20. We would like to thank Jaroslav Vrba, Alena Štrojsová and Feng Weisong for their methodical help and providing facilities to use epifluorescence microscope. Many thanks are given to Li Jianqiu for the sample preparations and to Liu Jingyuan for the review of this manuscript. The thanks also go to Li Lin, Zeng hui for their kind helps in sample collecting.

References

Boon, P. I., 1989. Organic matter degradation and nutrient regeneration in Australian freshwaters: I. Methods for exoenzyme assays in turbid aquatic environments. Archiv Fur Hydrobiologie 115: 339–359.

Buergi, H. R., H. Buehrer & B. Keller, 2003. Long-term changes in functional properties and biodiversity of plankton in Lake Greifensee (Switzerland) in response to phosphorus reduction. Aquatic Ecosystem Health and Management 6: 147–158.

Chróst, R. J., W. Siuda & G. Z. halemejko, 1984. Longterm studies on alkaline phosphatase activity (APA) in a lake with fish-aquaculture in relation to lake eutrophication and phosphorus cycle. Archiv Fur Hydrobiologie 70: 1–32.

Cronberg, G., 1982. Changes in the phytoplankton of lake Trummen induced by restoration. Hydrobiologia 86: 185–193.

Gage, M. A. & E. Gorham, 1985. Alkaline phosphatases activity and cellular phosphorus as an index of the phosphorus status of phytoplankton in Minnesota lakes. Freshwater Biology 15: 227–233.

Golterman, H. L., R. S. Clymo & M. A. M. Ohmstad, 1978. Methods for physical and chemical analysis of fresh waters. Blackwell Scientific Publications, Oxford, Handbook 8, 2nd edn., 111–117.

Gonzáles-Gil, S., B. Keafer, R. V. M. Jovine, A. Aquilera, S. Lu & D. M. Anderson, 1998. Detection and quantification of alkaline phosphatase in single cells of phosphorus-starved marine phytoplankton. Marine Ecology Progress Series 164: 21–35.

Graham, L. E. & L. W. Wilcox, 2000. Algae. Prentice hall, Inc., Upper Saddle River, USA.

Hu, H., R. Li, Y. Wei, C. Zhu, J. Chen & Z. Shi, 1979. Freshwater algae in China. Science Press, Peking (in Chinese).

James, W. F., J. W. Barko, H. L. Eakin & P. W. Sorge, 2002. Phosphorus budget and management strategies for an urban Wisconsin lake. Lake and Reservoir Management 18: 149–163.

Jeppesen, E., P. Kristensen, J. P. Jensen, M. Soendergaard, E. Mortensen & T. Lauridsen, 1990. Recovery resilience following a reduction in external phosphorus loading of shallow, eutrophic Danish lakes: duration, regulating factors and methods for overcoming resilience. 3. Int. Workshop on Ecosystem Research in Freshwater Environment Recovery, 26–29.

Karjalainen, H., S. Seppälä & M. Walls, 1997–1998. Nitrogen, phosphorus and Daphnia grazing in controlling phytoplankton biomass and composition–an experimental study. Hydrobiologia 363: 309–321.

Klotz, R. L., 1985. Influence of light on the alkaline phosphatase activity of Selenastrum capricornutum (Chlorophyceae) in streams. Canadian Journal of Fisheries and Aquatic Sciences 42: 384–388.

Knuttila, S., O.-P. Pietiläinen & L. Kauppi, 1994. Nutrient balances and phytoplankton dynamics in two agriculturally loaded shallow lakes. Hydrobiologia 275/276: 359–369.

Labry, C., A. Herbland & D. Delmas, 2002. The role of phosphorus on planktonic production of the Gironde plume waters in the Bay of Biscay. Journal of Plankton Research 24: 97–117.

Lessmann, D., A. Fyson & B. Nixdorf, 2003. Experimental eutrophication of a shallow acidic mining lake and effects on the phytoplankton. Hydrobiologia 506–509: 753–758.

Levine, M. A. & S. C. Whalen, 2001. Nutrient limitation of phytoplankton production in Alaskan Arctic foothill lakes. Hydrobiologia 455: 189–201.

Murphy, J. & P. Riley, 1962. A modified single solution method of the determination of phosphate in natural waters. Analytica Chimica Acta 27: 1–36.

Nedoma, J., A. Štrojsová, J. Vrba, J. Komárková & K. Šimek, 2003. Extracellular phosphatase activity of natural plankton studied with ELF97 phosphate: fluorescence quantification and labelling kinetics. Environmental Microbiology 5: 462–472.

Nuernberg, G. K, R. W. Bachmann, J. R. Jones, R. H. Peters & D. M. Soballe, 1995. Modeling phosphorus in lakes with possible sediment P release, case studies. Lake and Reservoir Management 11(2):176.

Ortega-Mayagoitial, E., C. Rojo & M. A. Rodrigo, 2003. Controlling factors of phytoplankton assemblages in wetlands: an experimental approach. Hydrobiologia 502: 177–186.

Overbeck, J., 1991. Early studies on ecto- and extracellular enzymes in aquatic environments. In Chróst, R. J. (ed.), Microbial enzymes in aquatic environments. Springer Verlag, New York, 29–59.

Pan, Y. D., R. Stevenson, V. Jan, S. J. Panchabi & C. J. Richardson, 2000. Changes in algal assemblages along observed and experimental phosphorus gradients in a subtropical wetland, USA. Freshwater Biology 44: 339–353.

Pillsbury, R. W., R. L. Lowe, Y. D. Pan & J. L. Greenwood, 2002. Changes in the benthic algal community and nutrient limitation in Saginaw Bay, Lake Huron, during the invasion of the zebra mussel (Dreissena polymorpha). Journal of the North American Benthological Society 21: 238–252.

Pollingher, U., 1986. Non-siliceous algae in a five meter core from Lake Kinneret (Israel). Hydrobiologia 143: 213–216.

Poulickova, A., L. Pechar & M. Kummel, 1998. Influence of sediment removal on fishpond phytoplankton. Archiv Fur Hydrobiologie Supplement 124: 107–120.

Qin, B., P. Xu, Q. Wu, L. Luo & Y. Zhang, 2007. Environmental issues of Lake Taihu, China. Hydrobiologia 581: 3–14.

Rengefors, K., K. Pettersson, T. Blenckner & D. M. Anderson, 2001. Species-specific alkaline phosphatase activity in freshwater spring phytoplankton: application of a novel method. Journal of Plankton Research 23: 435–443.

Roberts, E., J. Kroker, S. Koerner & A. Nicklisch, 2003. The role of periphyton during the re-colonization of a shallow lake with submerged macrophytes. Hydrobiologia 506–509: 525–530.

Sala, M. M. & H. Guede, 1995. Influence of algae and crustacean zooplankton on patterns of microbial hydrolytic enzyme activities—an experimental approach. Aquatic Microbial Ecology 48: 143–154.

Štrojsová, A., J. Vrba, J. Nedoma, J. Komárková & P. Znachor, 2003. Seasonal study on expression of extracellular phosphatases in the phytoplankton of an eutrophic reservoir. European Journal of Phycology 38: 295–306.

Vrba, J., J. Komárková & V. Vyhnálek, 1993. Enhanced activity of alkaline phosphatases—phytoplankton response to epilimnetic phosphorus depletion. Water Science and Technology 28: 15–24.

Hydrobiologia (2007) 581:297–303
DOI 10.1007/s10750-006-0498-0

EUTROPHICATION IN LAKES

Toxicity evaluation of Meiliang Bay, Lake Taihu, China—a drinking water source

Yuangao Chen · Haigang Chen · Yulin Wu · Zhaoli Li · Liwei Sun · Mengmeng Qu · Zhiming Kong

Abstract In this study, potential toxicity in organic pollutants in Meiliang Bay, Lake Taihu, a drinking water source, was investigated using the comet assay and zebrafish embryo test, and two control sites in Lake Changdang and the Changjiang (Yangtze) River were established. For the genotoxicity assay, results showed that organic extracts from water samples induce DNA damage on human lymphocytes and mouse testicular cells. A statistically significant difference ($p < 0.01$) was observed versus the solvent control, as shown by multiple comparisons at a dose of 100 ml tube^{-1}. The degree of DNA damage caused by Meiliang Bay water was most serious (human lymphocytes cells—184 arbitrary units (AU); 234 AU on mouse testicular cells). Organic extracts also affected zebrafish embryo development. Embryo coagulation, axis abnormality, slow absorbability of vitellicle, and multi-edema related to teratogenesis at 96 hpf were observed. In the high dose group, there was obvious edema in the hearts and vitellicles for most dysplastic embryos. Toxic potential in organic pollutants in drinking water sources from the Yangtze River and Lake Changdang were less serious than Meiliang Bay. Therefore, the drinking water source in Meiliang Bay was unsafe compared to the Changjiang River and Lake Changdang, and it is exigent that Meiliang Bay water quality should be ameliorated further.

Keywords Meiliang Bay of Lake Taihu · Drinking water source · Organic pollutant · Toxicity evaluation · Comet assay · Zebrafish embryo test

Guest editors: B. Qin, Z. Liu & K. Havens
Eutrophication of shallow lakes with special reference to Lake Taihu, China

Y. Chen · H. Chen · Y. Wu · Z. Li · L. Sun · M. Qu · Z. Kong (✉)
State Key Laboratory of Pollution Control and Resource Reuse, Environmental School of Nanjing University, Nanjing 210093, China
e-mail: zmkong@nju.edu.cn

Y. Chen
Nanjing Institute of Geography and Limnology, Chinese Academy of Sciences, Nanjing 210008, China

Introduction

There are many methods used to evaluate drinking water toxicity. Application of alkaline single-cell gel electrophoresis (SCGE), or comet assay, in environmental bio-monitoring to clinical applications, DNA repair processes and genetic toxicology, has been explored broadly (Singh et al., 1988; Tice et al., 2000). The comet assay for human lymphocytes has been used to investigate polluted surface water (Zhong et al., 2001; Rajaguru et al., 2002). Results indicate that the comet assay is useful for assessment of aquatic environments. The comet assay is a rapid, simple and sensitive visual technique for detecting potential double-strained DNA breaks in individual cells (Zhu, 1995; Rojas et al., 1999).

Male mice can be used as an animal model for toxicity research in drinking water. Since mice are mammals, results may demonstrate potential adverse effects on humans. Organic extracts may damage male reproductive systems with testis as the target organ (Tian et al., 2002). DNA damage to reproductive cells may induce DNA abnormality and miscarriages as well as developmental abnormalities and cancer susceptibility in offspring (Haines et al., 2001). Since mice spermatid cells are easy to separate and possess particular metabolic activation, they are good for evaluating genotoxicity of chemicals on human reproductive cells (Zhang et al., 2001).

Compared with the conventional acute fish test, the zebrafish embryo test has many advantages, such as low cost, fewer impact factors and high sensitivity. Therefore, it may be used as a standard method. The zebrafish embryo test also may provide information for chemical toxicity studies during different zebrafish embryo growth stages. The zebrafish embryo test has been used to evaluate toxic substance and toxicological mechanisms (Gellert & Heinrichsdorff, 2001; Zhu & Shi, 2002; Strmac et al., 2002).

In this study, toxicity of organic pollutants in Meiliang Bay, Lake Taihu, in Wuxi, Jiangsu province, China, was evaluated using the comet assay, on human peripheral blood lymphocytes and mouse testicular cells, and the zebrafish embryo test. Simultaneously, two control sites in the Yangtze River and Lake Changdang were established. The objective was to develop a simple modeling method for drinking water quality monitoring.

Materials and methods

Materials

Human peripheral blood samples were collected from two volunteers (healthy, non-smoking, adult male) by Nanjing Center Blood Bank (China).

Healthy male mice (Kunming strain) weighing 28–32 g were obtained from Animal Breeding Center of Nanjing University and used after one week of acclimatization. The animals were housed in plastic cages in a temperature-controlled room (25 ± 2°C) with a natural light-dark cycle and were given commercial food and water.

Zebrafish (*Danio rerio*) were obtained from a local pet store. A suitable number of healthy fish (females to males ratio was 1:2) were kept in a 100 l tank filled with water, which was prepared according to ISO (1996) (For per liter water, adding 294 mg $CaCl_2 \cdot H_2O$, 123 mg $MgSO_4 \cdot 7H_2O$, 65 mg $NaHCO_3$ and 6 mg KCl,). Water was maintained at 26 ± 1°C, DO > 7 mg l^{-1}, with a light/dark cycle of 14/10 h. Fish was fed daily.

Sampling procedures

Three water samples were collected from three sites (Fig. 1), the drinking water source in Meiliang Bay, Lake Taihu, in Wuxi (SL.1), the center of Lake Changdang in Jintan (SL.2), and the Changjiang River in Changzhou (SL.3). Water samples (100 l) were collected from each site at 0.5 m depth in December 2003. After being stored 24 h, each water sample was filtered with gauze and filter paper, to remove suspended material or sediment, and then passed though a column with non-polar neutral resin (XAD-2) to adsorb organic pollutants at a flow rate of 30–40 ml min^{-1}. The column was eluted with carbinol, acetone and dichloromethane. The elution was dried by blowing with nitrogen at 50°C and re-dissolved in 2.0 ml dimethysulphoxide (DMSO).

The comet assay on human peripheral lymphocyte cells

To obtain the peripheral lymphocytes, human blood samples were mixed with 3% glutin solution and subsided at 37°C for 30 min. The layer containing lymphocytes (0.2 ml) were transferred into a test tube with 0.8 ml phosphate-buffered saline (PBS). After centrifuging at 3,000 rpm for 3 min, the supernatant was removed carefully. The pellet containing lymphocytes was re-suspended after being exposed to different solutions containing the organics (i.e., each dosage equals

Fig. 1 Water sampling sites

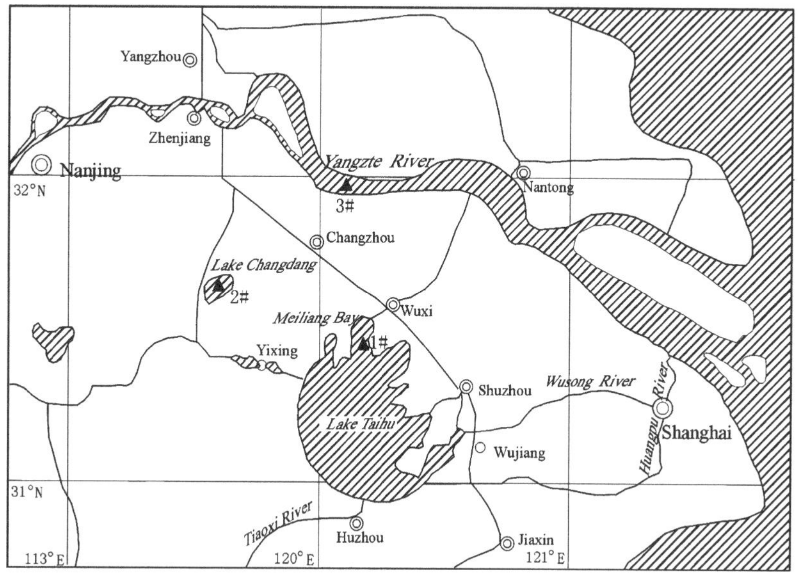

20 ml, 100 ml, and 500 ml source water) in the same buffer including 0.1 ml water sample extract and 0.9 ml PBS incubated for 1 h at 37°C. DMSO was used as a control. Viability of individual cells (>90%) was determined by the Trypan blue exclusion technique.

The comet assay was performed as described by Singh et al. (1988) with some modifications. Briefly, frosted microscope slides, on which cells were embedded in an agarose "sandwich", were submerged in cold lysis buffer (2.5 M NaCl, 100 mM EDTA, 10 mM Tris, adjusted to pH 10.0 with NaOH, 1% Triton X-100 and 10% DMSO both freshly added) at 4°C in the dark for at least 30 min to remove cellular proteins. Electrophoresis was conducted at 18°C with 25 V and 100 mA for 1 h. The experiment was done in red or yellow light to avoid DNA damage. DNA was stained with ethidium bromide, and slides were examined with a fluorescent microscope (BX41, Olympus, Japan). One observer evaluated the slides to minimize variability. Images were analyzed according to Collins et al. (1995). One hundred comets on each slide were scored visually as one of five classes, according to tail intensity, and given a value of 0, 1, 2, 3, or 4 (from undamaged, 0, to maximally damaged, 4). "Arbitrary units" (AU) were used to express DNA damage and was calculated as:

$$\text{Arbitrary units} = \sum_{i=0}^{4} i \times Ni$$

where Ni = frequency in class i; i = DNA damage class (0, 1, 2, 3, 4).

Comparisons of DNA damage induced by control and treated groups were analyzed using a one-way ANOVA using the SPSS software. Since studies have revealed that high dose exposure might damage DNA badly, a very high dose would result in a decline of visual DNA damage for the massive dissociative fraction to be missed in electrophoresis, and a very low dose made differences among groups invisible (Devaux et al., 1997). Therefore, in this study, genotoxicity of water samples was compared at 100 ml tube^{-1}. In addition, differences among groups were analyzed using the Duncan test with a significance level of 0.05.

The comet assay on mouse spermatid cells

Cell suspension from spermaries of adult male mice was performed according to Zhang et al. (2001). Briefly, mouse spermatid was extracted with 90% Dulbecco's Modified Eagle Medium (DMEM) for 1 h at 37°C. The dosage for tested water was the same as the human lymphocytes comet assay. Treated cells were centrifuged, and DMSO was used as a control. Individual cell viability (>90%) was determined by the Try-pan blue exclusion technique.

Zebrafish embryo test

Zebrafish eggs were obtained according to OECD (1998). Eggs were collected in 30 min and washed with tap water. A suitable number of eggs were transferred into prepared chemical concentrations. Eggs were observed and non-fertilized eggs were discarded. The experiment was considered valid when at least 90% of the embryos in the control were viable after 72 h. After hatching, embryos were transferred into wells in a 24-well multi-plate (1 embryo per well). Twenty wells were prepared with 2 ml water sample. Four wells were filled with egg water and were internal controls. DMSO concentration was not more than 0.25%. Tests were replicated at least three times (60 eggs per concentration and 5 concentrations tested for each sample) consecutively. Eggs were incubated at 26 ± 1°C with a light/dark cycle of 12/12 h.

Embryonic development was observed with an inverted microscope (TMD-EF, Nikon, Japan) and recorded daily. The toxicological endpoint was embryo mortality. In addition, differences among treated groups and the blank were analyzed with the χ^2 test.Table 1.

Results

Water samples from SL.1 and SL.2 (500 ml tube^{-1}) damaged DNA so badly that results could not be evaluated. Results from the human peripheral lymphocyte experiment show that all water samples extracts induced DNA damage, and a statistically significant difference ($p < 0.01$) was observed versus the solvent control. Results also showed that the DNA damage score increased with extract concentrations. Comparing genotoxicity of SL.1, 2 and 3 at 100 ml/tube^{-1}, the degree of DNA damage in human peripheral lymphocytes treated with SL.1 water was worst. A typical comet picture is shown in Figs. 2 and 3.

Results from the mouse spermatid experiment are shown in Table 2 and demonstrate that all water sample extracts induced DNA damage. A statistically significant difference ($p < 0.01$) was observed versus the solvent control group. With increasing dose, DNA damage became more extensive. Comparing the genotoxicity among the three water samples at 100 ml tube^{-1}, the order of the degree of DNA damage was SL.1 > SL.2 > SL.3.

Zebrafish embryos were affected by water sample organic extracts (SL.1, SL.2 and SL.3). Among them, SL.1 was the most toxic (Table 3). Compared with the blank control, SL.3 affected the 96 h-hatch-rate dramatically when organics were concentrated by 125. For SL.1 and SL.2, concentrated by 7.8 and 62.5, respectively, toxicity also was significant ($p < 0.01$).

Discussion

Jiangsu province has a developed chemical industry, and the chemical composition of organic

Table 1 DNA damages in human peripheral blood lymphocytes (mean ± SD) exposed to organics from different water samples

		Frequency (%)					
		Class 0	Class 1	Class 2	Class 3	Class 4	
SL.2	20	0.74 ± 1.27	86.61 ± 1.91	11.68 ± 0.17	0.97 ± 0.69	0	112.89 ± 0.66**
	100	0	42.06 ± 6.19	34.93 ± 8.17	20.00 ± 2.36	3.01 ± 1.30	183.95 ± 5.57**
	500	0	1.61 ± 1.21	51.84 ± 3.40	41.35 ± 4.58	5.20 ± 1.81	250.14 ± 1.95**
	20	29.56 ± 4.79	49.18 ± 2.14	21.25 ± 3.73	0	0	91.69 ± 8.32**
	100	0	37.34 ± 12.37	46.46 ± 10.36	16.04 ± 4.04	0.16 ± 0.35	179.01 ± 15.48**
	500	0	14.54 ± 6.86	40.68 ± 7.39	41.32 ± 5.16	3.46 ± 1.50	233.71 ± 9.39**
	20	10.07 ± 10.18	76.74 ± 8.47	13.20 ± 8.63	0	0	103.13 ± 16.87**
	100	0	66.44 ± 13.38	31.43 ± 13.18	2.13 ± 1.07	0	135.69 ± 13.64**
	500	0	33.75 ± 9.84	54.87 ± 12.28	10.21 ± 7.38	1.17 ± 1.43	178.80 ± 12.85**
Solvent control	0	73.02 ± 3.20	25.67 ± 3.16	1.30 ± 0.70	0	0	28.28 ± 3.39

** $p<0.01$

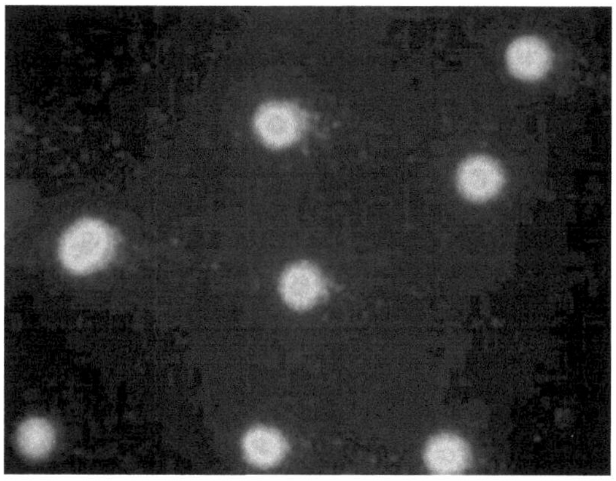

Fig. 2 The comet picture of DMSO on human peripheral blood lymphocytes (×400)

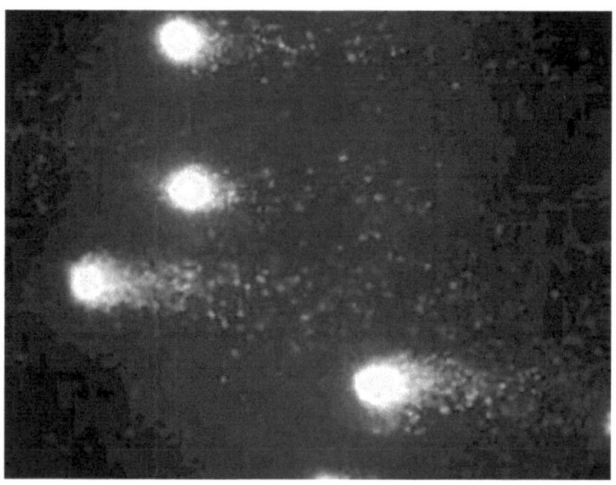

Fig. 3 The comet picture of a sample at the dose of 500 ml/tube^{-1} on human peripheral blood lymphocytes (×400)

extracts from the river and lake was mostly anthropogenic. There were more than 40 priority pollutants (such as pesticides, hydroxybenzenes and heterocyclics). The present study suggests that concentrated organics from water sample extracts were evident in genotoxins and could damage DNA of human peripheral blood lymphocytes and mouse spermatids. Additionally, mouse spermatid DNA damage was so severe in the high dose groups of samples SL.1 and 2 that heads and tails of some comets were separated from each other, which indicates that extracted organic pollutants may have promoted programmed cell death. Compared with human peripheral blood lymphocytes, mouse spermatids may be more sensitive to organic toxins. Mouse germ cells were more sensitive than somatic cells when evaluating potential harmful effects of organic pollutants in tap water (Wu et al., 1999). However, results of the comet assay showed DNA damage in the mouse spermatid control was more serious than human peripheral blood lymphocytes. Therefore, the isolation and culture of mouse spermatids were adjusted.

In the zebrafish embryo test, three aspects were considered when choosing the toxicological endpoints observed, including measurement, obvious toxic reaction and easy operation. The toxicological endpoint was the embryo hatch rate at 96 hpf (hour post fertilization). However, many kinds of dysplasia were observed, such as embryo coagulation and agensis correlated with death, axis abnormality, slow absorption of the vitellicle and

Table 2 DNA damages in mouse spermatid (mean ± SD) exposed to organics from different water samples

		Frequency (%)					
		Class 0	Class 1	Class 2	Class 3	Class 4	
	20	15.34 ± 1.64	48.70 ± 2.52	35.56 ± 3.50	0.40 ± 0.50	0	121.03 ± 4.76**
	100	0	16.93 ± 3.19	39.38 ± 3.55	36.20 ± 4.02	7.49 ± 1.23	234.25 ± 5.09**
	500						
	20	25.93 ± 1.75	45.02 ± 6.02	28.24 ± 6.32	0.80 ± 0.76	0	103.91 ± 8.43**
	100	5.56 ± 1.79	17.24 ± 2.57	37.16 ± 7.27	37.14 ± 3.87	2.91 ± 1.54	214.61 ± 5.89**
	500						
	20	36.44 ± 5.44	51.87 ± 5.81	11.69 ± 1.88	0	0	75.26 ± 5.71**
	100	1.91 ± 1.31	52.94 ± 6.45	41.33 ± 4.03	3.82 ± 2.53	0	147.05 ± 6.26**
	500	0	13.10 ± 3.78	43.39 ± 2.19	38.89 ± 6.50	4.61 ± 2.54	253.00 ± 8.47**
Solvent control	0	51.74 ± 5.25	47.53 ± 5.02	0.74 ± 1.27	0	0	49.00 ± 5.76

** $p<0.01$

Table 3 The 96-h-hatch rate of zebrafish embryo (mean ± SD) affected by different water samples

	Frequency (%)		
	SL.1	SL.2	SL.3
7.8125	66.8 ± 6.2**	93.3 ± 4.7	85.0 ± 14.7
15.625	65.0 ± 7.0**	85.0 ± 7.0	78.3 ± 6.2
31.25	36.7 ± 6.2**	81.7 ± 10.2	78.3 ± 11.8
62.5	0.0**	0.0**	90.0 ± 4.0
125	0.0**	0.0**	38.3 ± 28.9**
Blank control	87.7 ± 9.0		

** $p<0.01$

multi-edema related to teratogenesis at 96 hpf. In the high organics dose from SL.1 and 2, there was obvious edema in the hearts and vitellicles for most dysplastic embryos. The endpoint of teratogenesis was more sensitive than the death endpoint. Some studies showed that minnows had a similar toxicological endpoint for toxins (Strmac et al., 2002). Therefore, the nonspecific reaction to toxin in early life stages, such as acantha distortion, the slow absorption of the vitellicle and formation of edema are worthy of further study.

When exposed to toxin, the zebrafish embryo egg membrane protected the embryo. Selective permeability of the egg membrane changed with growth time and toxin differences. Due to underdevelopment of the embryo metabolizing system, interactions between the toxin and the enzyme system was difficult to observe in the test.

The report of The Environment Status Communique of Jiangsu Province in 2002 indicated that water quality in the Changjiang River was in the secondary category standard for surface water. However, water quality in Lake Taihu was worse than the fifth category standard. Lake Changdang is the third largest lake in Taihu basin, and its water quality approached the third category standard and became worse in recent years. From this study, results of the comet assay (human peripheral blood lymphocytes and mouse spermatid) and the zebrafish embryo test were consistent with the water quality of each water source. The three methods mentioned above could detect biologic toxicity of organic pollutants in water effectively. The research demonstrated that the drinking water source in Meiliang Bay was worst and should be ameliorated further prior to those in Lake Changdang and the Changjiang River.

Acknowledgements We would like to gratefully acknowledge financial support from Ministry of Science and Technology of China (No. 2002AA601013), National Natural Science Foundation of China (No. 40501078) and Chinese Academy of Sciences (No. KZCX1-SW-12, KZCX2-311).

References

Collins, A. R., A. G. Ma & S. J. Duthie, 1995. The kinetics of repair of oxidative DNA damage (strand breaks and oxidised pyrimidines) in human cells. Mutation Research 336: 69–77.

Devaux, A., M. Pesonen & G. Monod, 1997. Alkaline comet assay in rainbow trout hepatocytes. Toxicology in Vitro 11: 71–79.

Gellert, G. & J. Heinrichsdorff, 2001. Effect of age on the susceptibility of zerbrafish eggs to industrial wastewater. Water Research 35: 3754–3757.

Haines, G. A., J. H. Hendry & C. P. Daniel, 2001. Increased levels of comet-detected spermatozoa DNA damage following in vivo isotopic-or X-irradiation of spermatogonia. Mutation Research 495: 21–32.

OECD, 1998. Fish, short-term toxicity test on embryo and sac-fry stages. OECD Guideline for Testing of Chemicals, OECD TG212, Paris.

Rojas, E., M. C. Lopez & M. Valverde, 1999. Single cell gel electrophoresis assay: methodology and applications. Journal of Chromatography B 722: 225–254.

Rajaguru, P., L. Vidya & B. Baskarasethupathi, 2002. Genotoxicity evaluation of polluted ground water in human peripheral blood lymphocytes using the comet assay. Mutation Research 517: 29–37.

Singh, N. P., M. T. McCoy & R. R. Tic, 1988. A simple technique for quantity of low levels of DNA damage in individual cells. Experiment Cell Research 175: 184–191.

Strmac, M., A. Oberemm & T. Braunbeck, 2002. Effects of sediment elutes and extracts from differently polluted small rivers on zebrafish embryos and larvae. Journal of Fish Biology 61: 24–38.

Tian, H. J., D. S. Wu & R. Tao, 2002. Study of male reproductive toxicity induced by organic pollutants in drinking water. Acta Academiae Medicinae Militaris Tertiae 24(9): 1076–1078 (in Chinese).

Tice, R. R., E. Agurell & D. Anderson, 2000. Single cell gel/comet assay: Guidelines for in vitro and in vivo genetic toxicology testing. Environmental and Molecular Mutagenesis 35: 206–221.

Wu, K., C. H. Song & F. Liu, 1999. Evaluation of potential harmful effects of organic pollutants in tap water on mice in vivo. Chinese Journal of Preventive Medicine 33(3): 149–151 (in Chinese).

Zhong, Y., S. L. Feng & Z. M. Kong, 2001. Evaluating the genotoxicity of surface water of Yangzhong city, using the micronucleus test of *Vicia faba* and the comet assay. Bulletin of Environmental Contamination and Toxicology 67(2): 217–224.

Zhu, H. G., 1995. Comprehensive evaluation index of organic mutagens in water. Environment Science of Shanhai 14(10): 44–49 (in Chinese).

Zhang, Z. Z., Z. Z. Heng, Y. Liao & R. Zhao, 2001. Study on DNA damage induced by indirect mutagens using comet assay in testicular cells in vitro. Carcinogenesis, Teratogenesis and Mutagenesis 13(1): 4–7 (in Chinese).

Zhu, L. & S. J. Shi, 2002. Utilization of embryo development technique of zebrafish to evaluating toxicity on various chemicals. Chinese Journal of Application Ecology 13(2): 252–254 (in Chinese).

Hydrobiologia (2007) 581:305–318
DOI 10.1007/s10750-006-0509-1

EUTROPHICATION IN LAKES

Updating water quality targets for shallow Lake Balaton (Hungary), recovering from eutrophication

Vera Istvánovics · Adrienne Clement · László Somlyódy · András Specziár · László G.-Tóth · Judit Padisák

Abstract The paper presents an overview about recovery of shallow Lake Balaton from eutrophication by assessing quantitative and qualitative changes in phytoplankton, zooplankton, and chironomids as a function of load reduction. The aim was to update the present water quality targets. The proposed targeting scheme supplements the existing one with a range of lake-specific ecological criteria. We conclude that simple targets (desired phytoplankton biomass and permissible load) are the best choice during the initial stage of eutrophication management, but more complex schemes including ecological criteria are needed to trace recovery when re-organization of the ecosystem takes place.

Guest editors: B. Qin, Z. Liu & K. Havens
Eutrophication of shallow lakes with special reference to Lake Taihu, China

V. Istvánovics (✉) · A. Clement · L. Somlyódy
Department of Civil and Environmental Engineering, Budapest University of Technology and Economics, Műegyetem rkp. 3., Budapest 1111, Hungary
e-mail: istvera@goliat.eik.bme.hu

A. Specziár · L. G.-Tóth
Balaton Limnological Research Institute, Hungarian Academy of Sciences, Klebelsberg Kuno u. 3., Tihany 8237, Hungary

J. Padisák
Department of Limnology, University of Veszprém, Egyetem u. 10., Veszprem 8200, Hungary

Keywords Phosphorus load · Sediment behavior · Phytoplankton · Cyanobacteria · Zooplankton · Chironomid fauna

Introduction

Water quality targets serve as a measuring scale during eutrophication management that orients managers and the public in questions like "How much can the status of the lake be improved?", "What has been achieved by the implemented measures?", "Which are the main tasks of future management?" Target conditions are usually defined when intolerable economic and ecological consequences of eutrophication call for immediate action that is when the ecosystem had spectacularly transgressed steady state. In the best case, targets are set up using information about past steady state and changes observed during nutrient enrichment. The implicit initial assumption is that apart from some delay, the path of recovery will mirror the trajectory of eutrophication with a resulting steady state identical to the pre-eutrophication status of the lake. Full-scale recovery, however, has not been convincingly confirmed yet, and comprehensive studies of lake recovery suggest hysteretic effects due to both internal P load and biological resilience.

Sas (1989) and coworkers have recognized four stages during lake recovery. No decrease in algal biomass can be expected until external and internal P loads decrease below the level at which P regains its role in determining carrying capacity of the lake (Stage 1). In Stage 2 transparency shows a disproportionately large increase relative to the decrease in area-specific algal biomass due to dispersion of buoyant cyanobacteria and/or motile algae to greater depths with high nutrient concentration. This behavioural response to load reduction can be observed in deep lakes with stable stratification but not in shallow lakes. Once P has become growth limiting, the decrease of phytoplankton biomass is proportionate to the decrease in P supply (Stage 3). During Stage 4, a further reduction in algal biomass is coupled to floristic changes, most conspicuously to the retreat of cyanobacteria. Due to random fluctuations in the biomass and composition of phytoplankton, the various stages have no fixed boundaries. Although not considered in the study of Sas (1989), full recovery of a lake certainly requires one more step (Stage 5): regeneration of former macrovegetation and food web structure (Scheffer et al., 1993). Considering this general pattern of recovery as well as deviations of individual lakes from the "rule", occasional updating of water quality targets is inevitable to maintain the orienting function of these targets over the whole recovery process.

In the present study we overview the recovery of large (596 km^2), shallow (z_{mean} = 3.2 m) Lake Balaton in context of the above general model. Individual processes including the relationship between P load and phytoplankton biomass (Istvánovics & Somlyódy, 2001; Istvánovics et al., 2002), floristic changes of phytoplankton in response to restoration measures (Padisák & Reynolds, 1998), long-term changes in the biomass, composition and production of zooplankton (G.-Tóth, 2000), and the relationship between phytoplankton production and long-term dynamics of the chironomid fauna (Specziár & Vörös, 2001) have been published separately. The present overview was motivated by the practical aim of updating water quality targets considering both the observed response of the ecosystem to eutrophication management and realistic perspectives of further reduction in nutrient loads.

Data and methods

In order to detect trends and clarify patterns, we attempted to use as long data sets as possible. This could be achieved by supplementing the authors own data with critically evaluated historical records from both published sources and various unpublished archives. Whenever quantification was feasible, simple empirical relationships were applied.

Recovery of the lake was traced using one or both of the two independent approaches: (i) analysis of the trajectory of oligotrophication within the same lake area and (ii) comparison of the areas that had previously attained hypertrophy and mesotrophy/slight eutrophy. The advantage of the latter approach is that the influence of random variability caused by physical forcing is partly cancelled (Herodek, 1984). At the same time, poorly explored natural gradients within the lake may impose limitations to the application of this method.

Lake Balaton, its eutrophication and restoration

On the basis of large-scale circulation patterns (Shanahan et al., 1986), Lake Balaton can be divided into four basins (Fig. 1). Mean depth and surface area of the basins increase from west to east with a concomitant decrease in the area of the corresponding sub-watershed (Fig. 1). The largest tributary, the Zala River enters Basin 1, whereas the only outflow is the Sió Canal at the opposite end of the lake. The lake is extremely calcareous with high magnesian calcite making up 50–60% of sediments (Csermák & Máté, 2004). Turbidity is high particularly during summer due to both wind induced sediment resuspension and slow sedimentation of the precipitating carbonates (Istvánovics et al., 2004). Strong wave action restricts macrophyte growth to a narrow littoral zone (during the past 150 years, maximum 10% of lake area; Virág, 1998).

Specific morphometry gives rise to environmental gradients along the longitudinal axis from increasing water retention time to shifts in major ion composition and to decreasing area-specific nutrient loads. One can expect associated

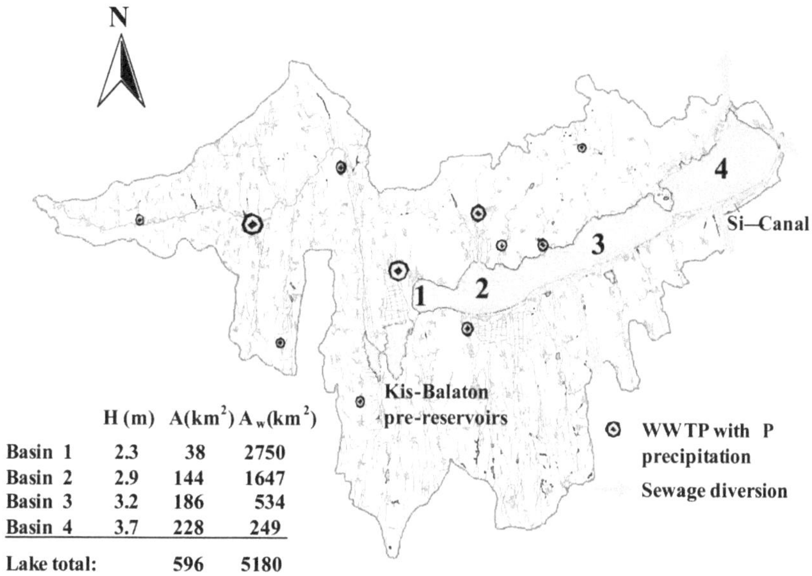

Fig. 1 Lake Balaton and its catchment. (H—average depth, A—surface area, Aw—area of sub-catchment, WWTP—waste water treatment plant)

differentiation in the biotic components under pristine conditions, but the available evidence is faint.

Eutrophication dramatically magnified many of the environmental gradients and particularly the biotic ones. By the end of the 1970s, area-specific P load in Basin 1 exceeded that in Basin 4 by a factor of 8 (Table 1, Somlyódy & Jolánkai, 1986). Primary production has increased 6–8-fold in Basin 1 and doubled in Basin 4 (Herodek, 1986). Blooms of N_2-fixing cyanobacteria developed regularly during summers in the western basins. Four years after its first appearance in 1978 (Padisák, 1997), invasive *Cylindrospermopsis raciborskii* (Woloszynska) Seenayya et Subba Raju became the dominant summer species in each basin (Herodek, 1986; Padisák & Reynolds, 1998; Istvánovics et al., 2002). Significant differences between the western and eastern areas have been reported in composition and production of

Table 1 Multi-annual mean loads (±sd) of total P in tons P year^{-1}

Source	Lake Balaton	Basin 1	Basin 2	Basin 3	Basin 4
Before eutrophication management (1980–1988)					
Geochemical background	18 ± 4	10 ± 2	6 ± 1	2	1
Sewage	61 ± 15	31 ± 7	10 ± 1	5 ± 4	16 ± 10
Urban runoff	56 ± 9	16 ± 4	19 ± 4	10 ± 1	10 ± 2
Agricultural (rural) runoff	137 ± 30	39 ± 17	42 ± 14	30 ± 5	26 ± 4
Atmosphearic	17 ± 2	1	4 ± 1	5 ± 1	7 ± 1
Total	289 ± 42	97 ± 23	80 ± 19	53 ± 4	60 ± 10
Through-flow			20 ± 4	21 ± 7	12 ± 6
After eutrophication management (1989–2002)					
Geochemical background	17 ± 6	9 ± 3	5 ± 2	2 ± 1	1
Sewage[a]	11 ± 3	7 ± 2	3 ± 1	1	0
Urban runoff	34 ± 11	9 ± 3	12 ± 4	6 ± 2	7 ± 3
Agricultural (rural) runoff	74 ± 27	20 ± 11	23 ± 8	14 ± 5	17 ± 6
Atmospheric	10 ± 5	1	2 ± 1	3 ± 2	4 ± 2
Total	145 ± 47	45 ± 19	46 ± 14	26 ± 8	29 ± 9
Through-flow			18 ± 9	21 ± 12	15 ± 11

a—value in 2002

the chironomid fauna (Dévai & Moldován, 1983; Ponyi et al., 1983) as well as in those of fish (Bíró & Vörös, 1990; Bíró, 1997).

In addition to its ecological values, Balaton is an internationally recognized recreational lake. Eutrophication caused significant economic loss in times when two thirds of Hungary's tourist income originated from the Balaton region. An inclusive eutrophication management plan was approved by the government in 1983. The aim was preventing further deterioration in water quality by 1990 and achieving the trophic conditions experienced during the early 1960s in the period 2005–2100 (Láng, 1986). Water quality targets and permissible phosphorus loads were also expressed in quantitative terms.

The most important management measures included (i) sewage diversion from shoreline settlements around Basin 4 and along the southern shore of Basin 3, (ii) introduction of chemical P precipitation at the large waste water treatment plants in the watershed, and (iii) pre-reservoir construction on the Zala River (Fig. 1). In the meantime, Hungarian agriculture has collapsed following the political changes in 1989. From the perspective of eutrophication, the most important consequence was a 90% drop in fertilizer application. Two periods of drought during 1990–1993 and 2000–2003 further decreased diffuse load of nutrients (Fig. 2). Multi-annual mean load of P decreased by about 50% in each basin (Clement et al., in press; Table 1). Considering, however, the additional load originating from through-flow, the reduction was only 30–40% with the exception of Basin 1 (Table 1).

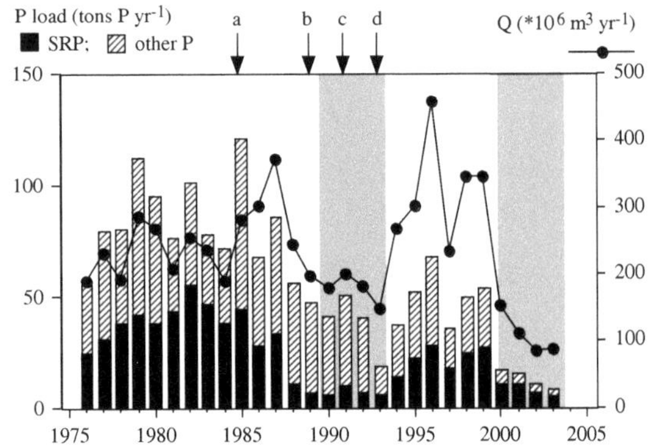

Fig. 2 Loads of soluble reactive P (SRP) and total P (sum of SRP and other P forms), and flow (Q) at the mouth section of the Zala River. (Arrows indicate timing of load reduction (Fig. 1) as follows: a—inundation of the Upper Kis-Balaton reservoir; b—sudden drop in fertilizer application due to the collapse of agriculture; c—introduction of chemical P removal at the sewage treatment plant of Zalaegerszeg, d—inundation of the Lower Kis-Balaton reservoir. Shaded areas indicate two periods of draught)

Results and discussion

The relationship between P load and algal biomass

In the western basins, annual mean biomass of algae decreased significantly after a delay of about one decade (Fig. 3). Although the last few summers were unusually warm and dry, the former adverse conditions have not returned. In the eastern basins, particularly in Basin 3 the improvement was marginal if any. In the period 1995–2003, annual average biomass of phytoplankton exceeded the prescribed range in 5 years (Fig. 3). The mean deviation was the lowest (25%) in Basin 1 and the highest (55%) in

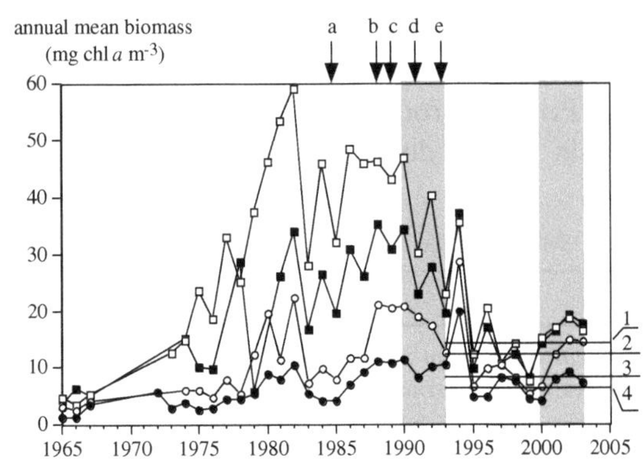

Fig. 3 Annual mean biomass of phytoplankton in the four basins of Lake Balaton. (Basin 1—open square, Basin 2—closed square, Basin 3—open circle, Basin 4—closed circle. Arrows indicate timing of load reduction (Fig. 1) as follows: a—inundation of the Upper Kis-Balaton reservoir; b—sewage diversion from the eastern basins; c—sudden drop in fertilizer application due to the collapse of agriculture; d—introduction of chemical P removal at the sewage treatment plant of Zalaegerszeg, e—inundation of the Lower Kis-Balaton reservoir. Shaded areas indicate two periods of draught. Numbered lines show water quality objectives to be achieved by 2010 in the four basins (cf. Table 2))

Basin 3. The conclusion is that in spite of the conspicuous improvement in the western areas, the lake is still far away from a full recovery.

Due to the large number of small, dynamic inflows (over 50, Fig. 1), annual load estimates are uncertain with the exception of Basin 1 where daily loads through the Zala River have been available from 1976 (Clement, 2000). Consequently, the relationship between algal biomass and phosphorus load can be traced at a higher resolution in this westernmost area than in other basins. The question is then how much the results are transferable to the other basins.

Pre-reservoir construction caused spectacular changes in the loading conditions of Basin 1 in addition to the reduction in absolute terms (Fig. 2). The most important ones were (i) an 80% decrease in the load of suspended solids, (ii) the increase in the calcium load related to total P and suspended solids loads by a factor of 2.5 and 4, respectively, and (iii) the decrease in the mean loading ratio of nitrate-N to phosphate-P below 2 (by weight) during summer. Changes in (i) and (ii) altered the apparent settling velocity of P and improved the composition of the freshly formed sediments (Istvánovics & Somlyódy, 2001), whereas (iii) may favor selection of N_2-fixing cyanobacteria.

Internal load of P is one of the basic processes that determine the path of recovery in shallow lakes (Marsden, 1989; Sas, 1989; Søndergaard et al., 1999). In well-oxygenated Lake Balaton with low concentrations of iron in the sediments (12–20 mg Fe g^{-1} dry mass), resuspension followed by P desoprtion is the predominant mechanism of transport (Boström et al., 1982; Lijklema et al., 1986). At long time scales relevant for recovery, transport can be considered constant provided that no significant changes occur in macrophyte coverage. In such lakes, long-term changes in the internal load depend on the concentration of mobile sediment P. These changes are rarely large enough to detect them directly (Søndergaard et al., 1999). Therefore we estimated long-term trends in mobile sediment P using the mass balance model of Lijklema and coworkers (1986) for the upper sediment layer. Since loads of Ca and suspended solids were available only for the Zala River, additional sources (about 5–10% of the total load) were neglected.

The model (Fig. 4) assumes that the "active" layer is fully mixed to a constant depth of h (m) due primarily to the burrowing activity of

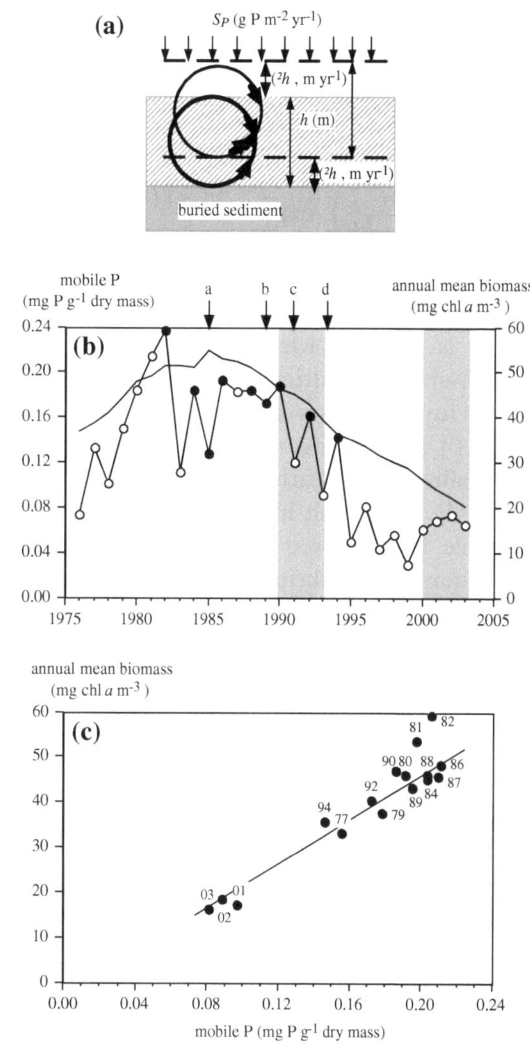

Fig. 4 Mass balance model of Lijklema et al. (1986) (**a**) and the estimated mobile P content of sediments (solid line) compared with annual mean biomass of phytoplankton (circle; **b**) in Basin 1. The relationship between mobile P content and phytoplankton biomass in years when internal P load approached its potential maximum (**c**). (In A closed circles denote years with large blooms of *Cylindrospermopsis raciborskii* (Fig. 6). Arrows indicate timing of load reduction (Fig. 1) as follows: a—inundation of the Upper Kis-Balaton reservoir; b—sudden drop in fertilizer application due to the collapse of agriculture; c—introduction of chemical P removal at the sewage treatment plant of Zalaegerszeg, d—inundation of the Lower Kis-Balaton reservoir. Shaded areas indicate two periods of draught. In B parameters of the regression line are $y = 234.94x$, $r^2 = 0.88$, $n = 16$)

zoobenthos. The rate of sedimentation (Δh, m year^{-1}) can be estimated from the mass balance of suspended solids and calcium. To keep the mixed depth constant, burial must equal the rate of sedimentation. Rate of P sedimentation (S_P, g P m^{-2} year^{-1}) can be derived from the mass balance of total P. The change in the P concentration of the active layer (P, g P m^{-3}) is

$$P = \left(P_0 - \frac{S_P}{\Delta h + kh}\right) \cdot \exp\left(-\frac{\Delta h + kh}{h} t\right) + \frac{S_P}{\Delta h + kh}$$

where P_0 is the initial concentration at $t = 0$ and k (year^{-1}) is the first-order "nutrient decay" rate. For conservative nutrients (like total P) $k = 0$, whereas for non-conservative ones (like total N or mobile P) $k > 0$. Diagenetic processes including re-crystallization of carbonates and exchange of Ca for Mg may result in gradual immobilization of mobile P in the sediments of Lake Balaton. The new equilibrium concentration is $P_{eq} = S_P/(\Delta h + kh)$. The rate of approaching to equilibrium is controlled by the exponent, $\tau = h/(\Delta h + kh)$.

Using the parameters calibrated in 1999 (P_0 = 67 g P m^{-3} = 0.15 mg P g^{-1} dry mass, h = 0.15 m, and k = 0.1 m year^{-1}; Istvánovics & Somlyódy, 2001), the rapid decrease in the mobile P pool is still in progress (Fig. 4). Considering that the concentration of mobile P sets an upper limit to the internal P load, the output of the model can be interpreted as the trend in the potential maximum of internal P load. This maximum has been fully exploited (i) during the initial period after the reduction in the external load, (ii) during the years of large blooms of *C. raciborskii* (there is a partial overlap between (i) and (ii)), and (iii) during the recent dry years when the weir of the Zala River was closed from about June to November. In other years physical constraints, biotic interactions or both might keep the biomass of phytoplankton below the highest achievable level.

In years when internal P load approached to the potential maximum, strong correlation ($r^2 = 0.88$, $n = 16$, Fig. 4) was observed between annual mean biomass of phytoplankton and concentration of mobile P. This suggests that the ultimate reason for the fast improvement in the western areas is the favorable long-term behavior of the highly calcareous sediments. The relationship can be used to predict future biomass of phytoplankton in Basin 1 under various loading scenarios.

The OECD approach (Vollenweider & Kerekes, 1982; Fig. 5) was used to explore the response of algal biomass to the reduction in external load of P in the various basins. Mean in-lake concentration of P (P, g P m^{-3}) was calculated from the area-specific P and hydraulic loads (l_{IN}, g P m^{-2} year^{-1} and q, m year^{-1}, respectively) taking into account both inflow and through-flow:

$$P_\lambda = \frac{l_{IN}}{q} \cdot \frac{1}{1 + 2\sqrt{\tau}}$$

where τ (year) is the theoretical water retention time.

Deteriorating water balance of the lake cancelled the effect of reduction in nutrient loads on P in the eastern basins (Fig. 5; Istvánovics et al., 2002). Pre- and post-management status of the various basins seemed to follow a common linear curve suggesting that the behavior of the different areas can be considered fairly similar for managerial purposes.

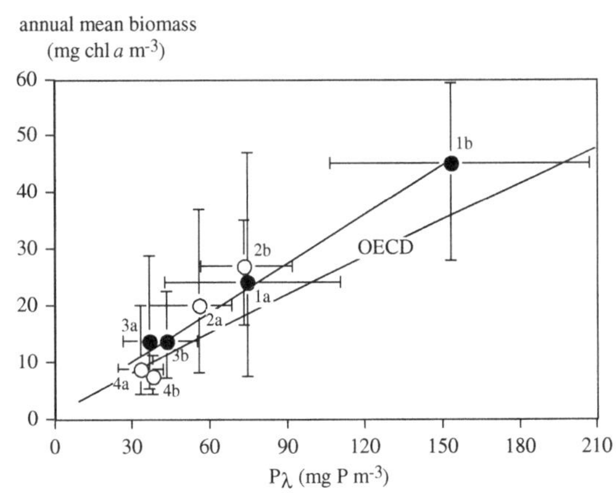

Fig. 5 Multi-annual mean biomass of phytoplankton as a function of the in-lake concentration of P (P_λ). (X and Y bars denote the minimum and maximum values during the period of averaging. Numbers refer to the four basins, "b" refers to the period before management (1980–1988), "a" refers to the period after management (1989–2003). The OECD model line and the linear trend for the four basins of Lake Balaton are given)

Changes in the composition of phytoplankton

In the eastern basins, phytoplankton composition and dynamics showed little difference during the 1970–1980s from those during the 1950s. *Cyclotella ocellata* Pantocsek and *C. compta* (Ehrenberg) Kützing dominated the spring bloom of diatoms. Main species of the summer maximum were *Ceratium hirundinella* (O.F. Müller) Schrank, meroplanktonic diatoms (*Aulacoseira granulata* (Ehrenberg) Simonsen, *A. granulata* var. *angustissima* (O. Müller) Simonsen and *Snowella lacustris* (Chodat) Komárek et Hindák. *Aphanizomenon klebahnii* Elenkin was also present but never dominant. In contrast to this, monomodal succession with a single large summer bloom characterized the phytoplankton in Basin 1 during the 1970s. Blooms of N_2-fixing cyanobacteria (mainly *Anabaena spiroides* Klebahn and *A. flos-aquae* (Lyngbye) Brébisson) might develop from as early as June and collapse as late as October. New cyanobacterial species were recorded almost each year. Floristic data collected during more than 70 years are clearly indicative of fast structural changes preceding the rapid increase in algal biomass.

Structural resilience could clearly be observed following the reduction in external loads. *Cylindrospermopsis raciborskii* continued to form large blooms in the western areas until 1994 (Fig. 6). A sudden shift occurred in 1995: the share of cyanobacteria has not exceeded 30% from this year and duration of the summer bloom shortened significantly (Fig. 6). Although the seasonal succession was monomodal in many years, the ratio of the summer to spring bloom tended to decrease.

In the eastern areas, the slight increase in annual mean biomass during the early post-management period (Fig. 3) coincided with the increasing share of *C. raciborskii* (Fig. 6). In the period 1988–1996, cyanobacteria contributed to the summer maximum by more than 20% in 7 years, whereas previously this occurred only in 1982 (Padisák, 1994; Padisák & Reynolds, 1998). The possible explanation is that blooms in the

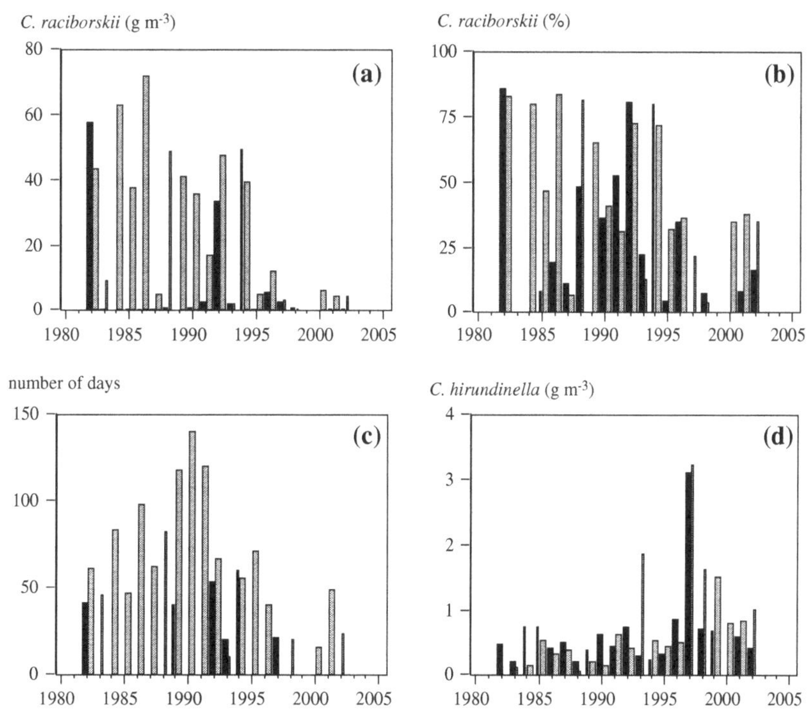

Fig. 6 Maximum biomass (**a**) and contribution to the total biomass during the cyanobacteria bloom (**b**) of *Cylindrospermopsis raciborskii*, number of days when total biomass exceeded 10 g m^{-3} (**c**), and maximum biomass of *Ceratium hirundinella* (**d**) in Basin 1 (dotted bar) and Basin 4 (closed bar). (No data were available in Basin 4 in 2000)

western areas may disperse in 3–4 weeks to inoculate the eastern ones (Istvánovics et al., 2002; Koncsos, 2003).

Apart from *C. raciborskii*, there are two species in Lake Balaton that deserve special attention due to their potential toxicity and/or ability to form excessive blooms. The one is *Aphanizomenon flos-aquae* (Lyngbye) Ralfs, which may form surface scum at the relatively low whole-column biomass of about 4 g m^{-3}. The other one is *Planktothrix agardhii* (Gomont) Anagnostidis et Komárek that bloomed first in the lake in 1989. Its share reached 20% and 50% of the maximum biomass in Basins 1 and 4, respectively. The first bloom might represent one of the unexpected consequences of pre-reservoir construction. By 1988, the upper reservoir (Fig. 1) developed into a hypertrophic pond with yearly mean chlorophyll concentration of 150 mg m^{-3} at the outflow. *P. agardhii* bloomed almost all around the year (Mátyás, 1996), and Basin 1 received a substantial *P. agardhii* "load" between 1988 and 1992. Although the reed-covered lower reservoir has fully eliminated this inoculation from 1993, the cyanobacterium established itself in the western areas of Lake Balaton and made up 30% of the maximum biomass in some years.

Ceratium hirundinella is one of the most successful canopy species of phytoplankton that was historically the prominent summer alga of Lake Balaton (Sebestyén, 1953). Its biomass remained relatively low and stable in Basin 4 (Fig. 6). In contrast to this, biomass of *C. hirundinella* was negatively correlated with that of *Cylindrospermopsis raciborskii* ($r = -0.59$) in Basin 1. The recent increase of *Ceratium hirundinella* (Fig. 6) is indicative of the beginning of structural recovery in the western areas.

In order to exploit the predictive power provided by compositional changes of phytoplankton, Padisák and coworkers (2003) elaborated a novel indicator, the *Q* value. Phytoplankton guilds/associations were identified in accordance with the phyto-sociological classification of Reynolds and coworkers (2002). Considering growth rates, life history, toxicity and distribution of species in 27 guilds, a water quality factor ranging from 2 to 10 was assigned to each functional group. A higher score indicates better quality. Five water quality classes were set up using the biomass-based relative frequency of the guilds weighted by the water quality factor. Presence of potentially toxic or bloom-forming species decreases the value of *Q* even if the total biomass is low. Vice versa, *Q* may attain a high value when the biomass is relatively high but bloom formation is unlikely in the absence of appropriate species.

The *Q* index was tested on phytoplankton time series from Lake Balaton at various temporal scales (Fig. 7). Annual mean biomass of phytoplankton was low in Basin 4 during the period 1983–1988. The *Q* index, however, qualified the area only as moderate or good. Indeed, annual mean biomass increased in the subsequent years

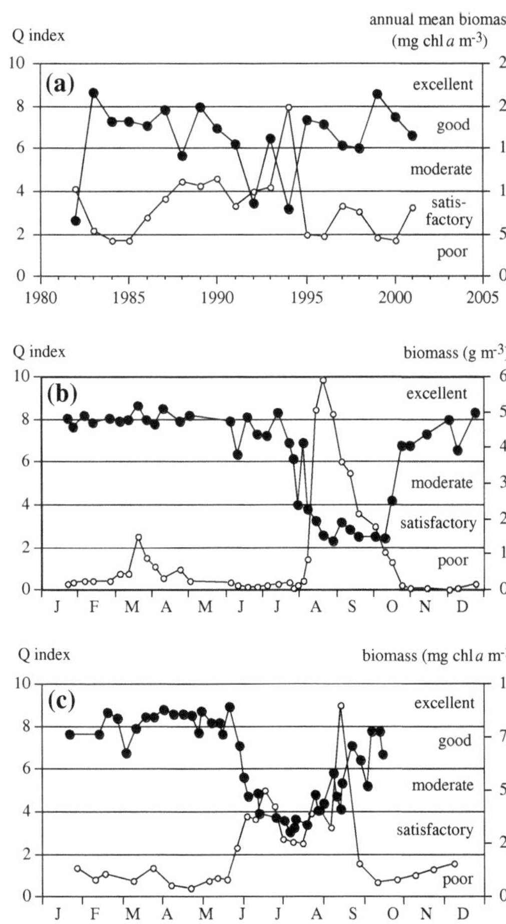

Fig. 7 The value of the *Q* index (closed circle) and the biomass of phytoplankton (open circle) in various years in Basin 4 (**a**), in 1994 in Basin 4 (**b**), and in 2002 in Basin 1 (**c**). (The *Q*-based classification of water quality is shown)

in spite of the reduction in nutrient loads (Fig. 3). At a finer scale, the predictive value of Q was even more striking. Low values of Q warned before the onset of the summer bloom of cyanobacteria that water quality problems were likely to occur.

Changes in the biomass and composition of crustacean zooplankton

Frequent sediment resuspension and intense carbonate precipitation makes Lake Balaton a highly stressful habitat for filter feeding zooplankton (G.-Tóth, 1992). Before eutrophication, *Eudiaptomus gracilis* (Kiefer, 1932) made up more than 80% of crustacean zooplankton. As an average, 8–10% of primary production was channeled to zooplankton. This is low in comparison to most other temperate lakes (G.-Tóth, 1991). The two important consequences are the prevalence of bottom–up regulation of zooplankton and low efficiency of energy utilization by fish (around 0.01%; Bíró & Vörös, 1982).

The moderate increase in the biomass of crustacean zooplankton by a factor of 1.5–3 was disproportionately low compared to the nearly order of magnitude increase in primary production in the western areas. The reasons of uncoupling might be manifold. (i) The share of inedible algae increased within the phytoplankton. The typically low community clearance rates decreased to negligible during blooms of filamentous cyanobacteria. (ii) Quality of ingestible algae might decrease due to frequent N deficiency. (iii) Elevated primary production further enhanced gross carbonate precipitation and thus counterbalanced the increase in the ratio of food to inorganic particles. In general, the most selective feeders (*Cyclops vicinus* (Uljanin, 1875), *Acanthocyclops vernalis* (Fischer, 1853), *A. vernalis* v. *robustus* (Sars, 1863)) benefited the most from eutrophication. Structural re-organization could be observed after 1996. Number of *E. gracilis* nauplii increased by a factor of 2–3 indicating the improved physiological condition of this species that took back its former leading position in the crustacean zooplankton.

Conclusively, crustacean zooplankton is hardly a suitable group for characterizing trophic status of Lake Balaton due to extreme physical constraints acting upon its composition and biomass.

Changes in the biomass and composition of the chironomid fauna

During the last 150–200 years, *Procladius choreus* (Meigen, 1804), *Tanypus punctipennis* (Meigen, 1818) and *Chironomus balatonicus* (Dévai, Wülker et Scholl, 1983) made up about 90% of the biomass and production of chironomid fauna in the profundal of Lake Balaton (Dévai & Moldován, 1983; Specziár & Vörös, 2001). Of them, carnivorous *P. choreus* was usually present in the highest densities.

Substantial structural and quantitative differences developed along the longitudinal axis of Lake Balaton during the period of eutrophication (Fig. 8; Dévai, 1990; Specziár & Vörös, 2001). The biomass remained low in Basin 4 (0.03–0.6 g dry mass m^{-2} depending on the season and the year), and *P. choreus* preserved its leading position. In the western areas, the biomass increased by a factor of 2–8 (0.2–4.25 g m^{-2})

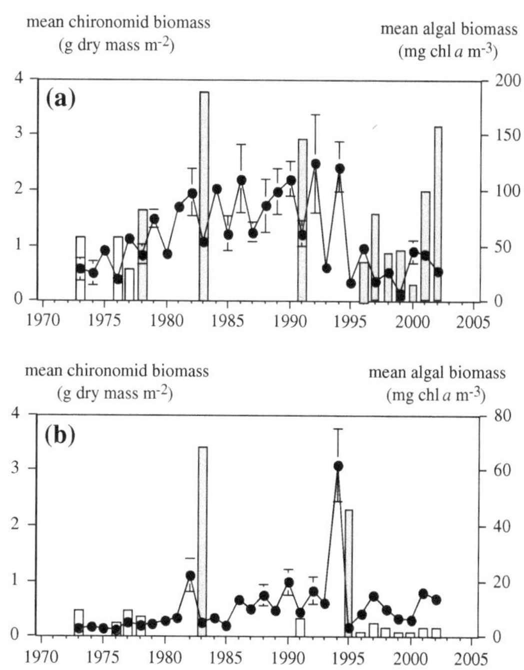

Fig. 8 Annual mean biomass of chironomid larvae (bars) and mean biomass (± sd) of phytoplankton (circle) during summer (July–September) in Basins 1 (**a**) and 4 (**b**)

indicating the relatively strong coupling between algal and chironomid production. The quantitative change was associated with a shift from *P. choreus* dominance to the dominance of phyto- and detritivorous *Ch. balatonicus*.

From the early post-management period, data are available only in 1991. Delayed response of phytoplankton seemed to postpone re-structuring of chironomid assemblages in Basin 1 (Fig. 8). From 1996, mean chironomid biomass tended to decrease but remained higher (0.3–3.1 g m^{-2}) than in the other basins (0.07–0.3 g m^{-2}). Considering the similar biomass of phytoplankton in Basins 1 and 2 (Fig. 3), the difference in the chironomid biomass between the two areas was surprising. The explanation lays in the divergent structure of chironomid assemblages. While *Ch. balatonicus* defended its dominance in Basin 1, *P. choreus* became once again the most abundant species in the other basins. This observation, along with structural changes during eutrophication suggests that high chironomid biomass (>0.5 g m^{-2}) is invariably associated with the dominance of *Ch. balatonicus*. Additional evidence also supports this contention. In Basin 4, two large chironomid peaks were observed following the bloom of *C. raciborskii* in 1982 and 1994. In both cases, *Ch. balatonicus* dominated the chironomid community (*cf.* Figs. 6 and 8).

The above pattern indicates that *Ch. balatonicus* may show a non-linear response to resource availability. To further explore this point, spring biomass of *Ch. balatonicus* was compared to phytoplankton biomass during the previous summer (Fig. 9). Above a threshold chlorophyll concentration of 20–30 mg m^{-3}, abundance of *Ch. balatonicus* exceeded 700–800 ind m^{-2} and the species dominated the chironomid fauna. Below this threshold, its abundance was close to the limit of detectability (≤1 ind m^{-2}). In the latter case, the importance of top–down control by fish increased (Tátrai, 1980; Bíró, 1997).

The shift from *P. choreus* to *Ch. balatonicus* dominance was shown to double (from 0.3–0.6% to 0.7–1.3%) the efficiency of energy transfer from primary producers to chironomids (Specziár & Vörös, 2001; Specziár, 2004). This, in turn, basically influenced population dynamics of several fish species including those that forage primarily for chironomids (*Abramis brama* (Linnaeus, 1758), *Gymnocephalus cernuus* (Linnaeus, 1758)) and those that temporarily supplement their diet with these organisms (*Rutilus rutilus* (Linnaeus, 1758), *Pelecus cultratus*

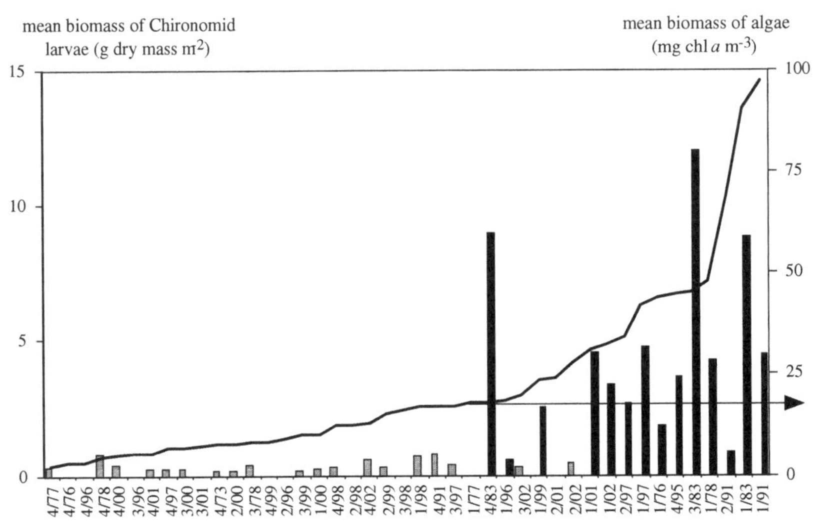

Fig. 9 Mean biomass of chironomid larvae during spring (March–April; bars) and mean biomass of phytoplankton (solid line) during the previous summer (July–September). (In the Z/XY label format of the abscissa Z refers to the basin and XY to the year of the chironomid data. Data were ordered by ascending phytoplankton biomass. Black bars denote years with the dominance of *Chironomus balatonicus*. Arrow indicates the approximate chlorophyll threshold above which *Ch. balatonicus* is likely to be present)

(Linnaeus, 1758), *Alburnus alburnus* (Linnaeus, 1758), *Cyprinus carpio* (Linnaeus, 1758), etc.). For example, growth of *Abramis brama* had been fast during the 1970s with a particularly strong population in Basin 1. In the 1990s, growth slowed down and the bream population decreased dramatically all along the lake (Bíró, 1997).

Updating water quality targets

As a first step, we examine whether the present water quality targets and permissible loads of phosphorus (Table 2) are realistic in the case of Basin 1. Maximum biomass of algae has been predicted for 2010 and 2025 (Table 3) from the relationship between the mobile P pool of the sediments (that is the potential maximum of internal P load) and yearly mean biomass of phytoplankton (Fig. 4). Three distinct load scenarios were defined using the observed loads during the past decade (Table 3). Scenario 1 represented the mean load over the whole period after the latest large management measure taken in 1992. Scenario 2 corresponded to the mean load under average hydro-meteorological conditions. Finally Scenario 3 was the mean load experienced during the draught in 2000–2003.

Table 2 Present water quality targets prescribed for 2010

Basin	Annual mean biomass (mg chl *a* m^{-3})	BAP load (tons P $year^{-1}$)
1	7–14	21
2	6–12	15
3	4–8	8
4	3–6	12

BAP—biologically available P

In order to achieve the present water quality targets until 2010, mean external load of P should be decreased immediately to the level of Scenario 3 (Table 3). We estimate that at the mouth section of the Zala River, 6–8 tons P $year^{-1}$ originate from sewage emission, whereas the diffuse load of biologically available P (BAP) ranges between 10 and 15 tons P $year^{-1}$ in average hydro-meteorological conditions. During the draught, pre-reservoirs released 1–5 tons P $year^{-1}$ of BAP due to increased internal P load. Thus, diffuse load dropped to 0–2 tons P $year^{-1}$, a value below the natural background load in years with average runoff (Table 1). Consequently, no combination of management measures can reduce the external load to the level of Scenario 3. Therefore the present water quality target is unlikely to be attained by 2010.

Taking into consideration planned development in sewerage, introducing an effluent standard of 0.5 g P m^{-3} in the whole watershed will reduce BAP load from point sources by no more than 30–40%. Such reduction would decrease the external load to the level of Scenario 1. The desired water quality could then be achieved by 2025 (Table 3). However, neither Scenario 1 seems to be realistic at the present time because it contains the strong effect of dry years on mean load (Table 3). In this way improved waste water treatment alone does not guarantee the expected improvement in water quality. In order to accomplish our objectives, diffuse load of BAP experienced under mean hydro-meteorological conditions (Scenario 2, Table 3) must be reduced by 3–4 tons P $year^{-1}$ in addition to improved waste water treatment. This reduction is to be

Table 3 Maximal expected value of annual mean algal biomass in Basin 1 assuming three constant load scenarios derived from past observations

Scenario	Period of past observation	TP load (tons P $year^{-1}$)	BAP load (tons P $year^{-1}$)	2010		2025	
				P_{mobil} (mg P g^{-1} dry mass)	Biomass (mg chl *a* m^{-3})	P_{mobil} (Mg P g^{-1} dry mass)	Biomass (mg chl *a* m^{-3})
1	1993–2003	37 ± 22	21 ± 12	0.066	16	0.056	13
2	1993–1999	50 ± 17	27 ± 10	0.077	18	0.074	17
3	2000–2003	15 ± 4	10 ± 2	0.047	11	0.022	5

TP—total phosphorus, BAP—biologically available phosphorus. P_{mobil}—mobil P content of the sediments. Estimates are based on the relationship in Fig. 4b. Loads of Ca and suspended solids were also assumed to be constant realized in a socio-economic environment where the unsustainable low level of fertilizer application is expected to rise, and abandoned agricultural areas are increasingly built in along the shoreline.

Considering the fairly similar behavior of various basins on the OECD plot (Fig. 5), permitted BAP loads can be expected to result in the desired water quality in each area (Table 2) by 2025.

As a second step, present water quality targets were supplemented with ecological criteria (Table 4). Replacing the simple scheme of Table 2 with a more complex classification system (Table 4) demands additional effort from the responsible authorities, which may reluctantly accept the new criteria. Nevertheless, there are at least three good reasons justifying our proposal. (i) The foregoing analysis indicated that recovery of Lake Balaton reached Stages 4 and 5. While mean biomass of phytoplankton may allow properly assessing progress of eutrophication management during Stages 1 and 3, the value of this indicator diminishes when structural changes take place during subsequent stages. (ii) Structural changes precede quantitative changes. Therefore properly elaborated structural indices may have considerable predictive power during Stages 4 and 5. (iii) The European Water Framework Directive urges classification of aquatic ecosystems using ecological indices.

As discussed above, the *Q* index of Padisák and coworkers (2003) is a suitable indicator of structural changes of phytoplankton that has been successfully tested in Lake Balaton at various time scales (Fig. 7). Thus, the index was incorporated as one of the ecological criteria (Table 4).

The long-term relationship between biomass of phytoplankton and that of chironomids (Fig. 8; Specziár & Vörös, 2001) was used to derive appropriate criteria concerning chironomid biomass (Table 4). A structural index, similar to the *Q* index has not been derived for the chironomid fauna yet. At the same time, presence or absence of *Chironomus balatonicus* was shown to basically influence overall efficiency of energy transfer from phytoplankton to fish. In this way abundance of this key species reflects important characteristics of ecosystem functioning. Moreover, among the 14 chironomid species inhabiting Lake Balaton and known to be capable of bloom formation, *Ch. balatonicus* was the single one that produced the actual blooms up to now. Its mass emergence attains an "intolerable" level when the biomass of chironomids exceeds 3–4 g m^{-2}. The threshold behavior of *Ch. balatonicus* (Fig. 9) allowed establishing ecological criteria in relation to the abundance of this species (Table 4).

Although we aimed at setting up the ecological targets in a consistent way, one cannot exclude that various indicators occasionally result in divergent classification. The question arises how an aggregated index of ecological status can be determined. One may accept the worst indication. Alternatively, the aggregated index may be composed from individual indices by appropriate weighting. The final choice must be made after evaluating the

Table 4 The proposed ecological target system to be realized by 2025

Basin	Ecological targets						Permissible TP load[a] (tons P $year^{-1}$)	Load reduction[g] (%)
	Phytoplankton			Chironomidae				
	Annual MB[b]	*Q* index[c]		Maximum MB[e]	*Ch. balatonicus*[f]			
	mg chl *a* m^{-3}	Value	Classification[d]	g dry mass m^{-2}	ind m^{-2}	Classification[d]		
1	14	6–8	Good	2.0	800	Good	37	17
2	12	6–8	Good	1.5	500	Good	21	54
3	8	8–10	Excellent	0.5	<1	Excellent	17	34
4	6	8–10	Excellent	0.5	<1	Excellent	22	24

MB—mean biomass. a—refers to the average of 5 years, yearly variability can reach ±50–60%. b—using the present monitoring frequency. c—in the period July–September. d—scaling: "excellent", "good", "moderate", "satisfactory", "poor". e—in the period March to June. f—maximum density of *Chironomus balatonicus* larvae. g—relative to the present load experienced in years of average runoff experience of experimental introduction of the new targets during the next few years.

Unlike the present scheme that expresses permissible load of P in terms of BAP, we propose to relay on total P loads (Table 4). Several conceptual and methodological problems introduce an additional uncertainty into the estimates of BAP load compared to that of total P load. Undertaking this higher uncertainty has been rational during the initial phase of eutrophication management, when the main effort had to be concentrated on the reduction of loads from point sources (Somlyódy & Jolánkai, 1986). Future management, however, must focus on the reduction of diffuse loads and therefore BAP can be replaced with total P. The proposed total P loads (Table 4) are equivalent to the BAP prescriptions of the present scheme (Table 2). The proposed values refer to loads during the years of average runoff. Depending on the hydrometeorological conditions, the deviation may reach ±50–60%.

Concluding remarks

During the initial stage of eutrophication management, the response of various shallow lakes converges to a single pattern: ones the sum of external and internal P loads reached capacity limiting levels, the phytoplankton biomass decreases in proportion to P supply. Obviously, long-term behavior of sediments determining the magnitude of internal load cause large variability with respect to timing and degree of recovery (Marsden, 1989; Sas, 1989). Nevertheless, a simple targeting scheme that includes the clear objectives (water quality to be restored) and the prerequisites of accomplishing these objectives (permissible loads) is satisfactory to orient both managers and the public. At this early stage, complicated targeting schemes should be avoided since they likely to make confusion.

The situation changes when recovery comes to the stage of re-organization of the aquatic ecosystem. During this stage, the simple targeting system should be replaced with one that integrates specific ecological features of the individual lakes. For example, biomass and composition of zooplankton can be good indicators of recovery in the majority of shallow lakes. In the specific geomorphological environment of Lake Balaton, however, this group fails to properly indicate trophic changes.

As demonstrated in our study, long-term observations must be available about biota in order to choose the most appropriate lake-specific indices and to define a consistent set of ecological criteria. Therefore we recommend including each major group that may a priori be considered to fulfill the role of an indicator in the given lake into the monitoring program from the very beginning of eutrophication management.

Acknowledgements This study was financially supported by the "Phytoplankton-on-line" project (No. EVK1-CT–1999-00037) and by the Ministry for Environment and Water (Hungary). We are greatly indebted to many colleagues who provided their published or unpublished data for this evaluation.

References

Bíró, P., 1997. Temporal variation in Lake Balaton and its fish populations. Ecology of Freshwater Fish 6: 196–216.

Bíró, P. & L. Vörös, 1982. Relationships between phytoplankton and fish yields in Lake Balaton. Hydrobiologia 97: 3–7.

Boström, B., M. Jansson & C. Forsberg, 1982. Phosphorus release from lake sediments. Archiv für Hydrobiologie Beiheft Ergebnisse der Limnologie 18: 5–59.

Clement, A., V. Istvánovics & L. Somlyódy, Evaluation of water quality of Lake Balaton. Vízügyi Közlemények (in Hungarian) (in press) .

Clement, A., 2000. Improving uncertain nutrient load estimates for Lake Balaton. Water Science and Technology 43: 279–286.

Csermák, K. & F. Máté, 2004. Soils of Lake Balaton. V.E. Georgikon Kar, Keszthely, Hungary (in Hungarian).

Dévai, Gy., 1990. Ecological background and importance of the change of chironomid fauna (Diptera: Chironomidae) in shallow Lake Balaton. Hydrobiologia 191: 189–198.

Dévai, Gy. & J. Moldován, 1983. An attempt to trace eutrophication in a shallow lake (Balaton, Hungary) using chironomids. Hydrobiologia 103: 169–175.

G.-Tóth, L., 1991. Respirative Electron Transport System (ETS) Activity and Carbon Metabolism of Net Zooplankton in Lake Balaton. In Bíró, P. (ed.), Hundred Years of Lake Balaton Research. Reproprint, Nemesvámos, 172–180 (in Hungarian).

G.-Tóth, L., 1992. Limiting effect of abioseston on food ingestion, postembryonic development time and

fecundity of daphnids in Lake Balaton (Hungary). Journal of Plankton Research 14: 435–446.
G.-Tóth, L., 2000. Feeding biology of planktonic crustaceans and their role in elimination of phytoplankton in Lake Balaton. DSc Thesis, Balaton Limnological Research Institute, Tihany (in Hungarian).
Herodek, S., 1984. The eutrophication of Lake Balaton: measurements, modeling and management. Internazionale Vereinigung für Theoretische und Angewandte Limnologie 22: 1087–1091.
Herodek, S., 1986. Phytoplankton Changes During Eutrophication and P and N Metabolism. In Somlyódy, L. & G. van Straten (eds), Modeling and Managing Shallow Lake Eutrophication. Springer-Verlag, Berlin, 183–204.
Istvánovics, V., A. Osztoics & M. Honti, 2004. Dynamics and ecological significance of daily internal load of phosphorus in shallow Lake Balaton, Hungary. Freshwater Biology 49: 232–252.
Istvánovics, V. & L. Somlyódy, 2001. Factors influencing lake recovery from eutrophication – the case of Basin 1 of Lake Balaton. Water Research 35: 729–735.
Istvánovics, V., L. Somlyódy & A. Clement, 2002. Cyanobacteria-mediated internal eutrophication in shallow Lake Balaton after load reduction. Water Research 36: 3314–3322.
Koncsos, L., 2003. Evaluation of the Achievement in Water Quality Management. In Mahunka, S. & J. Baczerowski (eds), Results of Lake Balaton Research in 2002. Hungarian Academy of Sciences, Budapest, 79–85 (in Hungarian).
Láng, I., 1986. Impact on Policymaking: Background to a Government Decision. In Somlyódy, L. & G. van Straten (eds), Modeling and Managing Shallow Lake Eutrophication. Springer-Verlag, Berlin, 110–122.
Lijklema, L., P. Gelencsér, F. Szilágyi & L. Somlyódy, 1986. Sediment and its Interaction with Water. In Somlyódy, L. & G. van Straten (eds), Modeling and Managing Shallow Lake Eutrophication. Springer-Verlag, Berlin, 156–183.
Marsden, M. W., 1989. Lake restoration by reducing external phosphorus loading: the influence of sediment phosphorus release. Freshwater Biology 21: 139–162.
Mátyás, K., 1996. Phytoplankton Studies in the Kis-Balaton Water Protection System between 1991 and 1995. In Pomogyi, P. (ed.), 2nd. Kis-Balaton Symposium, ISBN 963 02 9991 7. West-Transdanubian Water Authority, Keszthely, 119–130 (in Hungarian).
Padisák, J., 1994. Relationships between Short-term and Long-term Responses of Phytoplankton to Eutrophication of the Largest Shallow Lake in Central Europe (Balaton, Hungary). In Sund, H., H.-H. Stabel, W. Geller, Y. Xiaogan, Y. Kechang & S. Fengning (eds), Environmental Protection and Lake Ecosystem. Science & Technology Press, Beijing, 419–437.
Padisák, J., 1997. *Cylindrospermopsis raciborskii* (Woloszynska) Seenayya et Subba Raju, an expanding, highly adaptive cyanobacterial species: geographic distribution, population dynamics and autecology. Archive für Hydrobiologie (supplement) 107: 563–593.
Padisák, J. & C. S. Reynolds, 1998. Selection of phytoplankton associations in Lake Balaton, Hungary, in response to eutrophication and restoration measures, with special reference to the cyanoprokaryotes. Hydrobiologia 384: 41–53.
Padisák, J., É. Soróczki Pintér & Zs. Zámbóné Doma, 2003. Diversity and Spatial-temporal Patterns of Phytoplankton in Lake Balaton in 2002. (in Hungarian). In Mahunka, S. & J. Banczerowski (eds), Results of Lake Balaton Research in 2002, ISSN 1419-1075. MTA, Budapest, 35–42.
Ponyi, J., I. Tátrai & A. Frankó, 1983. Quantitative studies on Chironomidae and Oligochaeta in the benthos of Lake Balaton. Archiv für Hidrobiologie 97: 196–207.
Reynolds, C. S., V. Huszar, C. Kurk, L. Naselli-Flores & S. Melo, 2002. Towards a functional classification of the freshwater phytoplankton. Journal of Plankton Research 24: 417–428.
Sas, H., 1989. Lake Restoration and Reduction of Nutrient Loading: Expectations, Experiences, Extrapolations. Academia Verlag Richarz, St Augustin.
Scheffer, M., S. H. Hosper, M.-L. Meijer, B. Moss & E. Jeppesen, 1993. Alternative equilibria in shallow lakes. Trends in Ecology & Evolution 8: 275–279.
Sebestyén, O., 1953. Quantitative plankton studies in Lake Balaton. II. Decadal changes. Annales Instituti Biologici (Tihany) 21: 63–89 (in Hungarian).
Shanahan, P., D. R. F. Harleman & L. Somlyódy, 1986. Wind-induced Water Motion. In Somlyódy, L. & G. van Straten (eds), Modeling and Managing Shallow Lake Eutrophication. Springer-Verlag, Berlin, 204–255.
Somlyódy, L. & G. Jolánkai, 1986. Nutrient Loads. In Somlyódy, L. & G. van Straten (eds), Modeling and Managing Shallow Lake Eutrophication. Springer-Verlag, Berlin, 125–156.
Søndergaard, M., J. P. Jensen & E. Jeppesen, 1999. Internal phosphorus loading in shallow Danish lakes. Hydrobiologia 408/409: 145–152.
Specziár, A., 2004. Long-term Changes in the Chironomid Fauna and Feeding of Fish in Lake Balaton. In Fenyvesi, O. (ed.), Tudományos előadások 2004. MTA Veszprémi Területi Bizottsága, Veszprém, 69–85 (in Hungarian).
Specziár, A. & L. Vörös, 2001. Long-term dynamics of Lake Balaton's chironomid fauna and its dependence on the phytoplankton production. Archiv für Hydrobiologie 152: 119–142.
Tátrai, I., 1980. About feeding conditions of bream, (*Abramis brama* L.) in Lake Balaton. Developments in Hydrobiology 3: 81–86.
Virág, Á., 1998, The Past and the Present of Lake Balaton. (In Hungarian). Egri Nyomda Kft., Eger. ISBN 963 9060 216.
Vollenweider, R. A. & J. J. Kerekes, 1982. Background and Summary Results of the OECD Cooperative Programme on Eutrophication. OECD report, Paris.

Hydrobiologia (2007) 581:319–327
DOI 10.1007/s10750-006-0512-6

EUTROPHICATION IN LAKES

Removal of agricultural non-point source pollutants by ditch wetlands: implications for lake eutrophication control

Cuiling Jiang · Xiaoqiu Fan · Guangbo Cui · Yibin Zhang

Abstract Ditches grown with nature reed (*Phragmites communis* Trin) and wild rice (*Zizania latifolia* Turcz) were selected to study the removal capacity of agricultural non-point source pollutants so as to find a way to alleviate eutrophication in Lake Taihu. Ditches sediment from depths below 40 cm can accumulate organic matter and total nitrogen (TN). TN is correlated positively to organic matter in reed populated sediment and wild rice populated sediment. This suggests that the main composition of TN is organic nitrogen derived from plant decomposition. A significant negative relationship between TN and pH was found in reed and wild rice sediments. Seasonal harvest of helophyte vegetation is an effective method to remove N and P from wetlands. Organic matter and TN concentrations in water and sediments (0–20 cm) in areas where reeds were removed are lower than non-harvested areas (control). Reeds and wild rice have high uptake ability of nitrogen (N) and phosphorus (P). However, the low economic value of these plants will not stimulate voluntary harvest of farmers. *Zizania caduciflora* Turez Hand-mazt is a kind of vegetable widely cultivated in ditches around the lake. It can also absorb N and P effectively. Thus, large scale cultivation of *Z. caduciflora* to replace nature plants may improve water quality.

Keywords Ditch wetland · Helophyte vegetation · Non-point source pollutants

Guest editors: B. Qin, Z. Liu & K. Havens
Eutrophication of shallow lakes with special reference to Lake Taihu, China

C. Jiang (✉) · X. Fan · Y. Zhang
State Key Laboratory of Hydrology-Water Resources and Hydraulic Engineering, Hohai University, 1 Xi Kang Road, Nanjing 210098, China
e-mail: cljianghhu@163.com

C. Jiang · G. Cui
College of Water Resources and Environment, Hohai University, 1 Xi Kang Road, Nanjing 210098, China

Introduction

Lake Taihu is the third largest freshwater lake in China (Qin et al., 2007). Since the 1980s with rapid economic development in the Lake Taihu Basin, increasing nutrient loadings from point and non-point sources have resulted in water quality deterioration. Compared to oligotrophic conditions in the 1950s, Lake Taihu is now eutrophic (Huang et al., 2001). With enhanced control of point-source pollutants, non-point sources are the main influence on water quality of Lake Taihu. Excessive chemical fertilizer application led to large nitrogen (N) and phosphorus (P) increases (Roth & Fox, 1990; Angle et al., 1993; Withers et al., 2001). In Lake Taihu, the amount of fertilizers used (total pure nitrogen, phosphorus and kalium) has doubled from about 400 kg ha^{-1} in the 1980s to 800 kg ha^{-1} at the end of the 1990s (Gao et al., 2002). Uptake efficiency of chemical fertilizers by crops is 25–35% for N (Li & Li, 2001) and 10–20% for P. About 20–25% of N and 20–40% of P applied to crops annually is carried into ditches and channels with runoff and seepage (Xu, 1996; Huang et al., 2001).

Wetlands are effective at trapping and removing N and P by sediment adsorption, decomposition by microorganisms and uptake by helophyte vegetation (Raisin & Mitchell, 1995; Mitsch et al., 2000; Kang et al., 2002; Jordan et al., 2003). In many countries, constructed wetlands are used to treat point or non-point source pollutants (Fleischer et al., 1994; Spieles & Mitsch, 2000; Braskerud, 2002). In general, constructed wetlands are small, making them difficult to develop widely in China. However, there are irrigation and drainage networks that divide croplands in the lake basin. These ditches and channels are the first convergence points for agricultural non-point pollutants and conduits exporting nutrients to rivers and lakes. Wetland plants are flourishing in ditches and channels due to plentiful rainfall and suitable climate. Large emergent plants growing naturally in ditches and channels are reed (*Phragmites communis* Trin) and wild rice (*Zizania latifolia* Turcz), both of which have high capacity for N and P uptake. In the growing season, nutrients are assimilated and accumulated and become plant biomass. Reeds and wild rice have low economic value, and are difficult to harvest. In winter and spring, the plants decompose and release N and P, which may deteriorate water quality of rivers and the lake (Tanner, 2001).

Many studies have focused on the transport and transformation of N and P in constructed wetlands. Much attention was paid to N and P variations in wetland water (including inflow and outflow), but accumulation and transformation processes in sediments were ignored (Arheimer & Wittgren, 1994; Mitsch et al., 2000; Jan, 2002). Removal mechanisms of N and P in ditch wetlands with natural helophyte vegetation are not often discussed, and few, if any, economical methods to prevent secondary pollution have been proposed. The objective of this study is to identify factors affecting N and P transformations in ditches and formulate a plan to remove nutrients to alleviate surface water enrichment.

Materials and methods

Experimental site

The experimental site (32°14′ N, 119°52′ E) was located near Lake Taihu in a ditch wetland 600 m in length and 20–30 m in width (Fig. 1). Reeds and wild rice were growing abundantly in the ditch. The reed area was about 1.4 ha (560 m length × 25 m width) and wild rice was about 0.08 ha (40 m length × 20 m width). There is a concrete weir to keep water in the wetland. Only during flooding does water drain from east to west through the weir. About 1.8 ha of *Zizania caduciflora* Turez Hand-mazt was planted in another nearby ditch. The water table of the wetland in summer was about 40–80 cm deep. Dry and wet conditions alternated with the occurrence of rainfall in winter. During the experiment, only 0.23 ha of cropland was arranged to drain agricultural runoff into the wetland. Wheat and rice were grown in rotation. The wheat growing season was November to May of the next year and June to October for rice. Wheat and rice straws were returned to the soil after harvest. There were 787 kg ha^{-1} urea (46.2% N) and ammonia bicarbonate (17.1% N) and 1275 kg ha^{-1} compound fertilizer (9% N and 7% P) used on the cropland, equaling 607 kg ha^{-1} $year^{-1}$ of N and 90 kg ha^{-1} $year^{-1}$ of P.

Sampling and analyses

Soil samples were taken from cropland, reed and wild rice wetlands in September and December 2002 and March and June 2003 for analysis of organic matter, total nitrogen (TN), NH_4^+-N, NO_3^--N, soil moisture and pH. Two sample stations were selected in cropland and reed wetland, and one in wild rice wetland due to the small area (about 800 m^2). At each station, there were six layers of soil samples (0–10, 10–20,

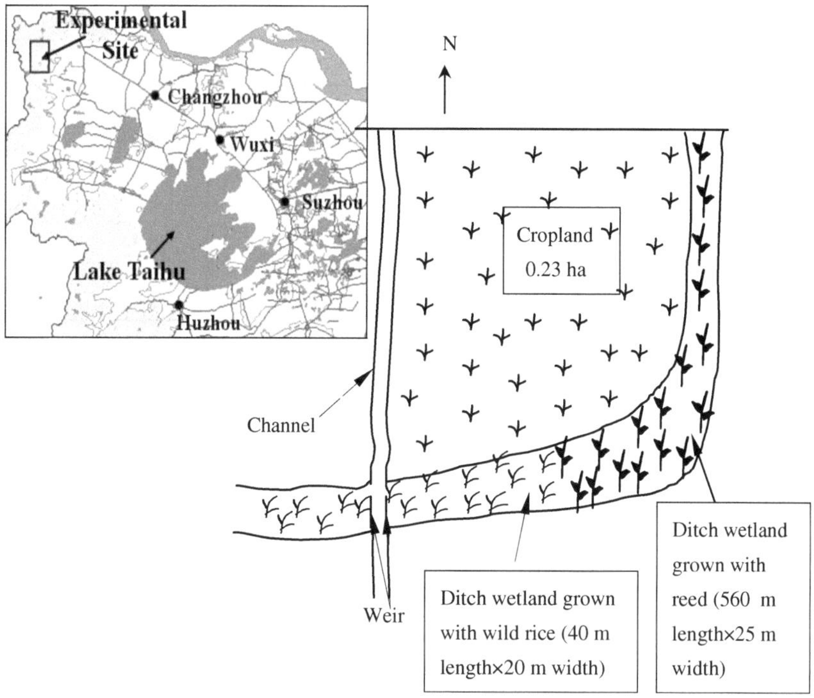

Fig. 1 Schematic illustration of ditch wetland in experimental site

20–40, 40–60, 60–80, 80–100 cm depth) collected using a 5 cm-diameter auger. During replication of sediment sampling in the reed wetland, one station was near the wild rice wetland (in the harvested plot after October) and another was about 150 m from the first (in the non-harvested plot). Two soil sampling stations in cropland were about 8–10 m from the wetland berm. Preliminary investigation showed little variation in nutrient contents between the two stations. Soil nutrient date reported in results is the average of the two stations.

Water samples in wild rice and reed wetlands were collected in September, November and December 2002, and March and June 2003 to analyze organic matter, TN, NH_4^+-N, NO_3^--N, and pH. Two or three water samples in each plot were taken and mixed. In October 2002, about 1/3 of the reed plant area adjacent to wild rice was harvested. Sediment and water samples in the harvested and non-harvested plot (control) were taken. Comparison sampling in the reed wetland was finished in March 2003.

Vegetation sampling was obtained from three randomly selected stations in growing seasons. Plant were sampled and separated into above and below ground fractions after transport to the laboratory. Plant samples were collected in September 2002, and June and September 2003 to analyze N and P in leaf, stem and root fractions. Characteristics of the helophyte plants are shown in Table 1.

Chemical analyses were conducted based on standard methodology (China Standard Press, 1998). Organic matter in water was determined by dichromate oxidation and $FeSO_4$ titration. NH_4^+-N concentration was analyzed using Nessler's reagent colorimetric method, and NO_3^--N concentration was measured using spectrophotometry with phenol disulfonic acid. For TN analyses, sample was acidified with H_2SO_4 to pH < 2 and subjected to Kjeldahl digestion and analysis by colorimetry. pH was measured with a glass electrode. All analyses in water were finished within 48 h.

For analyses of NH_4^+-N and NO_3^--N in soil, fresh samples were collected in pre-weighted aluminum cans and refrigerated immediately to prevent transformation of inorganic nitrogen. NH_4^+-N and NO_3^--N were extracted with 2 M

Table 1 Character of plants (average analyzed result of September 2002, June 2003 and September 2003)

Character	Reed	Wild rice	*Z. caduciflora*	Rice
Height (m)	2.3–3.1	1.47–1.5	2.1–2.2	0.83–0.85
Root length (cm)	27–62	35–47	38–50	10–18
Density (individual m^{-2})	200	184	64	192
Fresh weight of leaf and stem (g $individual^{-1}$)	61.36	21.01	73.34	7.57
Water content of plant (%)	55.56	80.30	76.34	58.12
Fresh weight (kg ha^{-1})	122,720	38,658	46,938	14,534
Dry weight (kg ha^{-1})	54,537	7,615	11,106	6087

KCl by shaking for 1 h with a reciprocal shaker. Determination of NH_4^+-N and NO_3^--N in extracts were the same as above.

Samples for organic matter, TN and pH were collected in plastic bags, air dried for about 1 month in the laboratory, sieved through a 2-mm stainless steel sieve, and kept in brown glass bottles. TN was analyzed using the semi-micro Kjeldahl method, and produced NH_4^+ was collected in a boric acid solution and titrated with H_2SO_4. Samples for pH analysis were extracted with 0.01 M $CaCl_2$ and measured with a glass electrode. Organic matter content was analyzed using potassium dichromate-sulfate oxidation method.

Plant samples, including above and below ground parts, were cleaned and oven-dried at 105°C for 1 h and 80°C for an additional hour. After being cooled, the samples were ground and kept in brown glass bottles. Nutrients in roots, leaves, and stems were analyzed separately. N and P were analyzed via H_2O_2/H_2SO_4 digestion and measured colorimetrically. Biomass was calculated based on fresh and dry weight and surface area of the sampling plot.

Results

Accumulation of organic matter and TN in cropland soil and wetland sediment

Organic matter in cropland soil likely comes from decomposition of crop remains since farmers no longer applied organic fertilizer. When rice and wheat were harvested, only usable parts were removed, and leftover straw was ploughed into soil, which became the organic matter and TN source. Above 60 cm, organic matter content and TN decreased with depth and varied little below 60 cm (Figs. 2, 3). TN was highly correlated with organic matter.

Organic matter and TN also were correlated in wetland sediment (Figs. 4, 5), with $r = 0.97$ ($n = 27$, significant level at $P < 0.001$) in the reed ditch and $r = 0.91$ ($n = 19$, $P < 0.001$) in the wild rice ditch. Therefore, TN in wetlands was composed of organic N from plant decomposition.

Organic matter and TN concentrations from 0 to 100 cm in rice and wheat soil decreased with depth, but, in reed and wild rice sediments, lowest concentrations occurred at 20–40 cm and increased below 40 cm (Figs. 2, 3). Therefore,

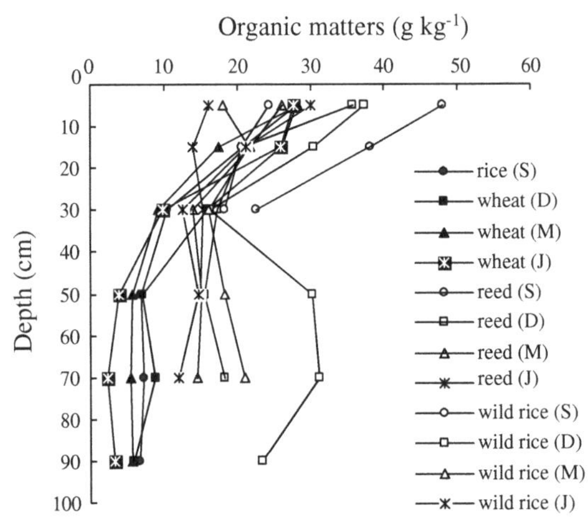

Fig. 2 Contents of organic matter in different layers of field soils (grown with rice and wheat) and wetland sediments (grown with reeds and wild rice) during the period 2002–2003. (S: in September 2002; D: in December 2002; M: in March 2003; J: in June 2003)

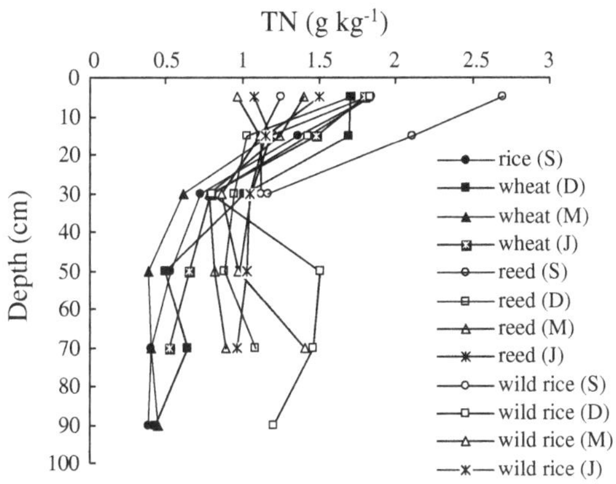

Fig. 3 Contents of TN in different layers of field soils (grown with rice and wheat) and wetland sediments (grown with reeds and wild rice) during the period 2002–2003. (S: in September 2002; D: in December 2002; M: in March 2003; J: in June 2003)

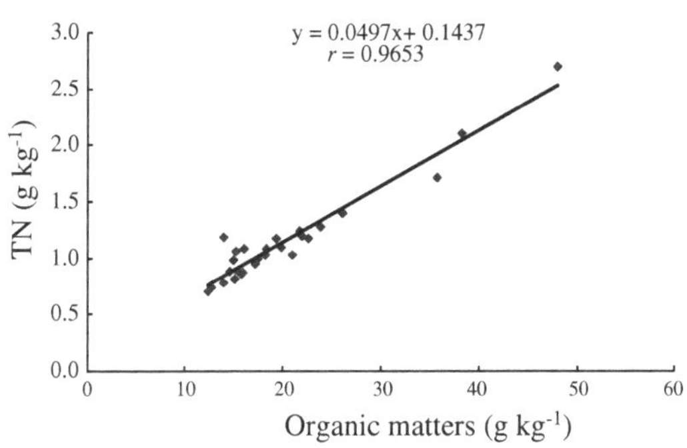

Fig. 4 Correlation between the contents of organic matter and TN in reed wetland sediment

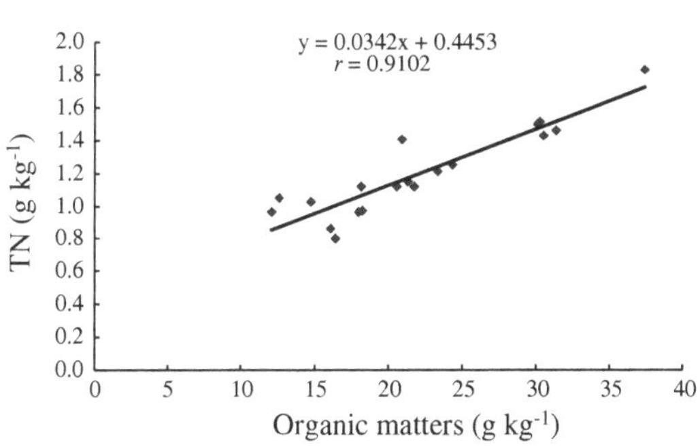

Fig. 5 Correlation between the contents of organic matter and TN in wild rice wetland sediment

organic matter and TN can accumulate in deep layers of wetland sediment.

In sediment above 40 cm, organic matter and TN varied with seasons. Variability in wet sediment was higher than dry soil suggesting that organic matter and TN in wetland surface layers decomposed more quickly due to dry and wet alternation.

Nutrients in harvested plot and in control

The reed wetland adjacent to wild rice was harvested in October 2002. After 20 days, water turbidity in the reaped area was lower than the non-harvested area (control). Organic matter and TN were 21.33 and 1.04 mg l^{-1}, respectively, in water of the reaped area, which were lower than those (28.4 and 1.22 mg l^{-1}, respectively) in the control. In December 2002 and March 2003, sediment organic matter and TN from 0 to 10 and 10 to 20 cm in the harvested area also were lower than the control (Fig. 6). Therefore, harvesting plants in autumn may reduce organic matter and TN accumulation in wetland water and surface sediment. However, NH_4^+-N and NO_3^--N in harvested and non-harvested plots were different from organic matter and TN (Fig. 7). Compared to the non-harvested area, inorganic N from 0 to 10 and 10 to 20 cm in harvested area was high (except low NO_3^--N in 10–20 cm in March 2003). Increased NH_4^+-N and NO_3^--N likely came from organic N decomposition.

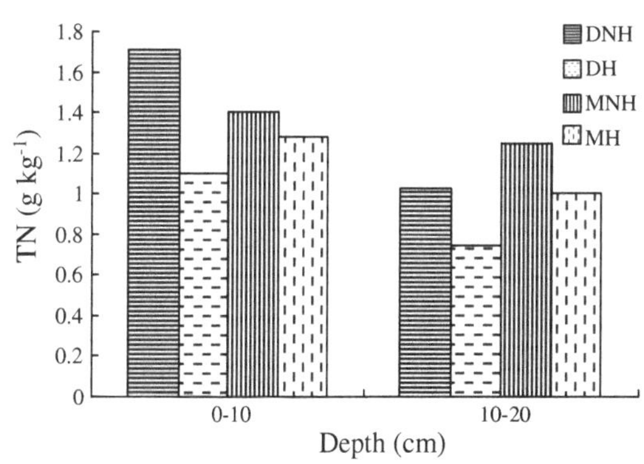

Fig. 6 Contents of TN in the depth of sediment from 0 to 20 cm in harvested and non-harvested plots of reed wetland (DNH: in non-harvested plot in December 2002; MNH: in non-harvested plot in March 2003; DH: in harvested plot in December 2002; MH: in harvested plot in March 2003)

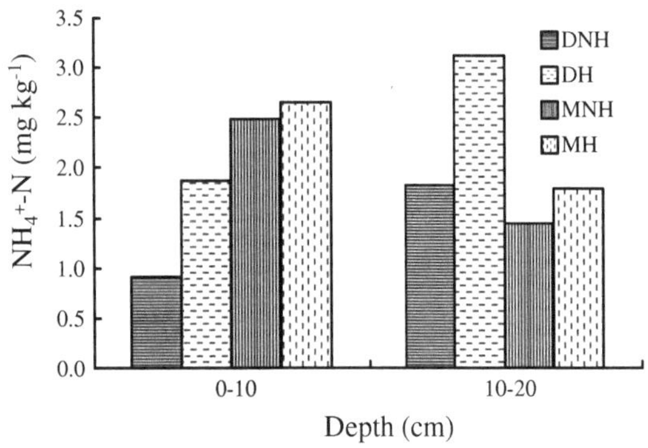

Fig. 7 Contents of NH_4^+-N in the depth of sediment from 0 to 20 cm in harvested and non-harvested plots of reed wetland (DNH: in non-harvested plot in December 2002; MNH: in non-harvested plot in March 2003; DH: in harvested plot in December 2002; MH: in harvested plot in March 2003)

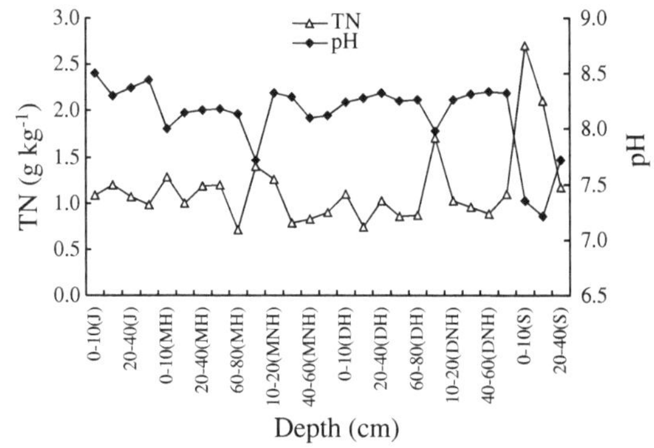

Fig. 8 TN and pH value in different layers of reed wetland sediment during the period 2002–2003. (J: in June 2003; MH: in harvested plot in March 2003; MNH: in non-harvested plot in March 2003; DH: in harvested plot in December 2002; DNH: in non-harvested plot in December 2002; S: in September 2002)

Significant relationships were found among water TN, NH_4^+-N and NO_3^--N in reaped and control plots in the reed wetland. The regression equation for TN and NH_4^+-N is $Y = 0.128X - 0.0338$, $r = 0.90$ ($n = 5$, significant level at $P < 0.05$, n is the number of samples obtained in harvested and non-harvested plots and in different times), and TN and NO_3^--N is $Y = 1.6935X - 0.7413$, $r = 0.96$ ($n = 5$, $P < 0.01$). These relationships suggest that TN mineralization is an important source of NH_4^+-N and NO_3^--N.

Relationship between TN and pH

Nutrient transformation in soil and sediment will cause variation of pH, which may promote chemical reactions. Organic matter and TN in surface soil were high and decreased with depth. In reed and wild rice sediments, organic matter and TN decreased in 0–40 cm, and then increased below 40 cm (Figs. 2, 3). However, pH variation was opposite (Figs. 8, 9). Lower pH coincided with higher organic matter and TN. TN in sediments and soil are negatively correlated with pH. The regression coefficient (r) is –0.78 ($n = 23$, $P < 0.001$) in soil, –0.82 ($n = 27$, $P < 0.001$) in reed sediment, and –0.69 ($n = 19$, $P < 0.01$) in wild rice sediment.

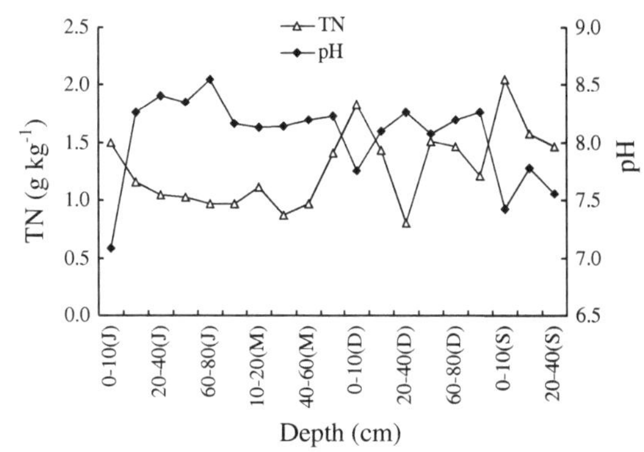

Fig. 9 TN and pH value in different layers of wild rice wetland sediment during the period 2002–2003 (J: in June 2003; M: in March 2003; D: in December 2002; S: in September 2002)

An effective method to remove N and P by plants

Plants are growing naturally in ditch wetlands in the Lake Taihu basin. In late June, reeds and wild rice are fully grown. Plant height, root length, density, fresh weight and water content were measured in September 2002 and June and September 2003 (Table 1). Mean dry reed and wild rice output were 54,537 and 7,615 kg ha^{-1} year^{-1}, respectively. Leaf and stem N and P concentrations are shown in Table 2. There were 818 kg ha^{-1} N and 103.6 kg ha^{-1} P

Table 2 Average contents of N and P in the part of plants above-ground and harvested amount of nutrients in a year

Tissue	*Zizania caduciflora*		Wild rice		Reed		Rice	
	N	P	N	P	N	P	N	P
Leaf and stem (g kg^{-1})	18.0	1.9	17.2	3.8	15.0	1.9	14.0	1.7
Harvested amount (kg ha^{-1})	200	21.1	131	28.9	818	103.6	85.2	10.3

removed after reeds harvest and 131 kg ha^{-1} N and 28.9 kg ha^{-1} P removed by wild rice in autumn. Nutrient uptake by above-ground reed parts is lower than wild rice, but its larger biomass provides high N and P removal capacity.

Zizania caduciflora Turez Hand-mazt is a kind of vegetable and also a perennial aquatic plant similar to wild rice in morphology and ecology. It can assimilate 200 kg ha^{-1} N and 21.1 kg ha^{-1} P in a year (Table 2). Compared with reeds and wild rice, average N content in above-ground parts of *Z. caduciflora* is higher, and P content is similar to reeds but lower than wild rice (Table 2). Therefore, planting *Zizania caduciflora* Turez Hand-mazt to replace natural reeds and wild rice may absorb N accumulated in ditch wetland sediment and alleviate eutrophication in lakes and rivers, but P removal is not as good as natural plants.

Discussion

Inorganic fertilizers used on crops were the primary N and P source in the experimental site. Organic matter and TN in cropland and wetland come from decomposition of plant remains, because farmers no longer use organic fertilizer. In general, inorganic N comprises <5% of TN (Rene et al., 1991; Bai et al., 2002). TN is correlated with organic matter in soil (Yang et al., 2002) and sediments (Figs. 4, 5). However, vertical and spatial distributions of organic matter and TN in cropland soil and wetlands sediment were different.

In rice and wheat soil, organic matter and TN from 0 to 100 cm decreased gradually with depth, but, in reed and wild rice sediments, lowest concentrations occurred in 20–40 cm and increased below 40 cm (Figs. 2, 3). Organic matter and TN decreased from 0 to 40 cm resulted from decomposition by large numbers of microorganisms in the root zone (Peder & Stefan, 1997). Compared with field soil, wetlands sediment contained more organic matter and TN below 40 cm. Therefore, wetlands can accumulate non-point source pollutants.

Organic matter and TN varied with season in surface sediment, more intensively than in soil, because of alternate aerobic and anaerobic conditions (Reddy & Patrick, 1975; Martin et al., 1997), which accelerated decomposition of organic matter and N in wetlands. In addition, cover and removal of plants also affected organic matter and N transformation in surface sediment. When reeds were harvested, improved transparency and aeration accelerated mineralization of organic matter and TN and led to a decrease in sediments from 0 to 20 cm (Fig. 6). Organic N mineralization was correlated with inorganic nitrogen (Tanner et al., 2002). NH_4^+-N and NO_3^--N from 0 to 20 cm in the sediment of harvested area were higher than the control. Significant relationships among TN, NH_4^+-N and NO_3^--N in reed wetland water were found. Therefore, organic nitrogen mineralization in the wetland was an important inorganic N source in addition to cropland runoff (Kovacic et al., 2000; Borin et al., 2001; Berit & Hans, 2002). Mineralization likely increased NH_4^+-N and NO_3^--N concentrations in wetland water temporarily, but TN in sediment decreased permanently. Therefore, harvest of dense plants removed pollutants from wetlands and accelerated organic matter and TN decomposition.

In general, high microorganism biomass occurs with high organic matter content and TN. Organic matter and TN decomposition by microorganisms changes pH. Organic matter produces organic acids during decomposition, and TN release H^+ when oxidized to NO_3^--N by ammonification and nitrification. These reactions result in decreased pH. Nitrification contributes to a pH decrease of two standard units according to DeSimone (1998).

On the basis of 607 kg ha^{-1} N and 90 kg ha^{-1} P fertilizer application in this region, and 20–25% N and 20–40% P loss rates according to Huang et al. (2001), 1 ha of reed wetland will remove N discharged from 5.4 to 8.0 ha of cropland and P discharged from 2.9 to 5.8 ha of field. Wild rice wetland can remove N from 0.9 to 1.3 ha of cropland and P from 0.8 to 1.6 ha of field annually. Therefore, plants are effective at removing agricultural non-point pollutants (Eriksson & Weisner, 1997; Li, 1997; Yang, 1998; Rickerl et al., 2000).

Despite high uptake ability of N and P by reeds and wild rice, wild plants have little economic value and cannot attract harvest of farmers. In winter and spring, the plants die and decompose to release N and P again, which causes secondary pollution. *Z. caduciflora* is a plant with value and is planted in the Lake Taihu basin. It has large biomass and high absorption capacity of N and P (Tables 1, 2). From August to September, *Z. caduciflora* is reproducing and requires high nutrient supply (Qiu et al., 2002). This is the same period for rice harvest, which requires fertilizing and irrigating. High nutrient concentrations in agricultural drainage meet the demand for rapid growth of *Z. caduciflora*. The growth period of *Z. caduciflora* is from March to September, similar to reeds and wild rice, coinciding with the growth period for wheat and rice rotated in crop fields. *Z. caduciflora* can absorb effectively N and P (Liu et al., 2002) discharged from field.

Acknowledgements We wish to express deep appreciation to associate Prof. L. Wang, Prof. A. J. M. Smits and A. Gashaw for valuable comments on the manuscript. We thank Junji Zhang for his practical assistance. The National Natural Science Foundation of China (No.50579018, 50239030) supported this study.

References

Angle, J. S., C. M. Gross, R. L. Hill & M. S. McIntosh, 1993. Soil nitrate concentrations under corn as affected by tillage, manure, and fertilizer application. Journal of Environmental Quality 22: 141–147.

Arheimer, B. & H. B. Wittgren, 1994. Modelling the effects of wetlands on regional nitrogen transport. Ambio 23: 378–386.

Bai, J., W. Deng & Y. Zhang, 2002. Spacial distribution characteristics of soil organic matter and nitrogen in the natural floodplain wetland. Environmental Science 23: 77–81 (in Chinese).

Berit, A. & B. W. Hans, 2002. Modelling nitrogen removal in potential wetlands at the catchment scale. Ecological Engineering 19: 63–80.

Borin, M., G. Bonaiti & L. Giardini, 2001. Controlled drainage and wetlands to reduce agricultural pollution: a lysimetric study. Journal of Environmental Quality 30: 1330–1340.

Braskerud, B. C., 2002. Factors affecting nitrogen retention in small constructed wetlands treating agricultural non-point source pollution. Ecological Engineering 18: 351–370.

China Standard Press, 1998. A Compilation of National Environmental Quality and Pollution Discharge Standards. China Standard Press, Beijing (in Chinese).

DeSimone, L. A., 1998. Nitrogen transport and transformations in a shallow aquifer receiving wastewater discharge: a mass balance approach. Water Resources Research 34: 271–285.

Eriksson, P. G. & S. E. B. Weisner, 1997. Nitrogen removal in a wastewater reservoir: the importance of denitrification by epiphytic biofilms on submersed vegetation. Journal of Environmental Quality 26: 905–910.

Fleischer, S., A. Gustafson, A. Joelsson, J. Pansar & L. Stibe, 1994. Nitrogen removal in created ponds. Ambio 23: 349–357.

Gao, C., J. Zhu & Y. Dou, 2002. Contribution of agricultural non-point source pollution to water quality deterioration in Taihu Lake watershed: recent trends and research priorities. Resources and Environment in the Yangtze Basin 11: 260–263 (in Chinese).

Huang, Y., C. Fang, P. Pu, J. Jiang & Q. Dai, 2001. Water Environment and Pollution Control of Lake Taihu. Science Publishing Company, Beijing, China, 259–264 (in Chinese).

Jan, V., 2002. The use of sub-surface constructed wetlands for wastewater treatment in the Czech Republic: 10 years experience. Ecological Engineering 18: 633–646.

Jordan, T. E., D. F. Whigham, K. H. Hofmockel & M. A. Pittek, 2003. Nutrient and sediment removal by a restored wetland receiving agricultural runoff. Journal of Environmental Quality 32: 1534–1547.

Kang, S., H. Kang, D. Ko & D. Lee, 2002. Nitrogen removal from a riverine wetland: a field survey and simulation study of *Phragmites japonica*. Ecological Engineering 18: 467–475.

Kovacic, D. A., M. B. David, L. E. Gentry, K. M. Starks & R. A. Cooke, 2000. Effectiveness of constructed wetlands in reducing nitrogen and phosphorus export from agricultural tile drainage. Journal of Environmental Quality 29: 1262–1274.

Li, S. & J. Li, 2001. Research progress on losses of fertilizer nitrogen. Agro-Environmental Protection 20: 377–379 (in Chinese).

Li, W., 1997. "Yellow water" in East Taihu Lake caused by *Zizania latifolia* and its prevention. Journal of Lake Sciences 9: 364–368 (in Chinese).

Liu, C., H. Hu & J. Zhang, 2002. Use of new type constructed wetland for treatment of low strength rural sewage. China Water and Wastewater 18: 1–4 (in Chinese).

Martin, H. W., D. B. Ivanoff, D. A. Graetz & K. R. Reddy, 1997. Water table effects on histosol drainage water carbon, nitrogen, and phosphorus. Journal of Environmental Quality 26: 1062–1071.

Mitsch, W. J., A. J. Horne & R. W. Nairn, 2000. Nitrogen and phosphorus retention in wetlands-ecological approaches to solving excess nutrient problems. Ecological Engineering 14: 1–7.

Peder, G. E. & E. B. W. Stefan, 1997. Nitrogen removal in a wastewater reservoir: the importance of denitrification by epiphytic biofilms on submersed vegetation. Journal of Environmental Quality 26: 905–910.

Qin, B. Q, P. Z. Xu, Q. L. Wu, L. C. Luo & Y. L. Zhang, 2007. Environmental issues of Lake Taihu, China. Hydrobiologia 581: 3–14.

Qiu, J., J. Jiang, B. Cao, L. Li, X. Xu & X. Han, 2002. Content and change trend of N, P, K in above ground organs of *Zizania latifolia*. Journal of Hubei Agricultural College 22: 22–125 (in Chinese).

Raisin, G. W. & D. S. Mitchell, 1995. The use of wetlands for the control of non-point source pollution. Water Science and Technology 32: 177–186.

Reddy, K. R. & W. H. Patrick, 1975. Effect of alternate aerobic and anaerobic conditions on redox potentical, organic matter decomposition and nitrogen loss in a flooded soil. Soil Biology & Biochemistry 7: 87–94.

Rene, L., Z. Malgorzata & B. Z. Joy, 1991. Nitrogen assessments in a constructed and a natural salt marsh of San Diego Bay. Ecological Applications 1: 40–51.

Rickerl, D. H., L. L. Janssen, & R. Woodland, 2000. Buffered wetlands in agricultural landscapes in the Prairie Pothole regions: environmental, agronomic, and economic evaluations. Journal of Soil and Water Conservation 55: 220–225.

Roth, G. W. & R. H. Fox, 1990. Soil nitrate accumulations following nitrogen-fertilized corn in Pennsylvania. Journal of Environmental Quality 19: 243–248.

Spieles, D. J. & W. J. Mitsch, 2000. The effects of season and hydrologic and chemical loading on nitrate retention in constructed wetlands: a comparison of low-and high-nutrient riverine systems. Ecological Engineering 14: 77–91.

Tanner, C. C., 2001. Growth and nutrient dynamics of soft-stem bulrush in constructed wetlands treating nutrient-rich wastewaters. Wetlands Ecology and Management 9: 49–73.

Tanner, C. C., R. H. Kadlec, M. M. Gibbs, J. P. S. Sukias & M. L. Nguyen, 2002. Nitrogen processing gradients in subsurface-flow treatment wetlands-influence of wastewater characteristics. Ecological Engineering 18: 499–520.

Withers, P. J. A., S. D. Clay & V. G. Breeze, 2001. Phosphorus transfer in runoff following application of fertilizer, manure, and sewage sludge. Journal of Environmental Quality 30: 180–188.

Xu, Q., 1996. A review on the status of non-point source pollution of chemical fertilizers and pesticides in China. Rural Eco-Environment 12: 39–43 (in Chinese).

Yang, Q., 1998. Ecological functions of aquatic vegetation in East Taihu Lake and its reasonable regulation. Journal of Lake Sciences 10: 76–72 (in Chinese).

Yang, Y., C. Tian, J. Sheng & Q. Wen, 2002. Spatial variability of soil organic matter, total nitrogen, phosphorus and potassium in cotton field. Agricultural Research in the Arid Areas 20: 26–30 (in Chinese).

MIX
Papier aus verantwortungsvollen Quellen
Paper from responsible sources
FSC® C105338

FSC
www.fsc.org

If you have any concerns about our products,
you can contact us on
ProductSafety@springernature.com

In case Publisher is established outside the EU,
the EU authorized representative is:
Springer Nature Customer Service Center GmbH
Europaplatz 3, 69115 Heidelberg, Germany

Printed by Libri Plureos GmbH
in Hamburg, Germany